Birkhäuser

Lecture Notes in Geosystems Mathematics and Computing

Lecture Notes in Geosystems Mathematics and Computing offers the opportunity to publish contributions at the interface between the geo-disciplines, mathematics, observational and computational strategies. Reflecting new developments and concepts, emerging topics, practical applications, as well as numerical strategies in high-performance computing (HPC), the series bridges real and virtual geoworld systems. The series supports the rapid communication between the Geoscientists, Mathematicians and HPComputer Scientists by publishing Tutorials, Collections to recent developments, proceedings of important meetings as well as outstanding PhD theses.

More information about this series at http://www.springer.com/series/15481

Alessio Fumagalli · Inga Berre ·
Luca Formaggia · Eirik Keilegavlen ·
Anna Scotti
Editors

Numerical Methods for Processes in Fractured Porous Media

Previously published in *GEM - International Journal on Geomathematics*, Volume 9, Issue 2, 2018, and Volume 10, Issue 1, 2019

Editors
Alessio Fumagalli
Torino, Italy

Luca Formaggia
Milan, Italy

Anna Scotti
Milan, Italy

Inga Berre
Bergen, Norway

Eirik Keilegavlen
Bergen, Norway

Lecture Notes in Geosystems Mathematics and Computing
ISBN 978-3-030-26940-1

Mathematics Subject Classification (2010): 65N08, 65N30, 65N38, 65N50, 65Y05, 65Y10, 65Z05

This book is published under the imprint Birkhäuser, www.birkhauser-science.com by the registered company Springer Nature Switzerland AG
The registered company address is: Gewerbestrasse 11, 6330 Cham, Switzerland

Preface

The research on mathematical and numerical modeling of processes in geological porous media is getting increasing interest because of the growing importance of underground exploitation. Issues like CO_2 sequestration, geothermal energy storage and production, more efficient exploitation of hydrocarbon resources, and environmental protection of aquifers call for a better understanding of subsurface flow, geomechanics, and their interaction. In all the aforementioned applications, fractures play an important role: they can enhance the permeability of a formation, and thus be beneficial for its exploitation, but may also represent a source of risk (i.e. leakage of pollutants).

Numerical modeling can be particularly useful in supporting geological studies since, unlike experiments, it is cheap, repeatable, and gives us full access to input data and results. However, the complexity of fractured porous media represents a challenge for numerical methods, mostly due to the geometric complexity, heterogeneity, and coupling of different phenomena.

This book aims at collecting state-of-the-art contributions on the topic of the numerical simulation of fractured porous media, focusing on flow and geomechanics. The first chapters discuss the application of some modern numerical methods to the problem of flow in fractured porous media, presenting different strategies for the discretization of the domain and the coupling between fracture network and porous matrix.

This book could be beneficial to computational scientists and numerical analysts who want to keep abreast of the developments of numerical discretization techniques for underground flow and geomechanics, as well as engineers and geologists interested in modern simulation techniques.

In the article by Köppel et al., the authors employ Lagrange multipliers to couple the n-dimensional domain with $n-1$ dimensional inclusions (fractures). This allows for independent finite element grids, thus providing more flexibility. Thanks to a suitable penalization term, the scheme can be proven to be stable and convergent.

The numerical method presented in the article by Burman et al. is based on a non-conforming discretization of the domain, i.e., fractures are allowed to cut the elements of the mesh. The unfitted finite element method based on a Nitsche type mortaring allows the authors to introduce a Laplace-Beltrami operator for the transport in the fracture. Optimal error estimates are shown along examples including bifurcating fractures.

Next, Chave et al. discuss a model for the passive transport of a solute in fractured porous media. The numerical method considered belongs to the new class of Hybrid High-Order. New energy based transmission conditions are introduced for the transport part by allowing solute jumps across the fracture.

In Berrone et al.'s article, the authors present a novel discretization strategy for discrete fracture matrix problems. Two methods are introduced: the first is based on finite and boundary element methods for the fracture and rock matrix, respectively, while the second deals with a full virtual element method approximation where arbitrary mesh elements can be used to accommodate geometrical complexity more easily.

A two-phase flow model is introduced in Aghili et al. where linear or non-linear transmission conditions at the matrix-fracture interfaces are proposed and compared. The discretization of the equations is based on a finite volume two-point-flux approximation, combined with a local non-linear solver to handle the transmission conditions to reduce the computational cost by local elimination of inter-facial unknowns.

The article by Gläser et al. introduces a hybrid-dimensional numerical model for non-isothermal two-phase flow in fractured porous media. The proposed model is intensively tested on synthetic cases for compressible fluids and strongly heterogeneous anisotropic full permeability tensor. These examples show good agreement with the corresponding equi-dimensional model in the case of highly conductive fractures.

The complex problem of two-codimensional coupling is introduced and analyzed in Cerroni et al.'s contribution, dealing with leak off or sink effects for multiple wells in porous media. The model considers the well as one-dimensional objects immersed in a three-dimensional rock matrix. The numerical scheme adopted in the discretization does not require conformity between the wells and the three-dimensional computational grid, increasing the flexibility of the framework.

To conclude this first part of chapters, a GPGPUs strategy is introduced and presented in Berrone et al. to speed up flow simulations in discrete fracture networks. The authors use the Nvidia Compute Unified Device Architecture to accelerate the linear algebra operations, the most computationally costly aspect in such problem.

The second group of chapters takes into account the problem of porous media and fracture mechanics coupled with fluid flow. It is indeed important to take into account the impact of fluid pressure on the deformation of geological porous media, and, conversely, the effect of deformation on porosity and permeability.

Ambartsumyan et al. introduce a model describing the flow in the fractures via the Stokes equations while Biot equations are used in the porous media. The global model is coupled with advection and diffusion of chemical species within the fluid. Lagrange multipliers are used to deal with the matrix-fracture interaction. Analysis of the problem as well as simulations validate the proposed approach.

In Mikelć et al., the authors study the propagation of hydraulic fractures using the fixed stress splitting method. The phase field approach is applied and the authors study in detail the derivation of an incremental formulation applied to hydraulic fracture in a poroelastic medium. The existence of a minimizer of the energy functional is established. Computational results from benchmark cases demonstrate the validity of the proposed approach.

Since diffusion-based models for fractured media are unable to reproduce important hydro-mechanical coupling phenomena, the work presented in Schmidt et al.'s article aims to overcome these difficulties by proposing a hybrid-dimensional formulation. The authors introduce an implicit weak coupling scheme, naturally able to perform non-conformal mesh calculations, and a strong coupling scheme in the form of interface elements.

The article by Yoshioka et al. concludes the list by discussing the numerical treatment of propagating fractures as embedded discontinuities. In this work, the authors compare the accuracy of two of the most popular smeared approaches along with an approach in which the solution space is locally enriched to capture a strong discontinuity. Differences and similarities are highlighted through examples for a toughness-dominated regime.

Torino, Italy — Alessio Fumagalli
Bergen, Norway — Inga Berre
Milan, Italy — Luca Formaggia
Bergen, Norway — Eirik Keilegavlen
Milan, Italy — Anna Scotti
June 2019

Contents

GEM - International Journal on Geomathematics
https://doi.org/10.1007/s13137-019-0117-7

ORIGINAL PAPER

A stabilized Lagrange multiplier finite-element method for flow in porous media with fractures

Markus Köppel[1] · **Vincent Martin**[2] · **Jean E. Roberts**[3]

Received: 6 April 2018 / Accepted: 10 November 2018

Abstract

In this work we introduce a stabilized, numerical method for a multidimensional, discrete-fracture model (DFM) for single-phase Darcy flow in fractured porous media. In the model, introduced in an earlier work, flow in the $(n-1)$-dimensional fracture domain is coupled with that in the n-dimensional bulk or matrix domain by the use of Lagrange multipliers. Thus the model permits a finite element discretization in which the meshes in the fracture and matrix domains are independent so that irregular meshing and in particular the generation of small elements can be avoided. In this paper we introduce in the numerical formulation, which is a saddle-point problem based on a primal, variational formulation for flow in the matrix domain and in the fracture system, a weakly consistent stabilizing term which penalizes discontinuities in the Lagrange multipliers. For this penalized scheme we show stability and prove convergence. With numerical experiments we analyze the performance of the method for various choices of the penalization parameter and compare with other numerical DFM's.

Keywords Discrete fracture model · Finite element method · Stabilized Lagrange multiplier method · Penalization · Nonconforming grids

Mathematics Subject Classification 35J50 · 35J57 · 65N12 · 65N85 · 76M10 · 76S05

✉ Markus Köppel
markus.koeppel@ians.uni-stuttgart.de

Vincent Martin
vincent.martin@utc.fr

Jean E. Roberts
jean-elizabeth.roberts@inria.fr

[1] Universtität Stuttgart, IANS, Pfaffenwaldring 57, 70569 Stuttgart, Germany

[2] Université de Technologie de Compiègne (UTC), LMAC, Rue du docteur Schweitzer CS 60319, 60203 Compiègne Cedex, France

[3] INRIA Paris, 2 Rue Simone Iff, 75589 Paris, France

1 Introduction

Fractures represent one of the most challenging heterogeneities for the approximation of fluid flow in porous media. Typically their lateral dimension is considerably smaller compared to their extensions in other directions. Moreover fractures may act as barriers to and/or conduits for fluid flow. Depending on the hydrogeological properties and the scale of consideration, the presence of fractures thus may lead to a significant change in the flow behavior in the subsurface. Because fault zones occur in many applications, such as CO_2 sequestration, underground storage of radioactive waste and enhanced oil recovery, the consideration of fractures in modeling of flow in porous media has received more and more attention in the last decades. A variety of different models have been proposed.

A common way to incorporate fractures in models is the discrete-fracture (DFM) approach, in which information concerning the fracture location in the domain of interest is required, and the fluid flow in the fracture as well as in the surrounding domain is calculated. In this context the fractures are often considered as $(n-1)$-dimensional objects within the surrounding n-dimensional matrix domain in order to avoid the generation of small elements of the spatial discretization grid. Such models have been studied, in e.g. Alboin et al. (2002) and Angot et al. (2009), assuming Darcy flow in both, fracture and matrix, parts of the domain. Other studies addressed Forchheimer flow in the fractures (Knabner and Roberts 2014) or Darcy-Brinkman flow (Lesinigo et al. 2011). Multiphase flow has also been considered, e.g. Ahmed et al. (2017), Brenner et al. (2015) and Hoteit and Firoozabadi (2008). Some articles deal with discrete fracture network (DFN) models, e.g. Berrone et al. (2014) and Pichot et al. (2012). Whereas some of these models are based on finite element methods (Baca et al. 1984), others use mixed or mixed-hybrid finite elements, Boon et al. (2018) and Martin et al. (2005), finite volume methods (Fumagalli et al. 2016; Karimi-Fard et al. 2004; Reichenberger et al. 2006), multi-point flux methods (Sandve et al. 2012), or mimetic finite difference methods (Antonietti et al. 2016b; Formaggia et al. 2018), or discontinuous Galerkin methods (Antonietti et al. 2016a; Massing 2017), to discretize the problem.

For discretization schemes, in what may be referred to as a matching fracture and matrix grid approach, the fracture mesh elements coincide with faces of the matrix mesh elements. However one may wish to discretize the fracture more finely in the case of a highly conductive fracture or more coarsely in the case of a barrier. Therefore it may be necessary to use methods allowing for non-matching grids; see e.g. Chave et al. (2018), Faille et al. (2016) and Frih et al. (2012). Still with these methods the matrix grid must be aligned with the fracture. By contrast, with nonconforming methods a fracture can cut through the interior of matrix elements because of an independent meshing of the corresponding domains. This can be achieved, for example, with locally enriched basis functions in the vicinity of the fracture to account for the resulting discontinuities, in what is commonly referred to as an extended finite element method (XFEM), e.g. in Fumagalli and Scotti (2013) and Schwenck et al. (2015).

This work presents an alternative nonconforming discretization scheme for a model, introduced in Köppel et al. (2018), for single-phase, Darcy flow in fractured porous media. The model uses Lagrange multiplier variables, which represent a local fluid exchange between fracture and matrix, in a primal variational formulation. The new numerical scheme, like that of Köppel et al. (2018), uses continuous piecewise linear or bilinear approximations for the pressure both in the matrix and in the fracture and piecewise constant functions to approximate the multipliers. Here however, following ideas of Burman and Hansbo (2010a), we add a stabilization term which penalizes jumps in the multipliers over regular portions of the fracture. The permeability in the fracture is assumed to be larger than that in the matrix. Hence the fluid pressure is continuous excluding the case of a geological barrier, which will be subject of future research. Because of the use of the multiplier this model allows for mutually independent grids of the matrix and the fracture, both discretized with continuous, piecewise-(bi)linear basis functions. As in Köppel et al. (2018), the Lagrange multiplier is discretized by means of discontinuous, piecewise-constant, basis functions, though here the multipliers are no longer associated with an independent but size-constrained grid but with a grid generated by intersections of the matrix grid with the fracture. Following Burman and Hansbo (2010a), we add a weakly consistent stabilizing term which penalizes the jumps of the discrete multipliers. This leads to a stabilization of the discrete saddle point system and thus reduces the condition numbers involved. In Sect. 2, we recall briefly the continuous formulation of the Lagrange multiplier method. Section 3 concerns the discrete formulation of the problem. We introduce a weakly consistent penalty term to stabilize the discrete system, prove the stability of the discrete formulation and its convergence under conditions on the regularity of the Lagrange multiplier. In Sect. 4, the theoretical findings are analyzed numerically by means of several numerical experiments, including two benchmarks from Flemisch et al. (2018), validating the method. Finally we conclude and discuss the proposed method in Sect. 5.

2 The continuous formulation for the Lagrange multiplier model

In this section we recall briefly the continuous model for the Lagrange-multiplier DFM, introduced in Köppel et al. (2018). Let Ω be a domain in $\mathbb{R}^2$, representing a porous medium and let $\gamma \subset \Omega$ be an one-dimensional surface representing a fracture. The extension to 3D does not pose real conceptual difficulties for the analysis though. Let $\mathbf{n}_\gamma$ denote one of the two possible continuous unit vector fields on γ, and let $\mathbf{K}$ and $\mathbf{K}_\gamma$ be symmetric, uniformly positive-definite, bounded, permeability tensor fields on Ω and γ respectively, with constants $C_{\mathbf{K}}^M$ and $C_{\mathbf{K}}^m > 0$ such that

$$\begin{aligned} &\|\mathbf{K}(x)\| \le C_{\mathbf{K}}^M, \quad \forall x \in \Omega, \qquad && C_{\mathbf{K}}^m \|\mathbf{v}\|_{0,\Omega} \le (\mathbf{K}\mathbf{v}, \mathbf{v})_\Omega, \quad \forall \mathbf{v} \in L^2(\Omega), \\ &\|\mathbf{K}_\gamma(x_\gamma)\| \le C_{\mathbf{K}}^M, \quad \forall x_\gamma \in \gamma, \qquad && C_{\mathbf{K}}^m \|\mathbf{v}_\gamma\|_{0,\gamma} \le \langle \mathbf{K}_\gamma \mathbf{v}_\gamma, \mathbf{v}_\gamma \rangle_\gamma, \quad \forall \mathbf{v}_\gamma L^2(\gamma), \end{aligned}$$

where we use the notation $(\cdot, \cdot)_\Omega$ and $\langle \cdot, \cdot \rangle_\gamma$ for the L^2 inner products on $L^2(\Omega)$ and $L^2(\gamma)$, respectively, and $\| \cdot \|_{0,\mathcal{O}}$ for the $L^2(\mathcal{O})$ norm on an open set $\mathcal{O} \subset \mathbb{R}^d$, $d = 1, 2$. Here $\|\mathbf{K}(x)\|$ denotes the operator norm as does $\|\mathbf{K}_\gamma(x_\gamma)\|$. For simplicity assume that γ is a line segment and that $\partial\gamma \subset \partial\Omega$. Flow in both Ω and γ is assumed to be governed by Darcy's law and the law of mass conservation, and for simplicity homogeneous Dirichlet boundary conditions are imposed on both $\partial\gamma$ and $\partial\Omega$. Fluid exchange between Ω and γ is through a source/sink term λ representing the discontinuity in the flux in Ω from one side of γ to the other. Letting p and p_γ represent the fluid pressure and f and f_γ external source terms in Ω and γ respectively, assuming sufficient regularity of f_γ, we may write the equations for the model as follows:

$$\begin{aligned} &\operatorname{div}(-\mathbf{K}\nabla p) - \lambda\,\delta_\gamma = f, && \text{in } \Omega,\\ &\operatorname{div}_\tau(-\mathbf{K}_\gamma \nabla_\tau p_\gamma) + \lambda = f_\gamma, && \text{in } \gamma,\\ &p|_\gamma = p_\gamma, && \text{in } \gamma,\\ &p = 0, && \text{on } \Gamma = \partial\Omega,\\ &p_\gamma = 0, && \text{on } \partial\gamma, \end{aligned} \tag{1}$$

where δ_γ denotes the Dirac measure on γ, and where the operators div_τ and ∇_τ denote the derivatives in the direction obtained by rotating $\mathbf{n}_\gamma$ through 90 degrees. For the variational formulation, the spaces V_Ω, V_γ, $\mathbf{V}$ and Λ are used:

$$V_\Omega = H_0^1(\Omega), \quad V_\gamma = H_0^1(\gamma), \quad \mathbf{V} = V_\Omega \times V_\gamma, \quad \text{and} \quad \Lambda = H_{00}^{-\frac{1}{2}}(\gamma). \tag{2}$$

We use the same notation $\langle\cdot,\ \cdot\rangle_\gamma$ for the duality pairing between $H_{00}^{-\frac{1}{2}}(\gamma)$ and $H_{00}^{\frac{1}{2}}(\gamma)$ as that which is used for the $L^2(\gamma)$–inner product when the functions are sufficiently regular. Now with the bilinear form $\mathcal{A}$ on $(\mathbf{V} \times \Lambda)^2$ defined by

$$\mathcal{A}(P, Q) = \int_\Omega \mathbf{K}\nabla p \cdot \nabla q + \int_\gamma \mathbf{K}_\gamma \nabla_\tau p_\gamma \cdot \nabla_\tau q_\gamma - \langle\lambda,\ q|_\gamma - q_\gamma\rangle_\gamma + \langle\mu,\ p|_\gamma - p_\gamma\rangle_\gamma,$$

for $P = (p, p_\gamma; \lambda)$ and $Q = (q, q_\gamma; \mu)$ in $\mathbf{V} \times \Lambda$, and the linear form ℓ on $\mathbf{V}$ defined by

$$\ell(q, q_\gamma) = \int_\Omega f q + \int_\gamma f_\gamma q_\gamma,$$

for $(q, q_\gamma) \in \mathbf{V}$, the variational formulation of (1) may be written as follows:

$$\begin{aligned} &\text{Find } P = (p, p_\gamma; \lambda) \in \mathbf{V} \times \Lambda \text{ such that}\\ &\mathcal{A}(P, Q) = \ell(q, q_\gamma), \quad \forall Q = (q, q_\gamma; \mu) \in \mathbf{V} \times \Lambda. \end{aligned} \tag{3}$$

In Köppel et al. (2018), it was proved that (3) has a unique solution. Note that λ can be interpreted as the jump in the flux across γ: $\lambda = [\![\mathbf{K}\nabla p \cdot \mathbf{n}_\gamma]\!]_\gamma$.

3 Discretization

Inspired by the work in Burman and Hansbo (2010a), we introduce a stabilized numerical discretization of problem (3) and show existence and uniqueness of the discrete solution as well as convergence. Recall that in Köppel et al. (2018), a different primal finite element method was used to discretize (3), one that uses different discretization spaces for the Lagrange multiplier and does not have a stabilization term. With the stabilized method we do not have the minimum size constraint on the support of the Lagrange multipliers.

3.1 A stabilized discrete formulation

We introduce independent finite element meshes, $\mathcal{T}_h$ and $\mathcal{T}_{h,\gamma}$, to define the approximation spaces $V_{h,\Omega} \subset V_\Omega$ and $V_{h,\gamma} \subset V_\gamma$. The mesh $\mathcal{T}_h$ on Ω is made up of triangles and/or rectangles, and $\mathcal{T}_{h,\gamma}$ is a mesh on γ. We assume that each of $\mathcal{T}_h$ and $\mathcal{T}_{h,\gamma}$ belongs to a uniformly regular family of discretizations. Let h and h_γ be the parameters associated with these families:

$$\begin{aligned} h &= \max_{T \in \mathcal{T}_h} h_T, \quad \text{where } h_T = \text{diam}(T), \\ h_\gamma &= \max_{t \in \mathcal{T}_{h,\gamma}} h_t, \quad \text{where } h_t = \text{diam}(t), \end{aligned}$$

For each $T \in \mathcal{T}_h$ let $\rho_T =$ the radius of the incircle of T, and let $\sigma_T = \frac{h_T}{\rho_T}$. Let $\rho_h = \min_{T \in \mathcal{T}_h} \rho_T$ and let $\sigma_h = \frac{h}{\rho_h}$. Let $\sigma_\Omega = \max_h \sigma_h$ be the upper bound guaranteed by uniform regularity.

There is naturally induced on γ a second mesh, which we will denote $\mathcal{T}_{h,\lambda}$ (as it will be associated with the space of discrete Lagrange multipliers), that consists of the segments $T \cap \gamma$ such that $T \in \mathcal{T}_h$, see Fig. 1:

$$\mathcal{T}_{h,\lambda} = \{s \subset \gamma \ : \ s = T \cap \gamma \text{ for some } T \in \mathcal{T}_h\}.$$

Let $\mathcal{F}_h$ denote the set of edges F of elements $T \in \mathcal{T}_h$, and $\mathcal{F}_{h,\lambda}$ the set of vertices f of elements $s \in \mathcal{T}_{h,\lambda}$ which do not lie on the boundary: $f \notin \partial\gamma$. The conforming approximation spaces $V_{h,\Omega}$ and $V_{h,\gamma}$ will consist of continuous functions that vanish on the boundary of Ω and γ, respectively. The functions in $V_{h,\gamma}$ will be piecewise linear subordinate to the mesh $\mathcal{T}_{h,\gamma}$ while those of $V_{h,\Omega}$, subordinate to the mesh $\mathcal{T}_h$ will be piecewise linear or bilinear depending on whether the element is a triangle or a rectangle:

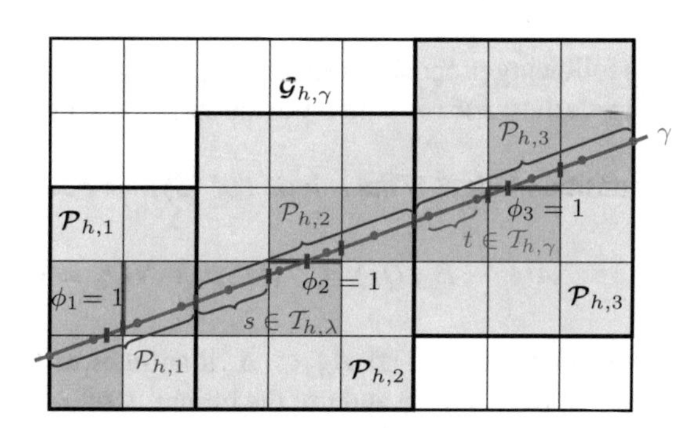

Fig. 1 Meshes $\mathcal{T}_h$, $\mathcal{T}_{h,\gamma}$ (elements t delimited by red dots) and $\mathcal{T}_{h,\lambda}$ (elements s delimited by blue segments). The domain $\mathcal{G}_{h,\gamma}$ around γ is in grey. The supports of the patch elements $\mathcal{P}_{h,i}$ (in blue) and $\boldsymbol{\mathcal{P}}_{h,i}$ (grey) are also depicted, with the chosen edge for ϕ_i (color figure online)

$$V_{h,\Omega} = \left\{ q \in H_0^1(\Omega) \; : \; \forall T \in \mathcal{T}_h, \; q\big|_T \in \begin{cases} \mathbb{P}^1(T) & \text{if } T \text{ is a triangle} \\ \mathbb{Q}^{1,1}(T) & \text{if } T \text{ is a rectangle} \end{cases} \right\},$$
$$V_{h,\gamma} = \left\{ q_\gamma \in H_0^1(\gamma) \; : \; \forall t \in \mathcal{T}_{h,\gamma}, \; q_\gamma\big|_t \in \mathbb{P}^1(t) \right\}, \text{ and } \boldsymbol{V}_h = V_{h,\Omega} \times V_{h,\gamma}. \tag{4}$$

The approximation space Λ_h for the Lagrange multiplier is defined as follows:

$$\Lambda_h = \{\lambda_h \in L^2(\gamma) \mid \lambda_h|_s \in \mathbb{P}^0(s), \; \forall s \in \mathcal{T}_{h,\lambda}\}. \tag{5}$$

Following Burman and Hansbo (2010a), we will introduce a stabilizing term $\mathcal{J}$ in the form of a bilinear operator on $\Lambda_h \times \Lambda_h$:

$$\mathcal{J}(\lambda_h, \mu_h) = \sum_{f \in \mathcal{F}_{h,\lambda}} \xi h^2 [\![\lambda_h]\!]_f [\![\mu_h]\!]_f, \tag{6}$$

where for $\mu_h \in \Lambda_h$, and $f \in \mathcal{F}_{h,\lambda}$, $[\![\mu_h]\!]_f$ denotes the jump in μ_h across the vertex f (i.e. along the fracture, and should not be confused with $[\![\![\cdot]\!]\!]_\gamma$ which is a jump normal to γ). Here, for simplicity, we assume that no edge $F \in \mathcal{F}_h$ lies along γ and that γ does not contain any vertex of the mesh $\mathcal{T}_h$. This ensures that $[\![\lambda_h]\!]_f$ is uniquely defined when $f \in \mathcal{F}_{h,\lambda}$ and $\lambda_h \in \Lambda_h$. Otherwise defining the jump term is more cumbersome, though it poses no real problem, and in fact, some of our numerical experiments treat such cases. We remark that we will at times use the notation $[\![\phi]\!]_f$ for functions ϕ not necessarily belonging to Λ_h but for which the jumps over the vertices $f \in \mathcal{F}_{h,\lambda}$ are well defined. Indeed, $\mathcal{J}(\cdot,\cdot)^{\frac{1}{2}}$ defines a semi-norm on Λ_h, and we have the Cauchy–Schwarz-like estimate

$$|\mathcal{J}(\lambda_h, \mu_h)| \leq \mathcal{J}(\lambda_h, \lambda_h)^{\frac{1}{2}} \; \mathcal{J}(\mu_h, \mu_h)^{\frac{1}{2}} \quad \forall \lambda_h, \mu_h \in \Lambda_h, \tag{7}$$

from the usual estimate $|\sum_f a_f b_f| \leq (\sum_f a_f^2)^{\frac{1}{2}} (\sum_f b_f^2)^{\frac{1}{2}}$ for $a_f, b_f \in \mathbb{R}$. The formulation of the discrete stabilized problem may be written as follows:

$$\begin{aligned} &\text{Find } P_h = (p_h, p_{\gamma,h}; \lambda_h) \in \boldsymbol{V}_h \times \Lambda_h \text{ such that} \\ &\mathcal{A}(P_h, Q_h) + \mathcal{J}(\lambda_h, \mu_h) = \ell(q_h, q_{\gamma,h}), \quad \forall Q_h = (q_h, q_{\gamma,h}; \mu_h) \in \boldsymbol{V}_h \times \Lambda_h. \end{aligned} \tag{8}$$

The following proposition states an approximate Galerkin orthogonality for (8) which will give the *weak consistency* of the method according to Burman and Hansbo (2010b).

Proposition 1 *If P is the solution of* (3) *and P_h the solution of* (8), *then*

$$\mathcal{A}(P - P_h, Q_h) = \mathcal{J}(\lambda_h, \mu_h) \;\; \forall Q_h = (q_h, q_{\gamma,h}; \mu_h) \in \boldsymbol{V}_h \times \Lambda_h. \tag{9}$$

Proof As $\boldsymbol{V}_h \subset \boldsymbol{V}$ and $\Lambda_h \subset \Lambda$, it suffices to take $Q = Q_h \in \boldsymbol{V}_h \times \Lambda_h$ in (3), substract (8) from (3) and use the bilinarity of $\mathcal{A}$ to obtain (9). □

3.2 Some discrete norms

We give the definition of some norms that will be useful for obtaining the approximation properties of the space $V_h \times \Lambda_h$. For $\zeta \in L^2(\gamma)$ and $h > 0$ we define the discrete norms

$$\|\zeta\|^2_{\frac{1}{2},h,\gamma} = \int_\gamma h^{-1}\zeta^2 = h^{-1}\|\zeta\|^2_{0,\gamma} \text{ and } \|\zeta\|^2_{-\frac{1}{2},h,\gamma} = \int_\gamma h\zeta^2 = h\|\zeta\|^2_{0,\gamma}, \tag{10}$$

and we recall the associated Cauchy–Schwarz type inequality

$$\langle \zeta, \eta \rangle_\gamma \leq \|\zeta\|_{-\frac{1}{2},h,\gamma}\|\eta\|_{\frac{1}{2},h,\gamma}, \qquad \forall\ \zeta \text{ and } \eta \in L^2(\gamma). \tag{11}$$

We will also use two more norms defined respectively for $Q = (q, q_\gamma; \mu) \in \mathbf{V} \times L^2(\gamma)$ for $Q_h = (q, q_\gamma; \mu_h) \in \mathbf{V} \times \Lambda_h$, and for $h > 0$ by

$$\begin{aligned}
|||Q|||^2_{0,h} &:= \|\nabla q\|^2_{0,\Omega} + \|\nabla_\tau q_\gamma\|^2_{0,\gamma} + \|\mu\|^2_{-\frac{1}{2},h,\gamma} + \|q|_\gamma - q_\gamma\|^2_{\frac{1}{2},h,\gamma} \\
|||Q_h|||^2_{1,h} &:= |||Q_h|||^2_{0,h} + \mathcal{J}(\mu_h, \mu_h).
\end{aligned}$$

That $|||\cdot|||_{0,h}$ indeed defines a norm on $\mathbf{V} \times L^2(\gamma)$ follows immediately from the Poincaré inequality. Thus $|||\cdot|||_{1,h}$ also defines a norm on $\mathbf{V} \times \Lambda_h$. That $\mathcal{A}$ is continuous in the $|||\cdot|||_{0,h}$ norm follows from the Cauchy–Schwarz inequality:

Proposition 2 *There exists a constant C_c, independent of h, such that if P and Q belong to $\mathbf{V} \times L^2(\gamma)$, then*

$$\mathcal{A}(P, Q) \leq C_c |||P|||_{0,h}\, |||Q|||_{0,h}. \tag{12}$$

Proof The Cauchy–Schwarz inequality, (7) and (11) yield (12) with $C_c = C^M_{\mathbf{K}}$. □

3.3 A subspace of Λ_h and some approximation lemmas

The family of inherited meshes $\mathcal{T}_{h,\lambda}$ on γ suffers from the fact that it is not uniformly regular: while for $s \in \mathcal{T}_{h,\lambda}$, its length $h_s \leq h$, there is not necessarily a $\sigma_\lambda > 0$, independent of h, such that $h_s \geq \frac{h}{\sigma_\lambda}$. For this reason we amalgamate elements of $\mathcal{T}_{h,\lambda}$ to obtain a supermesh $\mathcal{T}_{h,\mathcal{P}}$ of $\mathcal{T}_{h,\lambda}$ made up of patch-elements obtained by fusing two or more contiguous elements of $\mathcal{T}_{h,\lambda}$ to form n_h pairwise-disjoint patches, $\mathcal{P}_{h,i}$, $i = 1, \ldots n_h$, see Fig. 1. The patches are used for the analysis, but are not built in practice. The patches should be constructed in such a way that the length of each patch segment is bounded above and below by a multiple of h; i.e. there are positive constants c_1 and c_2, independent of h, such that

$$c_1 h \leq h_{\mathcal{P}_{h,i}} \leq c_2 h,\ i = 1, \ldots n_h, \tag{13}$$

where $h_{\mathcal{P}_{h,i}}$ denotes the length of the patch-segment $\mathcal{P}_{h,i}$. Let $h_{\mathcal{P}}$ be the maximum value of $h_{\mathcal{P}_{h,i}}$, $\mathcal{P}_{h,i} \in \mathcal{T}_{h,\mathcal{P}}$. An additional constraint on the patch construction will be given in Sect. 3.4 following the proof of Lemma 3. From the uniform regularity of $\boldsymbol{\mathcal{T}}_h$, the patch-segments can clearly be constructed so that the maximum number of elements $s \in \mathcal{T}_{h,\lambda}$ in a patch-element $\mathcal{P}_{h,i}$ is bounded above by some number $\bar{n}$ independent of h. The patches can be numbered in such a way that each of $\mathcal{P}_{h,1}$ and $\mathcal{P}_{h,n_h}$ has a vertex on the boundary of γ, and such that for $i = 1, \ldots, n_h - 1$, $\mathcal{P}_{h,i}$ and $\mathcal{P}_{h,i+1}$ have a vertex in common. Similarly, for each i; $i = 1, \ldots, n_h$, the patch $\mathcal{P}_{h,i}$ contains as subsets a certain number, n_i, of cells $s_{i,\ell} \in \mathcal{T}_{h,\lambda}$, $\ell = 1, \ldots n_i$ which we may assume are numbered such that the first and last cells have a vertex on $\partial\mathcal{P}_{h,i}$ and contiguous cells are numbered consecutively. Now define the space of patch-wise constant functions on γ

$$\mathcal{X}_h = \left\{ x_h \in L^2(\gamma) \ : \ x_h|_{\mathcal{P}_{h,i}} \in \mathbb{P}^0(\mathcal{P}_{h,i}),\ i = 1, \ldots, n_h \right\}.$$

(Please note that $\mathcal{X}_h$ is not meant to replace the multiplier space Λ_h but is to be used only in the demonstrations). Then let $\pi_{\mathcal{P}}: L^2(\gamma) \to \mathcal{X}_h$ be defined by $\eta \mapsto \pi_{\mathcal{P}}\eta$ where $\pi_{\mathcal{P}}\eta|_{\mathcal{P}_{h,i}} = \dfrac{1}{h_{\mathcal{P}_{h,i}}} \displaystyle\int_{\mathcal{P}_{h,i}} \eta$. As $\pi_{\mathcal{P}}$ is the $L^2(\gamma)$–projection operator from $L^2(\gamma)$ onto $\mathcal{X}_h$, for $\mu \in L^2(\gamma)$,

$$\langle \mu, \zeta_h \rangle_\gamma = \langle \pi_{\mathcal{P}}\mu, \zeta_h \rangle_\gamma, \qquad \forall \zeta_h \in \mathcal{X}_h, \tag{14}$$

$$\|\pi_{\mathcal{P}}\mu\|_{0,\gamma} \leq \|\mu\|_{0,\gamma}, \tag{15}$$

and if further $\mu \in H^1(\gamma)$, there is a constant $C_{\pi_{\mathcal{P}}}$, independent of h, such that the following Poincaré-Wirtinger-type inequality holds:

$$\|\mu - \pi_{\mathcal{P}}\mu\|_{0,\gamma} \leq C_{\pi_{\mathcal{P}}} h_{\mathcal{P}} \|\nabla_\tau \mu\|_{0,\gamma}. \tag{16}$$

Before stating approximation lemmas we define a mesh-dependent, thickened γ made up of the cells of $\boldsymbol{\mathcal{T}}_h$ crossed by γ plus an extra layer of cells on each side of γ. Let

$$\boldsymbol{\mathcal{S}}_h = \{T \in \boldsymbol{\mathcal{T}}_h \ : \ \exists T' \in \boldsymbol{\mathcal{T}}_h \text{ with } \overline{T} \cap \overline{T}' \neq \emptyset \text{ and } \overline{T}' \cap \gamma \neq \emptyset\},$$

and let $\mathcal{G}_{h,\gamma}$ be the interior of the union of the closures of the cells $T \in \boldsymbol{\mathcal{S}}_h$:

$$\mathcal{G}_{h,\gamma} = \text{Int}\Big(\bigcup_{T \in \boldsymbol{\mathcal{S}}_h} \overline{T} \Big). \tag{17}$$

The following two lemmas concern approximation in $\mathcal{X}_h$:

Lemma 1 *There exist constants* $C_1 \geq 1$ *and* $\tilde{C}_1 > 0$ *such that for* $(q_h, q_{\gamma,h}) \in \boldsymbol{V}_h$

$$\|(q_h|_\gamma - q_{\gamma,h}) - \pi_{\mathcal{P}}(q_h|_\gamma - q_{\gamma,h})\|^2_{\frac{1}{2},h,\gamma} \leq C_1 \Big(\|\nabla q_h\|^2_{0,\mathcal{G}_{h,\gamma}} + h \|\nabla_\tau q_{\gamma,h}\|^2_{0,\gamma} \Big),$$

$$\tilde{C}_1 \|q_h|_\gamma - q_{\gamma,h}\|^2_{\frac{1}{2},h,\gamma} - \|\nabla q_h\|^2_{0,\mathcal{G}_{h,\gamma}} - h\|\nabla_\tau q_{\gamma,h}\|^2_{0,\gamma} \le \|\pi_\mathcal{P}(q_h|_\gamma - q_{\gamma,h})\|^2_{\frac{1}{2},h,\gamma}.$$

Proof The second inequality follows directly from the first with $\tilde{C}_1 = \frac{1}{2C_1}$, so we only need to prove the first. For $q_{\gamma,h}$ in $V_{h,\gamma}$ and q_h in $V_{h,\Omega}$, we have $q_{\gamma,h} \in H^1(\gamma)$ and $q_h|_\gamma \in H^1(\gamma)$. Thus (16) and then (13) implies that

$$\begin{aligned}
\|q_{\gamma,h} - \pi_\mathcal{P} q_{\gamma,h}\|_{0,\gamma} &\le C_{\pi_\mathcal{P}} c_2 h \|\nabla_\tau(q_{\gamma,h})\|_{0,\gamma}, \\
\|q_h|_\gamma - \pi_\mathcal{P}(q_h|_\gamma)\|_{0,\gamma} &\le C_{\pi_\mathcal{P}} c_2 h \|\nabla_\tau(q_h|_\gamma)\|_{0,\gamma}.
\end{aligned}$$

It is obvious for a grid of triangles (when ∇q_h is piecewise constant) but also true for a grid of rectangles that there exists $C_\sigma > 0$ such that

$$\|\nabla_\tau(q_h|_\gamma)\|_{0,\gamma} \le C_\sigma h^{-\frac{1}{2}} \|\nabla_\tau q_h\|_{0,\mathcal{G}_{h,\gamma}} \le C_\sigma h^{-\frac{1}{2}} \|\nabla q_h\|_{0,\mathcal{G}_{h,\gamma}},$$

where C_σ depends on the uniform regularity constant σ_Ω. Combining the last three inequalities we obtain

$$\begin{aligned}
&\|(q_h|_\gamma - q_{\gamma,h}) - \pi_\mathcal{P}(q_h|_\gamma - q_{\gamma,h})\|^2_{\frac{1}{2},h,\gamma} \\
&\le 2h^{-1}\big(\|q_h|_\gamma - \pi_\mathcal{P}(q_h|_\gamma)\|^2_{0,\gamma} + \|q_{\gamma,h} - \pi_\mathcal{P} q_{\gamma,h}\|^2_{0,\gamma}\big) \\
&\le 2C^2_{\pi_\mathcal{P}} c_2^2 \big(C_\sigma^2 \|\nabla q_h\|^2_{0,\mathcal{G}_{h,\gamma}} + h\|\nabla_\tau q_{\gamma,h}\|^2_{0,\gamma}\big).
\end{aligned}$$

The lemma follows with $C_1 = \max\{1, 2C^2_{\pi_\mathcal{P}} c_2^2 \max\{C_\sigma^2, 1\}\}$. □

Lemma 2 *There is a positive constant C_2 such that if $\mu_h \in \Lambda_h$, then*

$$\|\mu_h - \pi_\mathcal{P}\mu_h\|^2_{-\frac{1}{2},h,\gamma} \le C_2 \mathcal{J}(\mu_h, \mu_h).$$

Proof As $\pi_\mathcal{P}$ is an L^2-projection, it suffices to show that there is a constant C_2 such that if $\mu_h \in \Lambda_h$ there is a function $\chi_h \in \mathcal{X}_h$ such that

$$\|\mu_h - \chi_h\|^2_{0,\gamma} \le C_2 h^{-1} \mathcal{J}(\mu_h, \mu_h).$$

Let $\mu_h \in \Lambda_h$. We construct $\chi_h \in \mathcal{X}_h$ such that the value χ_i of χ_h on the patch $\mathcal{P}_{h,i} \in \mathcal{T}_{h,\mathcal{P}}$ is the value $\mu_{i,1}$ of μ_h on the cell $s_{i,1} \in \mathcal{T}_{h,\lambda}$, i.e. $\chi_i = \mu_{i,1}$. Then

$$\begin{aligned}
\|\mu_h - \chi_h\|^2_{0,\gamma} &= \sum_{i=1}^{n_h} \|\mu_h - \chi_i\|^2_{0,\mathcal{P}_{h,i}} = \sum_{i=1}^{n_h}\sum_{\ell=2}^{n_i} \|\mu_h - \chi_i\|^2_{0,s_{i,\ell}} \\
&\le h_\lambda \sum_{i=1}^{n_h}\sum_{\ell=2}^{n_i} (\mu_{i,\ell} - \mu_{i,1})^2 \le h \sum_{i=1}^{n_h}\sum_{\ell=2}^{n_i} \left(\sum_{\nu=1}^{\ell-1} j_{i,\nu}\right)^2 \\
&\le h \sum_{i=1}^{n_h}\sum_{\ell=2}^{n_i} (\ell - 1) \sum_{\nu=1}^{\ell-1} j^2_{i,\nu} \le h \frac{\bar{n}^2}{2} \sum_{i=1}^{n_h}\sum_{\nu=1}^{n_i-1} j^2_{i,\nu},
\end{aligned}$$

where for $\ell = 1, \ldots, n_i$; $\mu_{i,\ell} = \mu_h|_{s_{i,\ell}}$; for $\nu = 1, \ldots, n_i - 1$; $j_{i,\nu} = \mu_{i,\nu+1} - \mu_{i,\nu}$ and $\bar{n}$ is an upper bound, independent of h, on the number of cells $s \in \mathcal{T}_{h,\lambda}$ contained in a patch $\mathcal{P}_h \in \mathcal{T}_{h,\mathcal{P}}$. The lemma follows with $C_2 = \frac{\bar{n}^2}{2\xi}$ since

$$\mathcal{J}(\mu_h, \mu_h) = \sum_{f \in \mathcal{F}_{h,\lambda}} \xi h^2 [\mu_h]_f^2 \geq \xi h^2 \sum_{i=1}^{n_h} \sum_{\nu=1}^{n_i-1} j_{i,\nu}^2.$$

□

3.4 A stability estimate

This section is devoted to the demonstration of the stability estimate for the formulation (8) given in Theorem 1. The proof relies on Lemmas 3 and 4.

Lemma 3 *There exist positive constants C_3, C_4 and C_5 independent of h, such that for $(p_h, p_{\gamma,h})$ in V_h there exists $\eta_h \in \Lambda_h$ satisfying:*

$$\begin{aligned}
&\langle \eta_h, p_h|_\gamma - p_{\gamma,h} \rangle_\gamma \geq C_3 \|\pi_{\mathcal{P}}(p_h|_\gamma - p_{\gamma,h})\|^2_{\frac{1}{2},h,\gamma},\\
&\|\eta_h\|^2_{-\frac{1}{2},h,\gamma} \leq C_4 \|\pi_{\mathcal{P}}(p_h|_\gamma - p_{\gamma,h})\|^2_{\frac{1}{2},h,\gamma},\\
&\mathcal{J}(\eta_h, \eta_h) \leq C_5 \|\pi_{\mathcal{P}}(p_h|_\gamma - p_{\gamma,h})\|^2_{\frac{1}{2},h,\gamma}.
\end{aligned}$$

Proof Given $(p_h, p_{\gamma,h})$ in V_h, define $\eta_h \in \mathcal{X}_h$ by $\eta_h = \frac{1}{h_{\mathcal{P}}} \pi_{\mathcal{P}}(p_h|_\gamma - p_{\gamma,h})$. Use Eqs. (14) and (13) to obtain the first inequality of the lemma:

$$\langle \eta_h, p_h|_\gamma - p_{\gamma,h} \rangle_\gamma = \|h_{\mathcal{P}}^{-\frac{1}{2}} \pi_{\mathcal{P}}(p_h|_\gamma - p_{\gamma,h})\|^2_{0,\gamma} \geq c_2^{-1} \|\pi_{\mathcal{P}}(p_h|_\gamma - p_{\gamma,h})\|^2_{\frac{1}{2},h,\gamma}.$$

For the second inequality, observe that

$$\|\eta_h\|^2_{-\frac{1}{2},h,\gamma} = h h_{\mathcal{P}}^{-2} \|\pi_{\mathcal{P}}(p_h|_\gamma - p_{\gamma,h})\|^2_{0,\gamma} \leq c_1^{-2} \|\pi_{\mathcal{P}}(p_h|_\gamma - p_{\gamma,h})\|^2_{\frac{1}{2},h,\gamma}.$$

For the third inequality, using the definition (6) of $\mathcal{J}$, the fact that η_h is constant on patches, using (13) and the definition of η_h, one obtains

$$\begin{aligned}
\mathcal{J}(\eta_h, \eta_h) &= \sum_{i=1}^{n_h-1} \xi h^2 \left(\eta_h|_{\mathcal{P}_{h,i}} - \eta_h|_{\mathcal{P}_{h,i+1}}\right)^2 \leq \sum_{i=1}^{n_h} 4\xi h^2 \left(\eta_h|_{\mathcal{P}_{h,i}}\right)^2\\
&\leq 4c_1^{-2} \xi \sum_{i=1}^{n_h} \left(\pi_{\mathcal{P}}(p_h|_\gamma - p_{\gamma,h})\big|_{\mathcal{P}_{h,i}}\right)^2 \leq 4c_1^{-3} \xi \|h^{-\frac{1}{2}} \pi_{\mathcal{P}}(p_h|_\gamma - p_{\gamma,h})\|^2_{0,\gamma}.
\end{aligned}$$

Now with $C_3 = c_2^{-1}$, $C_4 = c_1^{-2}$ and $C_5 = 4c_1^{-3}\xi$, the proof is completed. □

Before stating Lemma 4, still following Burman and Hansbo (2010a), we define a subspace $\mathcal{Y}_h$ of $V_{h,\Omega}$ consisting of certain functions having support "near" γ, [i.e. in $\mathcal{G}_{h,\gamma}$, see (17)]. Toward this end we partition $\mathcal{G}_{h,\gamma}$ into a set of nonoverlapping thickened patches $\boldsymbol{\mathcal{P}}_{h,i}$, $i = 1, \ldots, n_h$, where $\boldsymbol{\mathcal{P}}_{h,i}$ is made up of a choice of cells $T \in \boldsymbol{\mathcal{S}}_h$ such that $\mathcal{P}_{h,i} \subset \boldsymbol{\mathcal{P}}_{h,i}$, see Fig. 1. The 1D patches $\mathcal{P}_{h,i}$ should be constructed in such a way that to each thick patch $\boldsymbol{\mathcal{P}}_{h,i}$, we can associate a patch function $\phi_i \in V_{h,\Omega}$ such that for $i = 1, \ldots, n_h$ the patch function ϕ_i satisfies the following conditions:

- $0 \leq \phi_i(x) \leq 1, \quad \forall x \in \Omega,$
- ϕ_i vanishes outside $\boldsymbol{\mathcal{P}}_{h,i}$,
- ϕ_i is identically equal to 1 on some edge $F \in \mathcal{F}_h$ cut by γ,
- there are constants c_3 and c_4 independent of h and of i such that

$$c_3 \leq \frac{1}{h} \int_{\mathcal{P}_{h,i}} \phi_i \quad \text{and} \quad h|\nabla \phi_i(x)| \leq c_4, \ \text{for a. e. } x \in \Omega. \tag{18}$$

Because of the uniform regularity of $\boldsymbol{\mathcal{T}}_h$, it is always possible to construct the patches in such a way: it suffices to amalgamate enough elements $s \in \mathcal{T}_{h,\lambda}$. The subspace $\mathcal{Y}_h$ is the space generated by the functions ϕ_i; $i = 1, \ldots, n_h$. We note also that there is a constant c_5 independent of h such that

$$|\boldsymbol{\mathcal{P}}_{h,i}| \leq c_5 h^2_{\mathcal{P}_{h,i}}. \tag{19}$$

Lemma 4 *There exist positive constants C_6, C_7 and C_8, independent of h, such that for λ_h in Λ_h there is an element $r_h \in \mathcal{Y}_h \subset V_{h,\Omega}$ such that:*

$$\begin{aligned}
\langle \pi_{\mathcal{P}} \lambda_h,\, r_h \rangle_\gamma &\geq C_6 \|\pi_{\mathcal{P}} \lambda_h\|^2_{-\frac{1}{2},h,\gamma}, \\
\|r_h|_\gamma\|^2_{\frac{1}{2},h,\gamma} &\leq C_7 \|\lambda_h\|^2_{-\frac{1}{2},h,\gamma}, \\
\|\nabla r_h\|^2_{0,\Omega} &\leq C_8 \|\lambda_h\|^2_{-\frac{1}{2},h,\gamma}.
\end{aligned}$$

Proof Let $\lambda_h \in \Lambda_h$. We define $r_h \in \mathcal{Y}_h \subset V_{h,\Omega}$ by taking a linear combination of the patch functions as follows:

$$r_h = \sum_{i=1}^{n_h} C_{\mathcal{P}_i} \phi_i, \qquad \text{where} \qquad C_{\mathcal{P}_i} = \frac{\int_{\mathcal{P}_{h,i}} h_{\mathcal{P}} \lambda_h}{\int_{\mathcal{P}_{h,i}} \phi_i}. \tag{20}$$

We then have $r_h|_{\Omega \setminus \mathcal{G}_{h,\gamma}} = 0$, and recalling that ϕ_i vanishes outside of $\boldsymbol{\mathcal{P}}_{h,i}$, we also have

$$\int_{\mathcal{P}_{h,i}} r_h = \int_{\mathcal{P}_{h,i}} h_{\mathcal{P}} \lambda_h, \qquad \forall i = 1, \ldots, n_h. \tag{21}$$

One has $(\int_{\mathcal{P}_{h,i}} \phi_i)^{-2} \le c_3^{-2} h^{-2}$ from (18), and $(\int_{\mathcal{P}_{h,i}} \lambda_h)^2 \le h_{\mathcal{P}_{h,i}} \int_{\mathcal{P}_{h,i}} \lambda_h^2$ by the Cauchy–Schwarz inequality. Thus, using (13), we have the estimate

$$C_{\mathcal{P}_i}^2 \le c_2^3 c_3^{-2} \int_{\mathcal{P}_{h,i}} h\lambda_h^2 \qquad \forall i = 1, \ldots, n_h. \tag{22}$$

Now to check that r_h satisfies the first inequality of the lemma, since $\pi_{\mathcal{P}}\lambda_h$ is constant on $\mathcal{P}_{h,i}$, $i = 1, \ldots, n_h$, we can use (21), (14) and then (13) to obtain

$$\begin{aligned}\langle \pi_{\mathcal{P}}\lambda_h,\, r_h\rangle_\gamma &= \sum_{i=1}^{n_h} (\pi_{\mathcal{P}}\lambda_h)|_{\mathcal{P}_{h,i}} \int_{\mathcal{P}_{h,i}} r_h = \sum_{i=1}^{n_h} (\pi_{\mathcal{P}}\lambda_h)|_{\mathcal{P}_{h,i}} \int_{\mathcal{P}_{h,i}} h_{\mathcal{P}}\lambda_h \\ &= h_{\mathcal{P}} \langle \pi_{\mathcal{P}}\lambda_h,\, \lambda_h\rangle_\gamma = h_{\mathcal{P}} \int_\gamma (\pi_{\mathcal{P}}\lambda_h)^2 \ge c_1 \|\pi_{\mathcal{P}}\lambda_h\|^2_{-\frac{1}{2},h,\gamma}.\end{aligned}$$

For the second inequality we note that by the definition of r_h in (20), we have

$$\|r_h|_\gamma\|^2_{\frac{1}{2},h,\gamma} = h^{-1} \sum_{i=1}^{n_h} C_{\mathcal{P}_i}^2 \int_{\mathcal{P}_{h,i}} \phi_i^2 \le c_2 \sum_{i=1}^{n_h} C_{\mathcal{P}_i}^2$$

because (13), together with the fact that $|\phi_i(x)| \le 1$, $\forall x \in \gamma$, implies that $\int_{\mathcal{P}_{h,i}} \phi_i^2 \le h_{\mathcal{P}_{h,i}} \le c_2 h$. Thus using (22), one obtains

$$\|r_h|_\gamma\|^2_{\frac{1}{2},h,\gamma} \le \sum_{i=1}^{n_h} c_2^4 c_3^{-2} \int_{\mathcal{P}_{h,i}} h\lambda_h^2 = c_2^4 c_3^{-2} \|\lambda_h\|_{-\frac{1}{2},h,\gamma}.$$

For the third inequality, since the supports of the ϕ_i's are disjoint, we have

$$\|\nabla r_h\|^2_{0,\Omega} = \sum_{i=1}^{n_h} C_{\mathcal{P}_i}^2 \int_{\mathcal{P}_{h,i}} (\nabla\phi_i)^2 \le c_2^2 c_4^2 c_5 \sum_{i=1}^{n_h} C_{\mathcal{P}_i}^2 \le c_2^5 c_3^{-2} c_4^2 c_5 \sum_{i=1}^{n_h} \int_{\mathcal{P}_{h,i}} h\lambda_h^2,$$

where we have used (18), then (19) and (22). The lemma now follows with $C_6 = c_1$, $C_7 = c_2^4 c_3^{-2}$ and $C_8 = c_2^5 c_3^{-2} c_4^2 c_5$. □

We can now state a stability theorem for the problem given by (8).

Theorem 1 *Let $\xi > 0$. There is a positive constant θ, independent of h, such that if $P_h = (p_h, p_{\gamma,h}; \lambda_h) \in V_h \times \Lambda_h$, then*

$$\theta \||P_h\||_{1,h} \le \sup_{\substack{Q_h = (q_h, q_{\gamma,h};\, \mu_h) \\ Q_h \in V_h \times \Lambda_h}} \frac{\mathcal{A}(P_h, Q_h) + \mathcal{J}(\lambda_h, \mu_h)}{\||Q_h\||_{1,h}}. \tag{23}$$

Proof Clearly, it suffices to show that there exist positive constants θ_1 and θ_2 such that if $P_h = (p_h, p_{\gamma,h}, \lambda_h) \in \mathbf{V}_h \times \Lambda_h$ there exists $Q_h = (q_h, q_{\gamma,h}, \mu_h) \in \mathbf{V}_h \times \Lambda_h$ such that

$$\theta_1 |||Q_h|||_{1,h} \leq |||P_h|||_{1,h} \quad \text{and} \quad \theta_2 |||P_h|||_{1,h}^2 \leq \mathcal{A}(P_h, Q_h) + \mathcal{J}(\lambda_h, \mu_h). \tag{24}$$

Let $P_h = (p_h, p_{\gamma,h}; \lambda_h) \in \mathbf{V}_h \times \Lambda_h$. The idea is to put

$$Q_h = P_h + R_h, \quad \text{with } R_h = (-c_r r_h, 0; c_\eta \eta_h) \in \mathbf{V}_h \times \Lambda_h$$

with η_h and r_h as constructed in Lemmas 3 and 4, respectively, and with c_r and c_η positive constants to be determined in such a way that both equations of (24) are satisfied with θ_1 and θ_2 independent of the choice of P_h.

To obtain the first estimate of (24), we use Lemmas 3 and 4 with (15):

$$\begin{aligned}
|||Q_h|||_{1,h}^2 &= \|\nabla p_h - c_r \nabla r_h\|_{0,\Omega}^2 + \|\nabla_\tau p_{\gamma,h}\|_{0,\gamma}^2 + \|\lambda_h + c_\eta \eta_h\|_{-\frac{1}{2},h,\gamma}^2 \\
&\quad + \|p_h|_\gamma - c_r r_h|_\gamma - p_{\gamma,h}\|_{\frac{1}{2},h,\gamma}^2 + \mathcal{J}(\lambda_h + c_\eta \eta_h, \lambda_h + c_\eta \eta_h) \\
&\leq 2\left(|||P_h|||_{1,h}^2 + c_\eta^2 \left(\|\eta_h\|_{-\frac{1}{2},h,\gamma}^2 + \mathcal{J}(\eta_h, \eta_h)\right) + c_r^2 \left(\|\nabla r_h\|_{0,\Omega}^2 + \|r_h\|_{\frac{1}{2},h,\gamma}^2\right)\right) \\
&\leq 2\left(|||P_h|||_{1,h}^2 + c_\eta^2 (C_4 + C_5) \|(p_h|_\gamma - p_{\gamma,h})\|_{\frac{1}{2},h,\gamma}^2 + c_r^2 (C_7 + C_8) \|\lambda_h\|_{-\frac{1}{2},h,\gamma}^2\right).
\end{aligned}$$

Thus putting $\theta_1 = (2(1 + \max\{c_\eta^2(C_4 + C_5), c_r^2(C_7 + C_8)\}))^{-\frac{1}{2}}$ we obtain the first inequality of (24).

For the second estimate of (24), letting $\mu_h = \lambda_h + c_\eta \eta_h$, we have

$$\mathcal{A}(P_h, Q_h) + \mathcal{J}(\lambda_h, \mu_h) = \mathcal{A}(P_h, P_h) + \mathcal{J}(\lambda_h, \lambda_h) + \mathcal{A}(P_h, R_h) + c_\eta \mathcal{J}(\lambda_h, \eta_h). \tag{25}$$

Now we bound from below each term in (25). For the first two terms,

$$\begin{aligned}
\mathcal{A}(P_h, P_h) + \mathcal{J}(\lambda_h, \lambda_h) &= \|\mathbf{K}^{\frac{1}{2}} \nabla p_h\|_{0,\Omega}^2 + \|\mathbf{K}_\gamma^{\frac{1}{2}} \nabla_\tau p_{\gamma,h}\|_{0,\gamma}^2 + \mathcal{J}(\lambda_h, \lambda_h) \\
&\geq C_{\mathbf{K}}^m \left(\|\nabla p_h\|_{0,\Omega}^2 + \|\nabla_\tau p_{\gamma,h}\|_{0,\gamma}^2\right) + \mathcal{J}(\lambda_h, \lambda_h).
\end{aligned} \tag{26}$$

For the fourth term, using (7), Young's inequality, then the third inequality of Lemma 3 and (15), we obtain, for each $\varepsilon_\eta > 0$,

$$\begin{aligned}
c_\eta \mathcal{J}(\lambda_h, \eta_h) &\geq -c_\eta \mathcal{J}(\lambda_h, \lambda_h)^{\frac{1}{2}} \mathcal{J}(\eta_h, \eta_h)^{\frac{1}{2}} \geq -\tfrac{c_\eta}{\varepsilon_\eta} \mathcal{J}(\lambda_h, \lambda_h) - \tfrac{c_\eta \varepsilon_\eta}{4} \mathcal{J}(\eta_h, \eta_h) \\
&\geq -\tfrac{c_\eta}{\varepsilon_\eta} \mathcal{J}(\lambda_h, \lambda_h) - \tfrac{c_\eta \varepsilon_\eta}{4} C_5 \|p_h|_\gamma - p_{\gamma,h}\|_{\frac{1}{2},h,\gamma}^2.
\end{aligned} \tag{27}$$

There remains to bound the third term in (25):

$$\mathcal{A}(P_h, R_h) = -c_r \int_\Omega \mathbf{K} \nabla p_h \cdot \nabla r_h + c_r \langle \lambda_h, r_h|_\gamma \rangle_\gamma + c_\eta \langle \eta_h, p_h|_\gamma - p_{\gamma,h} \rangle_\gamma. \tag{28}$$

For the first term of the right hand side of (28) we have, for each $\varepsilon_r > 0$,

$$-c_r \int_{\Omega} \mathbf{K}\nabla p_h \cdot \nabla r_h \geq -\tfrac{c_r}{\varepsilon_r}\|\mathbf{K}^{\frac{1}{2}}\nabla p_h\|_{\Omega}^2 - \tfrac{c_r\varepsilon_r}{4}\|\mathbf{K}^{\frac{1}{2}}\nabla r_h\|_{\Omega}^2 \\ \geq -\tfrac{c_r}{\varepsilon_r} C_{\mathbf{K}}^M \|\nabla p_h\|_{\Omega}^2 - \tfrac{c_r\varepsilon_r}{4} C_{\mathbf{K}}^M C_8 \|\lambda_h\|_{-\frac{1}{2},h,\gamma}^2. \tag{29}$$

For the second term, using the Cauchy–Schwarz type inequality (11), the first inequality of Lemma 4, Young's inequality, the second inequality of Lemma 4 and Lemma 2, we obtain

$$\begin{aligned}
c_r\langle\lambda_h, r_h|_\gamma\rangle_\gamma &= c_r\langle\lambda_h - \pi_{\mathcal{P}}\lambda_h, r_h|_\gamma\rangle_\gamma + c_r\langle\pi_{\mathcal{P}}\lambda_h, r_h|_\gamma\rangle_\gamma \\
&\geq -\frac{c_r}{\varepsilon_r}\|\lambda_h - \pi_{\mathcal{P}}\lambda_h\|_{-\frac{1}{2},h,\gamma}^2 - \frac{c_r\varepsilon_r}{4}\|r_h|_\gamma\|_{\frac{1}{2},h,\gamma}^2 + c_r C_6\|\pi_{\mathcal{P}}\lambda_h\|_{-\frac{1}{2},h,\gamma}^2 \\
&\geq -\frac{c_r}{\varepsilon_r}\|\lambda_h - \pi_{\mathcal{P}}\lambda_h\|_{-\frac{1}{2},h,\gamma}^2 - \frac{c_r\varepsilon_r}{4} C_7\|\lambda_h\|_{-\frac{1}{2},h,\gamma}^2 \\
&\quad + c_r C_6\left(\frac{1}{2}\|\lambda_h\|_{-\frac{1}{2},h,\gamma}^2 - \|\lambda_h - \pi_{\mathcal{P}}\lambda_h\|_{-\frac{1}{2},h,\gamma}^2\right) \\
&\geq -c_r\left(C_6 + \frac{1}{\varepsilon_r}\right)\|\lambda_h - \pi_{\mathcal{P}}\lambda_h\|_{-\frac{1}{2},h,\gamma}^2 + c_r\left(\frac{1}{2}C_6 - \frac{\varepsilon_r}{4}C_7\right)\|\lambda_h\|_{-\frac{1}{2},h,\gamma}^2 \\
&\geq -c_r\left(C_6 + \frac{1}{\varepsilon_r}\right)C_2\mathcal{J}(\lambda_h, \lambda_h) + c_r\left(\frac{1}{2}C_6 - \frac{\varepsilon_r}{4}C_7\right)\|\lambda_h\|_{-\frac{1}{2},h,\gamma}^2,
\end{aligned} \tag{30}$$

where we have also used the inequality $\|a\|^2 \leq 2(\|a-b\|^2 + \|b\|^2)$ which gives $\|b\|^2 \geq \frac{1}{2}\|a\|^2 - \|a-b\|^2$. For the third term of the right hand side of (28) we use the first inequality of Lemma 3 and then Lemma 1 to obtain

$$c_\eta\langle\eta_h, p_h|_\gamma - p_{\gamma,h}\rangle_\gamma \geq c_\eta C_3\|\pi_{\mathcal{P}}(p_h|_\gamma - p_{\gamma,h})\|_{\frac{1}{2},h,\gamma}^2 \\ \geq c_\eta C_3\left(\tilde{C}_1\|p_h|_\gamma - p_{\gamma,h}\|_{\frac{1}{2},h,\gamma}^2 - \|\nabla p_h\|_{0,\mathcal{G}_{h,\gamma}}^2 - h\|\nabla_\tau p_{\gamma,h}\|_{0,\gamma}^2\right). \tag{31}$$

Thus adding Eqs. (26), (27), (29), (30) and (31), we have

$$\begin{aligned}
&\mathcal{A}(P_h, Q_h) + \mathcal{J}(\lambda_h, \mu_h) \geq \left(C_{\mathbf{K}}^m - c_\eta C_3 - \tfrac{c_r}{\varepsilon_r}C_{\mathbf{K}}^M\right)\|\nabla p_h\|_{0,\Omega}^2 \\
&+ (C_{\mathbf{K}}^m - c_\eta C_3 h)\|\nabla_\tau p_{\gamma,h}\|_{0,\gamma}^2 + c_\eta\left(\tilde{C}_1 C_3 - \tfrac{\varepsilon_\eta}{4}C_5\right)\|p_h|_\gamma - p_{\gamma,h}\|_{\frac{1}{2},h,\gamma}^2 \\
&+ c_r\left(\tfrac{C_6}{2} - \tfrac{\varepsilon_r}{4}(C_7 + C_{\mathbf{K}}^M C_8)\right)\|\lambda_h\|_{-\frac{1}{2},h,\gamma}^2 + \left(1 - \tfrac{c_\eta}{\varepsilon_\eta} - c_r C_2\left(C_6 + \tfrac{1}{\varepsilon_r}\right)\right)\mathcal{J}(\lambda_h, \lambda_h).
\end{aligned}$$

To complete the demonstration of the second estimate in (24), we have only to choose $\varepsilon_\eta, \varepsilon_r, c_r$ and c_η such that all of the constant factors above are positive. This can be done by choosing, for instance, $\varepsilon_r = \frac{C_6}{C_7 + C_{\mathbf{K}}^M C_8}$, $\varepsilon_\eta = \frac{2\tilde{C}_1 C_3}{C_5}$, $c_r = \min\{\frac{1}{4C_2(C_6 + \frac{1}{\varepsilon_r})}, \frac{\varepsilon_r}{4}\frac{C_{\mathbf{K}}^m}{C_{\mathbf{K}}^M}\}$ and $c_\eta = \min\{\frac{C_{\mathbf{K}}^m}{2hC_3}, \frac{C_{\mathbf{K}}^m}{4C_3}, \frac{\varepsilon_\eta}{4}\}$. We can then set $\theta_2 = \min\{\frac{1}{2}, \frac{1}{2}C_{\mathbf{K}}^m, \frac{1}{2}c_\eta\tilde{C}_1 C_3, \frac{1}{4}c_r C_6\}$. □

From Theorem 1, we deduce the following corollary.

Corollary 1 *For any* $\xi > 0$, *formulation* (8) *admits a unique solution* $P_h = (p_h, p_{\gamma,h}; \lambda_h) \in \boldsymbol{V}_h \times \Lambda_h$.

3.5 Convergence

Here we prove a Céa-type, best-approximation result and consider the question of convergence rates. The convergence depends on approximation results that are derived under assumptions concerning the regularity of problem (3).

Proposition 3 *Let* $P = (p, p_\gamma; \lambda) \in \boldsymbol{V} \times \Lambda$ *be the solution of* (3) *and let* $P_h = (p_h, p_{\gamma,h}; \lambda_h) \in \boldsymbol{V}_h \times \Lambda_h$ *be the solution of* (8). *Then there is a constant* $\theta_c > 0$ *independent of h such that for each* $Q_h = (q_h, q_{\gamma,h}; \mu_h) \in \boldsymbol{V}_h \times \Lambda_h$,

$$|||P - P_h|||_{0,h} + \mathcal{J}(\lambda_h, \lambda_h)^{\frac{1}{2}} \le \theta_c \left(|||P - Q_h|||_{0,h} + \mathcal{J}(\mu_h, \mu_h)^{\frac{1}{2}} \right). \tag{32}$$

Proof Let $Q_h = (q_h, q_{\gamma,h}; \mu_h) \in \boldsymbol{V}_h \times \Lambda_h$. From the triangle inequality we have

$$|||P - P_h|||_{0,h} + \mathcal{J}(\lambda_h, \lambda_h)^{\frac{1}{2}} \le |||P - Q_h|||_{0,h} + \mathcal{J}(\mu_h, \mu_h)^{\frac{1}{2}} + \sqrt{2}|||Q_h - P_h|||_{1,h}. \tag{33}$$

Because of (23) and (9), there exists $R_h = (r_h, r_{h,\gamma}; \eta_h) \in \boldsymbol{V}_h \times \Lambda_h$ with

$$\begin{aligned} \theta |||Q_h - P_h|||_{1,h} &\le \frac{\mathcal{A}(Q_h - P_h, R_h) + \mathcal{J}(\mu_h - \lambda_h, \eta_h)}{|||R_h|||_{1,h}} \\ &= \frac{\mathcal{A}(Q_h - P, R_h) + \mathcal{J}(\mu_h, \eta_h)}{|||R_h|||_{1,h}}. \end{aligned}$$

Thus using the continuity (12) and inequality (7), we obtain

$$\begin{aligned} \theta |||Q_h - P_h|||_{1,h} &\le C_c |||Q_h - P|||_{0,h} \frac{|||R_h|||_{0,h}}{|||R_h|||_{1,h}} + \mathcal{J}(\mu_h, \mu_h)^{\frac{1}{2}} \frac{\mathcal{J}(\eta_h, \eta_h)^{\frac{1}{2}}}{|||R_h|||_{1,h}} \\ &\le C_c |||Q_h - P|||_{0,h} + \mathcal{J}(\mu_h, \mu_h)^{\frac{1}{2}}. \end{aligned} \tag{34}$$

Then (32) follows from (33) and (34) with $\theta_c = \max\{1 + \frac{\sqrt{2}C_c}{\theta}, 1 + \frac{\sqrt{2}}{\theta}\}$. □

We now recall two trace inequalities. The proof of the first may be found in Costabel (1988, Lemma 3.6) [cf. also Ding (1996, Theorem 1)]. For $\mathcal{O}$ a Lipschitz domain, and $\alpha \in (\frac{1}{2}, \frac{3}{2})$, there is a constant $\widetilde{C}_{tr}$ such that

$$\|\phi|_{\partial\mathcal{O}}\|_{\alpha-\frac{1}{2},\partial\mathcal{O}} \le \widetilde{C}_{tr} \|\phi\|_{\alpha,\mathcal{O}} \qquad \forall \phi \in H^\alpha(\mathcal{O}). \tag{35}$$

The second is a multiplicative trace inequality which follows from Girault and Glowinski (1995, Lemma 2), or from Ainsworth (2007), see also Köppel et al. (2018,

Lemma 1). There is a constant $C_{tr} > 0$ such that if T a triangle or rectangle

$$\|q\|_{0,\gamma\cap T}^2 \leq C_{tr}(h^{-1}\|q\|_{0,T}^2 + h\|\nabla q\|_{0,T}^2), \quad \forall q \in H^1(T). \tag{36}$$

For approximation results we will use the Scott-Zhang interpolation operators associated with the approximation spaces $V_{h,\Omega}$ and $V_{h,\gamma}$:

$$\mathcal{I}_h^{\Omega} : H_0^1(\Omega) \longrightarrow V_{h,\Omega} \quad \text{and} \quad \mathcal{I}_h^{\gamma} : H_0^1(\gamma) \longrightarrow V_{h,\gamma}.$$

We have that if $\frac{1}{2} < \alpha \leq 2$ and $0 \leq \beta \leq \alpha$, then

$$\begin{aligned} \|\mathcal{I}_h^{\Omega} q - q\|_{\beta,\Omega} &\leq C_{\mathcal{I}_\Omega} h^{\alpha-\beta} |q|_{\alpha,\Omega}, && \forall q \in H^{\alpha}(\Omega), \\ \|\mathcal{I}_h^{\gamma} q_\gamma - q_\gamma\|_{\beta,\gamma} &\leq C_{\mathcal{I}_\gamma} h_\gamma^{\alpha-\beta} |q_\gamma|_{\alpha,\gamma}, && \forall q_\gamma \in H^{\alpha}(\gamma), \end{aligned} \tag{37}$$

where $|\cdot|_\alpha$ is an α-semi-norm; see Ern and Guermond (2004, Section 1.6.2).

Define π_h to be the $L^2(\Omega)$–projection onto $\mathbb{P}^0(\mathcal{T}_h)$, the space of piecewise constant functions on Ω subordinate to $\mathcal{T}_h$. We have

$$\|\pi_h q - q\|_{0,\Omega} \leq C_{\pi_h} h \|\nabla q\|_{0,\Omega}, \qquad \forall q \in H^1(\Omega). \tag{38}$$

We will make use of the following auxiliary problem: let Ω_1 and Ω_2 be the two subdomains of Ω obtained by splitting Ω along γ. For $\zeta \in H^{\frac{1}{2}}(\gamma)$ and for $j = 1, 2$, let $r_{\zeta,j} \in H^1(\Omega_j)$ be the solution of

$$\begin{aligned} -\Delta r_{\zeta,j} &= 0 && \text{in } \Omega_j, \\ r_{\zeta,j} &= \zeta && \text{on } \gamma, \\ \nabla r_{\zeta,j} \cdot \mathbf{n} &= 0 && \text{on } \partial\Omega_j \setminus \gamma. \end{aligned} \tag{39}$$

These problems are well posed $\forall \zeta \in H^{\frac{1}{2}}(\gamma)$ and for $j = 1, 2$, [cf. Galvis and Sarkis (2007, Lemma 2.1)] and r_ζ, defined by $r_\zeta\big|_{\Omega_j} = r_{\zeta,j}$, belongs to $H^1(\Omega)$ as $r_{\zeta,1}$ and $r_{\zeta,2}$ coincide on γ. We have

$$\|r_\zeta\|_{1,\Omega} \leq C_a \|\zeta\|_{\frac{1}{2},\gamma}. \tag{40}$$

Lemma 5 *There exists a constant C_9 such that if $\zeta \in H^{\frac{1}{2}}(\gamma)$, and $r_\zeta \in H^1(\Omega)$ is such that $r_\zeta|_{\Omega_1}$ and $r_\zeta|_{\Omega_2}$ are the solutions to* (39), *for $j = 1$ and 2, then*

$$\|\zeta - (\pi_h r_\zeta)\big|_\gamma\|_{-\frac{1}{2},h,\gamma} \leq C_9 h \|\zeta\|_{\frac{1}{2},\gamma} \quad \textit{and} \quad \mathcal{J}((\pi_h r_\zeta)|_\gamma, (\pi_h r_\zeta)|_\gamma)^{\frac{1}{2}} \leq C_9 h \|\zeta\|_{\frac{1}{2},\gamma}.$$

Proof Both of these estimates are based on the trace inequality (36) applied cell by cell. For the first estimate, inequality (36) is applied for each cell $T \in \mathcal{T}_h$ cut by γ, to $\gamma \cap T$ in relation to T: because ζ and r_ζ agree on γ and because $r_\zeta|_T$ and $(\pi_h r_\zeta)|_T$

belong to $H^1(T)$, using (10) and then using (36) for each cell cut by γ and summing over these cells and finally applying (38), we obtain

$$\begin{aligned}
&\|\zeta - (\pi_h r_\zeta)\big|_\gamma\|^2_{-\frac{1}{2},h,\gamma} = h\|(r_\zeta - \pi_h r_\zeta)\big|_\gamma\|^2_{0,\gamma} \\
&\leq C_{tr}\left(\|r_\zeta - \pi_h r_\zeta\|^2_{0,\mathcal{G}_{h,\gamma}} + h^2\|\nabla r_\zeta\|^2_{0,\mathcal{G}_{h,\gamma}}\right) \\
&\leq C_{tr}(1 + C^2_{\pi_h})h^2\|\nabla r_\zeta\|^2_{0,\mathcal{G}_{h,\gamma}}.
\end{aligned}$$

The first estimate, with $C_9 = \left(C_{tr}(1 + C^2_{\pi_h})\right)^{\frac{1}{2}} C_a$, now follows from (40).

For the second estimate, inequality (36) is applied for each edge $F \in \mathcal{F}_h$ cut by γ, to F in relation to each cell $T \in \mathcal{T}_h$ having F as an edge: using the definition of $\mathcal{J}$, the fact that, for each edge $F \in \mathcal{F}_h$, $[\![\pi_h r_\zeta]\!]_F$ is constant and $[\![r_\zeta]\!]_F = 0$ and $h \leq \frac{|F|}{\rho_h} h \leq \sigma_\Omega |F|$ and then using (36), (38) and (40) we obtain

$$\begin{aligned}
\mathcal{J}((\pi_h r_\zeta)|_\gamma, (\pi_h r_\zeta)|_\gamma) &= \sum_{f\in\mathcal{F}_{h,\lambda}} \xi h^2 [\![(\pi_h r_\zeta)|_\gamma]\!]^2_f \leq \xi\sigma_\Omega h \sum_{f\in\mathcal{F}_{h,\lambda}} \int_{F_f} [\![\pi_h r_\zeta]\!]^2_{F_f} \\
&= \xi\sigma_\Omega h \sum_{f\in\mathcal{F}_{h,\lambda}} \|[\![r_\zeta - \pi_h r_\zeta]\!]_{F_f}\|^2_{0,F_f} \\
&\leq 2\xi\sigma_\Omega \sum_{f\in\mathcal{F}_{h,\lambda}} \sum_{i=1}^{2} C_{tr}\left(\|r_\zeta - \pi_h r_\zeta\|^2_{0,T^i_{F_f}} + h^2\|\nabla r_\zeta\|^2_{0,T^i_{F_f}}\right) \\
&\leq 2\xi\sigma_\Omega \sum_{f\in\mathcal{F}_{h,\lambda}} \sum_{i=1}^{2} C_{tr}(1 + C^2_{\pi_h})h^2\|\nabla r_\zeta\|^2_{0,T^i_{F_f}} \\
&\leq 4\xi\sigma_\Omega C_{tr}(1 + C^2_{\pi_h})h^2\|\nabla r_\zeta\|^2_{0,\Omega}
\end{aligned}$$

where for $f \in \mathcal{F}_{h,\lambda}$, F_f denotes the edge in $\mathcal{F}_h$ containing f, and $T^1_{F_f}$ and $T^2_{F_f}$ are the two cells in $\mathcal{T}_h$ having F_f as an edge. The second estimate now follows as above with $C_9 = 2\left(\xi\sigma_\Omega C_{tr}(1 + C^2_{\pi_h})\right)^{\frac{1}{2}} C_a$. □

Lemma 6 *Assume that $p \in H^{1+\alpha}(\Omega)$ and $p|_{\Omega_j} \in H^2(\Omega_j)$ for $j = 1, 2$, $p_\gamma \in H^{1+\tilde{\alpha}}(\gamma)$, for some $\alpha, \tilde{\alpha} \in (\frac{1}{2}, \frac{3}{2})$, and $\lambda = [\![\mathbf{K}\nabla p \cdot \mathbf{n}_\gamma]\!]_\gamma \in H^{\frac{1}{2}}(\gamma)$.*

If $r_\lambda \in H^1(\Omega)$ is defined as in Lemma 5 with $\zeta = \lambda$, it follows that

$$\begin{aligned}
&\left|\left|\left|(p - \mathcal{I}^\Omega_h p, p_\gamma - \mathcal{I}^\gamma_h p_\gamma; \lambda - (\pi_h r_\lambda)|_\gamma)\right|\right|\right|_{1,h} \\
&\quad\leq C_{10}\big(h^\alpha\|p\|_{1+\alpha,\Omega} + h^{\tilde{\alpha}}_\gamma\|p_\gamma\|_{1+\tilde{\alpha},\gamma} + h\sum_{j=1,2}\|p\|_{2,\Omega_j}\big).
\end{aligned}$$

Proof Using the definition of the norm $|||\cdot|||_{0,h}$ we have

$$\begin{aligned}|||(p - \mathcal{I}_h^\Omega p, p_\gamma - \mathcal{I}_h^\gamma p_\gamma; \lambda - \pi_h r_\lambda)|||_{0,h}^2 &= \|\nabla(p - \mathcal{I}_h^\Omega p)\|_{0,\Omega}^2 \\ &+ \|\nabla_\tau(p_\gamma - \mathcal{I}_h^\gamma p_\gamma)\|_{0,\gamma}^2 + \|(p - \mathcal{I}_h^\Omega p)|_\gamma\|_{\frac{1}{2},h,\gamma}^2 \\ &+ \|p_\gamma - \mathcal{I}_h^\gamma p_\gamma\|_{\frac{1}{2},h,\gamma}^2 + \|\lambda - (\pi_h r_\lambda)|_\gamma\|_{-\frac{1}{2},h,\gamma}^2.\end{aligned}$$

The first two terms can be controlled using the interpolation estimates (37). For the third term, using (36), and again (37), we obtain

$$\begin{aligned}\|(p_h - \mathcal{I}_h^\Omega p_h)|_\gamma\|_{\frac{1}{2},h,\gamma}^2 &= \sum_{T \in \mathcal{S}_h} \|(p_h - \mathcal{I}_h^\Omega p_h)|_\gamma\|_{\frac{1}{2},h,T\cap\gamma}^2 \\ &\leq \sum_{T \in \mathcal{S}_h} C_{tr}\left(h^{-2}\|p_h - \mathcal{I}_h^\Omega p_h\|_{0,T}^2 + \|\nabla(p_h - \mathcal{I}_h^\Omega p_h)\|_{0,T}^2\right) \\ &\leq \sum_{T \in \mathcal{S}_h} 2C_{tr}C_{\mathcal{I}_\Omega}^2 h^{2\alpha}|p|_{1+\alpha,T}^2 \leq 2C_{tr}C_{\mathcal{I}_\Omega}^2 h^{2\alpha}|p|_{1+\alpha,\Omega}^2.\end{aligned}$$

To control the fourth term we need only (37)

$$\|p_\gamma - \mathcal{I}_h^\gamma p_\gamma\|_{\frac{1}{2},h_\gamma,\gamma} \leq h_\gamma^{-\frac{1}{2}} C_{\mathcal{I}_\gamma} h_\gamma^{1+\tilde{\alpha}} |p_\gamma|_{1+\tilde{\alpha},\gamma} = C_{\mathcal{I}_\gamma} h_\gamma^{\frac{1}{2}+\tilde{\alpha}} |p_\gamma|_{1+\tilde{\alpha},\gamma}.$$

For the fifth term we use Lemma 5 together with the definition of λ and (35):

$$\|\lambda\|_{\frac{1}{2},\gamma} = \|[\![\mathbf{K}\nabla p \cdot \mathbf{n}_\gamma]\!]_\gamma\|_{\frac{1}{2},\gamma} \leq \sum_{j=1,2} C_{\mathbf{K}}^M \widetilde{C}_{tr} \|\nabla p\|_{1,\Omega_j} \leq C_{\mathbf{K}}^M \widetilde{C}_{tr} \sum_{j=1,2} \|p\|_{2,\Omega_j}.$$

The stabilizing term $\mathcal{J}(\lambda_h, \lambda_h)^{\frac{1}{2}}$ is controlled similarly by the second estimate of Lemma 5. Thus the proof is completed. □

Theorem 2 *Again, let* $P = (p, p_\gamma; \lambda) \in V \times \Lambda$ *be the solution of* (3) *and let* $P_h = (p_h, p_{\gamma,h}; \lambda_h) \in \mathbf{V}_h \times \Lambda_h$ *be the solution of* (8), *and suppose the same regularity as in Lemma 6. Then there exists a constant C such that*

$$|||P - P_h|||_{0,h} + \mathcal{J}(\lambda_h, \lambda_h)^{\frac{1}{2}} \leq C\left(h^\alpha \|p\|_{1+\alpha,\Omega} + h_\gamma^{\tilde{\alpha}} \|p_\gamma\|_{1+\tilde{\alpha},\gamma} + h \sum_{j=1,2} \|p\|_{2,\Omega_j}\right).$$

Proof In Proposition 3 choose $q_h = \mathcal{I}_h^\Omega p$, $q_{\gamma,h} = \mathcal{I}_h^\gamma p_\gamma$ and $\mu_h = \pi_h r_\lambda$ where r_λ is defined as in Lemma 5 with $\zeta = \lambda$. Then the estimate follows directly from Lemma 6. □

Remark 1 Generally p is expected to belong to $H^{3/2-\varepsilon}(\Omega)$ for $\varepsilon > 0$, and it is reasonable to assume that p_γ belongs to $H^2(\gamma)$ (when f_γ is regular for instance) and that $p|_{\Omega_j} \in H^2(\Omega_j)$. Thus $(\mathbf{K}\nabla p|_{\Omega_j} \cdot \mathbf{n}_\gamma)|_\gamma$ belongs to $H^{\frac{1}{2}}(\gamma)$, and the jump λ is indeed in $H^{\frac{1}{2}}(\gamma)$. Thus Theorem 2 provides generally a suboptimal $\mathcal{O}(h^{\frac{1}{2}-\varepsilon})$ convergence. However a better rate of convergence of $\mathcal{O}(h)$ is frequently observed in practice.

4 Numerical results

This section is devoted to numerical experiments studying the proposed discretization (8). In particular we address accuracy, convergence and conditioning for various choices of the penalization parameter. A direct solver is used to solve the linear system. While in the theoretical considerations we have discussed only the case of a single fracture, we have looked at some numerical examples in a few more complex cases. These were treated in the simplest fashion: we assumed continuity of pressure and mass conservation at the intersection of fractures and no flow into or out of a fracture through a tip that is neither a tip of another fracture nor on the boundary of the domain. The simplest way to treat the multipliers is to assume that fractures don't cross but simply meet at their tips: 2 fractures crossing each other can be considered as 4 fractures with one tip in common. (Similarly a sharp angle at a point in a fracture would cause it to be considered as 2 fractures having this point as a tip.) In this way jumps in the multipliers at "crossing points" (or at "irregular points") are not penalized. Of course for a given data set of fractures it may be simpler to simply introduce extra multipliers at crossing points (and at irregular points). Moreover the winding number algorithm is used if the fracture lies along or contains vertices of the matrix mesh.

4.1 Case 1: a vertical fracture

The first setup is a two-dimensional, square domain $\Omega := [0, 1]^2$ with homogeneous Neumann conditions on the horizontal boundaries and nonhomogeneous Dirichlet conditions on the vertical boundaries ($p = 1$ on the left and $p = 4$ on the right). A vertical fracture γ with $\mathbf{K}_\gamma = 10$, $\mathbf{K} = \mathbf{I}$, is located in the middle of the matrix domain Ω with Dirichlet boundary values $p_\gamma = 1$ on the lower tip and $p_\gamma = 4$ on the upper tip. The test case and the pressure distribution are shown in Fig. 2. The simulations are performed for various mesh sizes h, h_γ and h_λ. However, due to the high permeability in the vertical fracture, we have taken $h_\gamma = h/2$ in order to obtain high accuracy at the fracture interface. In the simulations the fracture is either located on the edges of the rectangular matrix grid, referred to as *conforming*, or in the center of the rectangular elements, referred to as *nonconforming*.

On this basis we study the performance of the method. As the analytical solution is not known, a fine simulation of the stabilized Lagrange multiplier method with mesh size $h = 1/384$ is used as a reference solution. The influence of the penalty value ξ on accuracy, convergence and conditioning is analyzed in the conforming as well as the nonconforming case.

Figure 3 presents the numerical convergence analysis of the primary variables for $\xi = 0.5$ in both the cases of conforming and nonconforming meshes. As predicted by the theory and independently of the mesh configuration, the errors for p and p_γ converge linearly in the H^1 norm and the Lagrange multiplier converges linearly in the norm $\| \cdot \|_{-\frac{1}{2},\gamma,h}$. Depending on the mesh configuration the L^2 errors of the pressures converge with rates up to quadratic order: in the nonconforming case the matrix pressure converges linearly in the L^2 norm.

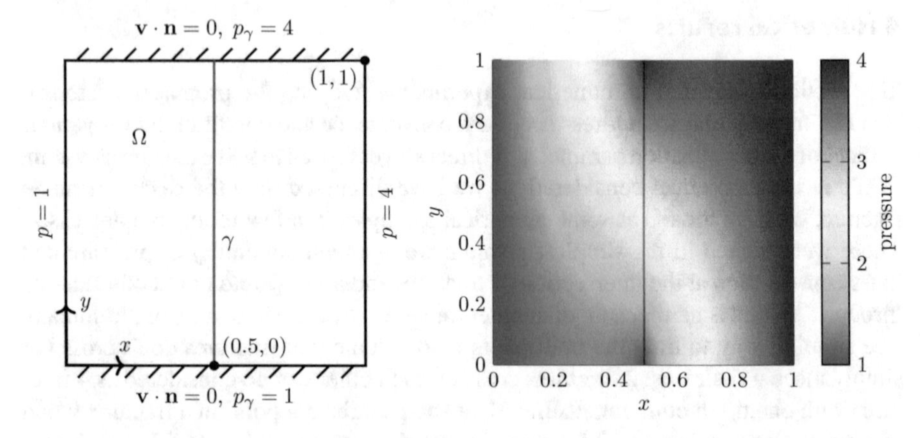

Fig. 2 Case 1 (a vertical fracture): general setting (left) and distribution of the pressures (right) of a simulation with $h = 1/24,\ h_\gamma = 1/48,\ \xi = 1$

Figures 4 and 5 display the approximation errors for matrix pressure, fracture pressure and the Lagrange multiplier for different penalty parameters ξ and different mesh sizes in the conforming and nonconforming cases. The plots indicate that the errors get smaller when the mesh is refined. The penalty parameter ξ has little influence on the H^1 errors for p and p_γ and seems to give slightly improved results for λ when $\xi \in [0.1,\ 1]$, see Fig. 4. The influence of ξ is more important for p and p_γ in the L^2 error, see Fig. 5. In the most interesting case (nonconforming), small to moderate penalty values ($\xi \leq 1$) seem optimal for the accuracy, since the penalty term aligns the multipliers at the expense of the coupling accuracy between matrix and fracture. In each plot, looking at the different curves for fixed ξ, one can determine the convergence rate, that is shown in Fig. 3 for $\xi = 0.5$. Globally, the smaller the mesh size h, the lesser the influence of the penalty parameter ξ. Furthermore an increase of the penalty parameter ξ generally leads to higher errors, which is a logical consequence of the construction of $\mathcal{J}$. We also observe in Fig. 5 that the nonconforming configuration of the mesh yields larger L^2 errors, errors that have a weaker dependence on ξ.

In addition to the nonconforming case in which the fracture cuts through the middle of a vertical strip of matrix mesh elements, we wish to study what happens as the matrix mesh edges approach the fracture line. Table 1 shows the condition numbers of the system matrix depending on the penalty parameter ξ and the distance between the matrix mesh edge and the fracture line for $h = 1/18,\ h_\gamma = 1/36$. The table indicates that the location of the fracture within the matrix grid does not influence significantly the conditioning of the system. However the condition number decreases with increasing values of the penalty parameter. This verifies the efficiency of the stabilizing term $\mathcal{J}$.

We conclude that the method (8) behaves as predicted by the theoretical results. The penalty parameter ξ should be chosen depending on accuracy and conditioning of the system matrix. The stabilizing term $\mathcal{J}$ penalizes the jumps of the Lagrange multiplier. The higher the penalty parameter ξ the more the discrete multipliers tend to a constant

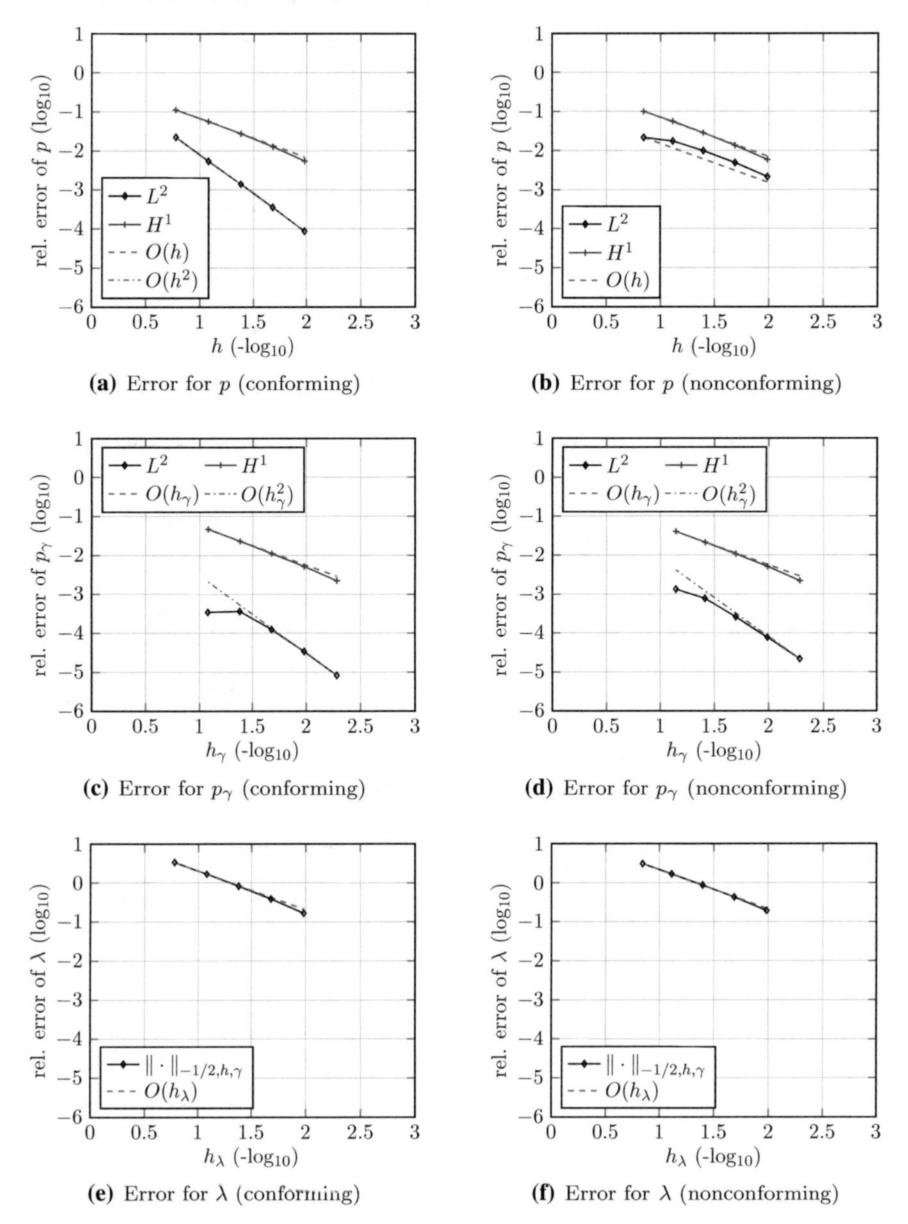

Fig. 3 Case 1 (a vertical fracture): convergence of matrix pressure p, fracture pressure p_γ and Lagrange multiplier λ for $\xi = 0.5$ in the conforming case (left) and the nonconforming case (right) depending on the mesh size

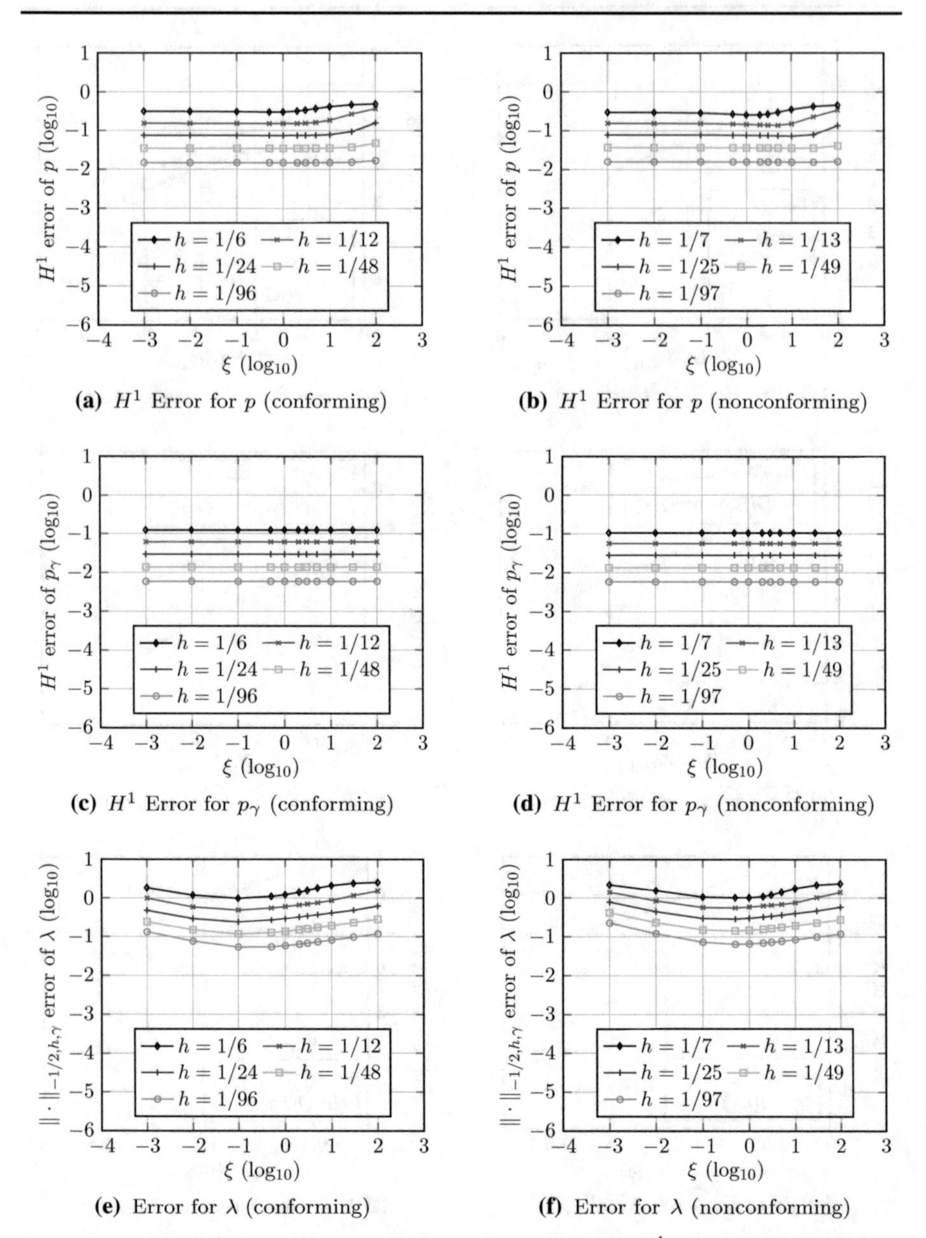

Fig. 4 Case 1 (a vertical fracture): influence of the penalty value ξ on the H^1 error for the matrix pressure p, the fracture pressure p_γ and on the $\|\cdot\|_{-\frac{1}{2},h,\gamma}$ error for the Lagrange multiplier λ in the conforming case (left) and in the nonconforming case (right) depending on the mesh refinement index h

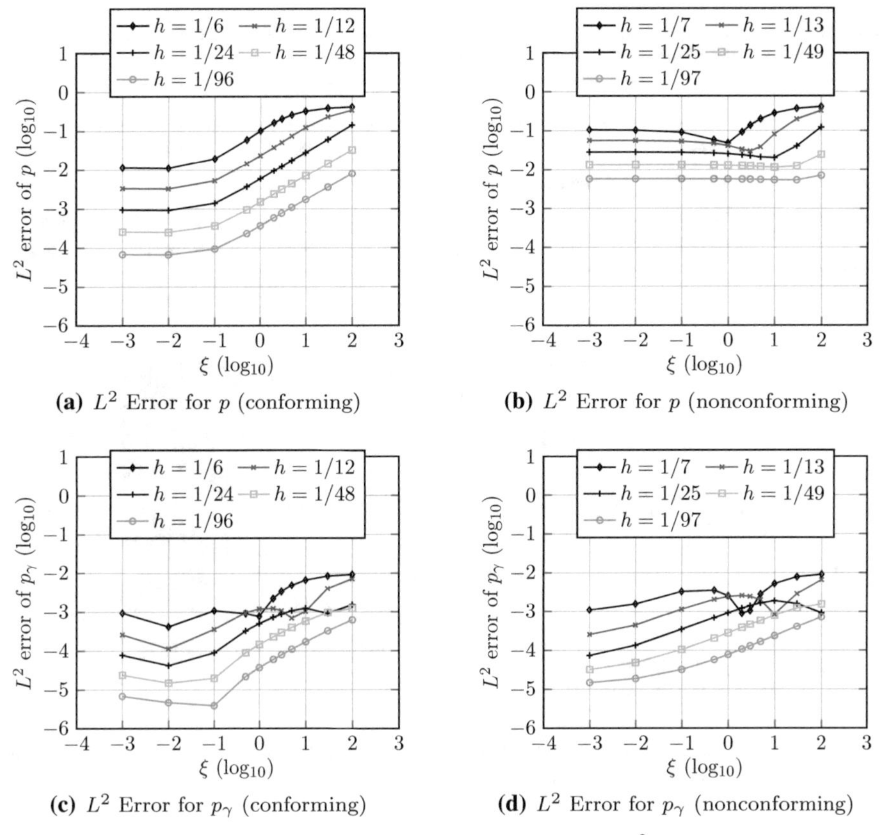

(a) L^2 Error for p (conforming)

(b) L^2 Error for p (nonconforming)

(c) L^2 Error for p_γ (conforming)

(d) L^2 Error for p_γ (nonconforming)

Fig. 5 Case 1 (a vertical fracture): influence of the penalty value ξ on the L^2 approximation error for the matrix pressure p and the fracture pressure p_γ in the conforming case (left) and in the nonconforming case (right) depending on the mesh refinement index h

value which affects the accuracy. However the results show clear convergence also for larger ξ. On the other hand high penalty parameters improve the conditioning.

Remark 2 We have not included comparisons with the non-penalized Lagrange multiplier method of Köppel et al. (2018) for lack of space. For this comparison see Köppel (2018).

4.2 Case 2: A fracture network

In this section a more complex test case with a regular fracture network, cf. Geiger et al. (2013) and Flemisch et al. (2018), is considered. The setup and the pressure distribution for a simulation with the stabilized method with nonconforming configuration and $h = 1/33$, $h_\gamma = 1/32$, $\xi = 1$, is illustrated in Fig. 6. All fractures of the test case are conductive and have a uniform permeability of $\mathbf{K}_\gamma = 1$. The unit square matrix domain is characterized by a permeability of $\mathbf{K} = \mathbf{I}$. Throughout this section the mesh size of the fracture will be in the same range as the mesh size of the matrix, i.e. $h_\gamma \approx h$.

Table 1 Case 1 (a vertical fracture): Condition numbers as a function of ξ and the distance between the matrix mesh edges and the fracture line for $h = 1/18,\ h_\gamma = 1/36$

Distance	$\xi = 0$	$\xi = 0.001$	$\xi = 0.01$	$\xi = 0.1$	$\xi = 1$	$\xi = 10$
$h/2$	4.06e+8	9.93e+7	1.19e+7	2.35e+6	9.19e+5	5.18e+5
$h/4$	3.31e+8	9.50e+7	1.18e+7	2.25e+6	8.88e+5	5.08e+5
$h/8$	2.69e+8	8.97e+7	1.18e+7	2.14e+6	8.53e+5	4.96e+5
0	2.14e+8	8.31e+7	1.18e+7	2.01e+6	8.08e+5	4.81e+5

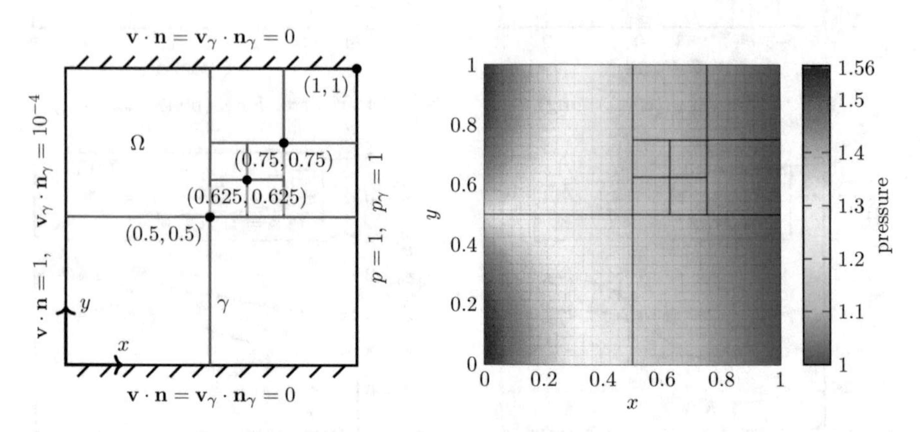

Fig. 6 Case 2 (a fracture network): general setup (left) and pressure distribution (right) of a nonconforming simulation with $h = 1/33,\ h_\gamma = 1/32,\ \xi = 1$

Table 2 Case 2 (a fracture network): Performance of the stabilized Lagrange multiplier method (SLM-FEM) compared to other DFM methods, cf. Flemisch et al. (2018), based on the number of degrees of freedom (d.o.f.), the number of matrix elements (#-matr), the number of fracture elements (#-frac), the matrix error (err_m), the fracture error (err_γ) and the condition number (cond)

Method	d.o.f.	#-matr	#-frac	err_m	err_γ	cond
SLM-FEM: $\xi = 0$	1374	1089 quads	112	1.0e−2	6.5e−3	6.9e+8
SLM-FEM: $\xi = 0.1$	1374	1089 quads	112	1.0e−2	6.5e−3	2.0e+6
SLM-FEM: $\xi = 1$	1374	1089 quads	112	9.9e−3	6.4e−3	7.7e+5
SLM-FEM: $\xi = 1000$	1374	1089 quads	112	6.4e−2	2.9e−2	1.3e+5
Box	577	1078 trias	74	1.1e−2	1.9e−4	2.2e+3
MPFA	1439	1348 trias	91	1.1e−2	4.5e−3	5.8e+4
EDFM	1501	1369 quads	132	6.5e−3	4.0e−3	5.6e+4
Flux-Mortar	3366	1280 trias	75	1.0e−2	6.9e−3	2.4e+6
P-XFEM	1650	961 quads	164	9.3e−3	7.3e−3	9.3e+9
D-XFEM	4474	1250 trias	126	9.6e−3	8.9e−3	1.2e+6

In Table 2 we compare the method (8) with several other available methods given in the benchmark study Flemisch et al. (2018) for single-phase flow in fractured porous

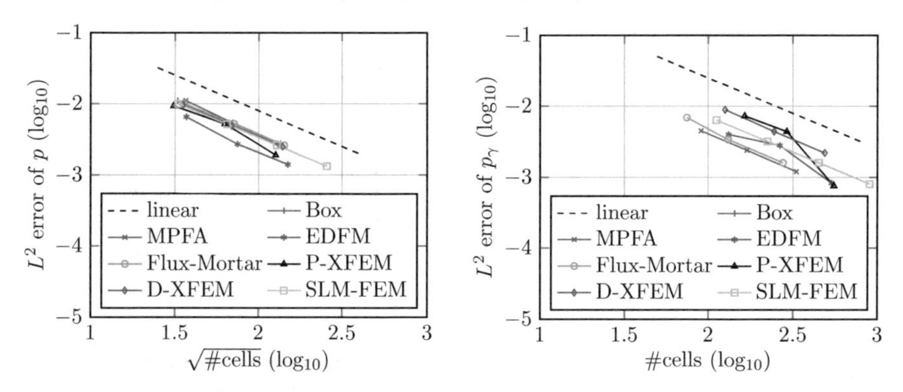

Fig. 7 Case 2 (a fracture network): convergence of matrix pressure p (left) and fracture pressure p_γ (right) of the stabilized Lagr. multiplier method (SLM-FEM) for $\xi = 1$ depending on the number of elements (#cells) compared to other methods, cf. Flemisch et al. (2018)

media. The reference solution is computed with a mimetic finite difference method (Brezzi et al. 2005) with a two-dimensional grid in the fracture as well as in the matrix domain. The table shows that the stabilized Lagrange multiplier method performs well. Intermediate values of the penalty parameter, i.e. $\xi \in [0.1, 1]$, yield a good balance between accuracy and conditioning. Note however that this range may change for more realistic permeabilities, see Sect. 4.3.

To study the numerical convergence of the stabilized method the configuration is refined three times by a factor of two ($h \in \{1/33, 1/65, 1/129, 1/257\}$) similar to Flemisch et al. (2018). The resulting convergence study of matrix and fracture pressure using the stabilized discretization with $\xi = 1$ is illustrated in Fig. 7. The figure shows that the stabilized method converges linearly in the matrix and in the fracture as the other methods of the benchmark study.

4.3 Case 3: A realistic case

The last numerical experiment represents a real set of fractures from an interpreted outcrop in the Sotra Island near Bergen in Norway of the benchmark study Flemisch et al. (2018). The domain is rectangular with uniform permeability $\mathbf{K} = 10^{-14}\text{m}^2$. It contains 64 fractures grouped in several networks with $\mathbf{K}_\gamma = 10^{-10}\text{m}^3$, see Fig. 8. The size of the domain is $700\,\text{m} \times 600\,\text{m}$ with homogeneous Neumann boundary conditions on top and bottom, a pressure of 10.1325 bar on the left and a pressure of 0 bar on the right.

The pressure distributions for simulations with $h = 10\,\text{m}$ and $\xi = 1$ and 100 are displayed in Fig. 9. A small undershoot can be observed in the vicinity of the upper right fracture connecting the top with the right boundary. In this region the pressures should be similar to the boundary condition on the right. Because of the piecewise linear basis functions and the high permeability contrast, the weak constraint of the equality of the pressures on the fracture interface leads to unphysical matrix pressures near the fracture. However the undershoot is smaller for the larger choice of ξ. Fig. 10

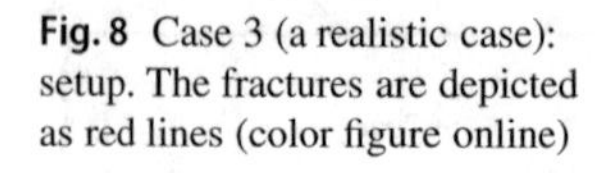

Fig. 8 Case 3 (a realistic case): setup. The fractures are depicted as red lines (color figure online)

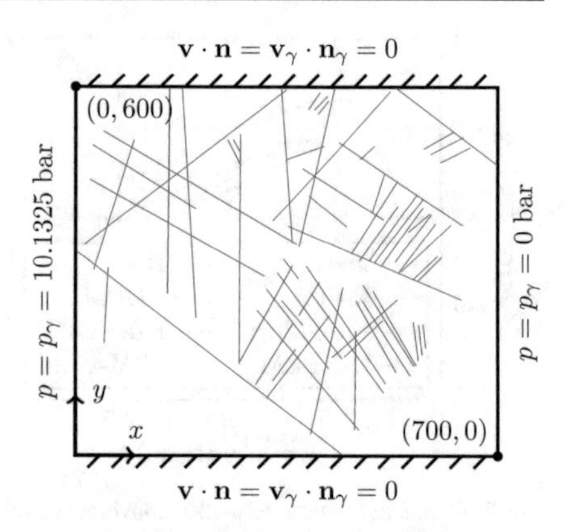

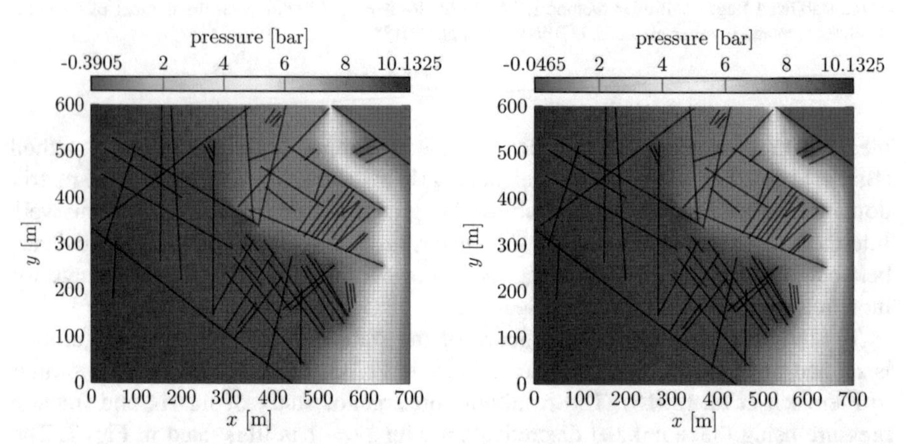

Fig. 9 Case 3 (a realistic case): pressure distribution. Simulation with $h = 10$ m and for $\xi = 1$ (left) and $\xi = 100$ (right). The lowest value is slightly different for the two figures

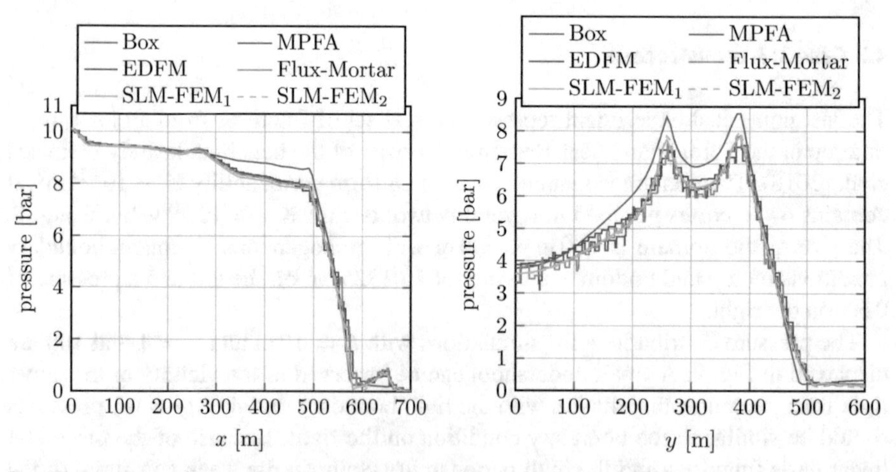

Fig. 10 Case 3 (a realistic case): pressure distribution along the line $y = 500$ m (left) and along the line $x = 625$ m (right) of the stabilized Lagrange multiplier method (with $h = 10$ m) in comparison with other methods, cf. Flemisch et al. (2018). SLM-FEM$_1$: $\xi = 1$. SLM-FEM$_2$: $\xi = 100$

Table 3 Benchmark 3 (a realistic case): performance of the stabilized Lagrange multiplier method (SLM-FEM) compared to other DFM methods based on the number of degrees of freedom (d.o.f.), the number of matrix elements (#-matr), the number of fracture elements (#-frac) and the condition number (cond), cf. Flemisch et al. (2018)

Method	d.o.f.	#-matr	#-frac	cond
SLM-FEM: $\xi = 0$	8081	5250 quads	1372	4.8e+9
SLM-FEM: $\xi = 1\text{e}{-4}$	8081	5250 quads	1372	3.5e+8
SLM-FEM: $\xi = 1$	8081	5250 quads	1372	6.1e+6
SLM-FEM: $\xi = 100$	8081	5250 quads	1372	6.2e+6
Box	5563	10807 trias	1386	9.3e+5
MPFA	8588	7614 trias	867	4.9e+6
EDFM	3599	2491 quads	1108	4.7e+6
Flux-Mortar	25,258	8319 trias	1317	2.2e+17

shows the distributions of the pressures at $y = 500$ m and $x = 625$ m of the stabilized Lagrange multiplier method compared to other methods of the benchmark study. The stabilized Lagrange multiplier method is in very good agreement with the pressure distributions of the other methods. A convergence analysis was not performed in this test case, as was noted in Flemisch et al. (2018), since it is a really difficult task to establish a full-dimensional reference solution with the mimetic finite difference method. The comparison in terms of the conditioning and degrees of freedom is given in Table 3 and shows good performances with moderate to large values of ξ.

Remark 3 In order to improve the conditioning in this realistic test case where the permeabilities are very small, a scaling was performed prior to the computations: the duality pairings $\langle \cdot, \cdot \rangle_\gamma$ of $\mathcal{A}$ defined in (2) and the stabilization term $\mathcal{J}$ defined in (6) were multiplied by the scalar $k = \mathbf{K}$. Thus the actual Lagrange multiplier unknown was $\tilde{\lambda} = k\lambda$, and the stabilization term $\mathcal{J}$ scales correctly with $\mathbf{K}$. It was then possible to divide the discrete system (8) by $\mathbf{K}_\gamma$, and obtain a system that is easier to solve. In the case when $\mathbf{K}$ is a tensor or varying in space, the same idea should be considered with an average value of its norm.

5 Conclusion

In this paper we presented a stabilized finite element discretization of a Lagrange multiplier model for single-phase Darcy flow in fractured porous media, cf. Köppel et al. (2018), where the multiplier represents a local exchange of the fluid between fracture and matrix domain allowing for the use of a mesh in the matrix that is not aligned with the fracture. The piecewise constant Lagrange multipliers of the stabilized discretization are defined on the intersections of the matrix elements with the fracture and, hence, are embedded on the fracture interface. In contrast to the method proposed in Köppel et al. (2018), a weakly consistent stability term penalizes the jumps of the consecutive multipliers to stabilize the discrete saddle point system. We proved stability and convergence of the discrete formulation following the ideas of Burman and Hansbo

(2010a). The numerical experiments are consistent with the theoretical results. They confirmed that with increasing values of the penalty parameter ξ the conditioning of the discrete system can be improved. On the other hand high penalty values deteriorate the accuracy of the approximation. Hence we recommend the use of intermediate penalty values to obtain optimal results. The particular choice of ξ depends on the test case considered. Despite the affected accuracy the results still show clear convergence even for large penalty parameters. In the numerical examples, when the coupling term λ is regular enough the errors of the matrix and fracture pressure converge linearly in the H^1 norm confirming the theoretical results. The Lagrange multiplier is characterized by linear rates of convergence in the discrete norm $\|\cdot\|_{-\frac{1}{2},h_\lambda,\gamma}$. These convergence rates are obtained independently of the possibly very irregular induced mesh for the Lagrange multiplier. The comparison with the benchmark results in Flemisch et al. (2018), leads to the conclusion that the penalized discretization is a good alternative to other models for the simulation of flow in fractured porous media.

Acknowledgements The authors would like to thank the German Research Foundation (DFG) for financial support of the project within the Cluster of Excellence in Simulation Technology (EXC 310/2) at the University of Stuttgart, and the two referees whose comments helped improving this paper.

References

Ahmed, E., Jaffré, J., Roberts, J.E.: A reduced fracture model for two-phase flow with different rock types. Math. Comput. Simul. **137**, 49–70 (2017)

Ainsworth, M.: A posteriori error estimation for discontinuous Galerkin finite element approximation. SIAM J. Numer. Anal. **45**(4), 1777–1798 (2007)

Alboin, C., Jaffré, J., Roberts, J.E., Serres, C.: Modeling fractures as interfaces for flow and transport in porous media. In: Fluid Flow and Transport in Porous Media: Mathematical and Numerical Treatment, Contemporary Mathematics, vol. 295, American Mathematical Society, Providence, RI, pp. 13–24 (2002)

Angot, P., Boyer, F., Hubert, F.: Asymptotic and numerical modelling of flows in fractured porous media. ESAIM: M2AN **43**(2), 239–275 (2009)

Antonietti, P.F., Facciolà, C., Russo, A., Verani, M.: Discontinuous Galerkin approximation of flows in fractured porous media on polytopic grids. Technical Report 22/2016, Politecnico di Milano (2016a)

Antonietti, P.F., Formaggia, L., Scotti, A., Verani, M., Verzott, N.: Mimetic finite difference approximation of flows in fractured porous media. ESAIM: M2AN **50**(3), 809–832 (2016b)

Baca, R.G., Arnett, R.C., Langford, D.W.: Modelling fluid flow in fractured-porous rock masses by finite-element techniques. Int. J. Numer. Methods Fluids **4**, 337–348 (1984)

Berrone, S., Pieraccini, S., Scialò, S.: An optimization approach for large scale simulations of discrete fracture network flows. J. Comput. Phys. **256**, 838–853 (2014)

Boon, W., Nordbotten, J., Yotov, I.: Robust discretization of flow in fractured porous media. SIAM J. Numer. Anal. **56**(4), 2203–2233 (2018)

Brenner, K., Groza, M., Guichard, C., Masson, R.: Vertex approximate gradient scheme for hybrid dimensional two-phase Darcy flows in fractured porous media. ESAIM Math. Model. Numer. Anal. **49**(2), 303–330 (2015)

Brezzi, F., Lipnikov, K., Simoncini, V.: A family of mimetic finite difference methods on polygonal and polyhedral meshes. Math. Models Methods Appl. Sci. **15**(10), 1533–1551 (2005)

Burman, E., Hansbo, P.: Fictitious domain finite element methods using cut elements: I. A stabilized Lagrange multiplier method. Comput. Methods Appl. Mech. Eng. **199**(41–44), 2680–2686 (2010a)

Burman, E., Hansbo, P.: Interior-penalty-stabilized Lagrange multiplier methods for the finite-element solution of elliptic interface problems. IMA J. Numer. Anal. **30**, 870–885 (2010b)

Chave, F.A., Di Pietro, D.A., Formaggia, L.: A hybrid high-order method for passive transport in fractured porous media (2018)/ https://hal.archives-ouvertes.fr/hal-01784181

Costabel, M.: Boundary integral operators on Lipschitz domains: elementary results. SIAM J. Math. Anal. **19**(3), 613–626 (1988)
Ding, Z.: A proof of the trace theorem of sobolev spaces on Lipschitz domains. Proc. Am. Math. Soc. **124**(2), 591–600 (1996)
Ern, A., Guermond, J.L.: Theory and Practice of Finite Elements, Applied Mathematical Sciences, vol. 159. Springer, New York (2004)
Faille, I., Fumagalli, A., Jaffré, J., Roberts, J.E.: Model reduction and discretization using hybrid finite volumes for flow in porous media containing faults. Comput. Geosci. **20**(2), 317–339 (2016)
Flemisch, B., Berre, I., Boon, W., Fumagalli, A., Schwenck, N., Scotti, A., Stefansson, I., Tatomir, A.: Benchmarks for single-phase flow in fractured porous media. Adv. Water Resour. **111**, 239–258 (2018)
Formaggia, L., Scotti, A., Sottocasa, F.: Analysis of a mimetic finite difference approximation of flows in fractured porous media. ESAIM: M2AN **52**(2), 595–630 (2018)
Frih, N., Martin, V., Roberts, J.E., Saâda, A.: Modeling fractures as interfaces with nonmatching grids. Comput. Geosci. **16**(4), 1043–1060 (2012)
Fumagalli, A., Scotti, A.: A numerical method for two-phase flow in fractured porous media with non-matching grids. Adv. Water Resour. **62**(Part C), 454–464. (Computational Methods in Geologic CO_2 Sequestration (2013))
Fumagalli, A., Pasquale, L., Zonca, S., Micheletti, S.: An upscaling procedure for fractured reservoirs with embedded grids. Water Resour. Res. **52**(8), 6506–6525 (2016)
Galvis, J., Sarkis, M.: Non-matching mortar discretization analysis for the coupling Stokes–Darcy equations. Electron. Trans. Numer. Anal. **26**, 350–384 (2007)
Geiger, S., Dentz, M., Neuweiler, I.: A novel multi-rate dual-porosity model for improved simulation of fractured and multi-porosity reservoirs. Soc. Petrol. Eng. J. **18**(4), 670–684 (2013)
Girault, V., Glowinski, R.: Error analysis of a fictitious domain method applied to a Dirichlet problem. Jpn. J. Ind. Appl. Math. **12**(3), 487 (1995)
Hoteit, H., Firoozabadi, A.: An efficient numerical model for incompressible two-phase flow in fractured media. Adv. Water Resour. **31**, 891–905 (2008)
Karimi-Fard, M., Durlofsky, L.J., Aziz, K.: An efficient discrete-fracture model applicable for general-purpose reservoir simulators. Soc. Petrol. Eng. J. **9**(2), 227–236 (2004)
Knabner, P., Roberts, J.E.: Mathematical analysis of a discrete fracture model coupling Darcy flow in the matrix with Darcy–Forchheimer flow in the fracture. ESAIM: Math. Model. Numer. Anal. **48**(5), 1451–1472 (2014)
Köppel, M.: Flow in Heterogeneous Porous Media: Fractures and Uncertainty Quantification. PhD thesis, University of Stuttgart, Germany (2018)
Köppel, M., Martin, V., Jaffré, J., Roberts, J.E.: A Lagrange multiplier method for a discrete fracture model for flow in porous media (2018 accepted). https://hal.archives-ouvertes.fr/hal-01700663
Lesinigo, M., D'Angelo, C., Quarteroni, A.: A multiscale Darcy–Brinkman model for fluid flow in fractured porous media. Numer. Math. **117**(4), 717–752 (2011)
Martin, V., Jaffré, J., Roberts, J.E.: Modeling fractures and barriers as interfaces for flow in porous media. SIAM J. Sci. Comput. **26**(5), 1667–1691 (2005)
Massing, A.: A Cut Discontinuous Galerkin method for coupled bulk-surface problems. (2017) ArXiv e-print. arXiv:1707.02153v1 [math.NA]
Pichot, G., Erhel, J., de Dreuzy, J.R.: A generalized mixed hybrid mortar method for solving flow in stochastic discrete fracture networks. SIAM J. Sci. Comput. **34**(1), B86–B105 (2012)
Reichenberger, V., Jakobs, H., Bastian, P., Helmig, R.: A mixed-dimensional finite volume method for two-phase flow in fractured porous media. Adv. Water Resour. **29**(7), 1020–1036 (2006)
Sandve, T.H., Berre, I., Nordbotten, J.M.: An efficient multi-point flux approximation method for Discrete Fracture–Matrix simulations. J. Comput. Phys. **231**(9), 3784–3800 (2012)
Schwenck, N., Flemisch, B., Helmig, R., Wohlmuth, B.I.: Dimensionally reduced flow models in fractured porous media: crossings and boundaries. Comput. Geosci. **19**(6), 1219–1230 (2015)

Publisher's Note Springer Nature remains neutral with regard to jurisdictional claims in published maps and institutional affiliations.

GEM - International Journal on Geomathematics
https://doi.org/10.1007/s13137-019-0120-z

ORIGINAL PAPER

A cut finite element method for elliptic bulk problems with embedded surfaces

Erik Burman[1] · Peter Hansbo[2] · Mats G. Larson[3] · David Samvin[2]

Received: 27 April 2018 / Accepted: 31 October 2018

Abstract

We propose an unfitted finite element method for flow in fractured porous media. The coupling across the fracture uses a Nitsche type mortaring, allowing for an accurate representation of the jump in the normal component of the gradient of the discrete solution across the fracture. The flow field in the fracture is modelled simultaneously, using the average of traces of the bulk variables on the fractures. In particular the Laplace–Beltrami operator for the transport in the fracture is included using the average of the projection on the tangential plane of the fracture of the trace of the bulk gradient. Optimal order error estimates are proven under suitable regularity assumptions on the domain geometry. The extension to the case of bifurcating fractures is discussed. Finally the theory is illustrated by a series of numerical examples.

Keywords Finite element · Unfitted · Embedded · Fractures

Mathematics Subject Classification 65M60 · 74S05 · 76S05

1 Introduction

We consider a model Darcy creeping flow problem with low permeability in the bulk and with embedded interfaces with high permeability. Our approach is based on the

✉ David Samvin
david.samvin@ju.se

Erik Burman
e.burman@ucl.ac.uk

Peter Hansbo
peter.hansbo@ju.se

Mats G. Larson
mats.larson@umu.se

[1] Mathematics, University College London, London, UK

[2] Mechanical Engineering, Jönköping University, Jönköping, Sweden

[3] Mathematics and Mathematical Statistics, Umeå University, Umeå, Sweden

Nitsche extended finite element of Hansbo and Hansbo (2002), which however did not include transport on the interface. Here, we follow Capatina et al. (2016) and let a suitable mean of the solution on the interface be affected by a transport equation see also Burman et al. (2015). We present a complete a priori analysis and consider the important extension to bifurcating fractures.

The flow model we use is essentially the one proposed in Capatina et al. (2016). More sophisticated models have been proposed, e.g., in Angot et al. (2009), Formaggia et al. (2014), Frih et al. (2008) and Martin et al. (2005), in particular allowing for jumps in the solution across the interfaces. To allow for such jumps, one can either align the mesh with the interfaces, as in, e.g., Berrone et al. (2017) and Hægland et al. (2009), or use extended finite element techniques, cf. Burman et al. (2015), Capatina et al. (2016), D'Angelo and Scotti (2012) and Del Pra et al. (2017).

In previous work Burman et al. (2017) we used a continuous approximation with the interface equations simply added to the bulk equation, which does not allow for jumps in the solution. This paper presents a more complex, but more accurate, discrete solution to the problem. To reduce the technical detail of the arguments we consider a semi-discretization of the problem where we assume that the integrals on the interface and the subdomains separated by the interface can be evaluated exactly. The results herein can be extended to the fully discrete setting, with a piecewise affine approximation of the fracture using the analysis detailed in Burman et al. (2016).

An outline of the paper is as follows: in Sect. 2 we formulate the model problem, its weak form, and investigate the regularity properties of the solution, in Sect. 3 we formulate the finite element method, in Sect. 4 we derive error estimates, in Sect. 5 we extend the approach to the case of bifurcating fractures, and in Sect. 6 we present numerical examples including a study of the convergence and a more applied example with a network of fractures.

2 The model problem

In this section we introduce our model problem. First we present the strong form of the equations and then we derive the weak form that is used for the finite element modelling. We discuss the regularity properties of the solution and show that if the fracture is sufficiently smooth the problem solution, restricted to the subdomains partitioning the global domain, has a regularity that allows for optimal approximation estimates for piecewise affine finite element methods (Fig. 1).

2.1 Strong and weak formulations

Let Ω be a convex polygonal domain in $\mathbb{R}^d$, with $d = 2$ or 3. Let Γ be a smooth embedded interface in Ω, which partitions Ω into two subdomains Ω_1 and Ω_2. We consider the problem: find the pressure $u : \Omega \to \mathbb{R}$ such that

$$-\nabla \cdot a\nabla u = f \qquad \text{in } \Omega_i, i = 1, 2 \tag{2.1}$$

$$-\nabla_\Gamma \cdot a_\Gamma \nabla_\Gamma u = f_\Gamma - [\![n \cdot a\nabla u]\!] \qquad \text{on } \Gamma \tag{2.2}$$

Fig. 1 Schematic figure of bifurcating fractures

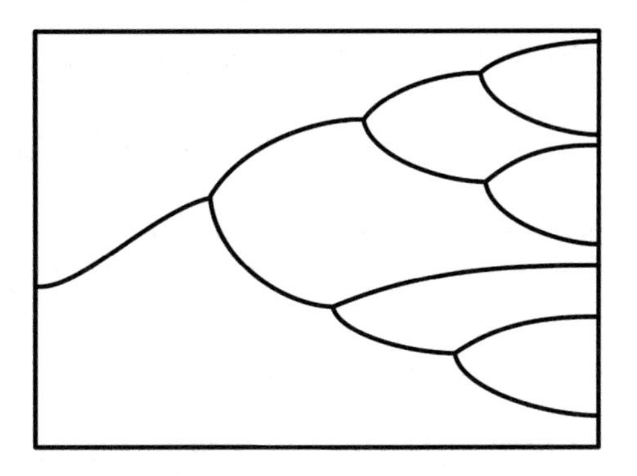

$$[u] = 0 \qquad \text{on } \Gamma \tag{2.3}$$
$$u = 0 \qquad \text{on } \partial\Omega \tag{2.4}$$

Here

$$[v] = v_1 - v_2, \qquad [\![n \cdot a\nabla v]\!] = n_1 \cdot a_1 \nabla v_1 + n_2 \cdot a_2 \nabla v_2 \tag{2.5}$$

where $v_i = v|_{H^1(\Omega_i)}$, n_i is the exterior unit normal to Ω_i, a_i are positive bounded permeability coefficients, for simplicity taken as constant, and $0 \leq a_\Gamma < \infty$ is a constant permeability coefficient on the interface. Note that it follows from (2.3) that v is continuous across Γ while from (2.2) we conclude that the normal flux is in general not continuous across Γ. Note also that taking $a_\Gamma = 0$ and $f_\Gamma = 0$ corresponds to a standard Poisson problem with possible jump in permeability coefficient across Γ.

To derive the weak formulation of the system we introduce the L^2-scalar product over a domain $X \subset \mathbb{R}^d$, or $X \subset \mathbb{R}^{d-1}$. For $u, v \in L^2(X)$ let

$$(u, v)_X = \int_X u\, v \,\mathrm{d}X \tag{2.6}$$

with the associated norm $\|u\|_X = (u, u)_X^{1/2}$. Multiplying (2.1) by

$$v \in V := \left\{ v \in H^1(\Omega); \quad v|_\Gamma \in H^1(\Gamma) \right\}, \tag{2.7}$$

integrating by parts over Ω_i, and using (2.2), we obtain

$$(f, v)_\Omega = -(\nabla \cdot a\nabla u, v)_{\Omega_1} - (\nabla \cdot a\nabla u, v)_{\Omega_2} \tag{2.8}$$
$$= (a\nabla u, \nabla v)_{\Omega_1} + (a\nabla u, \nabla v)_{\Omega_2} - ([\![n \cdot a\nabla u]\!], v)_\Gamma \tag{2.9}$$
$$= (a\nabla u, \nabla v)_\Omega - (f_\Gamma + \nabla_\Gamma \cdot a_\Gamma \nabla_\Gamma u, v)_\Gamma \tag{2.10}$$
$$= (a\nabla u, \nabla v)_\Omega + (a_\Gamma \nabla_\Gamma u, \nabla_\Gamma v)_\Gamma - (f_\Gamma, v)_\Gamma \tag{2.11}$$

We thus arrive at the weak formulation: find $u \in V$ such that

$$(a\nabla u, \nabla v)_\Omega + (a_\Gamma \nabla_\Gamma u, \nabla_\Gamma v)_\Gamma = (f, v)_\Omega + (f_\Gamma, v)_\Gamma \qquad \forall v \in V \tag{2.12}$$

Observing that V is a Hilbert space with scalar product

$$a(v, w) = (a\nabla v, \nabla w)_\Omega + (a_\Gamma \nabla_\Gamma v, \nabla_\Gamma w)_\Gamma \tag{2.13}$$

and associated norm $\|v\|_a^2 = a(v, v)$ it follows from the Lax–Milgram lemma that there is a unique solution to (2.12) in V for $f \in H^{-1}(\Omega)$ and $f_\Gamma \in H^{-1}(\Gamma)$.

2.2 Regularity properties

To prove optimality of our finite element method we need that the exact solution is sufficiently regular. However since the normal fluxes jump over the interface the solution can not have square integrable weak second derivatives. If the interface is smooth however we will prove that the solution restricted to the different subdomains Ω_1, Ω_2 and Γ is regular. The upshot of the unfitted finite element method is that this local regularity is sufficient for optimal order approximation. More precisely we have the elliptic regularity estimate

$$\boxed{\|u\|_{H^2(\Omega_1)} + \|u\|_{H^2(\Omega_2)} + \|u\|_{H^2(\Gamma)} \lesssim \|f\|_\Omega + \|f_\Gamma\|_\Gamma} \tag{2.14}$$

Proof Let $u_i \in H_0^1(\Omega_i)$ solve

$$(a_i \nabla u_i, \nabla v)_{\Omega_i} = (f, v)_{\Omega_i} \qquad \forall v \in H_0^1(\Omega_i) \tag{2.15}$$

Then we have

$$\|u_i\|_{H^2(\Omega_i)} \lesssim \|f\|_{\Omega_i} \qquad i = 1, 2 \tag{2.16}$$

Let $u = u_\Gamma + u_1 + u_2$ where u_Γ satisfies

$$-\nabla_\Gamma \cdot a_\Gamma \nabla_\Gamma u_\Gamma = f_\Gamma + [\![n \cdot a\nabla(u_\Gamma + u_1 + u_2)]\!] \tag{2.17}$$

$$= f_\Gamma + [\![n \cdot a\nabla u_\Gamma]\!] + n_1 \cdot a\nabla u_1 + n_2 \cdot a\nabla u_2 \qquad \text{on } \Gamma \tag{2.18}$$

and

$$-\nabla \cdot a\nabla u_\Gamma = 0 \qquad \text{on } \Omega_i, i = 1, 2 \tag{2.19}$$

Using (2.16) we conclude that

$$n_i \cdot a\nabla u_i|_\Gamma \in H^{1/2}(\Gamma) \qquad i = 1, 2 \tag{2.20}$$

Furthermore, using that $u_\Gamma \in H^1(\Gamma)$, which follows from the fact that $u_\Gamma \in V$ it follows that $u_\Gamma|_{\Omega_i} \in H^{3/2}(\Omega_i)$, $i = 1, 2$, and thus

$$[\![n \cdot a\nabla u_\Gamma]\!] \in H^{1/2}(\Gamma) \tag{2.21}$$

Since the right hand side of (2.18) is in $L^2(\Gamma)$ we may use elliptic regularity for the Laplace Beltrami operator to confirm that

$$u_\Gamma|_\Gamma \in H^2(\Gamma) \tag{2.22}$$

Collecting the bounds we obtain the refined regularity estimate

$$\|u_\Gamma\|_{H^2(\Gamma)} + \sum_{i=1}^{2} \left(\|u_\Gamma\|_{H^{5/2}(\Omega_i)} + \|u_i\|_{H^2(\Omega_i)} \right) \lesssim \|f\|_\Omega + \|f_\Gamma\|_\Gamma \tag{2.23}$$

where we note that we have stronger control of u_Γ on the subdomains. □

3 The finite element method

3.1 The mesh and finite element space

Let $\mathcal{T}_h$ be a quasi-uniform conformal mesh, consisting of shape regular elements with mesh parameter $h \in (0, h_0]$, on Ω and let

$$\mathcal{T}_{h,i} = \{T \in \mathcal{T}_h : T \cap \Omega_i \neq \emptyset\} \quad i = 1, 2 \tag{3.1}$$

be the active meshes associated with Ω_i, $i = 1, 2$. Let V_h be a finite element space consisting of continuous piecewise polynomials on $\mathcal{T}_h$ and define

$$V_{h,i} = V_h|_{\mathcal{T}_{h_i}} \qquad i = 1, 2 \tag{3.2}$$

and

$$W_h = V_{h,1} \oplus V_{h,2} \tag{3.3}$$

To $v = v_1 \oplus v_2 \in W_h$ we associate the function $\widetilde{v} \in L^2(\Omega)$ such that $\widetilde{v}|_{\Omega_i} = v_i|_{\Omega_i}$, $i = 1, 2$. In general, we simplify the notation and write $\widetilde{v} = v$. Finally, we use $\mathcal{T}_h(\Gamma)$ to denote the set of elements intersected by Γ.

3.2 Derivation of the method

To derive the finite element method we follow the same approach as when introducing the weak formulation, but taking care to handle the boundary integrals that appear due to the discontinuities in the approximation space.

Testing the exact problem with $v \in W_h$ and integrating by parts over Ω_1 and Ω_2 we find that

$$(f, v)_{\Omega_1} + (f, v)_{\Omega_2} \tag{3.4}$$

$$= (-\nabla \cdot a\nabla u, v)_{\Omega_1} + (-\nabla \cdot a\nabla u, v)_{\Omega_2} \tag{3.5}$$

$$= (a\nabla u, \nabla v)_\Omega - (\langle n \cdot a\nabla u \rangle, [v])_\Gamma - ([\![n \cdot a\nabla u]\!], \langle v \rangle_*)_\Gamma \tag{3.6}$$

$$= (a\nabla u, \nabla v)_\Omega - (\langle n \cdot a\nabla u\rangle, [v])_\Gamma - (\nabla_\Gamma \cdot a_\Gamma \nabla_\Gamma u, \langle v\rangle_*)_\Gamma - (f_\Gamma, \langle v\rangle_*)_\Gamma \tag{3.7}$$

$$= (a\nabla u, \nabla v)_\Omega - (\langle n \cdot a\nabla u\rangle, [v])_\Gamma + (a_\Gamma \nabla_\Gamma u, \nabla_\Gamma \langle v\rangle_*)_\Gamma - (f_\Gamma, \langle v\rangle_*)_\Gamma \tag{3.8}$$

$$= (a\nabla u, \nabla v)_\Omega - (\langle n \cdot a\nabla u\rangle, [v])_\Gamma - ([u], \langle n \cdot a\nabla v\rangle)_\Gamma \tag{3.9}$$

$$+(a_\Gamma \nabla_\Gamma u, \nabla_\Gamma \langle v\rangle_*)_\Gamma - (f_\Gamma, \langle v\rangle_*)_\Gamma \tag{3.10}$$

where in the last identity we symmetrized using the fact that $[u] = 0$. We also used the identity

$$[vw] = [v]\langle w\rangle + \langle v\rangle_*[w] \tag{3.11}$$

where the averages are defined by

$$\langle w\rangle = \kappa_1 w_1 + \kappa_2 w_2, \qquad \langle w\rangle_* = \kappa_2 w_1 + \kappa_1 w_2 \tag{3.12}$$

with $\kappa_1 + \kappa_2 = 1$ and $0 \leq \kappa_i \leq 1$.

Introducing the bilinear forms

$$a_\Omega(v, w) = (a\nabla v, \nabla w)_{\Omega_1} + (a\nabla v, \nabla w)_{\Omega_2} - (\langle n \cdot a\nabla v\rangle, [w])_\Gamma - ([v], \langle n \cdot a\nabla w\rangle)_\Gamma \tag{3.13}$$

$$a_{h,\Gamma}(v, w) = (a_\Gamma \nabla_\Gamma \langle v\rangle_*, \nabla_\Gamma \langle w\rangle_*)_\Gamma, \tag{3.14}$$

$$l_h(v) = (f, v)_\Omega + (f_\Gamma, \langle v\rangle_*)_\Gamma \tag{3.15}$$

the above formal derivation leads to the following consistent formulation for discontinuous test functions w. For $u \in W = H^1(\Omega) \cap H^{3/2}(\Omega_1) \cap H^{3/2}(\Omega_2) \cap H^1(\Gamma)$ the solution to (2.12) there holds

$$\boxed{a_\Omega(u, w) + a_{h,\Gamma}(u, w) = l_h(w) \qquad \forall w \in W_h} \tag{3.16}$$

Observe that we have modified $a_{h,\Gamma}$ by introducing the average $\langle v\rangle_*$ also in the left factor. This changes nothing when applied to a smooth solution, but will allow also to apply the form to the discontinuous discrete approximation space. The subscript h in the form indicates that it is modified to be well defined for the discontinuous approximation space. The definition of W is motivated by the fact that the trace terms should be well defined, for instance,

$$(\langle n \cdot a\nabla v\rangle, [w])_\Gamma \lesssim \left(\sum_{i=1}^2 \|v_i\|^2_{H^1(\partial\Omega_i)}\right)^{1/2} \left(\sum_{i=1}^2 \|w_i\|^2_{\partial\Omega_i}\right)^{1/2} \tag{3.17}$$

$$\lesssim \left(\sum_{i=1}^2 \|v_i\|^2_{H^{3/2}(\Omega_i)}\right)^{1/2} \left(\sum_{i=1}^2 \|w_i\|^2_{H^1(\Omega_i)}\right)^{1/2} \tag{3.18}$$

where we used the trace inequalities $\|v\|_{H^s(\partial\Omega_i)} \lesssim \|v\|_{H^{s+1/2}(\Omega_i)}$ for $s > 0$ and $\|w\|_{\partial\Omega_i} \lesssim \|w\|_{H^{1/2+\epsilon}(\Omega_i)} \lesssim \|w\|_{H^1(\Omega_i)}$ for $\epsilon > 0$.

3.3 The finite element method

The finite element method that we propose is based on the formulation (3.16). However, using this formulation as it stands does not lead to a robust approximation method. Indeed we need to ensure stability of the formulation through the addition of consistent penalty terms. First we need to enforce continuity of the discrete solution across Γ. To this end we introduce an augmented version of a_Ω,

$$a_h(v, w) = a_\Omega(v, w) + \beta h^{-1}([v], [w])_\Gamma$$

with β a positive parameter. Since the exact solution $u \in H^1(\Omega)$, there holds $a_\Omega(u, w) = a_h(u, w)$. Secondly, to obtain stability independently of how the interface cuts the computational mesh and for strongly varying permeabilities a_1, a_2 and a_Γ we also need some penalty terms in a neighbourhood of the interface. We define

$$s_h(v, w) = s_{h,1}(v_1, w_1) + s_{h,2}(v_2, w_2)$$

where

$$s_{h,i}(v_i, w_i) = \gamma h([n \cdot a\nabla v_i], [n \cdot a\nabla w_i])_{\mathcal{F}_{h,i}} \qquad i = 1, 2 \tag{3.19}$$

where γ is a positive parameters and $\mathcal{F}_{h,i}$ is the set of interior faces in $\mathcal{T}_{h,i}$ that belongs to an element $T \in \mathcal{T}_{h,i}$ which intersects Γ, see Fig. 6. Observe that for $u \in H^2(\Omega_1 \cup \Omega_2)$, $s_h(u, v) = 0$ for all $v \in W_h$.

Collecting the above bilinear forms in

$$A_h(v, w) = a_h(v, w) + s_h(v, w) + a_{h,\Gamma}(v, w) \tag{3.20}$$

the finite element method reads:

$$\boxed{\text{Find } u_h \in W_h \text{ such that: } A_h(u_h, v) = l_h(v) \qquad \forall v \in W_h} \tag{3.21}$$

4 Analysis of the method

In this section we derive the basic error estimates that the solution of the formulation (3.21) satisfies. The technical detail is kept to a minimum to improve readability. In particular, we assume that the bilinear forms can be computed exactly and that Γ fulfils the conditions of Hansbo and Hansbo (2002). For a more complete exposition in a similar context we refer to Burman et al. (2016).

4.1 Properties of the bilinear form

For the analysis it is convenient to define the following energy norm

$$|||v|||_h^2 = \sum_{i=1}^{2} \left(\|a_i^{1/2} \nabla v\|_{\Omega_i}^2 + |v|_{s_i}^2 \right) + c_a h \|\langle n \cdot a \nabla v \rangle\|_\Gamma^2 + \beta h^{-1} \|[v]\|_\Gamma^2 + \|a_\Gamma \nabla_\Gamma \langle v \rangle_* \|_\Gamma^2 \tag{4.1}$$

where $|v|_{s_i} = s_i(v, v)^{1/2}$ and c_a is a constant fulfilling $c_a \sim a_{\min}^{-1}$, cf. Lemma 4.1 below.

Lemma 4.1 *The form A_h, defined in* (3.20), *satisfies the following bounds:*

- *A_h is continuous*

$$\boxed{A_h(v, w) \lesssim |||v|||_h |||w|||_h \qquad v, w \in W \cup W_h} \tag{4.2}$$

  where W was introduced in (3.16).
- *A_h is coercive on W_h,*

$$\boxed{|||v|||_h^2 \lesssim A_h(v, v) \qquad v \in W_h} \tag{4.3}$$

  provided β is large enough.

Proof The first estimate (4.2) follows directly from the Cauchy–Schwarz inequality. To show (4.3) we recall the following inequalities:

$$\|a_i^{1/2} \nabla v\|_{\mathcal{T}_{h,i}}^2 \lesssim \|a_i^{1/2} \nabla v\|_{\Omega_i}^2 + |v|_{s_{h,i}}^2 \quad \text{(see Burman and Hansbo 2012)} \tag{4.4}$$

$$h \|\langle n \cdot a \nabla v \rangle\|_\Gamma^2 \lesssim \sum_{i=1}^{2} \|\kappa_i a_i \nabla v\|_{\mathcal{T}_{h,i}(\Gamma)}^2 \quad \text{(see Hansbo and Hansbo 2002)} \tag{4.5}$$

In (4.5) we used the notation $\mathcal{T}_{h,i}(\Gamma) := \{T \in \mathcal{T}_{h,i} : T \cap \Gamma \neq \emptyset\}$. To prove the claim observe that for all $v \in W_h$

$$\begin{aligned} A_h(v, v) = &\sum_{i=1}^{2} \left(\|a_i^{1/2} \nabla v\|_{\Omega_i}^2 + |v|_{s_i}^2 \right) + \beta h^{-1} \|[v]\|_\Gamma^2 \\ &+ \|a_\Gamma \nabla_\Gamma \langle v \rangle_* \|_\Gamma^2 - 2 \left(\langle n \cdot a \nabla v \rangle, [v] \right)_\Gamma \end{aligned} \tag{4.6}$$

Using (4.4) and (4.5) we obtain the following bound on the fluxes

$$h \|\langle n \cdot a \nabla v \rangle\|_\Gamma^2 \leq C \sum_{i=1}^{2} \kappa_i a_i \left(\|a_i^{1/2} \nabla v\|_{\Omega_i}^2 + |v|_{s_{h,i}}^2 \right) \tag{4.7}$$

Now assume that $\kappa_i a_i \leq a_{\min} := \min_{i \in \{1,2\}} a_i$, for instance one may take $\kappa_1 = a_2/(a_1 + a_2)$ and $\kappa_2 = a_1/(a_1 + a_2)$ then

$$2 \left(\langle n \cdot a \nabla v \rangle, [v] \right)_\Gamma \leq 2 a_{\min}^{-1/2} h^{1/2} \|\langle n \cdot a \nabla v \rangle\|_\Gamma a_{\min}^{1/2} h^{-1/2} \|[v]\|_\Gamma \tag{4.8}$$

$$\leq \varepsilon h a_{\min}^{-1} \| \langle n \cdot a \nabla v \rangle \|_{\Gamma}^2 + a_{\min} h^{-1} \varepsilon^{-1} \| [v] \|_{\Gamma}^2 \tag{4.9}$$

$$\leq C\varepsilon \sum_{i=1}^{2} \left(\| a_i^{1/2} \nabla v \|_{\Omega_i}^2 + |v|_{s_{h,i}}^2 \right) + a_{\min} h^{-1} \varepsilon^{-1} \| [v] \|_{\Gamma}^2 \tag{4.10}$$

It follows that

$$A_h(v, v) \geq (1 - C\varepsilon) \sum_{i=1}^{2} \left(\| a_i^{1/2} \nabla v \|_{\Omega_i}^2 + |v|_{s_i}^2 \right)$$
$$+ (\beta - a_{\min}/\varepsilon)\, h^{-1} \| [v] \|_{\Gamma}^2 + \| a_{\Gamma}^{1/2} \nabla_{\Gamma} \langle v \rangle_* \|_{\Gamma}^2 \tag{4.11}$$

The bound (4.3) now follows taking $\varepsilon = 1/(2C)$ and $\beta > 2Ca_{\min}$ and by applying once again (4.7), taking $c_a \sim a_{\min}^{-1}$. □

A consequence of the bound (4.3) is the existence of a unique solution to (3.21).

Lemma 4.2 *The linear system defined by the formulation* (3.21) *is invertible.*

Proof Follows from Lax–Milgram's lemma. □

4.2 Interpolation

For $\delta > 0$ let $E_i : H^s(\Omega_i) \to H^s(\Omega)$ be a continuous extension operator $s > 0$. We define the interpolation operator

$$\pi_h : L^2(\Omega) \ni v \mapsto \pi_{h,1} v_1 \oplus \pi_{h,2} v_2 \in V_{h,1} \oplus V_{h,2} = W_h \tag{4.12}$$

where $\pi_{h,i} : L^2(\Omega_i) \ni v_i \mapsto \pi_{h,i}^{SZ} E_i v_i \in V_{h,i}$, $i = 1, 2$, and π_h^{SZ} is the Scott–Zhang interpolation operator. We then have the interpolation error estimate

$$\boxed{|||u - \pi_h u|||_h \lesssim h\Big(\|u\|_{H^2(\Omega_1)} + \|u\|_{H^2(\Omega_2)} + \|u\|_{L^\infty_\delta(H^2(\Gamma_t))} \Big)} \tag{4.13}$$

where, with ρ_Γ the signed distance function associated with Γ,

$$\Gamma_t = \{ x \in \Omega : \rho_\Gamma(x) = t \}, \qquad |t| \leq \delta \tag{4.14}$$

and

$$\|v\|_{L^\infty_\delta(H^s(\Gamma_t))} = \sup_{|t| \leq \delta} \|v\|_{(H^s(\Gamma_t))} \tag{4.15}$$

Proof To prove the estimate (4.13) we use a trace-inequality on functions in $H^1(\mathcal{T}_h(\Gamma))$ (i.e., with $\| \cdot \|_{\mathcal{T}_h(\Gamma)}$ the broken H^1-norm over the elements intersected by Γ),

$$\|v_i\|_\Gamma \lesssim h^{-1/2} \|E_i v_i\|_{\mathcal{T}_h(\Gamma)} + h^{1/2} \|\nabla E_i v_i\|_{\mathcal{T}_h(\Gamma)} \tag{4.16}$$

see Hansbo and Hansbo (2002), then interpolation on $\mathcal{T}_h(\Gamma)$ and finally we use the stability of the extension operator E_i. First observe that by using the trace inequality (4.16) we obtain, with $v = u - \pi_h u$

$$\|(\beta h)^{-1/2}[v]\|_\Gamma + c_a h\|\langle n \cdot a\nabla v\rangle\|_\Gamma \lesssim \sum_{i=1}^{2}\Big(h^{-1}\|v_i\|_{\mathcal{T}_h(\Gamma)} + \|\nabla v_i\|_{\mathcal{T}_h(\Gamma)}) + h\|\nabla^2 v_i\|_{\mathcal{T}_h(\Gamma)})\Big) \tag{4.17}$$

Using standard interpolation for the Scott–Zhang interpolation operator we get the bound

$$\begin{aligned}\|(\beta h)^{-1/2}[v]\|_\Gamma + c_a h\|\langle n \cdot a\nabla v\rangle\|_\Gamma &\lesssim h\sum_{i=1}^{2}|E_i u_i|_{H^2(\mathcal{T}_h(\Gamma))}\\ &\lesssim h\sum_{i=1}^{2}|u_i|_{H^2(\Omega_i)}\end{aligned} \tag{4.18}$$

where we used the stability of the extension operator in the last inequality. The bound

$$|u - \pi_h u|_{s_i} \lesssim h\sum_{i=1}^{2}|a_i^{1/2}u_i|_{H^2(\Omega_i)} \tag{4.19}$$

follows similarly using element-wise trace inequalities follows by interpolation (c.f. Burman and Hansbo 2012). The interpolation error estimate for the terms due to the Laplace–Beltrami operator on Γ is a bit more delicate. We use a trace inequality to conclude that

$$\|a_\Gamma^{1/2}\nabla_\Gamma\langle u - \pi_h u\rangle_\star\|_\Gamma^2 \lesssim \sum_{i=1}^{2}\|a_\Gamma^{1/2}\nabla_\Gamma(u_i - \pi_{h,i}u_i)\|_\Gamma^2 \tag{4.20}$$

$$\lesssim \sum_{i=1}^{2} h^{-1}\|\nabla(u_i - \pi_{h,i}u_i)\|_{\mathcal{T}_h(\Gamma)}^2 + h\|\nabla^2(u_i - \pi_{h,i}u_i)\|_{\mathcal{T}_h(\Gamma)}^2 \tag{4.21}$$

$$\lesssim \sum_{i=1}^{2} h\|\nabla^2 u_i\|_{\mathcal{T}_h(\Gamma)}^2 \tag{4.22}$$

$$\lesssim \delta h\|u\|_{L_\delta^\infty(H^2(\Gamma_t))}^2 \tag{4.23}$$

Observing that we may take $\delta \sim h$ the estimate follows. □

Comparing (4.13) with (2.23) we see that we have a small mismatch between the regularity that we can prove and that required to achieve optimal convergence. In view of this we need to assume a slightly more regular solution for the H^1-error estimates below. The sub optimal regularity also interferes in the L^2-error estimates. Here we need to use (2.23) on the dual solution and in this case the additional regularity of the estimate (4.13) is not available. Instead we need to find the largest $\zeta \in [0, 1]$ such that $|||u - \pi_h u|||_h \lesssim h^\zeta \sum_{i=1}^2 \|u\|_{H^2(\Omega_i)}$, which will result in a suboptimality by a power of $1 - \zeta$ in the convergence order in the L^2-norm. Revisiting the analysis above up to (4.22) we see that

$$|||u - \pi_h u|||_h \lesssim h \sum_{i=1}^{2} |u_i|_{H^2(\Omega_i)} + \sum_{i=1}^{2} h^{-1/2} \|\nabla(u_i - \pi_{h,i} u_i)\|_{\mathcal{T}_h(\Gamma)} + h^{1/2} \|\nabla^2(u_i - \pi_{h,i} u_i)\|_{\mathcal{T}_h(\Gamma)} \tag{4.24}$$

$$\lesssim \left(h + h^{1/2}\right) \sum_{i=1}^{2} |u_i|_{H^2(\Omega_i)} \tag{4.25}$$

4.3 Error estimates

Theorem 4.1 *If u is the solution to* (2.1)–(2.4)*, satisfying $u \in H^2(\Omega_1 \cup \Omega_2) \cup L^\infty_\delta(H^2(\Gamma_t))$, and u_h is the finite element approximation defined by* (3.21)*, then*

$$|||u - u_h|||_h \lesssim h\Big(\|u\|_{H^2(\Omega_1)} + \|u\|_{H^2(\Omega_2)} + \|u\|_{L^\infty_\delta(H^2(\Gamma_t))}\Big) \tag{4.26}$$

$$\|u - u_h\|_\Omega + \|u - u_h\|_\Gamma \lesssim h^{3/2}\Big(\|u\|_{H^2(\Omega_1)} + \|u\|_{H^2(\Omega_2)} + \|u\|_{L^\infty_\delta(H^2(\Gamma_t))}\Big) \tag{4.27}$$

Proof (4.26). Splitting the error and using the interpolation error estimate we have

$$|||u - u_h|||_h \leq |||u - \pi_h u|||_h + |||\pi_h u - u_h|||_h \tag{4.28}$$

Using coercivity (4.3), Galerkin orthogonality and continuity (4.2) the second term can be estimated as follows

$$|||\pi_h u - u_h|||_h^2 \lesssim A_h(\pi_h u - u_h, \pi_h u - u_h) \tag{4.29}$$

$$= A_h(\pi_h u - u, \pi_h u - u_h) \tag{4.30}$$

$$\leq |||\pi_h u - u|||_h |||\pi_h u - u_h|||_h \tag{4.31}$$

and thus applying the approximation result (4.13) we conclude that

$$|||u - u_h|||_h \lesssim |||u - \pi_h u|||_h \tag{4.32}$$

$$\lesssim h\Big(\|u\|_{H^2(\Omega_1)} + \|u\|_{H^2(\Omega_2)} + \|u\|_{L^\infty_\delta(H^2(\Gamma_t))}\Big) \tag{4.33}$$

(4.27). Consider the dual problem

$$A(v, \phi) = (v, \psi)_\Omega + (v, \psi_\Gamma) \quad \forall v \in V \tag{4.34}$$

and recall that by (2.23) we have the elliptic regularity

$$\sum_{i=1}^{2} \|\phi\|_{H^2(\Omega_i)} + \|\phi_\Gamma\|_{H^2(\Gamma)} \lesssim \sum_{i=1}^{2} \|\psi_i\|_{\Omega_i} + \|\psi_\Gamma\|_\Gamma \tag{4.35}$$

Setting $v = e = u - u_h$ and using Galerkin orthogonality, followed by the continuity (4.2) and the suboptimal approximation estimate (4.24) on $|||\phi - \pi_h\phi|||_h$ we get

$$(e, \psi)_\Omega + (e, \psi_\Gamma)_\Gamma = A_h(e, \phi) \tag{4.36}$$

$$= A_h(e, \phi - \pi_h\phi) \tag{4.37}$$

$$\leq |||e|||_h |||\phi - \pi_h\phi|||_h \tag{4.38}$$

$$\lesssim |||e|||_h h^{1/2} \left(\sum_{i=1}^{2} \|\phi\|_{H^2(\Omega_i} + \|\phi_\Gamma\|_{H^2(\Gamma)} \right) \tag{4.39}$$

$$\lesssim h^{1/2} |||e|||_h \left(\sum_{i=1}^{2} \|\psi_i\|_{\Omega_i} + \|\psi_\Gamma\|_\Gamma \right). \tag{4.40}$$

In the last step we used the elliptic regularity estimate (4.35) for the dual problem. Setting $\psi_i = e_i/\|e_i\|_{\Omega_i}$ and $\psi_\Gamma = e_\Gamma/\|e_\Gamma\|_\Gamma$ estimate (4.27) follows. □

Remark 4.1 As noted before the error estimate in the L^2-norm is suboptimal with a power 1/2. To improve on this estimate we would need to sharpen the regularities required for the approximation estimate (4.13). This appears to be highly non-trivial since the interpolation of u and u_Γ can not be separated when both are interpolated using the bulk unknowns. Therefore we did not manage to exploit the stronger control that we have on the harmonic extension of u_Γ in (2.23). Note however that if separate fields are used on the fracture and in the bulk domains we would recover optimal order convergence in L^2.

Remark 4.2 Using the stronger control of the regularity of the harmonic extension provided by (2.23) we may however establish an optimal order L^2 error estimate for the solution on Γ,

$$\boxed{\|u - u_h\|_\Gamma \lesssim h^2 \Big(\|u\|_{H^2(\Omega_1)} + \|u\|_{H^2(\Omega_2)} + \|u\|_{L^\infty_\delta(H^2(\Gamma_t))} \Big)} \tag{2.14}$$

5 Extension to bifurcating fractures

In the case most common in applications, fractures bifurcate, leading to networks of interfaces in the bulk. It is straightforward to include this case in the method above

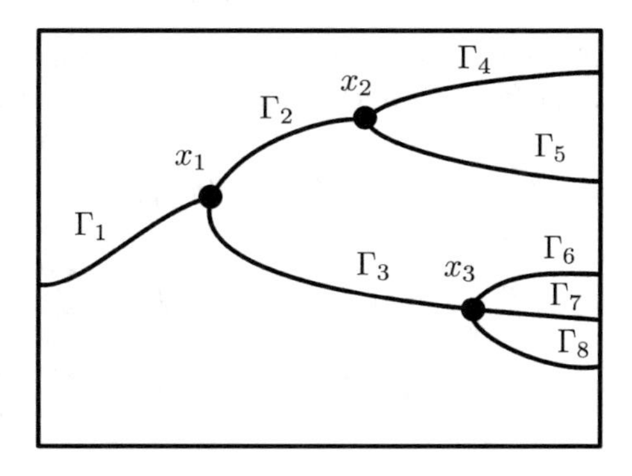

Fig. 2 Notation for bifurcating fractures

and we will discuss the method with bifurcating fractures below. The analysis can also be extended under suitable regularity assumptions, but becomes increasingly technical. We leave the analysis of the methods modelling flow in fractured media with bifurcating interfaces to future work.

5.1 The model problem

Description of the domain Let us for simplicity consider a two dimensional problem with a one dimensional interface. We define the following:

- Let the interface Γ be described as a planar graph with nodes $\mathcal{N} = \{x_i\}_{i \in I_N}$ and edges $\mathcal{E} = \{\Gamma_j\}_{j \in I_E}$, where I_N, I_E are finite index sets, and each Γ_j is a smooth curve between two nodes with indexes $I_N(j)$. Note that edges only meet in nodes.
- For each $i \in I_N$ we let $I_E(i)$ be the set of indexes corresponding to edges for which x_i is a node. For each $i \in I_N$ we let $I_E(i)$ be the set of indexes j such that x_i is an end point of Γ_j, see Fig. 2.
- The graph Γ defines a partition of Ω into N subdomains Ω_i, $i = 1, \dots, N$.

The Kirchhoff condition The governing equations are given by (2.1)–(2.4) together with two conditions at each of the nodes $x_i \in \mathcal{N}$, the continuity condition

$$u_{\Gamma_k}(x_i) = u_{\Gamma_l}(x_i) \qquad \forall k, l \in I_E(i) \tag{5.1}$$

and the Kirchhoff condition

$$\sum_{j \in I_E(i)} (t_{\Gamma_j} \cdot a_{\Gamma_j} \nabla_{\Gamma_j} u_{\Gamma_j})|_{x_j} = 0 \tag{5.2}$$

where $t_{\Gamma_j}(x_i)$ is the exterior tangent unit vector to Γ_j at x_i. Note that in the special case when a node x_i is an end point of only one curve the Kirchhoff condition becomes a homogeneous Neumann condition.

5.2 The finite element method

Forms associated with the bifurcating interface Let $V_\Gamma = \{v \in C(\Gamma) : v \in H^1(\Gamma_j),\ j \in I_E\}$ and $V = H^1_0(\Omega) \cap V_\Gamma$. We proceed as in the derivation (2.8)–(2.11) of the weak problem (2.12) in the standard case. However, when we use Green's formula on Γ we proceed segment by segment as follows

$$\sum_{j\in I_E} -\left(\nabla_{\Gamma_j}\cdot a_{\Gamma_j}\nabla_{\Gamma_j}u_j, \langle v_j\rangle_*\right)_{\Gamma_j}$$

$$= \sum_{j\in I_E}\left(a_{\Gamma_j}\nabla_{\Gamma_j}u, \nabla_{\Gamma_j}\langle v\rangle_*\right)_{\Gamma_j} - \sum_{j\in I_E}\sum_{i\in I_N(j)}\left(t_i\cdot a_{\Gamma_j}\nabla_{\Gamma_j}u, \langle v\rangle_*\right)_{x_i} \tag{5.3}$$

$$= \sum_{j\in I_E}\left(a_{\Gamma_j}\nabla_{\Gamma_j}u, \nabla_{\Gamma_j}\langle v\rangle_*\right)_{\Gamma_j} - \sum_{i\in I_N}\sum_{j\in I_E(i)}\left(t_i\cdot a_{\Gamma_j}\nabla_{\Gamma_j}u, \langle v\rangle_* - \langle\langle v\rangle_*\rangle_i\right)_{x_i} \tag{5.4}$$

where we changed the order of summation and used the Kirchhoff condition (5.2) to subtract the nodal average

$$\langle v\rangle_i = \sum_{j\in I_E(i)}\kappa_j^\Gamma v_j(x_i) \tag{5.5}$$

where $0 < \kappa_i^\Gamma$, and $\sum_{j\in I_E(i)}\kappa_j^\Gamma = 1$. Note that when a node x_i is an end point of only one curve the contribution from x_i is zero, because in that case we have $\langle\langle v\rangle_*\rangle_i|_{x_i} - \langle v\rangle_* = 0$ since there is only one element in $I_E(i)$, and thus we get the standard weak enforcement of the homogeneous Neumann condition.

Symmetrizing and adding a penalty term we obtain the form

$$\begin{aligned} a_{h,\Gamma}(v,w) = &\sum_{j\in I_E}\left(a_{\Gamma_j}\nabla_{\Gamma_j}\langle v\rangle_*, \nabla_{\Gamma_j}\langle w\rangle_*\right)_{\Gamma_j} \\ &-\sum_{i\in I_N}\sum_{j\in I_E(i)}\left(t_j\cdot a_{\Gamma_j}\nabla_{\Gamma_j}\langle v\rangle_*, \langle w\rangle_* - \langle\langle v\rangle_*\rangle_i\right)_{x_i} \\ &-\sum_{i\in I_N}\sum_{j\in I_E(i)}\left(\langle v\rangle_* - \langle\langle v\rangle_*\rangle_i, t_j\cdot a_{\Gamma_j}\nabla_{\Gamma_j}\langle w\rangle\right)_{x_i} \\ &+\sum_{i\in I_N}\sum_{j\in I_E(i)}\beta^\Gamma h^{-1}\left(\langle v\rangle_* - \langle\langle v\rangle_*\rangle_i, \langle w\rangle_* - \langle\langle w\rangle_*\rangle_i\right)_{x_i} \end{aligned} \tag{5.6}$$

where β^Γ is a stabilisation parameter with the same function as β. A similar derivation can be performed for a two dimensional bifurcating fracture embedded into $\mathbb{R}^3$, see Hansbo et al. (2017) for further details.

To ensure coercivity we add a stabilization term of the form

$$s_{h,\Gamma}(v,w) = \sum_{j\in I_E} s_{h,\Gamma_j}(v,w) \tag{5.7}$$

where

$$s_{h,\Gamma_j}(v,w) = \left([\nabla_{\Gamma_j}\langle v\rangle_*], [\nabla_{\Gamma_j}\langle w\rangle_*]\right)_{\mathcal{X}_h(\Gamma_j)} \tag{5.8}$$

and $\mathcal{X}_h(\Gamma_j)$ is the set of points

$$\Gamma_j \cap \mathcal{F}_h(x_i) \tag{5.9}$$

where $\mathcal{F}_h(x_i)$ is the set of interior faces in the patch of elements $\mathcal{N}_h(T(x_i))$ and $T(x_i)$ is an element such that $x_i \in T$.

We finally define the form $A_{h,\Gamma}$ associated with the bifurcating crack by

$$A_{h,\Gamma}(v,w) = a_{h,\Gamma}(v,w) + s_{h,\Gamma}(v,w) \qquad \forall v \in W_h \tag{5.10}$$

The method Define

$$W_h = \bigoplus_{i=1}^{N} V_{h,i} \tag{5.11}$$

where $V_{h,i} = V_h|_{\mathcal{T}_{h,i}}$. The method takes the form: find $u_h \in W_h$ such that

$$A_h(u_h, v) = l_h(v) \quad \forall v \in W_h \tag{5.12}$$

where

$$A_h(v,w) = \sum_{i=1}^{N} A_{h,i}(v,w) + A_{h,\Gamma}(v,w) \tag{5.13}$$

and

$$A_{h,i}(v,w) = a_{h,i}(v,w) + s_{h,i}(v,w) \tag{5.14}$$

6 Numerical examples

6.1 Implementation details

We will employ piecewise linear triangles and extend the implementation approach proposed in Hansbo and Hansbo (2002) to include also bifurcating fractures. Recall that $\mathcal{T}_h(\Gamma)$ denotes the set of elements intersected by Γ, where each side of the intersection belongs to Ω_1 and Ω_2, respectively. For each element in $T_i \in \mathcal{T}_h(\Gamma)$, we assign elements $T_{i,1} \in \mathcal{T}_{h,1}$ and $T_{i,2} \in \mathcal{T}_{h,2}$ by overlapping the existing element $T_i \in \mathcal{T}_h(\Gamma)$ using the *same* nodes from the original triangulation. Elements $T_{i,1}$ and $T_{i,2}$ coincide geometrically, see Fig. 3. To ensure continuity, we used the same process on the neighboring elements and checked if new nodes had already been assigned. For each bifurcation point, two approaches can be adapted. Either by letting the bifurcation point coincide with a node or by the less straight-forward approach to overlap the existing element $T_i \in \mathcal{T}_h(\Gamma)$ into $T_{i,1}$, $T_{i,2}$ and $T_{i,3}$, see Fig. 4. For simplicity of implementation, we have here chosen to let the bifurcating point coincide with a node. The triangles $T_i \notin \mathcal{T}_h(\Gamma)$ were handled in the usual way. The stabilization (3.19) was only applied to the *cut sides* of the elements which in all examples was sufficient for stability.

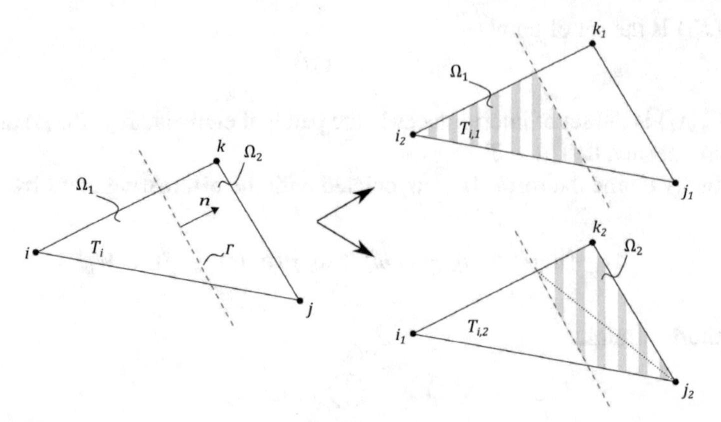

Fig. 3 The split of a triangle without bifurcation point

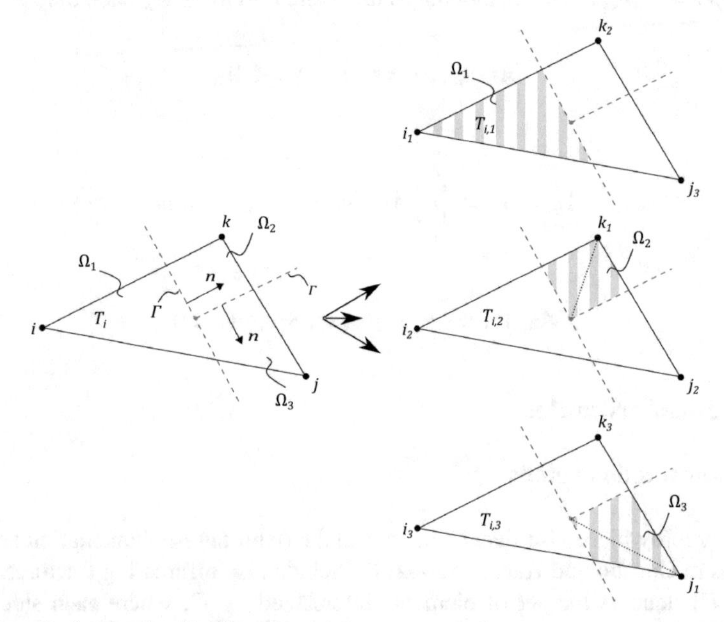

Fig. 4 The split of a triangle with bifurcation point

6.2 Example 1: No flow in fracture

We consider an example on $\Omega = (0, 1) \times (0, 1)$, from Hansbo and Hansbo (2002). We solved the example with an added bifurcation point. For the added fracture, we denote the diffusion coefficient by a_{Γ_1}. The exact solution is given by

$$u(x, y) = \begin{cases} \frac{r^2}{a_1}, & \text{if } r \leqslant r_0 \\ \frac{r^2}{a_2} - \frac{r_0^2}{a_2} + \frac{r_0^2}{a_1}, & \text{if } r > r_0 \end{cases} \tag{6.1}$$

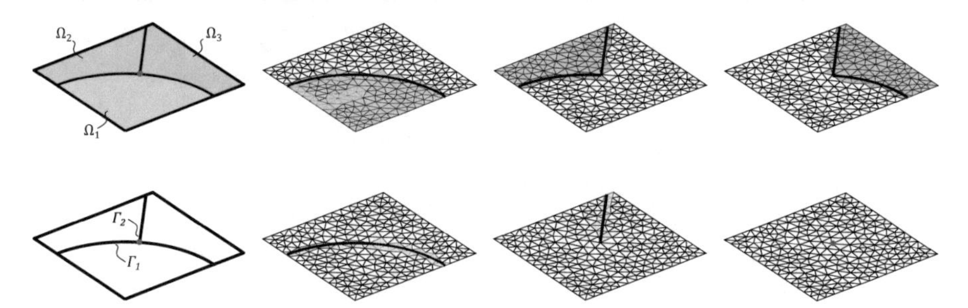

Fig. 5 Active meshes with two embedded fractures, Example 1

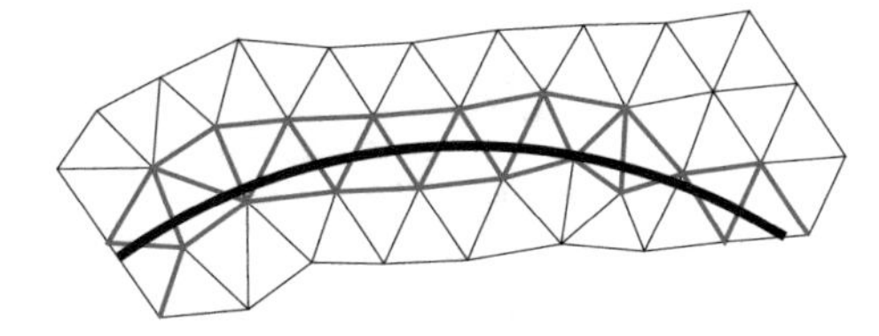

Fig. 6 The red edges indicates the selection for computing stabilization terms asscording to (3.19)

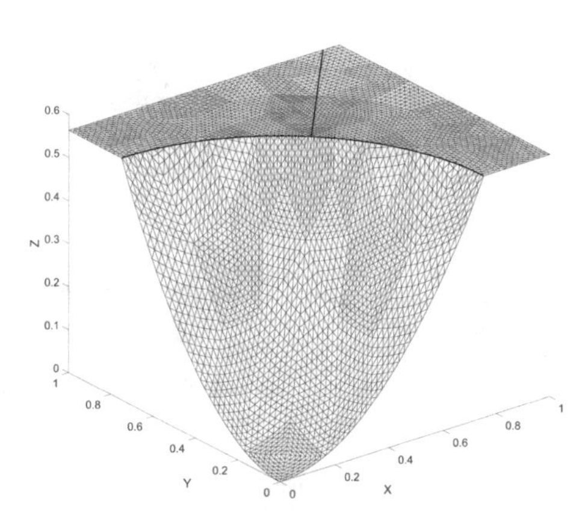

Fig. 7 Elevation of the approximate solution with two embedded fractures, Example 1 (color figure online)

where $r = \sqrt{x^2 + y^2}$. We chose $r_0 = 3/4$, $a_1 = 1$, $a_2 = 1000$ and $a_\Gamma = a_{\Gamma_1} = 0$, with a right-hand side $f = -4$ and $f_\Gamma = 0$. The boundary conditions were symmetry boundaries at $x = 0$ and $y = 0$ and Dirichlet boundary conditions corresponding to the exact solution at $x = 1$ and $y = 1$. This example is outlined in Figs. 5 and 6. We give the elevation of the approximate solution in Fig. 7, on the last mesh in a sequence. The corresponding convergence of the L_2-norm and the energy-norm is given in Fig. 8.

6.3 Example 2: Flow in the fracture

We considered a two-dimensional example on the domain $\Omega = \left(1, e^{5/4}\right) \times \left(1, e^{5/4}\right)$, from Burman et al. (2017). We solved the example with an additional fracture added, see Fig. 9. The exact solution is given by

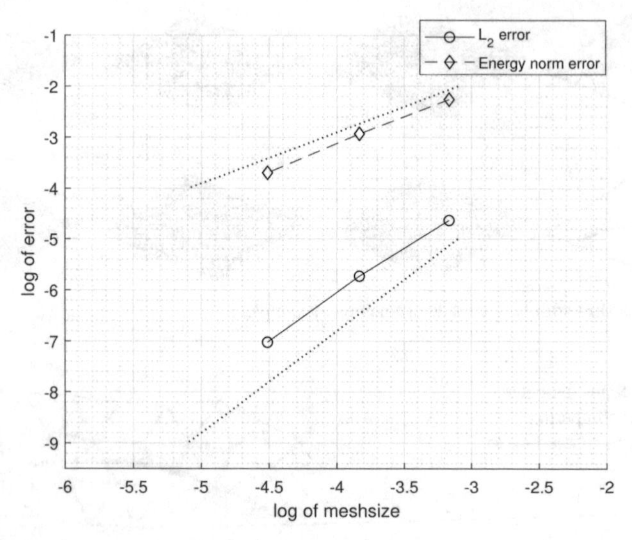

Fig. 8 L_2-norm and energy-norm convergence using natural logarithm with two embedded fractures, Example 1. Dotted lines signify optimal convergence. Inclination 1:1 for energy-norm and 2:1 for L_2-norm

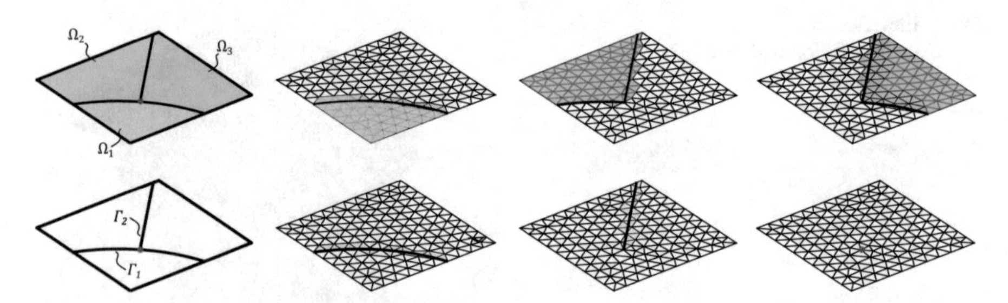

Fig. 9 Active meshes with two embedded fractures, Example 2

$$
u_1 = \frac{\log(r)}{5}(4+e) \quad \text{for} \quad 1 < r < e,
$$

$$
u_2 = \frac{4-4e}{5}\left(\log(r) - \frac{5}{4}\right) + 1 \quad \text{for} \quad e < r < e^{5/4},
$$

where $\sqrt{x^2+y^2} := r = e$. We chose $a_1 = a_2 = a_\Gamma = 1$ and the right hand side to $f = f_\Gamma = 0$. For the added crack we chose $a_{\Gamma_1} = 0$. The Dirichlet boundary conditions corresponding to the exact solution at $x, y = 1$ and $x, y = e^{5/4}$. In Fig. 10, we give the elevation of the approximate solution. The corresponding L_2-norm convergence and the energy-norm is given in Fig. 11.

6.4 Example 3: Flow in bifurcating fractures

We consider an example with two bifurcating points. The fractures are modeled using higher order curves. In Fig. 12 we show the fractures and construction of individual

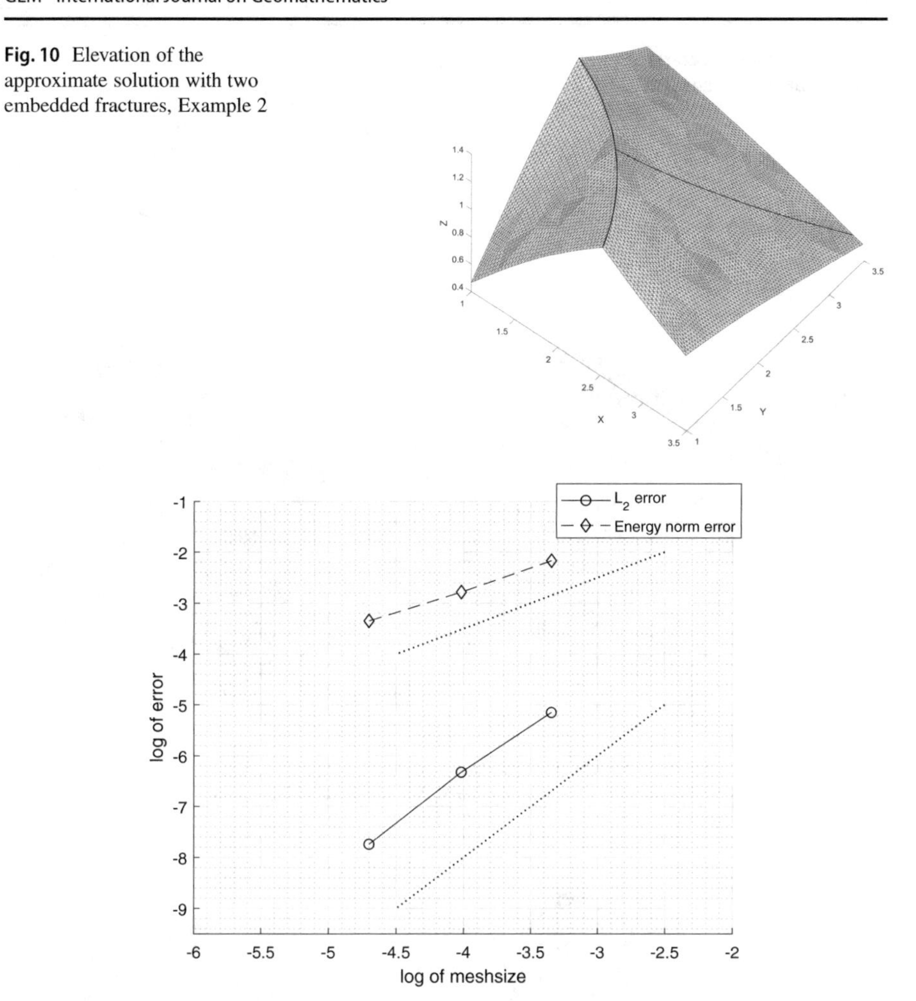

Fig. 10 Elevation of the approximate solution with two embedded fractures, Example 2

Fig. 11 L_2-norm and energy-norm convergence using natural logarithm with two embedded fractures, Example 2. Dotted lines signify optimal convergence. Inclination 1:1 for energy-norm and 2:1 for L_2-norm

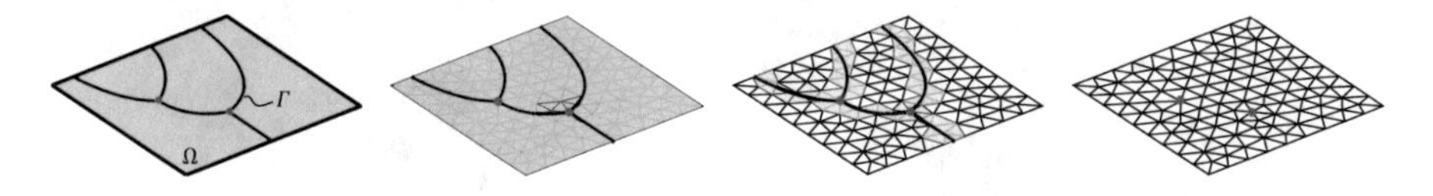

Fig. 12 Active meshes with two bifurcating points, Example 3

elements. On the domain $\Omega = (0, 1) \times (0, 1)$, we chose $a_1 = a_2 = 1$, $f_\Omega = 1$ and $f_\Gamma = 0$. We impose the Dirichlet boundary conditions $u = 0$ at $x, y = 0$ and $u = 0$ at $x, y = 1$. For the diffusion coefficient, we denote a_{Γ_i} for each fracture where $a_{\Gamma_i} \in \{0, 100\}$ and assign an individual value for each Γ_i, see Fig. 13. We report six different solutions by allowing different fractures to be active. The first result have been obtained with $a_{\Gamma_i} = 0$, thus no flow in the fractures is allowed, see Fig. 14.

Fig. 13 Embedded fractures with assigned Γ, Example 3

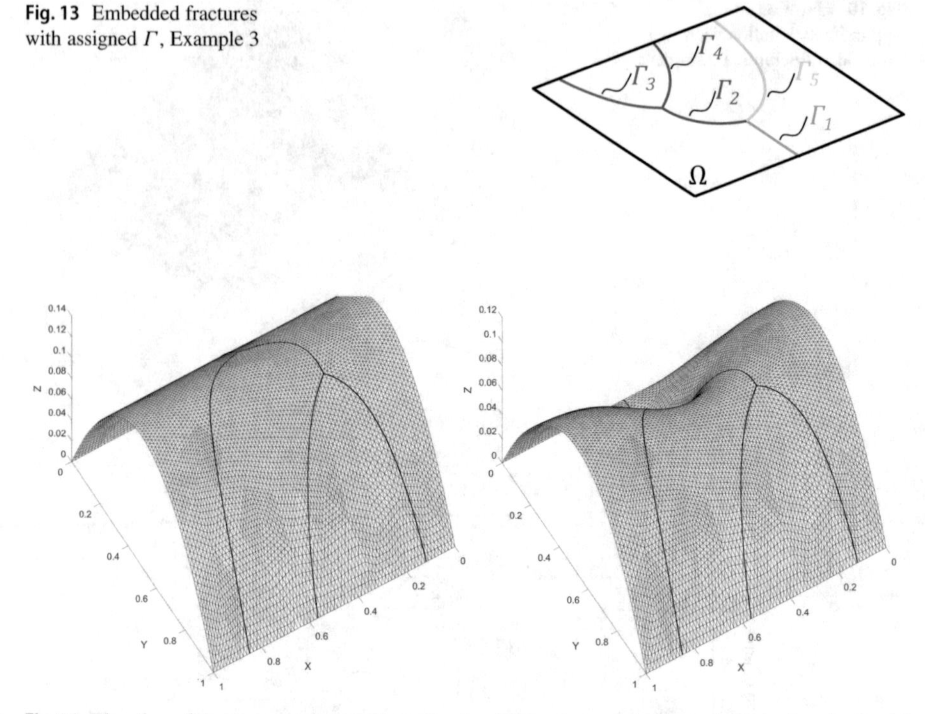

Fig. 14 Elevation of the approximate solution using two bifurcating points, Example 3. Assigned value to the left figure: $a_{\Gamma_1} = a_{\Gamma_2} = a_{\Gamma_3} = a_{\Gamma_4} = a_{\Gamma_5} = 0$, and assigned values to the right figure: $a_{\Gamma_1} = 100$ and $a_{\Gamma_2} = a_{\Gamma_3} = a_{\Gamma_4} = a_{\Gamma_5} = 0$

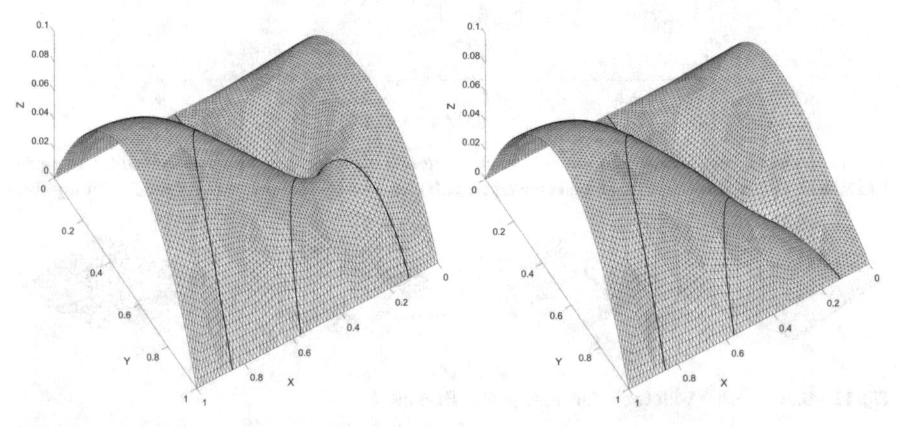

Fig. 15 Elevation of the approximate solution using two bifurcating points, Example 3. Assigned value to the left figure: $a_{\Gamma_1} = a_{\Gamma_2} = 100$ and $a_{\Gamma_3} = a_{\Gamma_4} = a_{\Gamma_5} = 0$, and assigned values to the right figure: $a_{\Gamma_1} = a_{\Gamma_2} = a_{\Gamma_3} = 100$ and $a_{\Gamma_4} = a_{\Gamma_5} = 0$

Further, each solution has one additional fracture activated by changing the diffusion coefficent $a_{\Gamma_i} = 100$, see Figs. 14, 15 and 16. The last solution is presented with flow in all fractures.

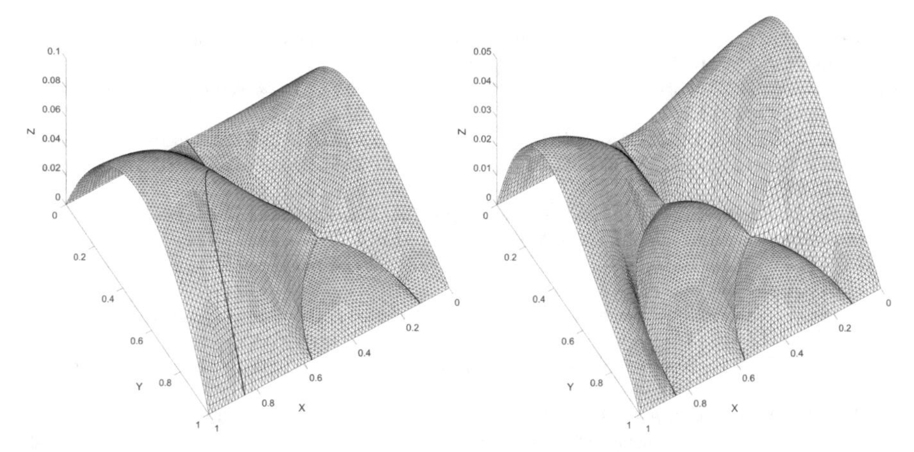

Fig. 16 Elevation of the approximate solution using two bifurcating points, Example 3. Assigned value to the left figure: $a_{\Gamma_1} = a_{\Gamma_2} = a_{\Gamma_3} = a_{\Gamma_4} = 100$ and $a_{\Gamma_5} = 0$, and assigned values to the right figure: $a_{\Gamma_1} = a_{\Gamma_2} = a_{\Gamma_3} = a_{\Gamma_4} = a_{\Gamma_5} = 100$

7 Concluding remarks

We proposed a discontinuous finite element method using a one–field approach to modelling Darcy flow in a cracked medium. The pressure in the crack was modelled as an average of pressure on either side of the crack which, unlike our previous work (Burman et al. 2017), allows for pressure jumps across the crack. In particular, the case of bifurcating fractures was considered. Optimal order error estimates were proven and backed up by numerical experiments. Extension to other flow models in the crack have been considered in Burman et al. (2018).

Acknowledgements This research was supported in part by the Swedish Foundation for Strategic Research Grant No. AM13-0029, the Swedish Research Council Grants Nos. 2013-4708, 2017-03911, and the Swedish Research Programme Essence. EB was supported in part by the EPSRC Grant EP/P01576X/1.

References

Angot, P., Boyer, F., Hubert, F.: Asymptotic and numerical modelling of flows in fractured porous media. ESAIM: Math. Model. Numer. Anal. **43**(2), 239–275 (2009)

Berrone, S., Pieraccini, S., Scialò, S.: Flow simulations in porous media with immersed intersecting fractures. J. Comput. Phys. **345**, 768–791 (2017)

Burman, E., Hansbo, P.: Fictitious domain finite element methods using cut elements: II. A stabilized Nitsche method. Appl. Numer. Math. **62**(4), 328–341 (2012)

Burman, E., Claus, S., Hansbo, P., Larson, M.G., Massing, A.: CutFEM: discretizing geometry and partial differential equations. Int. J. Numer. Methods Eng. **104**(7), 472–501 (2015)

Burman, E., Hansbo, P., Larson, M.G., Zahedi, S.: Cut finite element methods for coupled bulk-surface problems. Numer. Math. **133**(2), 203–231 (2016)

Burman, E., Hansbo, P., Larson, M.G.: A simple finite element method for elliptic bulk problems with embedded surfaces. ArXiv e-prints, Sept. (2017)
Burman, E., Hansbo, P., Larson, M.G, Larsson, K.: Cut finite elements for convection in fractured domains. Comput. Fluids (2018)
Capatina, D., Luce, R., El-Otmany, H., Barrau, N.: Nitsche's extended finite element method for a fracture model in porous media. Appl. Anal. **95**(10), 2224–2242 (2016). https://doi.org/10.1016/j.compfluid.2018.07.022
D'Angelo, C., Scotti, A.: A mixed finite element method for Darcy flow in fractured porous media with non-matching grids. ESAIM: Math. Model. Numer. Anal. **46**(2), 465–489 (2012)
Del Pra, M., Fumagalli, A., Scotti, A.: Well posedness of fully coupled fracture/bulk Darcy flow with XFEM. SIAM J. Numer. Anal. **55**(2), 785–811 (2017)
Formaggia, L., Fumagalli, A., Scotti, A., Ruffo, P.: A reduced model for Darcy's problem in networks of fractures. ESAIM: Math. Model. Numer. Anal. **48**(4), 1089–1116 (2014)
Frih, N., Roberts, J.E., Saada, A.: Modeling fractures as interfaces: a model for Forchheimer fractures. Comput. Geosci. **12**(1), 91–104 (2008)
Hægland, H., Assteerawatt, A., Dahle, H.K., Eigestad, G.T., Helmig, R.: Comparison of cell- and vertex-centered discretization methods for flow in a two-dimensional discrete-fracture–matrix system. Adv. Water Resour. **32**(12), 1740–1755 (2009)
Hansbo, A., Hansbo, P.: An unfitted finite element method, based on Nitsche's method, for elliptic interface problems. Comput. Methods Appl. Mech. Eng. **191**(47–48), 5537–5552 (2002)
Hansbo, P., Jonsson, T., Larson, M.G., Larsson, K.: A Nitsche method for elliptic problems on composite surfaces. Comput. Methods Appl. Mech. Eng. **326**, 505–525 (2017)
Martin, V., Jaffré, J., Roberts, J.E.: Modeling fractures and barriers as interfaces for flow in porous media. SIAM J. Sci. Comput. **26**(5), 1667–1691 (2005)

GEM - International Journal on Geomathematics
https://doi.org/10.1007/s13137-019-0114-x

ORIGINAL PAPER

A Hybrid High-Order method for passive transport in fractured porous media

Florent Chave[1,2] · **Daniele A. Di Pietro**[1] · **Luca Formaggia**[2]

Received: 20 April 2018 / Accepted: 25 September 2018

Abstract

In this work, we propose a model for the passive transport of a solute in a fractured porous medium, for which we develop a Hybrid High-Order (HHO) space discretization. We consider, for the sake of simplicity, the case where the flow problem is fully decoupled from the transport problem. The novel transmission conditions in our model mimic at the discrete level the property that the advection terms do not contribute to the energy balance. This choice enables us to handle the case where the concentration of the solute jumps across the fracture. The HHO discretization hinges on a mixed formulation in the bulk region and on a primal formulation inside the fracture for the flow problem, and on a primal formulation both in the bulk region and inside the fracture for the transport problem. Relevant features of the method include the treatment of nonconforming discretizations of the fracture, as well as the support of arbitrary approximation orders on fairly general meshes.

Keywords Hybrid High-Order methods · Finite volume methods · Finite element methods · Fractured porous media · Darcy flow · Miscible displacement · Passive transport

Mathematics Subject Classification 76S05 · 65N08 · 65N30

1 Introduction

Over the last decades, the research on fluid flows in fractured porous media has received a great amount of attention because of its relevance in many areas of the geosciences,

The second author acknowledges the partial support of Agence Nationale de la Recherche grant HHOMM (Ref. ANR-15-CE40-0005-01). The third author acknowledges the support of INdaM-GNCS under the program Progetti 2017. The authors also acknowledge the support of the Vinci Programme of Université Franco Italienne.

✉ Florent Chave
florent.chave@umontpellier.fr

Extended author information available on the last page of the article

ranging from ground-water hydrology to hydrocarbon exploitation. Fractures in the subsurface are indeed ubiquitous, and can be caused by tectonic forces, changes of temperature, drying processes, by leaching in the plane of stratification, or by schistosity. Depending on the material that has accumulated within the fractures, they may act as conduits or barriers, and thus affect the flow patterns in a substantial way. For instance, it has been observed that fractures near boreholes tend to increase the productivity of wells during oil recovery. In the context of geological isolation of radioactive waste, the presence of fractures in the disposal areas due to, for example, tunnel excavation, can drastically accelerate the migration process of radionuclides.

A common feature of fractures in porous media is the variety of length scales. While the presence of smaller fractures may be accounted for by using homogenization or other upscaling techniques, fractures with larger extension have to be modelled explicitly, and there are several possible ways to incorporate their presence. Our focus is here on the approach developed in Martin et al. (2005), where a reduced model for the flow in the fracture is obtained by an averaging process, and the fracture is treated as an interface inside the bulk region. The fracture is assumed to be filled of debris, so that the flow therein can still be modelled by Darcy's law. The problem is closed by interface conditions that relate the average and jump of the bulk pressure to the normal flux and pressure in the fracture. In Chave et al. (2018) we have designed and analysed a Hybrid High-Order (HHO) method to discretize this model, and proved stability and order $O(h^{k+1})$ convergence of the discretization error measured in an energy-like norm, with h denoting the meshsize and $k \geq 0$ the polynomial degree. This method is based on a mixed formulation for the bulk coupled with a primal method for the fracture. This choice is motivated by the fact that the unknowns of the method are those that naturally appear in the coupling conditions (4), namely the normal component of the bulk flux and the fracture pressure. For a review of other formulations, we refer the reader to Flemisch et al. (2016). Concerning the equivalence of mixed and primal HHO methods, see Aghili et al. (2015) and Boffi and Di Pietro (2018). We also refer the reader to Di Pietro and Ern (2015), Cockburn et al. (2016) and also Di Pietro and Tittarelli (2018, Section 3.2.5) concerning flux formulations of HHO methods, which highlight their local conservation properties. Several other discretization schemes have been proposed for this type of models; see, e.g., Brenner et al. (2018), Angot et al. (2009), Antonietti et al. (2016a), Benedetto et al. (2014), Boon et al. (2018), Scotti et al. (2017), D'Angelo and Scotti (2012), Fumagalli and Keilegavlen (2018) and Berrone et al. (2017) and references therein. Other works where fractures are treated as interfaces include Bastian et al. (1999), Angot et al. (1999) and Faille et al. (2002).

The literature on passive transport in fractured porous media and related problems is, however, more scarce. In Gross et al. (2015), the authors study a system of advection–diffusion equations where the jump of the diffusive bulk flux acts as a source term inside the fracture. In the coupling conditions, only the diffusive part of the total bulk flux is considered. The discretization is based on the Unfitted Finite Element method, for which well-posedness and $O(h^k)$ convergence in the energy-norm are proved. In Chernyshenko et al. (2016), a Finite Volume method is combined with a Trace Finite Element method to solve a transport problem in the bulk region and inside the fracture, with the jump of the total bulk flux acting as a source term in the surface problem and under the assumption that the concentration is continuous at the interface. Convergence in $O(h)$ is numerically observed for the energy-norm of the discretization error. A similar problem is studied in Alboin et al. (2002). In Fumagalli and Scotti (2011), the authors use an averaging technique similar to Martin et al. (2005) in order to derive coupling conditions for a transport problem which allow the concentration to jump across the fracture. This enables them to model high concentration gradients near the fracture resulting from highly heterogeneous diffusivity. The problem is discretized by eXtended Finite Elements (XFEM), and numerical evidence is provided. Yet another approach is represented by Discrete Fracture Networks (DFNs) models, where the bulk surrounding fractures is considered as impervious, so that the flow can only occur through the fracture planes and across their intersections; see, e.g., Berrone et al. (2016), where authors propose a system of unsteady advection–diffusion in DFNs.

In this work, we consider the passive transport of a solute driven by a velocity field solution of a (decoupled) Darcy problem. We present two novel contributions:

(i) first, we propose new coupling conditions between the bulk region and fracture inspired by energy-based arguments, following the general ideas developed by Formaggia et al. (2013) in a different context. Crucially, these transmission conditions allow the solute concentration to jump across the fracture;
(ii) second, we propose a novel HHO discretization of this new model where the Darcy velocity field results from an HHO approximation of the flow problem in the spirit of Chave et al. (2018). The discretization is designed to incorporate the new transmission conditions and to reproduce at the discrete level the energy argument from which they originate.

The main source of inspiration for the discretization of the advection terms in the bulk region and inside the fracture is Di Pietro et al. (2015), where the authors develop an HHO method that is proven to be robust across the entire range $[0, +\infty]$ of local Péclet numbers and that supports locally degenerate diffusion. The adaptation of the analysis techniques developed in this reference to the present case seems possible, and will make the object of a future, theoretically oriented work. Concerning the coupling of the flow and transport problems, we take inspiration from Anderson and Droniou (2018), where an HHO discretization of miscible displacements in non-fractured porous media described by the Peaceman model is considered. Therein, in order to obtain a well-posed discrete problem, the flow problem has to be solved using polynomials of degree twice as high as the transport problem. In our work, we find that a similar condition is required to ensure the coercivity of the transport bilinear form; see Remark 11 for further details. A thorough numerical investigation is carried out to demonstrate the order of convergence of the method and showcase its performance on physical test cases.

The material is organized as follows. In Sect. 2 we describe the equations that govern the model in the steady case along with their weak formulation. In Sect. 3 we discuss the discrete setting. In Sect. 4, we formulate the HHO space approximation and hint to the generalization to the unsteady case. Section 5 contains a complete panel of steady and unsteady numerical tests, including a numerical study of the convergence properties of the method and more physical test cases corresponding to conductive and impermeable fractures.

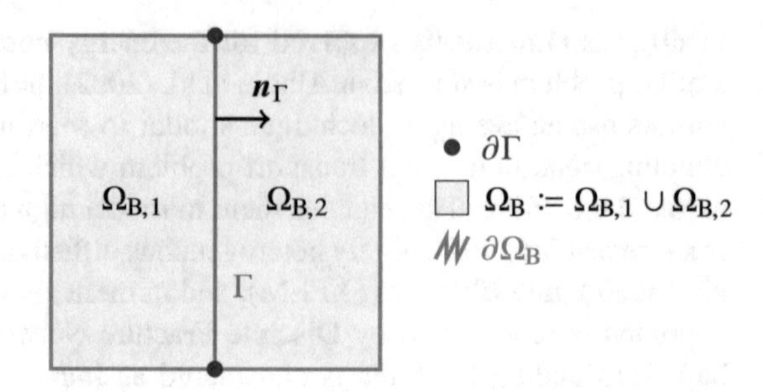

Fig. 1 Illustration of the notation introduced in Sect. 2.1

2 The differential model

In this section we introduce the strong and weak formulations of the flow and passive transport problems in the steady case. For the sake of simplicity, the presentation focuses on the two-dimensional case with a single fracture.

2.1 Notation

We consider a porous medium saturated by an incompressible fluid that occupies a space region $\Omega \subset \mathbb{R}^2$ traversed by a fracture Γ. We assume that Ω is an open, bounded, connected, polygonal set with Lipschitz boundary $\partial\Omega$. The fracture Γ is represented by an open line segment of nonzero length which cuts Ω into two disjoint connected polygonal subdomains $\Omega_{B,1}$ and $\Omega_{B,2}$ with Lipschitz boundary. The set $\Omega_B := \Omega\backslash\overline{\Gamma} = \Omega_{B,1} \cup \Omega_{B,2}$ corresponds to the bulk region. We denote by $\partial\Omega_B := \bigcup_{i=1}^{2}(\partial\Omega_{B,i} \setminus \overline{\Gamma})$ the external boundary of the bulk region and by $\boldsymbol{n}_{\partial\Omega}$ the unit normal vector on $\partial\Omega_B$ pointing out of Ω_B. For $i \in \{1, 2\}$, we let $\partial\Omega_{B,i} := \partial\Omega_B \cap \overline{\Omega_{B,i}}$ denote the external boundary of the subdomain $\Omega_{B,i}$. The boundary of the fracture Γ is denoted by $\partial\Gamma$, and the corresponding outward unit tangential vector is $\boldsymbol{\tau}_{\partial\Gamma}$. Finally, $\boldsymbol{n}_\Gamma$ denotes the unit normal vector to Γ pointing out of $\Omega_{B,1}$. This notation is illustrated in Fig. 1.

For any function φ sufficiently regular to admit a (possibly two-valued) trace on Γ, we define the jump and average operators such that

$$[\![\varphi]\!]_\Gamma := (\varphi_1 - \varphi_2)_{|\Gamma}, \qquad \{\!\{\varphi\}\!\}_\Gamma := \left(\frac{\varphi_1 + \varphi_2}{2}\right)_{|\Gamma},$$

where $\varphi_i := \varphi_{|\Omega_{B,i}}$ denotes the restriction of φ to the subdomain $\Omega_{B,i} \subset \Omega_B$. When applied to vector-valued functions, these operators act component-wise.

Finally, for any $X \subset \overline{\Omega}$, we denote by $(\cdot, \cdot)_X$ and $\|\cdot\|_X$ the usual inner product and norm of $L^2(X)$ or $L^2(X)^2$, according to the context.

2.2 Darcy flow

We now formulate the equations that govern the flow in the saturated, fractured porous medium and discuss a weak formulation inspired by Antonietti et al. (2016b) and Del Pra et al. (2017).

2.2.1 Governing equations

In the bulk region Ω_B and in the fracture Γ, we model the fluid flow by Darcy's law in mixed and primal form, respectively, so that the bulk Darcy velocity $\boldsymbol{u} : \Omega_B \to \mathbb{R}^2$, the bulk pressure $p : \Omega_B \to \mathbb{R}$, and the fracture pressure $p_\Gamma : \Gamma \to \mathbb{R}$ satisfy

$$\begin{aligned}
\boldsymbol{u} + \boldsymbol{K}\nabla p &= 0 && \text{in } \Omega_B, && \text{(1a)}\\
\nabla \cdot \boldsymbol{u} &= f && \text{in } \Omega_B, && \text{(1b)}\\
-\nabla_\tau \cdot (K_\Gamma \nabla_\tau p_\Gamma) &= \ell_\Gamma f_\Gamma + [\![\boldsymbol{u}]\!]_\Gamma \cdot \boldsymbol{n}_\Gamma && \text{in } \Gamma, && \text{(1c)}\\
\boldsymbol{u} \cdot \boldsymbol{n}_{\partial\Omega} &= 0 && \text{on } \partial\Omega_B, && \text{(1d)}\\
-K_\Gamma \nabla_\tau p_\Gamma \cdot \boldsymbol{\tau}_{\partial\Gamma} &= 0 && \text{on } \partial\Gamma, && \text{(1e)}\\
\int_\Gamma p_\Gamma &= 0, && && \text{(1f)}
\end{aligned}$$

where $f \in L^2(\Omega_B)$ and $f_\Gamma \in L^2(\Gamma)$ denote source or sink terms, $\boldsymbol{K} : \Omega_B \to \mathbb{R}^{2\times 2}$ the bulk permeability tensor, and we have set $K_\Gamma := \kappa_\Gamma^\tau \ell_\Gamma$, with $\kappa_\Gamma^\tau : \Gamma \to \mathbb{R}$ denoting the tangential permeability inside the fracture and $\ell_\Gamma : \Gamma \to \mathbb{R}$ the fracture thickness. In (1c) and (1e), ∇_τ and $\nabla_\tau\cdot$ denote the tangential gradient and divergence operators along Γ, respectively. We assume that $\boldsymbol{K}$ is symmetric, piecewise constant on a finite polygonal partition

$$\mathcal{P}_B = \{\omega_{B,i} \ : \ i \in I_B\} \tag{2}$$

of Ω_B, and uniformly elliptic, so that there exist two strictly positive real numbers $\underline{K}_B$ and $\overline{K}_B$ such that, for almost every $\boldsymbol{x} \in \Omega_B$ and all $z \in \mathbb{R}^2$ with $|z| = 1$,

$$0 < \underline{K}_B \le \boldsymbol{K}(\boldsymbol{x})z \cdot z \le \overline{K}_B.$$

The quantities κ_Γ^τ and ℓ_Γ are also assumed piecewise constant on a finite partition

$$\mathcal{P}_\Gamma = \{\omega_{\Gamma,i} \ : \ i \in I_\Gamma\} \tag{3}$$

of Γ, and such that there exist strictly positive real numbers $\underline{\ell}_\Gamma$,$\overline{\ell}_\Gamma$ $\underline{K}_\Gamma$,$\overline{K}_\Gamma$ such that, for almost every $\boldsymbol{x} \in \Gamma$,

$$0 < \underline{\ell}_\Gamma \le \ell_\Gamma(\boldsymbol{x}) \le \overline{\ell}_\Gamma, \qquad 0 < \underline{K}_\Gamma \le K_\Gamma(\boldsymbol{x}) \le \overline{K}_\Gamma.$$

To close the problem, we add the following transmission conditions across the fracture:

$$\begin{aligned}
\lambda_\Gamma \{\!\{\boldsymbol{u}\}\!\}_\Gamma \cdot \boldsymbol{n}_\Gamma &= [\![p]\!]_\Gamma && \text{on } \Gamma,\\
\lambda_\Gamma^\xi [\![\boldsymbol{u}]\!]_\Gamma \cdot \boldsymbol{n}_\Gamma &= \{\!\{p\}\!\}_\Gamma - p_\Gamma && \text{on } \Gamma,
\end{aligned} \tag{4}$$

where, denoting by $\xi \in \left(\frac{1}{2}, 1\right]$ a user-dependent model parameter, we have set

$$\lambda_\Gamma := \frac{\ell_\Gamma}{\kappa_\Gamma^n}, \qquad \lambda_\Gamma^\xi := \lambda_\Gamma \left(\frac{\xi}{2} - \frac{1}{4}\right).$$

Here, $\kappa_\Gamma^n : \Gamma \to \mathbb{R}$ represents the normal permeability inside the fracture, which is assumed piecewise constant on the partition $\mathcal{P}_\Gamma$ of Γ and such that, for almost every $\boldsymbol{x} \in \Gamma$,

$$0 < \underline{\lambda}_\Gamma \leq \lambda_\Gamma(\boldsymbol{x}) \leq \overline{\lambda}_\Gamma,$$

for two given strictly positive real numbers $\underline{\lambda}_\Gamma$ and $\overline{\lambda}_\Gamma$.

Remark 1 (Compatibility condition) Since homogeneous Neumann boundary conditions are considered on both the bulk and fracture boundaries (cf. (1d), (1e)), the flow through the porous medium is entirely driven by the source terms f and f_Γ, which typically model injection or production wells according to their sign. Decomposing f and f_Γ into their positive and negative parts, i.e., writing $f = f^+ - f^-$ and $f_\Gamma = f_\Gamma^+ - f_\Gamma^-$ with $f^\pm := \frac{|f| \pm f}{2}$ and $f_\Gamma^\pm := \frac{|f_\Gamma| \pm f_\Gamma}{2}$, we need to further assume the following compatibility condition in order to ensure that a global mass balance is satisfied:

$$\int_{\Omega_B} f^+ + \int_\Gamma \ell_\Gamma f_\Gamma^+ = \int_{\Omega_B} f^- + \int_\Gamma \ell_\Gamma f_\Gamma^-, \tag{5}$$

which translates the fact that all the fluid that enters the domain through injection wells must exit the domain through production wells. In this configuration, the fracture pressure p_Γ is defined up to a constant that is fixed by the zero-average constraint (1f). The bulk pressure, on the other hand, is uniquely defined without additional conditions owing to the coupling conditions (4).

Remark 2 (Boundary conditions) The model can be adapted to incorporate all the usual boundary conditions. One can consider, e.g., non-homogeneous Neumann boundary conditions on the bulk and, in the case where the fracture hits the domain boundary $\partial\Omega$, non-homogeneous Neumann boundary conditions on its tip. On the other hand, if the fracture boundaries lie in the interior of the domain Ω, a no-flow condition is required, where suitable compatibility conditions have to be enforced in each case. We do not delve further into this topic here, as the extension of the proposed method is relatively standard.

2.2.2 Weak formulation

We define the space $\boldsymbol{H}(\text{div}; \Omega_B)$, spanned by vector-valued functions on Ω_B whose restriction to every bulk subregion $\Omega_{B,i}$, $i \in \{1, 2\}$, is in $\boldsymbol{H}(\text{div}; \Omega_{B,i})$. The Darcy velocity space is

$$\boldsymbol{U} := \left\{ \boldsymbol{u} \in \boldsymbol{H}(\text{div}; \Omega_B) \ : \ \boldsymbol{u} \cdot \boldsymbol{n}_{\partial\Omega} = 0 \text{ on } \partial\Omega_B \text{ and } (\boldsymbol{u}_1 \cdot \boldsymbol{n}_\Gamma, \boldsymbol{u}_2 \cdot \boldsymbol{n}_\Gamma) \in L^2(\Gamma)^2 \right\}.$$

The fracture pressure space is $P_\Gamma := H^1(\Gamma) \cap L_0^2(\Gamma)$, with $L_0^2(\Gamma)$ spanned by square-integrable functions with zero mean value on Γ. We define the bilinear forms $a_K^\xi : \boldsymbol{U} \times \boldsymbol{U} \to \mathbb{R}$, $a_K^\Gamma : H^1(\Gamma) \times H^1(\Gamma) \to \mathbb{R}$, $b : \boldsymbol{U} \times L^2(\Omega_B) \to \mathbb{R}$ and $d : \boldsymbol{U} \times L^2(\Gamma) \to \mathbb{R}$ such that

$$a_K^\xi(\boldsymbol{u}, \boldsymbol{q}) := (\boldsymbol{K}^{-1}\boldsymbol{u}, \boldsymbol{q})_{\Omega_B} + (\lambda_\Gamma^\xi [\![\boldsymbol{u}]\!]_\Gamma \cdot \boldsymbol{n}_\Gamma, [\![\boldsymbol{q}]\!]_\Gamma \cdot \boldsymbol{n}_\Gamma)_\Gamma + (\lambda_\Gamma \{\!\{\boldsymbol{u}\}\!\}_\Gamma \cdot \boldsymbol{n}_\Gamma, \{\!\{\boldsymbol{q}\}\!\}_\Gamma \cdot \boldsymbol{n}_\Gamma)_\Gamma,$$
$$a_K^\Gamma(p_\Gamma, z_\Gamma) := (K_\Gamma \nabla_\tau p_\Gamma, \nabla_\tau z_\Gamma)_\Gamma, \qquad b(\boldsymbol{u}, z) := (\nabla \cdot \boldsymbol{u}, z)_{\Omega_B},$$
$$d(\boldsymbol{u}, z_\Gamma) := ([\![\boldsymbol{u}]\!]_\Gamma \cdot \boldsymbol{n}_\Gamma, z_\Gamma)_\Gamma,$$

as well as the global bilinear form $\mathcal{A}_\xi^{\text{flow}} : (\boldsymbol{U} \times L^2(\Omega_B) \times H^1(\Gamma)) \times (\boldsymbol{U} \times L^2(\Omega_B) \times H^1(\Gamma)) \to \mathbb{R}$ such that

$$\mathcal{A}_\xi^{\text{flow}}((\boldsymbol{u}, p, p_\Gamma), (\boldsymbol{q}, z, z_\Gamma)) := a_K^\xi(\boldsymbol{u}, \boldsymbol{q}) + b(\boldsymbol{u}, z) - b(\boldsymbol{q}, p) + d(\boldsymbol{q}, p_\Gamma) - d(\boldsymbol{u}, z_\Gamma) + a_K^\Gamma(p_\Gamma, z_\Gamma).$$

With these spaces and bilinear forms, the weak formulation of problem (1)–(4) reads: Find $(\boldsymbol{u}, p, p_\Gamma) \in \boldsymbol{U} \times L^2(\Omega_B) \times P_\Gamma$ such that, for all $(\boldsymbol{q}, z, z_\Gamma) \in \boldsymbol{U} \times L^2(\Omega_B) \times H^1(\Gamma)$,

$$\mathcal{A}_\xi^{\text{flow}}((\boldsymbol{u}, p, p_\Gamma), (\boldsymbol{q}, z, z_\Gamma)) = (f, z)_{\Omega_B} + (\ell_\Gamma f_\Gamma, z_\Gamma)_\Gamma. \tag{6}$$

The well-posedness of problem (6) with mixed boundary conditions is studied in Antonietti et al. (2016b); cf. also Del Pra et al. (2017), Scotti et al. (2017) and references therein.

2.3 Passive transport

We next formulate the equations that govern the passive transport of a solute by the Darcy flow solution of problem (1)–(4). For the sake of simplicity, we focus on the case where the transport problem is fully decoupled. This section contains the first main contribution of this paper, namely novel transmission conditions that enable the treatment of discontinuous solute concentrations across the fracture.

2.3.1 Bulk region

Denoting by $c : \Omega_B \to \mathbb{R}$ the concentration of the solute in the bulk and by $\boldsymbol{D} : \Omega_B \to \mathbb{R}^{2\times 2}$ the symmetric, uniformly elliptic bulk diffusion–dispersion tensor, the passive transport of the solute in the bulk region is governed by the following equations:

$$\nabla \cdot (\boldsymbol{u}c - \boldsymbol{D}\nabla c) + f^- c = f^+ \widehat{c} \quad \text{in } \Omega_B, \tag{7a}$$
$$-\boldsymbol{D}\nabla c \cdot \boldsymbol{n}_{\partial\Omega} = 0 \quad \text{on } \partial\Omega_B, \tag{7b}$$

where the term $f^- c$ acts as a sink, while the term $f^+ \widehat{c}$, with $\widehat{c} : \Omega_B \to \mathbb{R}$ denoting the concentration of solute as it is injected, acts as a source. We assume that both $\boldsymbol{D}$ and $\widehat{c}$ are piecewise constant on the polygonal partition $\mathcal{P}_B$ of Ω_B (see 2), and that there exist two strictly positive real numbers $\underline{D}_B$ and $\overline{D}_B$ such that, for almost every $\boldsymbol{x} \in \Omega_B$ and all $\boldsymbol{z} \in \mathbb{R}^2$ such that $|\boldsymbol{z}| = 1$,

$$0 \le \widehat{c}(\boldsymbol{x}) \le 1, \qquad 0 < \underline{D}_B \le \boldsymbol{D}(\boldsymbol{x})\boldsymbol{z} \cdot \boldsymbol{z} \le \overline{D}_B.$$

More generally $\boldsymbol{D}$ can depend on $\boldsymbol{u}$. While the theoretical results provided hereafter focus on the case of $\boldsymbol{D}$ independent from $\boldsymbol{u}$, this dependence has been considered in some numerical experiments presented in Sect. 5.

2.3.2 Fracture

We define the Darcy velocity $\boldsymbol{u}_\Gamma : \Gamma \to \mathbb{R}^2$ inside the fracture such that $\boldsymbol{u}_\Gamma := -K_\Gamma \nabla_\tau p_\Gamma$ where $p_\Gamma : \Gamma \to \mathbb{R}$ is the fracture pressure solution of problem (1)–(4). Denoting by $c_\Gamma : \Gamma \to \mathbb{R}$ the concentration of the solute inside the fracture, and letting $D_\Gamma := \mathcal{D}_\Gamma^\tau \ell_\Gamma$ with $\mathcal{D}_\Gamma^\tau : \Gamma \to \mathbb{R}$ denoting the (strictly positive almost everywhere) tangential diffusion–dispersion coefficient of the fracture, the governing equations for the transport problem inside the fracture are:

$$\nabla_\tau \cdot (\boldsymbol{u}_\Gamma c_\Gamma - D_\Gamma \nabla_\tau c_\Gamma) + \ell_\Gamma f_\Gamma^- c_\Gamma = \ell_\Gamma f_\Gamma^+ \widehat{c_\Gamma} + [\![\boldsymbol{u}c - \boldsymbol{D}\nabla c]\!]_\Gamma \cdot \boldsymbol{n}_\Gamma \quad \text{in } \Gamma, \tag{8a}$$

$$-D_\Gamma \nabla_\tau c_\Gamma \cdot \boldsymbol{\tau}_{\partial\Gamma} = 0 \quad \text{on } \partial\Gamma, \tag{8b}$$

where again $f_\Gamma^- c$ acts as a sink term while $f_\Gamma^+ \widehat{c_\Gamma}$ acts as a source, with $\widehat{c_\Gamma} : \Gamma \to \mathbb{R}$ denoting the concentration of solute as it is injected into the fracture. For the sake of simplicity, we assume in what follows that both $\widehat{c_\Gamma}$ and D_Γ are piecewise constant on the partition $\mathcal{P}_\Gamma$ of Γ (see 3), and such that there exist two strictly positive real numbers $\underline{D}_\Gamma$ and $\overline{D}_\Gamma$ such that, for almost every $\boldsymbol{x} \in \Gamma$,

$$0 \le \widehat{c_\Gamma}(\boldsymbol{x}) \le 1, \qquad 0 < \underline{D}_\Gamma \le D_\Gamma(\boldsymbol{x}) \le \overline{D}_\Gamma.$$

Remark 3 (Bulk and fracture boundary conditions) Considering no-flow boundary conditions on the bulk and fracture flux (1d) and (1e) entails a slight simplification, since we do not have to deal with the decomposition of the bulk or fracture boundary into their respective inflow or outflow parts. Other boundary conditions can be considered, but this topic will not be further developed here for the sake of brevity.

2.3.3 Transmission conditions

To derive transmission conditions for the hybrid dimensional passive transport problem, we have followed a technique similar to that used in Formaggia et al. (2013) and Di Pietro et al. (2015) in a completely different context. We started from the observation that in the unreduced model, where the fracture is not reduced to an internal interface, the transport operator does not contribute to the energy balance, aside from a possible contribution at the boundary (which is zero if we prescribe zero normal Darcy velocity). Therefore, we want to obtain also in the reduced model an energy estimate where the transport term behaves similarly. In particular, in the energy estimate for the hybrid dimensional problem, the terms related to transport in the coupling conditions have to cancel out and give no contribution to the energy. This is crucial, since energy estimates are a key ingredient for the coercivity of the differential problem. This is far from trivial, however we will show that this is possible if the following conditions are used (see Theorem 5):

$$\begin{aligned}
\{\!\{\boldsymbol{u}c - \boldsymbol{D}\nabla c\}\!\}_\Gamma \cdot \boldsymbol{n}_\Gamma &= \beta_\Gamma [\![c]\!]_\Gamma + (\{\!\{\boldsymbol{u}\}\!\}_\Gamma \cdot \boldsymbol{n}_\Gamma)\{\!\{c\}\!\}_\Gamma + \frac{1}{8}([\![\boldsymbol{u}]\!]_\Gamma \cdot \boldsymbol{n}_\Gamma)[\![c]\!]_\Gamma && \text{on } \Gamma, \\
[\![\boldsymbol{u}c - \boldsymbol{D}\nabla c]\!]_\Gamma \cdot \boldsymbol{n}_\Gamma &= \beta_\Gamma^\xi(\{\!\{c\}\!\}_\Gamma - c_\Gamma) + \frac{1}{2}([\![\boldsymbol{u}]\!]_\Gamma \cdot \boldsymbol{n}_\Gamma)(\{\!\{c\}\!\}_\Gamma + c_\Gamma) && \text{on } \Gamma,
\end{aligned} \tag{9}$$

where ξ is the user-dependent model parameter introduced in Sect. 2.2.1, and we have set

$$\beta_\Gamma := \frac{\mathcal{D}_\Gamma^n}{\ell_\Gamma}, \qquad \beta_\Gamma^\xi := \beta_\Gamma \left(\frac{\xi}{2} - \frac{1}{4}\right)^{-1}.$$

The term $\mathcal{D}_\Gamma^n : \Gamma \to \mathbb{R}$ represents the normal diffusion–dispersion coefficient of the fracture, which is assumed piecewise constant on the partition $\mathcal{P}_\Gamma$ of Γ (see 3), strictly positive almost everywhere on Γ, and such that, for almost every $\boldsymbol{x} \in \Gamma$,

$$0 < \underline{\beta}_\Gamma \leq \beta_\Gamma(\boldsymbol{x}) \leq \overline{\beta}_\Gamma,$$

for two given strictly positive real numbers $\underline{\beta}_\Gamma$ and $\overline{\beta}_\Gamma$.

Remark 4 (Limit cases) We may notice that in the case of a pure diffusion–dispersion problem in the bulk and in the fracture, corresponding to the case $\boldsymbol{u} = \boldsymbol{0}$, the transmission conditions (9) reduce to

$$\{\!\{-\boldsymbol{D}\nabla c\}\!\}_\Gamma \cdot \boldsymbol{n}_\Gamma = \beta_\Gamma [\![c]\!]_\Gamma \qquad \text{on } \Gamma, \tag{10a}$$

$$[\![-\boldsymbol{D}\nabla c]\!]_\Gamma \cdot \boldsymbol{n}_\Gamma = \beta_\Gamma^\xi(\{\!\{c\}\!\}_\Gamma - c_\Gamma) \qquad \text{on } \Gamma. \tag{10b}$$

The first equation (10a) stipulates that the diffusive flux across the fracture is proportional to the difference of concentration at the two sides of the fracture, while the second equation (10b) stipulates that the exchange between the bulk and the fracture is proportional to the difference between the average concentration across the fracture and the concentation at the interior of the fracture. On the other hand, for a pure transport problem (which corresponds to the case when all the diffusion–dispersion coefficients are zero), simple algebraic manipulations show that we obtain $c_1 = c_2 = c_\Gamma$ across the fracture Γ, provided that $[\![\boldsymbol{u}]\!]_\Gamma \cdot \boldsymbol{n}_\Gamma \neq 0$. This is reasonable, since in this limit case concentration is just transmitted across the interface. In intermediate cases, the transmission conditions (9) as designed so as to guarantee the satisfaction of the energy inequality, as we will show in Sect. 2.3.5.

2.3.4 Weak formulation

Let $H^1(\Omega_\mathrm{B})$ denote the broken Sobolev space spanned by scalar-valued functions on Ω_B whose restriction to every bulk subregion $\Omega_{\mathrm{B},i}$, $i \in \{1, 2\}$, is in $H^1(\Omega_{\mathrm{B},i})$. We define the molecular diffusion bilinear form $a_D : H^1(\Omega_\mathrm{B}) \times H^1(\Omega_\mathrm{B}) \to \mathbb{R}$, the advection–reaction bilinear form $a_{\boldsymbol{u},f} : H^1(\Omega_B) \times H^1(\Omega_B) \to \mathbb{R}$, and the diffusion–advection–reaction bilinear form $a : H^1(\Omega_B) \times H^1(\Omega_B)$ such that

$$a_D(c,z) := \int_{\Omega_B} \boldsymbol{D}\nabla c \cdot \nabla z, \qquad a_{\boldsymbol{u},f}(c,z) := \int_{\Omega_B} \Big(-c(\boldsymbol{u}\cdot\nabla z) + f^- cz \Big), \tag{11}$$
$$a(c,z) := a_D(c,z) + a_{\boldsymbol{u},f}(c,z).$$

We also define their fracture-based counterparts $a_D^\Gamma : H^1(\Gamma) \times H^1(\Gamma) \to \mathbb{R}$, $a_{\boldsymbol{u},f}^\Gamma : H^1(\Gamma) \times H^1(\Gamma) \to \mathbb{R}$ and $a_\Gamma : H^1(\Gamma) \times H^1(\Gamma) \to \mathbb{R}$ such that

$$\begin{aligned} a_D^\Gamma(c_\Gamma, z_\Gamma) &:= \int_\Gamma D_\Gamma \nabla_\tau c_\Gamma \cdot \nabla_\tau z_\Gamma, \\ a_{\boldsymbol{u},f}^\Gamma(c_\Gamma, z_\Gamma) &:= \int_\Gamma \Big(-c_\Gamma(\boldsymbol{u}_\Gamma \cdot \nabla_\tau z_\Gamma) + \ell_\Gamma f_\Gamma^- c_\Gamma z_\Gamma \Big), \\ a_\Gamma(c_\Gamma, z_\Gamma) &:= a_D^\Gamma(c_\Gamma, z_\Gamma) + a_{\boldsymbol{u},f}^\Gamma(c_\Gamma, z_\Gamma). \end{aligned} \tag{12}$$

The global bilinear form $\mathcal{A}_\xi^{\text{transp}} : \big(H^1(\Omega_B) \times H^1(\Gamma)\big) \times \big(H^1(\Omega_B) \times H^1(\Gamma)\big) \to \mathbb{R}$, that additionally takes into account terms that stem from the coupling equations, is defined as follows:

$$\begin{aligned} \mathcal{A}_\xi^{\text{transp}}((c, c_\Gamma), (z, z_\Gamma)) &:= a(c,z) + a_\Gamma(c_\Gamma, z_\Gamma) + \int_\Gamma \beta_\Gamma^\xi (\{\!\{c\}\!\}_\Gamma - c_\Gamma)(\{\!\{z\}\!\}_\Gamma - z_\Gamma) \\ &+ \int_\Gamma \Big(\beta_\Gamma [\![c]\!]_\Gamma [\![z]\!]_\Gamma + \frac{1}{2}([\![\boldsymbol{u}]\!]_\Gamma \cdot \boldsymbol{n}_\Gamma)(\{\!\{c\}\!\}_\Gamma + c_\Gamma)(\{\!\{z\}\!\}_\Gamma - z_\Gamma) \Big) \\ &+ \int_\Gamma \Big((\{\!\{\boldsymbol{u}\}\!\}_\Gamma \cdot \boldsymbol{n}_\Gamma)\{\!\{c\}\!\}_\Gamma [\![z]\!]_\Gamma + \frac{1}{8}([\![\boldsymbol{u}]\!]_\Gamma \cdot \boldsymbol{n}_\Gamma)[\![c]\!]_\Gamma [\![z]\!]_\Gamma \Big). \end{aligned} \tag{13}$$

With these spaces and bilinear forms, the weak formulation of problem (7)–(8)–(9) reads: Find $(c, c_\Gamma) \in H^1(\Omega_B) \times H^1(\Gamma)$ such that, for all $(z, z_\Gamma) \in H^1(\Omega_B) \times H^1(\Gamma)$

$$\mathcal{A}_\xi^{\text{transp}}((c, c_\Gamma), (z, z_\Gamma)) = (f^+\widehat{c}, z)_{\Omega_B} + (\ell_\Gamma f_\Gamma^+ \widehat{c_\Gamma}, z_\Gamma)_\Gamma. \tag{14}$$

2.3.5 Coercivity

In the following theorem, we prove the coercivity of the global transport bilinear form defined by (13) and show that, thanks to the new transmission conditions (8), the advective terms do not dissipate energy. This result is the key ingredient to derive a stability result for problem (14).

Theorem 5 (Coercivity) *Let* $\xi > 1/2$. *Then, for all* $(z, z_\Gamma) \in H^1(\Omega_B) \times H^1(\Gamma)$, *it holds*

$$\begin{aligned} \mathcal{A}_\xi^{\text{transp}}((z, z_\Gamma), (z, z_\Gamma)) &= \|\boldsymbol{D}^{1/2}\nabla z\|_{\Omega_B}^2 + \|D_\Gamma^{1/2}\nabla_\tau z_\Gamma\|_\Gamma^2 + \|\chi_B^{1/2} z\|_{\Omega_B}^2 + \|\chi_\Gamma^{1/2} z_\Gamma\|_\Gamma^2 \\ &+ \|(\beta_\Gamma^\xi)^{1/2}(\{\!\{z\}\!\}_\Gamma - z_\Gamma)\|_\Gamma^2 + \|(\beta_\Gamma)^{1/2}[\![z]\!]_\Gamma\|_\Gamma^2, \end{aligned} \tag{15}$$

with $\chi_{\mathrm{B}} := \dfrac{|f|}{2}$ *and* $\chi_\Gamma := \dfrac{\ell_\Gamma |f_\Gamma|}{2}$.

Remark 6 (Energy balance) Equation (15) can be interpreted as a global energy balance. The transmission conditions (9) are designed so that the advective terms do not contribute to this balance. Additionally, if $z = z_\Gamma$ across Γ, also all terms related to the diffusion–dispersion across the fracture, collected in the second line of (15), disappear.

Proof Let $(z, z_\Gamma) \in H^1(\Omega_{\mathrm{B}}) \times H^1(\Gamma)$. By definition of the global bilinear form $\mathcal{A}_\xi^{\mathrm{transp}}$ (13), it holds

$$
\begin{aligned}
\mathcal{A}_\xi^{\mathrm{transp}}((z, z_\Gamma), (z, z_\Gamma)) &= a(z, z) + a_\Gamma(z_\Gamma, z_\Gamma) + \|(\beta_\Gamma^\xi)^{1/2}(\{\!\{z\}\!\}_\Gamma - z_\Gamma)\|_\Gamma^2 \\
&\quad + \|(\beta_\Gamma)^{1/2}[\![z]\!]_\Gamma\|_\Gamma^2 \\
&\quad + \int_\Gamma \frac{1}{2}([\![\boldsymbol{u}]\!]_\Gamma \cdot \boldsymbol{n}_\Gamma)(\{\!\{z\}\!\}_\Gamma + z_\Gamma)(\{\!\{z\}\!\}_\Gamma - z_\Gamma) \\
&\quad + \int_\Gamma \Big((\{\!\{\boldsymbol{u}\}\!\}_\Gamma \cdot \boldsymbol{n}_\Gamma)\{\!\{z\}\!\}_\Gamma [\![z]\!]_\Gamma + \frac{1}{8}([\![\boldsymbol{u}]\!]_\Gamma \cdot \boldsymbol{n}_\Gamma)[\![z]\!]_\Gamma^2 \Big),
\end{aligned}
\tag{16}
$$

Using the definitions (11) and (12) of the bilinear forms a and a_Γ, we obtain

$$
a(z, z) = \|\boldsymbol{D}^{1/2}\nabla z\|_{\Omega_{\mathrm{B}}}^2 + a_{\boldsymbol{u},f}(z, z), \qquad a_\Gamma(z_\Gamma, z_\Gamma) = \|D_\Gamma^{1/2}\nabla_\tau z_\Gamma\|_\Gamma^2 + a_{\boldsymbol{u},f}^\Gamma(z_\Gamma, z_\Gamma).
\tag{17}
$$

Expanding the bilinear form $a_{\boldsymbol{u},f}$ according to its definition (11), we get

$$
\begin{aligned}
a_{\boldsymbol{u},f}(z, z) &= \int_{\Omega_{\mathrm{B}}} \Big(-z(\boldsymbol{u} \cdot \nabla z) + f^- z^2 \Big) \\
&= \int_{\Omega_{\mathrm{B}}} \Big(-\boldsymbol{u} \cdot \nabla(\frac{z^2}{2}) + f^- z^2 \Big) \\
&= \int_{\Omega_{\mathrm{B}}} \Big(\frac{1}{2}(\nabla \cdot \boldsymbol{u}) z^2 + f^- z^2 \Big) - \frac{1}{2} \int_\Gamma [\![\boldsymbol{u} z^2]\!]_\Gamma \cdot \boldsymbol{n}_\Gamma \\
&= \|\chi_{\mathrm{B}}^{1/2} z\|_{\Omega_{\mathrm{B}}}^2 - \frac{1}{2} \int_\Gamma \Big([\![\boldsymbol{u}]\!]_\Gamma \cdot \boldsymbol{n}_\Gamma \{\!\{z^2\}\!\}_\Gamma + \{\!\{\boldsymbol{u}\}\!\}_\Gamma \cdot \boldsymbol{n}_\Gamma [\![z^2]\!]_\Gamma \Big),
\end{aligned}
\tag{18}
$$

where we have used an integration by parts together with the boundary condition (1d) to pass to the third line while to pass to the fourth line, we have used (1b) to write $\frac{1}{2}(\nabla \cdot \boldsymbol{u}) + f^- = \frac{f}{2} + f^- = \frac{|f|}{2}$ followed by the relation

$$
[\![ab]\!]_\Gamma = [\![a]\!]_\Gamma \{\!\{b\}\!\}_\Gamma + \{\!\{a\}\!\}_\Gamma [\![b]\!]_\Gamma.
\tag{19}
$$

Similarly, expanding $a_{\boldsymbol{u},f}^\Gamma$ according to its definition (12), we find

$$
\begin{aligned}
a^{\Gamma}_{\boldsymbol{u},f}(z_\Gamma, z_\Gamma) &= \int_\Gamma \left(-z_\Gamma(\boldsymbol{u}_\Gamma \cdot \nabla_\tau z_\Gamma) + \ell_\Gamma f_\Gamma^- z_\Gamma^2 \right) \\
&= \int_\Gamma \left(-\boldsymbol{u}_\Gamma \cdot \nabla(\frac{z_\Gamma^2}{2}) + \ell_\Gamma f_\Gamma^- z_\Gamma^2 \right) \\
&= \int_\Gamma \left(\frac{1}{2}(\nabla_\tau \cdot \boldsymbol{u}_\Gamma) z_\Gamma^2 + \ell_\Gamma f_\Gamma^- z_\Gamma^2 \right) \\
&= \int_\Gamma \left(\frac{1}{2}(\ell_\Gamma f_\Gamma + [\![\boldsymbol{u}]\!]_\Gamma \cdot \boldsymbol{n}_\Gamma) z_\Gamma^2 + \ell_\Gamma f_\Gamma^- z_\Gamma^2 \right) \\
&= \|\chi_\Gamma^{1/2} z_\Gamma\|_\Gamma^2 + \frac{1}{2}\int_\Gamma ([\![\boldsymbol{u}]\!]_\Gamma \cdot \boldsymbol{n}_\Gamma) z_\Gamma^2,
\end{aligned} \tag{20}
$$

where we have, at first, integrated by parts and used (1e) to pass to the third line, then we have used (1c) after recalling that $\boldsymbol{u}_\Gamma := -K_\Gamma \nabla_\tau p_\Gamma$ to pass to the fourth line, and invoked the definition of χ_Γ to conclude. Plugging (17), (18) and (20) into (16), we obtain

$$
\begin{aligned}
&\mathcal{A}_\xi^{\text{transp}}((z, z_\Gamma), (z, z_\Gamma)) \\
&= \|\boldsymbol{D}^{1/2}\nabla z\|_{\Omega_B}^2 + \|D_\Gamma^{1/2}\nabla_\tau z_\Gamma\|_\Gamma^2 + \|\chi_B^{1/2} z\|_{\Omega_B}^2 + \|\chi_\Gamma^{1/2} z_\Gamma\|_\Gamma^2 \\
&+ \|(\beta_\Gamma^\xi)^{1/2}(\{\!\{z\}\!\}_\Gamma - z_\Gamma)\|_\Gamma^2 + \|(\beta_\Gamma)^{1/2}[\![z]\!]_\Gamma\|_\Gamma^2 \\
&+ \int_\Gamma \left(-\frac{1}{2}([\![\boldsymbol{u}]\!]_\Gamma \cdot \boldsymbol{n}_\Gamma)\{\!\{z^2\}\!\}_\Gamma + (\{\!\{\boldsymbol{u}\}\!\}_\Gamma \cdot \boldsymbol{n}_\Gamma)\left(\cancel{\{\!\{z\}\!\}_\Gamma [\![z]\!]_\Gamma - \frac{1}{2}[\![z^2]\!]_\Gamma} \right) \right) \\
&+ \int_\Gamma \frac{1}{2}\left(([\![\boldsymbol{u}]\!]_\Gamma \cdot \boldsymbol{n}_\Gamma) z_\Gamma^2 + ([\![\boldsymbol{u}]\!]_\Gamma \cdot \boldsymbol{n}_\Gamma)(\{\!\{z\}\!\}_\Gamma + z_\Gamma)(\{\!\{z\}\!\}_\Gamma - z_\Gamma) \right) \\
&+ \int_\Gamma \frac{1}{8}([\![\boldsymbol{u}]\!]_\Gamma \cdot \boldsymbol{n}_\Gamma)[\![z]\!]_\Gamma^2,
\end{aligned}
$$

where, to cancel the last term in the third line, we have used formula (19) with $a = b = z$ to infer $\frac{1}{2}[\![z^2]\!]_\Gamma = \{\!\{z\}\!\}_\Gamma [\![z]\!]_\Gamma$. Rearranging the terms on Γ, we arrive at

$$
\begin{aligned}
\mathcal{A}_\xi^{\text{transp}}((z, z_\Gamma), (z, z_\Gamma)) &= \|\boldsymbol{D}^{1/2}\nabla z\|_{\Omega_B}^2 + \|D_\Gamma^{1/2}\nabla_\tau z_\Gamma\|_\Gamma^2 + \|\chi_B^{1/2} z\|_{\Omega_B}^2 + \|\chi_\Gamma^{1/2} z_\Gamma\|_\Gamma^2 \\
&+ \|(\beta_\Gamma^\xi)^{1/2}(\{\!\{z\}\!\}_\Gamma - z_\Gamma)\|_\Gamma^2 + \|(\beta_\Gamma)^{1/2}[\![z]\!]_\Gamma\|_\Gamma^2 \\
&+ \int_\Gamma \frac{1}{2}([\![\boldsymbol{u}]\!]_\Gamma \cdot \boldsymbol{n}_\Gamma)\left(\cancel{z_\Gamma^2} - \{\!\{z^2\}\!\}_\Gamma + \{\!\{z\}\!\}_\Gamma^2 - \cancel{z_\Gamma^2} + \frac{1}{4}[\![z]\!]_\Gamma^2 \right).
\end{aligned} \tag{21}
$$

Using the formula

$$
\{\!\{ab\}\!\}_\Gamma = \{\!\{a\}\!\}_\Gamma \{\!\{b\}\!\}_\Gamma + \frac{1}{4}[\![a]\!]_\Gamma [\![b]\!]_\Gamma
$$

with $a = b = z$ to write $\{\!\{z^2\}\!\}_\Gamma = \{\!\{z\}\!\}_\Gamma^2 + \frac{1}{4}[\![z]\!]_\Gamma^2$ in the last line of (21), (15) follows. □

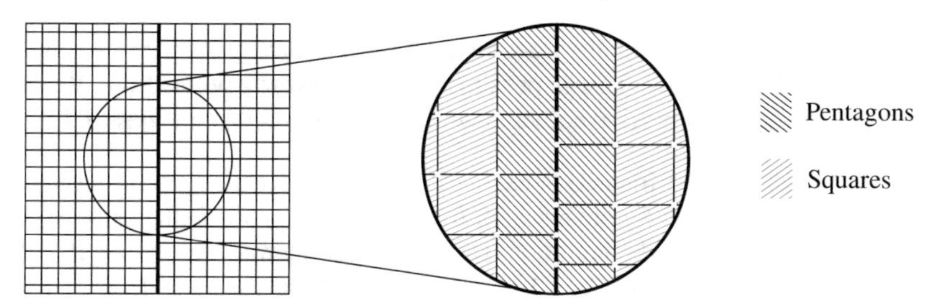

Fig. 2 Treatment of nonconforming fracture discretizations

3 Discrete setting

The HHO method is built upon a polygonal mesh of the domain Ω defined prescribing a set of mesh elements $\mathcal{T}_h$ and a set of mesh faces $\mathcal{F}_h$.

The set of mesh elements $\mathcal{T}_h$ is a finite collection of open disjoint polygons with nonzero area such that $\overline{\Omega} = \bigcup_{T \in \mathcal{T}_h} \overline{T}$ and $h = \max_{T \in \mathcal{T}_h} h_T$, with h_T denoting the diameter of T. We also denote by ∂T the boundary of a mesh element $T \in \mathcal{T}_h$. The set of mesh faces $\mathcal{F}_h$ is a finite collection of open disjoint line segments in $\overline{\Omega}$ with nonzero length such that, for all $F \in \mathcal{F}_h$, (i) either there exist two distinct mesh elements $T_1, T_2 \in \mathcal{T}_h$ such that $F \subset \partial T_1 \cap \partial T_2$ (and F is called an interface) or (ii) there exist a (unique) mesh element $T \in \mathcal{T}_h$ such that $F \subset \partial T \cap \partial\Omega$ (and F is called a boundary face). We assume that $\mathcal{F}_h$ partitions the mesh skeleton in the sense that $\bigcup_{T \in \mathcal{T}_h} \partial T = \bigcup_{F \in \mathcal{F}_h} \overline{F}$.

Remark 7 (Mesh faces) Despite working in two space dimensions, we use the terminology "face" over "edge" in order to (i) be consistent with the standard HHO nomenclature and (ii) stress the fact that faces *need not* coincide with polygonal edges (but can be subsets thereof); see also Remark 8.

We denote by $\mathcal{F}_h^{\mathrm{i}}$ the set of all interfaces and by $\mathcal{F}_h^{\mathrm{b}}$ the set of all boundary faces, so that $\mathcal{F}_h = \mathcal{F}_h^{\mathrm{i}} \cup \mathcal{F}_h^{\mathrm{b}}$. The length of a face $F \in \mathcal{F}_h$ is denoted by h_F. For any mesh element $T \in \mathcal{T}_h$, $\mathcal{F}_T$ is the set of faces that lie on ∂T and, for any $F \in \mathcal{F}_T$, $\boldsymbol{n}_{TF}$ is the unit normal to F pointing out of T. Symmetrically, for any $F \in \mathcal{F}_h$, $\mathcal{T}_F$ is the set containing the mesh elements sharing the face F (two if F is an interface, one if F is a boundary face).

To account for the presence of the fracture, we make the following assumption.

Assumption 1 *(Geometric compliance with the fracture)* The mesh is compliant with the fracture, i.e., there exists a subset $\mathcal{F}_h^{\Gamma} \subset \mathcal{F}_h^{\mathrm{i}}$ such that $\overline{\Gamma} = \bigcup_{F \in \mathcal{F}_h^{\Gamma}} \overline{F}$. As a result, $\mathcal{F}_h^{\Gamma}$ is a (1-dimensional) mesh of the fracture.

Remark 8 (Polygonal meshes and geometric compliance with the fracture) Fulfilling Assumption 1 does not pose particular problems in the context of polygonal methods, even when the fracture discretization is nonconforming in the classical sense. Consider, e.g., the situation illustrated in Fig. 2, where the fracture lies at the intersection of two nonmatching Cartesian submeshes. In this case, no special treatment is required if the mesh elements in contact with the fracture are treated as pentagons with two coplanar faces instead of rectangles. This is possible since, as already pointed out, the set of mesh faces $\mathcal{F}_h$ does not need to coincide with the set of polygonal edges of $\mathcal{T}_h$.

The set of vertices of the fracture is denoted by $\mathcal{V}_h$ and, for all $F \in \mathcal{F}_h^\Gamma$, we denote by $\mathcal{V}_F$ the vertices of F. Symmetrically, for any $V \in \mathcal{V}_h$, $\mathcal{F}_V$ is the set containing the fracture faces sharing the vertex V (two if V is an internal vertex, one if V is on the boundary on the fracture). For all $F \in \mathcal{F}_h^\Gamma$ and all $V \in \mathcal{V}_F$, $\boldsymbol{\tau}_{FV}$ denotes the unit vector tangent to the fracture and oriented so that it points out of F from V. Finally, $\mathcal{V}_h^{\rm i}$ is the set containing the internal vertices and $\mathcal{V}_h^{\rm b}$ is the set containing the points in $\partial\Gamma$, so that $\mathcal{V}_h = \mathcal{V}_h^{\rm i} \cup \mathcal{V}_h^{\rm b}$.

To avoid dealing with jumps of the problem data inside mesh elements, as well as on boundary and fracture faces, we additionally make the following assumption.

Assumption 2 *(Compliance with the problem data)* The mesh is compliant with the data, i.e.: (i) for each mesh element $T \in \mathcal{T}_h$, there exists a unique sudomain $\omega_{\rm B} \in \mathcal{P}_{\rm B}$ (see 2) such that $T \subset \omega_{\rm B}$; (ii) for each fracture face $F \in \mathcal{F}_h^\Gamma$, there is a unique subdomain $\omega_\Gamma \in \mathcal{P}_\Gamma$ (see 3) such that $F \subset \omega_\Gamma$.

4 The hybrid high-order method

In this section, we formulate the HHO discretization of problems (6) (Darcy flow) and (14) (steady passive transport).

4.1 Darcy flow

We start with the discretization of problem (6), which is closely inspired by Chave et al. (2018). Through this section, we denote by $l \geq 0$ a fixed integer polynomial degree.

4.1.1 Discrete bulk Darcy velocity unkonwns, bulk Darcy velocity reconstruction, and permeability-weighted product of Darcy velocities

Let an element $T \in \mathcal{T}_h$ be fixed, and denote by $\boldsymbol{K}_T$ the (constant) restriction to T of the bulk permeability. For any integer $m \geq 0$, set

$$\boldsymbol{U}_T^m := \boldsymbol{K}_T \nabla \mathbb{P}^m(T), \tag{22}$$

with $\mathbb{P}^m(T)$ denoting the space spanned by the restriction to T of two-variate polynomials of total degree up to m. We define the following space of fully discontinuous bulk Darcy velocity unknowns:

$$\underline{\boldsymbol{U}}_h^l := \Big\{ \underline{\boldsymbol{q}}_h := (\boldsymbol{q}_T, (q_{TF})_{F\in\mathcal{F}_T})_{T\in\mathcal{T}_h} \ : \ \text{for all } T \in \mathcal{T}_h,\ \boldsymbol{q}_T \in \boldsymbol{U}_T^l \text{ and } q_{TF} \in \mathbb{P}^l(F) \text{ for all } F \in \mathcal{F}_T \Big\}.$$

For any $T \in \mathcal{T}_h$, the element-based unknown $\boldsymbol{q}_T$ represents the Darcy velocity inside the element, while the face-based unknown q_{TF}, $F \in \mathcal{F}_T$, represents the normal Darcy velocity exiting T through F. Furthermore, we denote by $\underline{\boldsymbol{U}}_T^l$ the restriction of $\widetilde{\underline{\boldsymbol{U}}}_h^l$ to T and, for any $\underline{\boldsymbol{q}}_h \in \widetilde{\underline{\boldsymbol{U}}}_h^l$, we let $\underline{\boldsymbol{q}}_T := (\boldsymbol{q}_T, (q_{TF})_{F \in \mathcal{F}_T}) \in \underline{\boldsymbol{U}}_T^l$. The following subspace of $\widetilde{\underline{\boldsymbol{U}}}_h^l$ strongly incorporates the continuity of Darcy velocity unknowns at each interface $F \in \mathcal{F}_h^{\mathrm{i}} \setminus \mathcal{F}_h^{\Gamma}$ contained in the bulk region, as well as the homogeneous Neumann boundary condition on $\partial\Omega_{\mathrm{B}}$:

$$\underline{\boldsymbol{U}}_h^l := \left\{ \underline{\boldsymbol{q}}_h \in \widetilde{\underline{\boldsymbol{U}}}_h^l \;:\; [\![\underline{\boldsymbol{q}}_h]\!]_F = 0 \text{ for all } F \in \mathcal{F}_h^{\mathrm{i}} \setminus \mathcal{F}_h^{\Gamma} \text{ and } q_F = 0 \text{ for all } F \in \mathcal{F}_h^{\mathrm{b}} \right\}, \tag{23}$$

where, for all $F \in \mathcal{F}_h^{\mathrm{b}}$, we have set $q_F := q_{TF}$ with T denoting the unique mesh element such that $F \in \mathcal{F}_T$ and, for all $F \in \mathcal{F}_h^{\mathrm{i}}$, we have defined the jump operator such that, for any $\underline{\boldsymbol{q}}_h \in \widetilde{\underline{\boldsymbol{U}}}_h^l$,

$$[\![\underline{\boldsymbol{q}}_h]\!]_F := \sum_{T \in \mathcal{T}_F} q_{TF}.$$

For all $T \in \mathcal{T}_h$, we define the local discrete Darcy velocity reconstruction operator $\boldsymbol{F}_T^{l+1} : \underline{\boldsymbol{U}}_T^l \to \boldsymbol{U}_T^{l+1}$ (see 22) such that, for all $\underline{\boldsymbol{q}}_T = (\boldsymbol{q}_T, (q_{TF})_{F \in \mathcal{F}_T}) \in \underline{\boldsymbol{U}}_T^l$, $\boldsymbol{F}_T^{l+1}\underline{\boldsymbol{q}}_T$ solves

$$\int_T \boldsymbol{F}_T^{l+1}\underline{\boldsymbol{q}}_T \cdot \nabla w_T = \int_T \boldsymbol{q}_T \cdot \nabla \pi_T^l w_T + \sum_{F \in \mathcal{F}_T} \int_F q_{TF}(w_T - \pi_T^l w_T) \quad \forall w_T \in \mathbb{P}^{l+1}(T), \tag{24}$$

with $\pi_T^l : L^1(T) \to \mathbb{P}^l(T)$ denoting the L^2-orthogonal projector on $\mathbb{P}^l(T)$; see, e.g., Di Pietro and Droniou (2017, Appendix A.2). Notice that the quantity $\boldsymbol{F}_T^{l+1}\underline{\boldsymbol{q}}_T$ provides a representation of the Darcy velocity inside T one degree higher than the element-based unknown $\boldsymbol{q}_T$. It can be checked that condition (24) defines a unique element of $\boldsymbol{U}_T^{l+1}$, and that it is equivalent to Chave et al. (2018, Eq. 19) with discrete divergence operator expanded according to its definition.

Based on this Darcy velocity reconstruction operator, we define the global permeability-weighted product of Darcy velocities $a_{K,h}^l : \underline{\boldsymbol{U}}_h^l \times \underline{\boldsymbol{U}}_h^l \to \mathbb{R}$ such that, for all $(\underline{\boldsymbol{u}}_h, \underline{\boldsymbol{q}}_h) \in \underline{\boldsymbol{U}}_h^l \times \underline{\boldsymbol{U}}_h^l$,

$$a_{K,h}^l(\underline{\boldsymbol{u}}_h, \underline{\boldsymbol{q}}_h) := \sum_{T \in \mathcal{T}_h} \left(\int_T \boldsymbol{K}_T^{-1} \boldsymbol{F}_T^{l+1}\underline{\boldsymbol{u}}_T \cdot \boldsymbol{F}_T^{l+1}\underline{\boldsymbol{q}}_T + s_{K,T}^l(\underline{\boldsymbol{u}}_T, \underline{\boldsymbol{q}}_T) \right). \tag{25}$$

Here, the first term is the Galerkin contribution responsible for consistency while, for all $T \in \mathcal{T}_h$, $s_{K,T}^l : \underline{\boldsymbol{U}}_T^l \times \underline{\boldsymbol{U}}_T^l \to \mathbb{R}$ is the stabilization bilinear form such that, for all $(\underline{\boldsymbol{u}}_T, \underline{\boldsymbol{q}}_T) \in \underline{\boldsymbol{U}}_T^l \times \underline{\boldsymbol{U}}_T^l$,

$$s_{K,T}^l(\underline{\boldsymbol{u}}_T, \underline{\boldsymbol{q}}_T) := \sum_{F \in \mathcal{F}_T} \int_F \frac{h_F}{K_{TF}} (\boldsymbol{F}_T^{l+1}\underline{\boldsymbol{u}}_T \cdot \boldsymbol{n}_{TF} - u_{TF})(\boldsymbol{F}_T^{l+1}\underline{\boldsymbol{q}}_T \cdot \boldsymbol{n}_{TF} - q_{TF}),$$

where, for all $F \in \mathcal{F}_T$, we have set $K_{TF} := \boldsymbol{K}_T \boldsymbol{n}_{TF} \cdot \boldsymbol{n}_{TF}$.

4.1.2 Discrete fracture pressure unknowns, fracture pressure reconstruction, and tangential diffusion bilinear form

The space of discrete fracture pressure unknowns is given by

$$\underline{P}^l_{\Gamma,h} := \Big\{ \underline{z}^\Gamma_h := ((z^\Gamma_F)_{F\in\mathcal{F}^\Gamma_h}, (z^\Gamma_V)_{V\in\mathcal{V}_h}) \;:\; z^\Gamma_F \in \mathbb{P}^l(F) \text{ for all } F \in \mathcal{F}^\Gamma_h \text{ and } z^\Gamma_V \in \mathbb{R} \text{ for all } V \in \mathcal{V}_h \Big\}. \tag{26}$$

For all $F \in \mathcal{F}^\Gamma_h$, we denote by $\underline{P}^l_{\Gamma,F}$ the restriction of $\underline{P}^l_{\Gamma,h}$ to F, and set $\underline{z}^\Gamma_F := (z^\Gamma_F, (z^\Gamma_V)_{V\in\mathcal{V}_F}) \in \underline{P}^l_{\Gamma,F}$. We also introduce the following subspace which embeds the zero-mean value constraint:

$$\underline{P}^l_{\Gamma,h,0} := \left\{ \underline{z}^\Gamma_h \in \underline{P}^l_{\Gamma,h} \;:\; \int_\Gamma z^\Gamma_h = 0 \right\}, \tag{27}$$

where $z^\Gamma_h \in \mathbb{P}^l(\mathcal{F}^\Gamma_h)$ is the broken polynomial function on $\mathcal{F}^\Gamma_h$ such that $(z^\Gamma_h)_{|F} := z^\Gamma_F$ for all $F \in \mathcal{F}^\Gamma_h$.

Let $F \in \mathcal{F}^\Gamma_h$ and denote by K_F the (constant) restriction to F of the fracture permeability. We define the local fracture pressure reconstruction operator $r^{l+1}_{K,F} : \underline{P}^l_{\Gamma,F} \to \mathbb{P}^{l+1}(F)$ such that, for all $\underline{z}^\Gamma_F = (z^\Gamma_F, (z^\Gamma_V)_{\mathcal{V}_F}) \in \underline{P}^l_{\Gamma,F}$, $r^{l+1}_{K,F}\underline{z}^\Gamma_F$ is such that, for all $w^\Gamma_F \in \mathbb{P}^{l+1}(F)$,

$$\int_F K_F \nabla_\tau r^{l+1}_{K,F}\underline{z}^\Gamma_F \cdot \nabla_\tau w^\Gamma_F = -\int_F \left(z^\Gamma_F \nabla_\tau \cdot (K_F \nabla_\tau w^\Gamma_F) \right) + \sum_{V\in\mathcal{V}_F} z^\Gamma_V (K_F \nabla_\tau w^\Gamma_F)(V) \cdot \boldsymbol{\tau}_{FV}. \tag{28}$$

This relation defines a unique element $\nabla_\tau r^{l+1}_{K,F}\underline{z}^\Gamma_F$, hence a polynomial $r^{l+1}_{K,F}\underline{z}^\Gamma_F \in \mathbb{P}^{l+1}(F)$ up to an additive constant, which we fix by additionally imposing that

$$\int_F \left(r^{l+1}_{K,F}\underline{z}^\Gamma_F - z^\Gamma_F \right) = 0.$$

The reconstruction $r^{l+1}_{K,F}\underline{z}^\Gamma_F$ provides inside F a representation of the fracture pressure one degree higher than the element-based fracture unknown z^Γ_F.

We can now define the tangential diffusion bilinear form $a^{\Gamma,l}_{K,h} : \underline{P}^l_{\Gamma,h} \times \underline{P}^l_{\Gamma,h} \to \mathbb{R}$ such that

$$a^{\Gamma,l}_{K,h}(\underline{z}^\Gamma_h, \underline{q}^\Gamma_h) := \sum_{F\in\mathcal{F}^\Gamma_h} \left(\int_F K_F \nabla_\tau r^{l+1}_{K,F}\underline{z}^\Gamma_F \cdot \nabla_\tau r^{l+1}_{K,F}\underline{q}^\Gamma_F + s^{\Gamma,l}_{K,F}(\underline{z}^\Gamma_F, \underline{q}^\Gamma_F) \right), \tag{29}$$

where the first term is the Galerkin contribution responsible for consistency, while $s^{\Gamma,l}_{K,F} : \underline{P}^l_{\Gamma,F} \times \underline{P}^l_{\Gamma,F} \to \mathbb{R}$ is the stabilization bilinear form such that, for all $(\underline{z}^\Gamma_F, \underline{q}^\Gamma_F) \in \underline{P}^l_{\Gamma,F} \times \underline{P}^l_{\Gamma,F}$,

$$s_{K,F}^{\Gamma,l}(\underline{z}_F^\Gamma, \underline{q}_F^\Gamma) := \sum_{V \in \mathcal{V}_F} \frac{K_F}{h_F}(R_{K,F}^{l+1}\underline{z}_F^\Gamma(V) - z_V^\Gamma)(R_{K,F}^{l+1}\underline{q}_F^\Gamma(V) - q_V^\Gamma), \tag{30}$$

with $R_{K,F}^{l+1} : \underline{P}_{\Gamma,F}^l \to \mathbb{P}^{l+1}(F)$ such that, for all $\underline{z}_F^\Gamma \in \underline{P}_{\Gamma,F}^l$, $R_{K,F}^{l+1}\underline{z}_F^\Gamma := z_F^\Gamma + (r_{K,F}^{l+1}\underline{z}_F^\Gamma - \pi_F^l r_{K,F}^{l+1}\underline{z}_F^\Gamma)$.

4.1.3 Discrete flow problem

Let an integer $k \geq 0$ be fixed. Following Anderson and Droniou (2018), in order to have a sufficiently accurate representation of the Darcy velocity when writing the HHO approximation of degree k of the transport problem (14), we solve the flow problem (6) with an HHO approximation of degree $2k$. Thus, the bulk velocity, bulk pressure, and fracture pressure will be sought, respectively, in $\underline{\boldsymbol{U}}_h^{2k}$ (see 23), $P_{\mathrm{B},h}^{2k} := \mathbb{P}^{2k}(\mathcal{T}_h)$ (the space of broken polynomials of total degree $\leq 2k$ over $\mathcal{T}_h$), and $\underline{P}_{\Gamma,h}^{2k}$ (see 27). The discrete counterparts of the continuous bilinear forms defined in Sect. 2.2 are the bilinear forms $a_{K,h}^{\xi,2k} : \underline{\boldsymbol{U}}_h^{2k} \times \underline{\boldsymbol{U}}_h^{2k} \to \mathbb{R}$, $b_h^{2k} : \underline{\boldsymbol{U}}_h^{2k} \times P_{\mathrm{B},h}^{2k} \to \mathbb{R}$, $d_h^{2k} : \underline{\boldsymbol{U}}_h^{2k} \times \underline{P}_{\Gamma,h}^{2k} \to \mathbb{R}$ such that

$$a_{K,h}^{\xi,2k}(\underline{\boldsymbol{u}}_h, \underline{\boldsymbol{q}}_h) := a_{K,h}^{2k}(\underline{\boldsymbol{u}}_h, \underline{\boldsymbol{q}}_h) + \sum_{F \in \mathcal{F}_h^\Gamma} \Big((\lambda_F^\xi [\![\underline{\boldsymbol{u}}_h]\!]_F, [\![\underline{\boldsymbol{q}}_h]\!]_F)_F + (\lambda_F \{\!\{\underline{\boldsymbol{u}}_h\}\!\}_F, \{\!\{\underline{\boldsymbol{q}}_h\}\!\}_F)_F \Big),$$

$$b_h^{2k}(\underline{\boldsymbol{u}}_h, p_h) := \sum_{T \in \mathcal{T}_h} \left(-\int_T \boldsymbol{u}_T \cdot \nabla p_T + \sum_{F \in \mathcal{F}_T} \int_F u_{TF}\, p_T \right),$$

$$d_h^{2k}(\underline{\boldsymbol{u}}_h, \underline{p}_h^\Gamma) := \sum_{F \in \mathcal{F}_h^\Gamma} \int_F [\![\underline{\boldsymbol{u}}_h]\!]_F\, p_F^\Gamma,$$

where the bilinear forms $a_{K,h}^{2k}$ and $a_{K,h}^{\Gamma,2k}$ are defined by (25) and (29), respectively, and, for all $p_h \in P_{\mathrm{B},h}^{2k}$ and all $T \in \mathcal{T}_h$, we have set $p_T := p_{h|T}$.

Letting $\mathcal{A}_{\xi,h,2k}^{\text{flow}} : (\underline{\boldsymbol{U}}_h^{2k} \times P_{\mathrm{B},h}^{2k} \times \underline{P}_{\Gamma,h}^{2k}) \times (\underline{\boldsymbol{U}}_h^{2k} \times P_{\mathrm{B},h}^{2k} \times \underline{P}_{\Gamma,h}^{2k}) \to \mathbb{R}$ be the global bilinear form such that

$$\begin{aligned}\mathcal{A}_{\xi,h,2k}^{\text{flow}}((\underline{\boldsymbol{u}}_h, p_h, \underline{p}_h^\Gamma), (\underline{\boldsymbol{q}}_h, z_h, \underline{z}_h^\Gamma)) := a_{K,h}^{\xi,2k}(\underline{\boldsymbol{u}}_h, \underline{\boldsymbol{q}}_h) + b_h^{2k}(\underline{\boldsymbol{u}}_h, z_h) - b_h^{2k}(\underline{\boldsymbol{q}}_h, p_h)\\ + d_h^{2k}(\underline{\boldsymbol{q}}_h, \underline{p}_h^\Gamma) - d_h^{2k}(\underline{\boldsymbol{u}}_h, \underline{z}_h^\Gamma) + a_{K,h}^{\Gamma,2k}(\underline{p}_h^\Gamma, \underline{z}_h^\Gamma),\end{aligned}$$

the HHO discretization of problem (6) reads: Find $(\underline{\boldsymbol{u}}_h, p_h, \underline{p}_h^\Gamma) \in \underline{\boldsymbol{U}}_h^{2k} \times P_{\mathrm{B},h}^{2k} \times \underline{P}_{\Gamma,h,0}^{2k}$ such that, for all $(\underline{\boldsymbol{q}}_h, z_h, \underline{z}_h^\Gamma) \in \underline{\boldsymbol{U}}_h^{2k} \times P_{\mathrm{B},h}^{2k} \times \underline{P}_{\Gamma,h}^{2k}$,

$$\mathcal{A}_{\xi,h,2k}^{\text{flow}}((\underline{\boldsymbol{u}}_h, p_h, \underline{p}_h^\Gamma), (\underline{\boldsymbol{q}}_h, z_h, \underline{z}_h^\Gamma)) = (f, z_h)_{\Omega_{\mathrm{B}}} + (\ell_\Gamma f_\Gamma, z_h^\Gamma)_\Gamma. \tag{31}$$

4.2 Passive transport

We now formulate the HHO discretization of the steady passive transport problem (14). In what follows, the polynomial degree k is the same as in Sect. 4.1.3.

4.2.1 Discrete bulk concentration unknowns, bulk concentration reconstruction, and molecular diffusion bilinear form

We define the fully discontinuous space of bulk concentration unknowns as follows:

$$\widetilde{\underline{P}}_{\mathrm{B},h}^{k} := \Big\{ \underline{z}_h = (z_T, (z_{TF})_{F\in\mathcal{F}_T})_{T\in\mathcal{T}_h} \;:\; \text{for all } T \in \mathcal{T}_h,\ z_T \in \mathbb{P}^k(T) \text{ and } z_{TF} \in \mathbb{P}^k(F) \text{ for all } F \in \mathcal{F}_T \Big\}.$$

For all $T \in \mathcal{T}_h$, we denote by $\underline{P}_{\mathrm{B},T}^{k}$ the restriction of $\widetilde{\underline{P}}_{\mathrm{B},h}^{k}$ to T, and we set $\underline{z}_T = (z_T, (z_{TF})_{F\in\mathcal{F}_T}) \in \underline{P}_{\mathrm{B},T}^{k}$. For any interface $F \in \mathcal{F}_h^{\mathrm{i}}$ shared by distinct elements $T_1, T_2 \in \mathcal{T}_F$, we introduce the jump and average operators such that, for any $\underline{z}_h \in \widetilde{\underline{P}}_{\mathrm{B},h}^{k}$,

$$[\![\underline{z}_h]\!]_F = z_{T_1F} - z_{T_2F}, \qquad \{\!\{\underline{z}_h\}\!\}_F = \frac{z_{T_1F} + z_{T_2F}}{2}. \tag{32}$$

The following subspace of $\widetilde{\underline{P}}_{\mathrm{B},h}^{k}$ strongly incorporates the continuity of concentration unknowns across interfaces contained in the bulk region:

$$\underline{P}_{\mathrm{B},h}^{k} := \Big\{ \underline{z}_h \in \widetilde{\underline{P}}_{\mathrm{B},h}^{k} \;:\; [\![\underline{z}_h]\!]_F = 0 \text{ for all } F \in \mathcal{F}_h^{\mathrm{i}} \setminus \mathcal{F}_h^{\Gamma} \Big\}. \tag{33}$$

Let now an element $T \in \mathcal{T}_h$ be fixed, and denote by $\boldsymbol{D}_T$ the restriction to T of the bulk diffusion–dispersion tensor. We define the bulk concentration reconstruction operator $r_{D,T}^{k+1} : \underline{P}_{\mathrm{B},T}^{k} \to \mathbb{P}^{k+1}(T)$ such that, for all $\underline{z}_T = (z_T, (z_{TF})_{F\in\mathcal{F}_T}) \in \underline{P}_{\mathrm{B},T}^{k}$, $r_{D,T}^{k+1}\underline{z}_T$ solves

$$\int_T \boldsymbol{D}_T \nabla r_{D,T}^{k+1}\underline{z}_T \cdot \nabla w_T = -\int_T \Big(z_T \nabla\cdot(\boldsymbol{D}_T\nabla w_T) \Big) + \sum_{F\in\mathcal{F}_T} \int_F z_{TF} (\boldsymbol{D}_T \nabla w_T \cdot \boldsymbol{n}_{TF}) \quad \forall w_T \in \mathbb{P}^{k+1}(T).$$

This condition defines $r_{D,T}^{k+1}\underline{z}_T$ up to a constant, which we fix by additionally imposing that

$$\int_T \Big(r_{D,T}^{k+1}\underline{z}_T - z_T \Big) = 0.$$

The polynomial $r_{D,T}^{k+1}\underline{z}_T$ provides a representation of the concentration inside T one degree higher than the element-based unknown z_T.

We are now ready to define a global molecular diffusion bilinear form closely inspired by Di Pietro et al. (2014). More precisely, we let $a_{D,h}^k : \underline{P}_{\mathrm{B},h}^k \times \underline{P}_{\mathrm{B},h}^k \to \mathbb{R}$ be such that, for all $(\underline{c}_h, \underline{z}_h) \in \underline{P}_{\mathrm{B},h}^k \times \underline{P}_{\mathrm{B},h}^k$,

$$a_{D,h}^k(\underline{c}_h, \underline{z}_h) := \sum_{T \in \mathcal{T}_h} \left(\int_T \boldsymbol{D}_T \nabla r_{D,T}^{k+1} \underline{c}_T \cdot \nabla r_{D,T}^{k+1} \underline{z}_T + s_{D,T}^k(\underline{c}_T, \underline{z}_T) \right),$$

where the first term is the Galerkin contribution responsible for consistency, while $s_{D,T}^k : \underline{P}_{\mathrm{B},T}^k \times \underline{P}_{\mathrm{B},T}^k \to \mathbb{R}$ is the stabilization bilinear form such that, for all $(\underline{c}_T, \underline{z}_T) \in \underline{P}_{\mathrm{B},T}^k \times \underline{P}_{\mathrm{B},T}^k$,

$$s_{D,T}^k(\underline{c}_T, \underline{z}_T) := \sum_{F \in \mathcal{F}_T} \int_F \frac{D_{TF}}{h_F} (R_{D,T}^{k+1} \underline{c}_T - c_{TF})(R_{D,T}^{k+1} \underline{z}_T - z_{TF}),$$

with $D_{TF} := \boldsymbol{D}_T \boldsymbol{n}_{TF} \cdot \boldsymbol{n}_{TF}$ for all $F \in \mathcal{F}_T$ and $R_{D,T}^{k+1} : \underline{P}_{\mathrm{B},T}^k \to \mathbb{P}^{k+1}(T)$ is such that, for all $\underline{z}_T \in \underline{P}_{\mathrm{B},T}^k$, $R_{D,T}^{k+1} \underline{z}_T := z_T + (r_{D,T}^{k+1} \underline{z}_T - \pi_T^k r_{D,T}^{k+1} \underline{z}_T)$.

4.2.2 Fracture concentration unknowns, fracture concentration reconstruction, and molecular diffusion bilinear form

The fracture concentration is sought in the space $\underline{P}_{\Gamma,h}^k$ defined by (26) with $l = k$. For all $F \in \mathcal{F}_h^\Gamma$, we define the fracture concentration reconstruction operator $r_{D,F}^{k+1} : \underline{P}_{\Gamma,F}^k \to \mathbb{P}^{k+1}(F)$ as in (28) setting $l = k$ and replacing K_F by $D_F := D_{\Gamma|F}$. Similary, we denote by $a_{D,h}^{\Gamma,k} : \underline{P}_{\Gamma,h}^k \times \underline{P}_{\Gamma,h}^k \to \mathbb{R}$ the tangential molecular diffusion bilinear form defined as (29)–(30) with $l = k$ and K_F replaced by D_F.

4.2.3 Darcy velocities and advection–reaction bilinear forms in the bulk region and in the fracture

In order to discretize the advection–reaction terms that appear in the passive transport problem, we need suitable representations of the Darcy velocity both in the bulk region and inside the fracture.

Denote by $(\underline{\boldsymbol{u}}_h, p_h, \underline{p}_h^\Gamma) \in \underline{\boldsymbol{U}}_h^{2k} \times P_{\mathrm{B},h}^{2k} \times \underline{P}_{\Gamma,h,0}^{2k}$ the solution of the discrete flow problem (31). For any $T \in \mathcal{T}_h$, taking in (31) $\underline{\boldsymbol{q}}_h = \underline{\boldsymbol{0}}$, z_h such that $z_{T'} = 0$ for all $T' \in \mathcal{T}_h \setminus \{T\}$ while z_T spans $\mathbb{P}^{2k}(T)$, and $\underline{z}_h^\Gamma = \underline{0}$, we infer the following local balance for the discrete bulk Darcy velocity:

$$\int_T -\boldsymbol{u}_T \cdot \nabla z_T + \sum_{F \in \mathcal{F}_T} \int_F u_{TF} z_T = \int_T f z_T \qquad \forall z_T \in \mathbb{P}^{2k}(T). \tag{34}$$

Additionally, by definition (23) of $\underline{\boldsymbol{U}}_h^{2k}$, the Darcy velocity thus defined has continuous normal components across interfaces contained in the bulk in the sense that $[\![\underline{\boldsymbol{u}}_h]\!]_F = 0$

for all $F \in \mathcal{F}_h^{\mathrm{i}} \setminus \mathcal{F}_h^{\Gamma}$ where the jump operator is defined applying (32) componentwise. Thus, $\underline{\boldsymbol{u}}_h$ is the natural candidate to play the role of the Darcy velocity in the bulk region.

Let now a fracture face $F \in \mathcal{F}_h^{\Gamma}$ be fixed, and define the fracture Darcy velocity $\underline{\boldsymbol{u}}_F^{\Gamma} = (\boldsymbol{u}_F^{\Gamma}, (u_{FV}^{\Gamma})_{V \in \mathcal{V}_F})$ such that

$$\boldsymbol{u}_F^{\Gamma} := -K_F \nabla_\tau r_{K,F}^{2k+1} \underline{p}_F^{\Gamma} \text{ and, for all } V \in \mathcal{V}_F,$$

$$u_{FV}^{\Gamma} := \begin{cases} \boldsymbol{u}_F^{\Gamma}(V) \cdot \boldsymbol{\tau}_{FV} + \gamma_{FV}^{\mathrm{num}}(\underline{p}_F^{\Gamma}) & \text{if } V \in \mathcal{V}_h^{\mathrm{i}} \\ 0 & \text{if } V \in \mathcal{V}_h^{\mathrm{b}}, \end{cases}$$

where, for all $V \in \mathcal{V}_F$, $\gamma_{FV}^{\mathrm{num}} : \underline{P}_{\Gamma,F}^{2k} \to \mathbb{R}$ is the boundary residual operator defined as in Di Pietro and Tittarelli (2018, Lemma 3). With this choice for the fracture Darcy velocity, the following local balance holds for all $F \in \mathcal{F}_h^{\Gamma}$:

$$-\int_F \boldsymbol{u}_F^{\Gamma} \cdot \nabla_\tau z_F^{\Gamma} + \sum_{V \in \mathcal{V}_F} u_{FV}^{\Gamma}(z_F^{\Gamma}(V) - z_V^{\Gamma}) = \int_F \left(\ell_\Gamma f_\Gamma + [\![\underline{\boldsymbol{u}}_h]\!]_F\right) z_F^{\Gamma} \quad \forall \underline{z}_F^{\Gamma} \in \underline{P}_{\Gamma,F}^{2k}. \tag{35}$$

Moreover, the discrete fracture Darcy velocity is continuous across internal vertices, that is to say,

$$\sum_{F \in \mathcal{F}_V} u_{FV}^{\Gamma} = 0 \text{ for all } V \in \mathcal{V}_h^{\mathrm{i}}. \tag{36}$$

$\underline{\boldsymbol{u}}_F^{\Gamma}$ is therefore the natural candidate to play the role of the Darcy velocity inside the fracture.

We now have all the ingredients to define discrete counterparts of the advective terms in the bulk region and inside the fracture. More precisely, closely following Di Pietro et al. (2015), we define the advection–reaction bilinear forms $a_{\boldsymbol{u},f,h}^{k} : \underline{P}_{\mathrm{B},h}^{k} \times \underline{P}_{\mathrm{B},h}^{k} \to \mathbb{R}$ and $a_{\boldsymbol{u},f,h}^{\Gamma,k} : \underline{P}_{\Gamma,h}^{k} \times \underline{P}_{\Gamma,h}^{k}$ such that

$$\begin{aligned} a_{\boldsymbol{u},f,h}^{k}(\underline{c}_h, \underline{z}_h) &:= \sum_{T \in \mathcal{T}_h} \Bigg(\int_T c_T(-\boldsymbol{u}_T \cdot \nabla z_T + f^- z_T) + \sum_{F \in \mathcal{F}_T} \int_F u_{TF} c_T (z_T - z_{TF}) \\ &\qquad + s_{\boldsymbol{u},T}^{k}(\underline{c}_T, \underline{z}_T) \Bigg), \\ a_{\boldsymbol{u},f,h}^{\Gamma,k}(\underline{c}_h^{\Gamma}, \underline{z}_h^{\Gamma}) &:= \sum_{F \in \mathcal{F}_h^{\Gamma}} \Bigg(\int_F c_F^{\Gamma}(-\boldsymbol{u}_F^{\Gamma} \cdot \nabla_\tau z_F^{\Gamma} + \ell_\Gamma f_\Gamma^- z_F^{\Gamma}) + \sum_{V \in \mathcal{V}_F} u_{FV}^{\Gamma} c_F^{\Gamma}(V)(z_F^{\Gamma}(V) - z_V^{\Gamma}) \\ &\qquad + s_{\boldsymbol{u},F}^{\Gamma,k}(\underline{c}_F^{\Gamma}, \underline{z}_F^{\Gamma}) \Bigg), \end{aligned} \tag{37}$$

where, for all $T \in \mathcal{T}_h$ and all $F \in \mathcal{F}_h^{\Gamma}$, $s_{\boldsymbol{u},T}^{k} : \underline{P}_{\mathrm{B},T}^{k} \times \underline{P}_{\mathrm{B},T}^{k} \to \mathbb{R}$ and $s_{\boldsymbol{u},F}^{\Gamma,k} : \underline{P}_{\Gamma,F}^{k} \times \underline{P}_{\Gamma,F}^{k} \to \mathbb{R}$ are the upwind stabilization bilinear forms respectively in the bulk and inside the fracture such that

$$
\begin{aligned}
s_{\boldsymbol{u},T}^{k}(\underline{c}_T, \underline{z}_T) &:= \sum_{F\in\mathcal{F}_T} \int_F \frac{|u_{TF}| - u_{TF}}{2}(c_T - c_{TF})(z_T - z_{TF}),\\
s_{\boldsymbol{u},F}^{\Gamma,k}(\underline{c}_F^{\Gamma}, \underline{z}_F^{\Gamma}) &:= \sum_{V\in\mathcal{V}_F} \frac{|u_{FV}^{\Gamma}| - u_{FV}^{\Gamma}}{2}(c_F^{\Gamma}(V) - c_V^{\Gamma})(z_F^{\Gamma}(V) - z_V^{\Gamma}).
\end{aligned}
\tag{38}
$$

4.2.4 Passive transport problem

We are now ready to state the HHO discretization of the steady transport problem (14). At the discrete level, the counterpart of the continuous bilinear form defined in (13) is the bilinear form $\mathcal{A}_{\xi,h,k}^{\text{transp}} : \left(\underline{P}_{\text{B},h}^{k} \times \underline{P}_{\Gamma,h}^{k}\right) \times \left(\underline{P}_{\text{B},h}^{k} \times \underline{P}_{\Gamma,h}^{k}\right) \to \mathbb{R}$ such that

$$
\begin{aligned}
&\mathcal{A}_{\xi,h,k}^{\text{transp}}((\underline{c}_h, \underline{c}_h^{\Gamma}), (\underline{z}_h, \underline{z}_F^{\Gamma}))\\
&\quad := a_{D,h}^{k}(\underline{c}_h, \underline{z}_h) + a_{\boldsymbol{u},f,h}^{k}(\underline{c}_h, \underline{z}_h) + a_{D,h}^{\Gamma,k}(\underline{c}_h^{\Gamma}, \underline{z}_h^{\Gamma}) + a_{\boldsymbol{u},f,h}^{\Gamma,k}(\underline{c}_h^{\Gamma}, \underline{z}_h^{\Gamma})\\
&\qquad + \sum_{F\in\mathcal{F}_h^{\Gamma}} \int_F \left(\beta_F^{\xi}(\{\!\{\underline{c}_h\}\!\}_F - c_F^{\Gamma})(\{\!\{\underline{z}_h\}\!\}_F - z_F^{\Gamma}) + \beta_F [\![\underline{c}_h]\!]_F [\![\underline{z}_h]\!]_F\right)\\
&\qquad + \sum_{F\in\mathcal{F}_h^{\Gamma}} \int_F \left(\{\!\{\underline{\boldsymbol{u}}_h\}\!\}_F \{\!\{\underline{c}_h\}\!\}_F [\![\underline{z}_h]\!]_F + \frac{1}{8}[\![\underline{\boldsymbol{u}}_h]\!]_F [\![\underline{c}_h]\!]_F [\![\underline{z}_h]\!]_F\right)\\
&\qquad + \sum_{F\in\mathcal{F}_h^{\Gamma}} \int_F \frac{1}{2}[\![\underline{\boldsymbol{u}}_h]\!]_F (\{\!\{\underline{c}_h\}\!\}_F + c_F^{\Gamma})(\{\!\{\underline{z}_h\}\!\}_F - z_F^{\Gamma}),
\end{aligned}
\tag{39}
$$

where the role of the terms in the last three lines is to enforce the transmission conditions (9) on Γ. The HHO discretization of problem (14) then reads: Find $(\underline{c}_h, \underline{c}_h^{\Gamma}) \in \underline{P}_{\text{B},h}^{k} \times \underline{P}_{\Gamma,h}^{k}$ such that

$$
\mathcal{A}_{\xi,h,k}^{\text{transp}}((\underline{c}_h, \underline{c}_h^{\Gamma}), (\underline{z}_h, \underline{z}_F^{\Gamma}))= \int_{\Omega_{\text{B}}} f^{+}\widehat{c}z_h + \int_{\Gamma} \ell_{\Gamma} f_{\Gamma}^{+}\widehat{c}_{\Gamma} z_h^{\Gamma} \quad \forall (\underline{z}_h, \underline{z}_h^{\Gamma}) \in \underline{P}_{\text{B},h}^{k} \times \underline{P}_{\Gamma,h}^{k}.
\tag{40}
$$

We now prove the discrete counterpart of the Theorem 5.

Theorem 9 (Discrete coercivity) *Let $\xi > 1/2$. Then, for all $(\underline{z}_h, \underline{z}_h^{\Gamma}) \in \underline{P}_{\text{B},h}^{k} \times \underline{P}_{\Gamma,h}^{k}$, it holds*

$$
\begin{aligned}
\mathcal{A}_{\xi,h,k}^{\text{transp}}((\underline{z}_h, \underline{z}_h^{\Gamma}), (\underline{z}_h, \underline{z}_F^{\Gamma})) &= a_{D,h}^{k}(\underline{z}_h, \underline{z}_h) + a_{D,h}^{\Gamma,k}(\underline{z}_h^{\Gamma}, \underline{z}_h^{\Gamma})\\
&\quad + \sum_{T\in\mathcal{T}_h} \left(\|\chi_{\text{B},T}^{1/2} z_T\|_T^2 + \sum_{F\in\mathcal{F}_T} \frac{1}{2} \||u_{TF}|^{1/2}(z_T - z_{TF})\|_F^2 \right)\\
&\quad + \sum_{F\in\mathcal{F}_h^{\Gamma}} \left(\|\chi_{\Gamma,F}^{1/2} z_F^{\Gamma}\|_F^2 + \sum_{V\in\mathcal{V}_F} \frac{1}{2} |u_{FV}^{\Gamma}|(z_F^{\Gamma}(V) - z_V^{\Gamma})^2 \right)
\end{aligned}
$$

$$+\sum_{F\in\mathcal{F}_h^\Gamma}\left(\beta_F^\xi\|\{\!\{\underline{z}_h\}\!\}_F - z_F^\Gamma\|_F^2 + \beta_F\|[\![\underline{c}_h]\!]_F\|_F^2\right), \quad (41)$$

where, for all $T \in \mathcal{T}_h$ and all $F \in \mathcal{F}_h^\Gamma$, $\chi_{\mathrm{B},T} := (\chi_\mathrm{B})_{|T}$ and $\chi_{\Gamma,F} := (\chi_\Gamma)_{|F}$, respectively.

Remark 10 (Upwind contributions) Unlike the continuous case (see Theorem 5), we have in the second and third lines of the energy balance (41) upwind-related contributions of bulk and fracture region, respectively. These could be removed at the price of having coercivity in a weaker norm.

Proof The proof is similar to the one of the Theorem 5. Let $(\underline{z}_h, \underline{z}_h^\Gamma) \in \underline{P}_{\mathrm{B},h}^k \times \underline{P}_{\Gamma,h}^k$ be fixed and set $(\underline{\mathfrak{z}}_h, \underline{\mathfrak{z}}_h^\Gamma) \in \underline{P}_{\mathrm{B},h}^{2k} \times \underline{P}_{\Gamma,h}^{2k}$ such that,

$$\begin{aligned}
\forall T \in \mathcal{T}_h, \qquad (\underline{\mathfrak{z}}_h)_{|T} &= \underline{\mathfrak{z}}_T = (\mathfrak{z}_T, (\mathfrak{z}_{TF})_{F\in\mathcal{F}_T}) := (\frac{z_T^2}{2}, (\frac{z_{TF}^2}{2})_{F\in\mathcal{F}_T}),\\
\forall F \in \mathcal{F}_h^\Gamma, \qquad (\underline{\mathfrak{z}}_h^\Gamma)_{|F} &= \underline{\mathfrak{z}}_F^\Gamma = (\mathfrak{z}_F^\Gamma, (\mathfrak{z}_V^\Gamma)_{V\in\mathcal{V}_F}) := (\frac{(z_F^\Gamma)^2}{2}, (\frac{(z_V^\Gamma)^2}{2})_{V\in\mathcal{V}_F}).
\end{aligned} \quad (42)$$

Using the definition of the global bilinear form $\mathcal{A}_{\xi,h,k}^{\mathrm{transp}}$ (39) with $(\underline{c}_h, \underline{c}_h^\Gamma) = (\underline{z}_h, \underline{z}_h^\Gamma)$, we immediately obtain the terms in the first and last line of (41). Let now $\mathcal{I}_1 := a_{\boldsymbol{u},f,h}^k(\underline{z}_h, \underline{z}_h)$, $\mathcal{I}_2 := a_{\boldsymbol{u},f,h}^{\Gamma,k}(\underline{z}_h^\Gamma, \underline{z}_h^\Gamma)$, and let $\mathcal{I}_3$ gather the remaining coupling terms, that is to say, the two last lines on the right-hand side of (39) with $\underline{c}_h = \underline{z}_h$ and $c_F^\Gamma = z_F^\Gamma$ for all $F \in \mathcal{F}_h^\Gamma$. Expanding $\mathcal{I}_1$ and $\mathcal{I}_2$ according to their respective definitions (37), and recalling the definitions of the stabilization bilinear forms $s_{\boldsymbol{u},T}^k$ and $s_{\boldsymbol{u},F}^k$ (38), it is inferred that

$$\mathcal{I}_1 = \sum_{T\in\mathcal{T}_h}\left(\int_T\left(-\boldsymbol{u}_T\cdot\nabla\mathfrak{z}_T + f^- z_T^2\right) + \sum_{F\in\mathcal{F}_T}\int_F\left(u_{TF}(\mathfrak{z}_T - \mathfrak{z}_{TF}) + \frac{1}{2}|u_{TF}|(z_T - z_{TF})^2\right)\right), \quad (43a)$$

$$\mathcal{I}_2 = \sum_{F\in\mathcal{F}_h^\Gamma}\left(\int_F\left(-\boldsymbol{u}_F^\Gamma\cdot\nabla_\tau\mathfrak{z}_F^\Gamma + \ell_\Gamma f_\Gamma^- {z_F^\Gamma}^2\right) + \sum_{V\in\mathcal{V}_F}\left(u_{FV}^\Gamma(\mathfrak{z}_F^\Gamma(V) - \mathfrak{z}_V^\Gamma) + \frac{1}{2}|u_{FV}^\Gamma|(z_F^\Gamma(V) - z_V^\Gamma)^2\right)\right). \quad (43b)$$

Using the local balances (34) in the bulk and (35) inside the fracture (that hold since $\mathfrak{z}_T \in \mathbb{P}^{2k}(T)$ for all $T \in \mathcal{T}_h$ and $\underline{\mathfrak{z}}_F^\Gamma \in \underline{P}_{\Gamma,F}^{2k}$ for all $F \in \mathcal{F}_h^\Gamma$) together with the fact that

$$\sum_{T\in\mathcal{T}_h}\sum_{F\in\mathcal{F}_T} u_{TF}\mathfrak{z}_{TF} = \sum_{F\in\mathcal{F}_h^\Gamma}[\![\boldsymbol{u}_h\mathfrak{z}_h]\!]_F,$$

which follows from $(\underline{\boldsymbol{u}}_h, \underline{\mathfrak{z}}_h) \in \underline{\boldsymbol{U}}_h^{2k} \times \underline{P}_{\mathrm{B},h}^{2k}$, we finally get from (43a) and (43b)

$$\mathcal{I}_1 = \sum_{T\in\mathcal{T}_h} \left(\|\chi_{\mathrm{B},T}^{1/2} z_T\|_T^2 + \sum_{F\in\mathcal{F}_T} \frac{1}{2} \| |u_{TF}|^{1/2}(z_T - z_{TF})\|_F^2 \right) - \sum_{F\in\mathcal{F}_h^\Gamma} \int_F [\![\underline{\boldsymbol{u}}_h \underline{\mathfrak{z}}_h]\!]_F, \tag{44a}$$

$$\mathcal{I}_2 = \sum_{F\in\mathcal{F}_h^\Gamma} \left(\|\chi_{\Gamma,F}^{1/2} z_F^\Gamma\|_F^2 + \sum_{V\in\mathcal{V}_F} \frac{1}{2} |u_{FV}^\Gamma| (z_F^\Gamma(V) - z_V^\Gamma)^2 \right) + \sum_{F\in\mathcal{F}_h^\Gamma} \int_F [\![\underline{\boldsymbol{u}}_h]\!]_F \mathfrak{z}_F^\Gamma. \tag{44b}$$

To conclude, it suffices to prove that the sum of the last term in the right-hand side of (44a) and the last term in the right-hand side of (44b) and $\mathcal{I}_3$ is equal to zero. Using (19) to infer first that $[\![\underline{\boldsymbol{u}}_h \underline{\mathfrak{z}}_h]\!]_F = [\![\underline{\boldsymbol{u}}_h]\!]_F \{\{\underline{\mathfrak{z}}_h\}\}_F + \{\{\underline{\boldsymbol{u}}_h\}\}_F [\![\underline{\mathfrak{z}}_h]\!]_F$ and then that $[\![\underline{\mathfrak{z}}_h]\!]_F = [\![\underline{z}_h]\!]_F \{\{\underline{z}_h\}\}_F$, we get

$$\sum_{F\in\mathcal{F}_h^\Gamma} \int_F \left([\![\underline{\boldsymbol{u}}_h]\!]_F \mathfrak{z}_F^\Gamma - [\![\underline{\boldsymbol{u}}_h \underline{\mathfrak{z}}_h]\!]_F \right) + \mathcal{I}_3 = \sum_{F\in\mathcal{F}_h^\Gamma} \int_F \left([\![\underline{\boldsymbol{u}}_h]\!]_F ({}^1\!/\!{}_2 \{\{\underline{z}_h\}\}_F^2 + {}^1\!/\!{}_8 [\![\underline{z}_h]\!]_F^2 - \{\{\underline{\mathfrak{z}}_h\}\}_F) \right),$$

that concludes the proof since $\{\{\underline{\mathfrak{z}}_h\}\}_F = \frac{1}{2}(\{\{\underline{z}_h\}\}_F^2 + \frac{1}{4}[\![\underline{z}_h]\!]_F^2)$. □

Remark 11 (Polynomial degree and local conservation) The use of polynomials of degree $2k$ to solve the discrete flow problem (40) is required in the proof of Theorem 9. Indeed, to pass from (43) to (44), the argument is that both the local balances (34) and (35) are valid when we use as test functions $\underline{\mathfrak{z}}_h \in \underline{P}_{\mathrm{B},h}^{2k}$ and $\underline{\mathfrak{z}}_h^\Gamma \in \underline{P}_{\Gamma,h}^{2k}$ defined by (42).

4.3 Extension to the unsteady case

In the numerical tests of Sects. 5.2, 5.3 below, we consider the physically relevant situation of unsteady passive transport with a steady Darcy velocity field. The extension of the HHO scheme (40) to this situation is briefly discussed in what follows.

The transport problem can be extended to the unsteady case by assuming that the unknowns depend on time and adding the unsteady contributions $\phi \mathrm{d}_t c$ and $\ell_\Gamma \phi_\Gamma \mathrm{d}_t c_\Gamma$ in, respectively, (7a) and (8a), where $\phi : \Omega_\mathrm{B} \to \mathbb{R}$ and $\phi_\Gamma : \Gamma \to \mathbb{R}$ stand, respectively, for the porosity in the bulk region and in the fracture such that $0 < \phi < 1$ and $0 < \phi_\Gamma < 1$. In the numerical tests, we assume that these quantities are piecewise constant on the partitions $\mathcal{P}_\mathrm{B}$ and $\mathcal{P}_\Gamma$ (see 2, 3), respectively. More generally, the porosities could also depend on time. Initial conditions for the bulk and the fracture concentration $c(t = 0, \cdot) = c^0(\cdot)$ and $c_\Gamma(t = 0, \cdot) = c_\Gamma^0(\cdot)$ close the problem. The functions $\widehat{c}$ and $\widehat{c_\Gamma}$ that represent the concentration of solute as it is injected in, respectively, the bulk and the fracture, will also be allowed to depend on time.

To discretize in time, we consider for sake of simplicity a uniform partition $(t^n)_{0\le n\le N}$ of the time interval $[0, t_F]$ with $t^0 = 0$, $t^N = t_F$ the final time of computation, and $t^n - t^{n-1} = \delta t$ the constant time step for all $1 \le n \le N$. For any sufficiently regular function of time φ taking values in a vector space V, we denote by $\varphi^n \in V$ its value at discrete time t^n and we introduce the backward differencing operator δ_t such that, for all $1 \le n \le N$,

$$\delta_t \varphi^n := \frac{\varphi^n - \varphi^{n-1}}{\delta t} \in V.$$

With this notation, the discrete problem reads: For all $1 \le n \le N$, find $(\underline{c}_h^n, \underline{c}_h^{\Gamma,n}) \in \underline{P}_{\mathrm{B},h}^k \times \underline{P}_{\Gamma,h}^k$ such that, for all $(\underline{z}_h, \underline{z}_h^\Gamma) \in \underline{P}_{\mathrm{B},h}^k \times \underline{P}_{\Gamma,h}^k$,

$$\begin{aligned}\int_{\Omega_\mathrm{B}} \phi \delta_t c_h^n z_h + \int_\Gamma \ell_\Gamma \phi_\Gamma \delta_t c_h^{\Gamma,n} z_h^\Gamma + \mathcal{A}_{\xi,h,k}^{\mathrm{transp}}((\underline{c}_h^n, \underline{c}_h^{\Gamma,n}), (\underline{z}_h, \underline{z}_F^\Gamma)) \\ = \int_{\Omega_\mathrm{B}} f^+ \widehat{c}^n z_h + \int_\Gamma \ell_\Gamma f_\Gamma^+ \widehat{c}_\Gamma^n z_h^\Gamma. \end{aligned} \tag{45}$$

The initial condition is discretized taking c_h^0 and $c_h^{\Gamma,0}$ equal to the L^2-orthogonal projections on $\mathbb{P}^k(\mathcal{T}_h)$ and $\mathbb{P}^l(\mathcal{F}_h^\Gamma)$ of c^0 and c_Γ^0, respectively. Notice that it is not necessary to prescribe face values for the concentration in the bulk region, nor vertex values for the concentration in the fracture, as these discrete unknowns do not appear in the discretization of the time derivative.

5 Numerical results

This section contains an extensive numerical validation of the HHO method. We first study numerically the convergence rates achieved by the method, and then propose two more physical test cases in which fractures act as barriers or conduits, depending on the value of the permeability parameters.

5.1 Convergence for a steady problem

We start by a numerical study of the convergence rates of the method for both the flow problem (31) and the steady passive transport problem (40).

5.1.1 Analytical solution

We approximate problems (31) and (40) on the square domain $\Omega = (0, 1)^2$ crossed by the fracture $\Gamma = \{\boldsymbol{x} \in \Omega : x_1 = 0.5\}$, and set $\ell_\Gamma = 0.01$ and $\xi = 3/4$. For the flow problem, we consider the exact solutions corresponding to the bulk and fracture pressures

$$p(\boldsymbol{x}) := \begin{cases} \cos(2x_1)\cos(\pi x_2) & \text{if } x_1 < 0.5 \\ \cos(\pi x_1)\cos(\pi x_2) & \text{if } x_1 > 0.5 \end{cases}, \qquad p_\Gamma(\boldsymbol{x}) := \{\!\{c\}\!\}_\Gamma - \lambda_\Gamma [\![\boldsymbol{u}]\!]_\Gamma \cdot \boldsymbol{n}_\Gamma,$$

and let $\boldsymbol{u}_{|\Omega_{\mathrm{B},i}} = -\boldsymbol{K}\nabla p_{|\Omega_{\mathrm{B},i}}$ for $i \in \{1, 2\}$ and $\boldsymbol{u}_\Gamma = -K_\Gamma \nabla_\tau p_\Gamma$, with $\kappa_\Gamma^\tau = 1$, $\kappa_\Gamma^n = 0.01$ and

$$\boldsymbol{K} := \frac{\cos(1)}{\sin(1) + \pi/2} \begin{bmatrix} \kappa_\Gamma^n/(2\ell_\Gamma) & 0 \\ 0 & 1 \end{bmatrix}, \qquad K_\Gamma := \kappa_\Gamma^\tau \ell_\Gamma.$$

For the steady passive transport problem, the exact solutions corresponding to the bulk and fracture concentrations are given by

$$c(\boldsymbol{x}) := \begin{cases} \exp\left(2/\pi \cos(\pi x_1)\left(\beta_\Gamma - \frac{1}{8}\cos(\pi x_2)\frac{\kappa_\Gamma^n}{\ell_\Gamma}\frac{\cos(1)(\sin(1)-\pi/2)}{\sin(1)+\pi/2}\right)\right) & \text{if } x_1 < 0.5 \\ \exp\left(2/\pi (\cos(\pi x_1) - \pi)\left(\beta_\Gamma - \frac{1}{8}\cos(\pi x_2)\frac{\kappa_\Gamma^n}{\ell_\Gamma}\frac{\cos(1)(\sin(1)-\pi/2)}{\sin(1)+\pi/2}\right)\right) & \text{if } x_1 > 0.5 \end{cases},$$

$$c_\Gamma(\boldsymbol{x}) := \frac{[\![\boldsymbol{u}c - \boldsymbol{D}\nabla c]\!]_\Gamma \cdot \boldsymbol{n}_\Gamma - \{\!\{c\}\!\}_\Gamma \left(1/2[\![\boldsymbol{u}]\!]_\Gamma \cdot \boldsymbol{n}_\Gamma + \beta_\Gamma^\xi\right)}{1/2[\![\boldsymbol{u}]\!]_\Gamma \cdot \boldsymbol{n}_\Gamma - \beta_\Gamma^\xi}, \tag{46}$$

with $\boldsymbol{D} = \boldsymbol{I}_2$, the identity matrix of $\mathbb{R}^{2\times 2}$, $\mathcal{D}_\Gamma^\tau = 1$ and $\mathcal{D}_\Gamma^n = 0.01$. The source terms f, f_Γ are inferred from (1b) and (1c), respectively. The right-hand sides of (7a) and (8a) are also modified by introducing nonzero terms in accordance with the expressions of c and c_Γ; see (46). It can be checked that, with this choice of analytical solutions, the jump and average of p, $\boldsymbol{u}$, c, $\boldsymbol{D}\nabla c$ are not identically zero on the fracture, which enables us to test the weak enforcement of the transmission conditions (4) for the flow problem and (9) for the steady passive transport problem.

5.1.2 Error measures

On the spaces of discrete bulk unknowns $\underline{\boldsymbol{U}}_h^{2k}$ and $\underline{P}_h^k$, we define the norms $\|\cdot\|_{U,h}$ and $\|\cdot\|_{D,h}$ such that, for all $\underline{\boldsymbol{q}}_h \in \underline{\boldsymbol{U}}_h^{2k}$ and all $\underline{z}_h \in \underline{P}_h^k$,

$$\|\underline{\boldsymbol{q}}_h\|_{U,h}^2 := \sum_{T\in\mathcal{T}_h} (\overline{K}_{\mathrm{B},T})^{-1} \left(\|\boldsymbol{q}_T\|_T^2 + \sum_{F\in\mathcal{F}_T} h_F \|q_{TF}\|_F^2 \right),$$

$$\|\underline{z}_h\|_{D,h}^2 := \sum_{T\in\mathcal{T}_h} \varrho_{\boldsymbol{D},T}^{-1} \Big(\|\boldsymbol{D}_T^{1/2}\nabla z_T\|_T^2 + \sum_{F\in\mathcal{F}_T} \frac{D_{TF}}{h_F}\|z_T - z_{TF}\|_F^2 \Big),$$

where, for any $T \in \mathcal{T}_h$, $\overline{K}_{\mathrm{B},T}$ is the largest eigenvalue of the (constant) permeability tensor $\boldsymbol{K}_T$, while $\varrho_{\boldsymbol{D},T} := \overline{D}_{\mathrm{B},T}/\underline{D}_{\mathrm{B},T}$ is the bulk anisotropy ratio with $\overline{D}_{\mathrm{B},T}$, $\underline{D}_{\mathrm{B},T} > 0$ denoting, respectively, the largest and smallest eigenvalue of the (constant) local bulk diffusion–dispersion tensor $\boldsymbol{D}_T$.

On the spaces of discrete fracture unknowns $\underline{P}_{\Gamma,h}^{2k}$ and $\underline{P}_{\Gamma,h}^k$ we define the norms $\|\cdot\|_{\Gamma,K,h}$ and $\|\cdot\|_{\Gamma,D,h}$ such that, for all $\underline{v}_h^\Gamma \in \underline{P}_{\Gamma,h}^{2k}$ and all $\underline{z}_h^\Gamma \in \underline{P}_{\Gamma,h}^k$,

$$\|\underline{v}_h^\Gamma\|_{\Gamma,K,h}^2 := \sum_{F\in\mathcal{F}_h^\Gamma} \left(\|K_F^{1/2}\nabla_\tau v_F^\Gamma\|_F^2 + \sum_{V\in\mathcal{V}_F} \frac{K_F}{h_F}(v_F^\Gamma(V) - v_V^\Gamma)^2 \right),$$

$$\|\underline{z}_h^\Gamma\|_{\Gamma,D,h}^2 := \sum_{F\in\mathcal{F}_h^\Gamma} \left(\|D_F^{1/2}\nabla_\tau z_F^\Gamma\|_F^2 + \sum_{V\in\mathcal{V}_F} \frac{D_F}{h_F}(z_F^\Gamma(V) - z_V^\Gamma)^2 \right).$$

For the flow problem, we monitor the following errors defined as the difference between the numerical solution and suitable projections of the exact solution:

$$\|\underline{\boldsymbol{u}}_h - \underline{\boldsymbol{I}}_h^{2k}\boldsymbol{u}\|_{U,h}, \quad \|\underline{p}_h^\Gamma - \underline{I}_h^{2k} p_\Gamma\|_{\Gamma,K,h}, \quad \|p_h - \pi_h^{2k} p\|_{L^2(\Omega_\mathrm{B})}, \quad \|p_h^\Gamma - \pi_{\Gamma,h}^{2k} p_\Gamma\|_{L^2(\Gamma)}, \tag{47}$$

where $\underline{\boldsymbol{I}}_h^{2k}\boldsymbol{u} := (\boldsymbol{K}_T\nabla y_T, (\pi_F^{2k}(\boldsymbol{u}\cdot\boldsymbol{n}_{TF})_{F\in\mathcal{F}_T})_{T\in\mathcal{T}_h}$ with $y_T \in \mathbb{P}^{2k}(T)$ is such that $\int_T (\boldsymbol{K}_T\nabla y_T - \boldsymbol{u})\cdot\nabla v_T = 0$ for all $v_T \in \mathbb{P}^{2k}(T)$, $\underline{I}_{\Gamma,h}^{2k} p_\Gamma := ((\pi_F^{2k} p_{\Gamma|F})_{F\in\mathcal{F}_h^\Gamma}, (p_\Gamma(V))_{V\in\mathcal{V}_h})$ with π_F^{2k} denoting the L^2-orthogonal projector on $\mathbb{P}^{2k}(F)$, and $\pi_h^{2k} p$ and $\pi_{\Gamma,h}^{2k} p_\Gamma$ denote, respectively, the L^2-orthogonal projections of p and p_Γ on $P_{\mathrm{B},h}^{2k}$ and $P_{\Gamma,h}^{2k}$.

Similarly, for the steady passive transport problem we consider the following error measures:

$$\|\underline{c}_h - \underline{I}_h^k c\|_{D,h}, \quad \|\underline{c}_h^\Gamma - \underline{I}_{\Gamma,h}^k c_\Gamma\|_{\Gamma,D,h}, \quad \|c_h - \pi_h^k c\|_{L^2(\Omega_\mathrm{B})}, \quad \|c_h^\Gamma - \pi_{\Gamma,h}^k c_\Gamma\|_{L^2(\Gamma)}, \tag{48}$$

where $\underline{I}_h^k c := ((\pi_T^k c_{|T})_{T\in\mathcal{T}_h}, (\pi_F^k(c_{|F}))_{F\in\mathcal{F}_h})$ with π_T^k and π_F^k denoting, respectively, the L^2-orthogonal projectors on $\mathbb{P}^k(T)$ and $\mathbb{P}^k(F)$, $\underline{I}_{\Gamma,h}^k c_\Gamma := ((\pi_F^k c_{\Gamma|F})_{F\in\mathcal{F}_h^\Gamma}, (c_\Gamma(V))_{V\in\mathcal{V}_h})$, and $\pi_h^k c$ and $\pi_{\Gamma,h}^k c_\Gamma$ denote, respectively, the L^2-orthogonal projections of c and c_Γ on $P_{\mathrm{B},h}^k$ and $P_{\Gamma,h}^k$.

5.1.3 Results

We consider the triangular, Cartesian and nonconforming mesh families of Fig. 3.

In Fig. 4, we display the errors (47) for the flow problem as functions of the meshsize. The flow problem (31) is solved using polynomials two times higher than for the passive transport problem, so higher convergence rates than for the passive transport problem are to be expected. More specifically, on the triangular mesh we observe convergence in h^{2k+1} of the discretization error measured in the energy-like norms $\|\underline{\boldsymbol{u}}_h - \underline{\boldsymbol{I}}_h^{2k}\boldsymbol{u}\|_{\boldsymbol{P},h}$ and $\|\underline{p}_h^\Gamma - \underline{I}_h^{2k} p_\Gamma\|_{\Gamma,K,h}$, and convergence in h^{2k+2} for the error measured in the L^2-norms $\|p_h - \pi_h^{2k} p\|_{L^2(\Omega_\mathrm{B})}$ and $\|p_h^\Gamma - \pi_{\Gamma,h}^{2k} p_\Gamma\|_{L^2(\Gamma)}$. Slightly better convergence rates are observed on Cartesian and nonconforming meshes, as already noticed in Chave et al. (2018).

For the steady passive transport problem (40), we plot in Fig. 5 the errors (48) as functions of the meshsize. For both the energy-like norms of the error $\|\underline{c}_h - \underline{I}_h^k c\|_{D,h}$ and $\|\underline{c}_h^\Gamma - \underline{I}_{\Gamma,h}^k c_\Gamma\|_{\Gamma,D,h}$, we obain convergence in h^{k+1}. For the L^2-norms of the error $\|c_h - \pi_h^k c\|_{L^2(\Omega_\mathrm{B})}$ and $\|c_h^\Gamma - \pi_{\Gamma,h}^k c\|_{L^2(\Gamma)}$, on the other hand, we obtain convergence

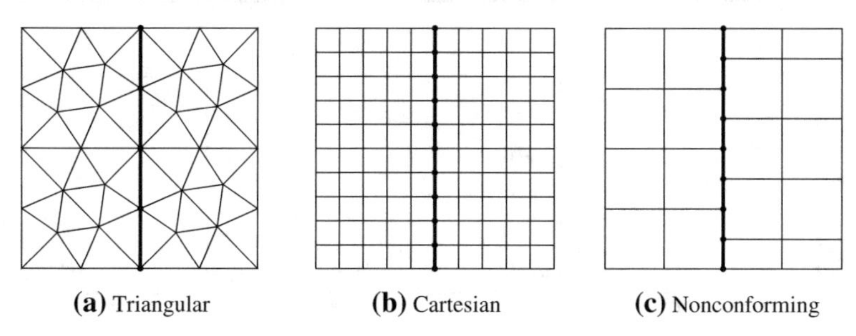

(a) Triangular (b) Cartesian (c) Nonconforming

Fig. 3 Mesh families for the numerical tests

in h^{k+2} using piecewise linear or quadratic polynomials, and for the case $k = 0$ in a fracture, we remark a stagnation of convergence around 10^{-3}. This phenomenon will be investigate in further works.

5.2 Unsteady transport with impermeable fractures

We next consider a physical test case modelling the unsteady passive displacement of a solute in a porous medium in which the fractures act as barriers.

The configuration is depicted in Figure 6a. More specifically, the computational domain is the unit square $\Omega = (0, 1)^2$, with fractures of constant thickness $\ell_\Gamma = 10^{-2}$ corresponding to

$$\Gamma = \{\boldsymbol{x}=(x_1, x_2)\in\Omega \ : \ (x_1<0.75 \text{ and } x_2 \in \{0.25, 0.75\}) \text{ or } (x_1>0.25 \text{ and } x_2=0.5)\}.$$

The injection well is located in $(0.5, 0)$, the production one in $(0.5, 1)$, and both are modeled by the source term f defined such that

$$f(\boldsymbol{x}) = \frac{1}{2}\Bigg(\tanh\left(200(0.025 - \sqrt{(x_1 - 0.5)^2 + x_2^2}\right) \\ - \tanh\left(200(0.025 - \sqrt{(x_1 - 0.5)^2 + (x_2 - 1)^2}\right)\Bigg).$$

The fracture source term f_Γ is set to 0. It can be checked that the average of f in Ω_{B} is zero, so the compatibility condition (5) is verified. We set the user parameter $\xi = 0.75$.

Concerning the flow problem, we select the values of the permeability in the bulk and in the fracture so as to obtain impermeable fractures. More specifically, in the bulk we set $\boldsymbol{K} = 10^{-3}\boldsymbol{I}$, while in the fractures the tangential and normal permeability are, respectively, $\kappa_\Gamma^\tau = 10^{-3}$ and $\kappa_\Gamma^n = 10^{-6}$. In Fig. 6a, we display the bulk pressure p obtained with such parameters and the plot over $x_1 = 0.5$. We can clearly see that the pressure jumps across the fractures and decreases from the injection to the production well.

We consider the unsteady passive transport problem (45), set the final time $t_F = 100$ and the time step $\delta t = 1$. At $t = 0$, there is not solute in the bulk nor in the fractures.

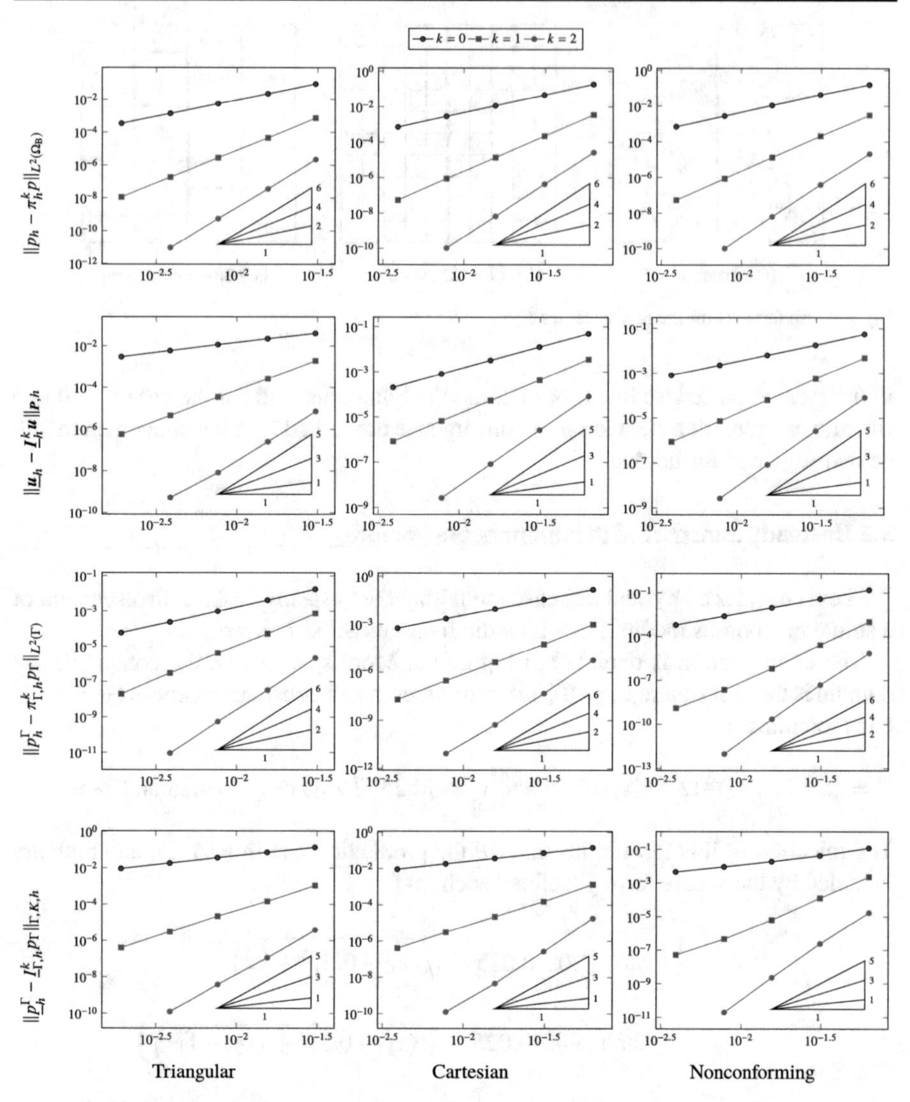

Fig. 4 Convergence results for the test case of Sect. 5.1. Errors (47) for the flow problem v. h on the triangular, Cartesian and nonconforming mesh families of Fig. 3

The concentration of injected solute in the bulk is given, for all $\boldsymbol{x} \in \Omega_B$, by $\widehat{c}(t, \boldsymbol{x}) = 1$ if $t < 30$ and $\widehat{c}(t, \boldsymbol{x}) = 0$ otherwise. Since we do not have wells in the fracture, we set $\widehat{c}_\Gamma \equiv 0$. The porosity in the bulk and in the fracture is set to $\phi = \phi_\Gamma = 10^{-1}$. Following Anderson and Droniou (2018); Peaceman (1966), the diffusion–dispersion tensor in the bulk is defined locally for all $T \in \mathcal{T}_h$ such that

$$\boldsymbol{D}_T = \phi d_{\mathrm{m}} \boldsymbol{I}_2 + \phi |\boldsymbol{F}_T^{2k+1} \underline{\boldsymbol{u}}_T| (d_{\mathrm{l}} \boldsymbol{E}_{\boldsymbol{u},T} + d_{\mathrm{t}} (\boldsymbol{I}_2 - \boldsymbol{E}_{\boldsymbol{u},T})),$$

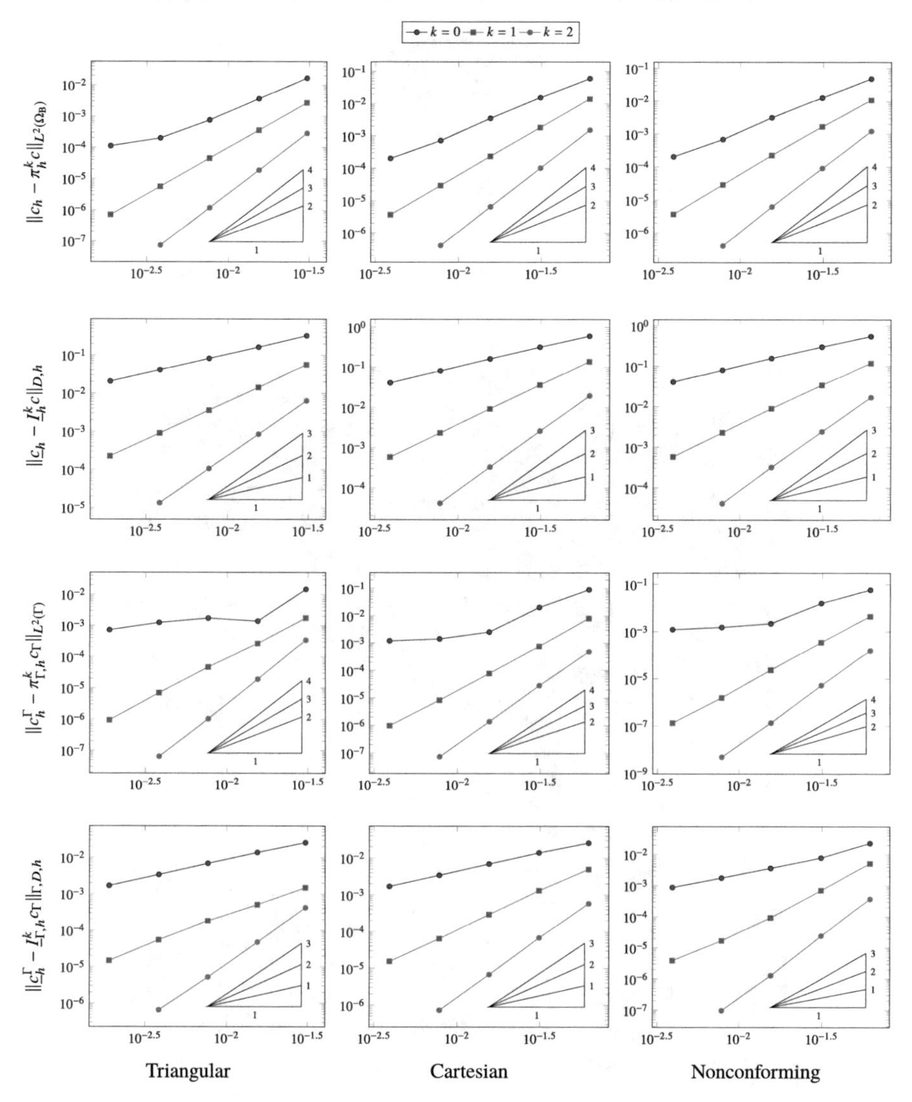

Fig. 5 Convergence results for the test case of Sect. 5.1. Errors (48) for the passive transport problem v. h on the triangular, Cartesian and nonconforming mesh families of Fig. 3

where $|\boldsymbol{F}_T^{2k+1}\underline{\boldsymbol{u}}_T|$ is the Euclidean norm of $\boldsymbol{F}_T^{2k+1}\underline{\boldsymbol{u}}_T$, $\boldsymbol{E}_{\boldsymbol{u},T} := |\boldsymbol{F}_T^{2k+1}\underline{\boldsymbol{u}}_T|^{-2} \left(\boldsymbol{F}_T^{2k+1}\underline{\boldsymbol{u}}_T \otimes \boldsymbol{F}_T^{2k+1}\underline{\boldsymbol{u}}_T\right)$, while $d_\mathrm{m} = 10^{-5}$, $d_\mathrm{l} = 1$ and $d_\mathrm{t} = 10^{-2}$ denote, respectively, the molecular diffusion, longitudinal, and transverse dispersion coefficients. Notice that the high-order reconstruction of the Darcy velocity is needed to define $\boldsymbol{D}_T$ since, if using constant elements $k = 0$, we do not have cell-based DOFs for the flux. The fracture counterpart of the diffusion–dispersion coefficient is defined, for all $F \in \mathcal{F}_h^\Gamma$, as follows

(a) Domain configuration (left), bulk pressure p (middle) and bulk pressure profile over $x_1 = 0.5$ (right).

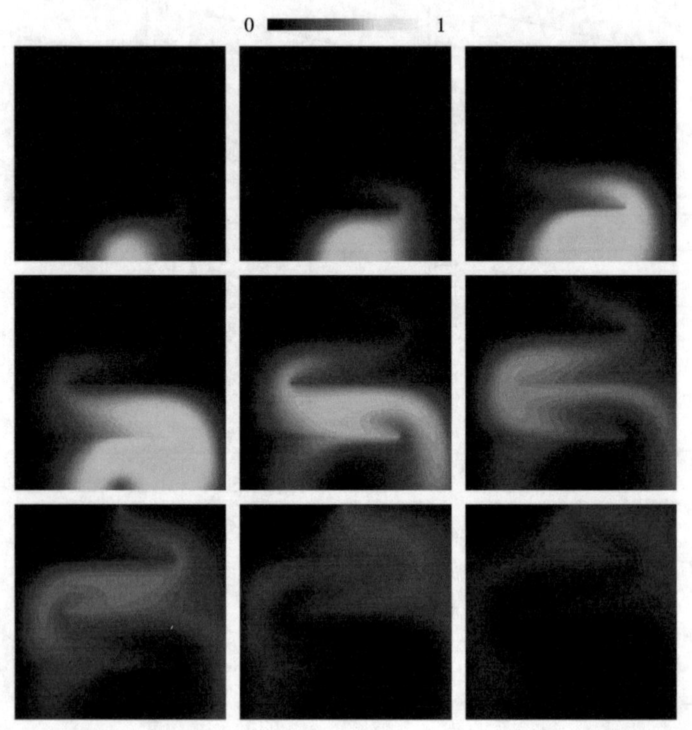

(b) Snapshots of the bulk concentration c at times (from left to right, top to bottom): $t =$ 5, 10, 20, 30, 40, 50, 60, 80, 100.

Fig. 6 Configuration and numerical results for the test of Sect. 5.2 (unsteady transport with impermeable fractures)

$$D_F = \ell_\Gamma \phi_\Gamma d_\mathrm{m}^\Gamma \boldsymbol{I}_2 + \phi_\Gamma |\boldsymbol{u}_F^\Gamma| (d_\mathrm{l}^\Gamma \boldsymbol{E}_{\boldsymbol{u},F} + d_\mathrm{t}^\Gamma (\boldsymbol{I}_2 - \boldsymbol{E}_{\boldsymbol{u},F})),$$

with $\boldsymbol{E}_{\boldsymbol{u},F} := |\boldsymbol{u}_F^\Gamma|^{-2} \left(\boldsymbol{u}_F^\Gamma \otimes \boldsymbol{u}_F^\Gamma\right)$ and where $d_\mathrm{m}^\Gamma = 10^{-5}$, $d_\mathrm{l}^\Gamma = 1$ and $d_\mathrm{t}^\Gamma = 10^{-2}$ denote, respectively, the fracture molecular diffusion, longitudinal, and transverse dispersion coefficients. We set the normal diffusion–dispersion coefficient of the fracture $\mathcal{D}_\Gamma^n$ equal to 1. A more in-depth investigation of the meaning of this term is postponed to a future work.

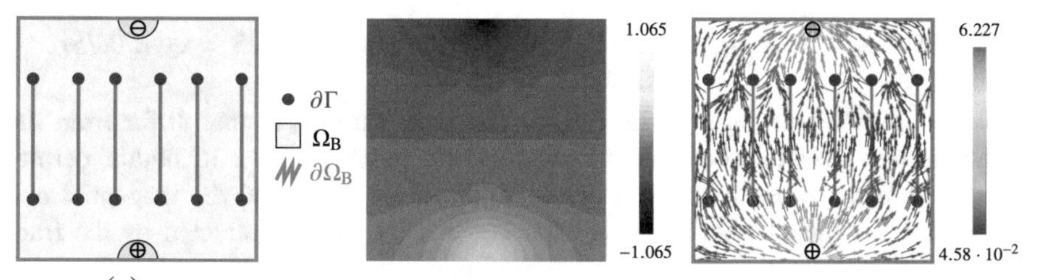

(a) Domain configuration (left), bulk pressure p (middle), and Darcy velocity $\boldsymbol{u}$ (right).

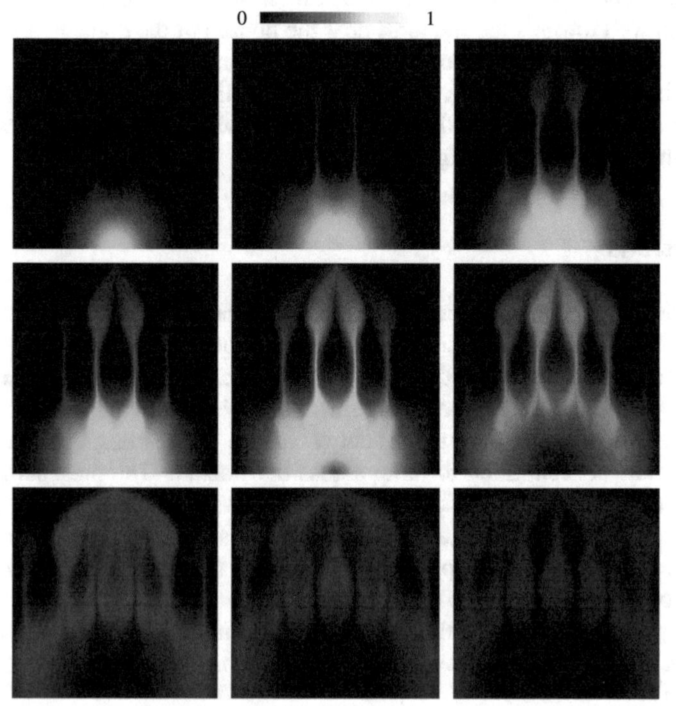

(b) Snapshots of the bulk concentration c at times (from left to right, top to bottom): t = 5, 10, 15, 20, 30, 40, 60, 80, 100.

Fig. 7 Configuration and numerical results for the test of Sect. 5.3 (unsteady transport with permeable fractures)

We run the test case on the Cartesian mesh depicted in Fig. 3b of meshsize $h = 7.81 \cdot 10^{-3}$ with $k = 2$. In Fig. 6b, we display the bulk concentration at different time t. As expected, the solute follows the corridors designed by the fractures that act as barriers and goes from the injection to the production well.

5.3 Unsteady transport with permeable fractures

We next focus on the case where the fractures act as conduits. The domain is still the square unit $\Omega = (0, 1)^2$, the fractures of constant thickness $\ell_\Gamma = 10^{-2}$ are located in

$$\Gamma = \{\boldsymbol{x} \in \Omega \,:\, x_1 \in \{2/32, 8/32, 13/32, 19/32, 24/32, 30/32\} \text{ and } 0.25 < x_2 < 0.75\}.$$

The configuration is depicted in Fig. 7a. The only parameters that differ from the previous test case of Sect. 5.2 are the fracture permeabilities: to obtain permeable fractures, we set the normal permeability $\kappa_\Gamma^n = 10^{-3}$ and the tangential one $\kappa_\Gamma^\tau = 10^{-1}$. With this choice, it is expected that the flow is attracted by the fractures.

In Fig. 7a, we display the bulk pressure p and Darcy velocity $\boldsymbol{u}$ where, for the latter, the color scale correspond to the value of the magnitude. As expected, the flow is from the injection well towards the fractures near the bottom of the domain, and from the fractures to the production well near the top of the domain.

In Fig. 7b, we display the bulk concentration c at different times. We can distinctly see that the solute channeled by the fractures flows towards the production well faster than the solute in the surrounding bulk medium.

6 Conclusions

We conclude this paper by pointing out its main contributions and discuss perspectives of further works. We have introduced a new reduced model for the passive transport of a solute in fractured porous media driven by Darcy velocites. To derive the transmission conditions, we used an energy-based argument such that, as in the unreduced model, transport terms do not contribute to the energy balance. These transmission conditions allow the solute concentration to jump across the fracture. The presentation of the model and its discretization are done considering the steady case, while the extension to the unsteady case is presented and used for numerical experiments. In future works, we will investigate further the physical meaning of the fracture normal diffusion–dispersion coefficient $\mathcal{D}_\Gamma^n$ in the new transmission conditions (9), and we will carry out the complete analysis of the discrete formulation, including its well-posedness and the study of the convergence properties of the HHO method adapting the techniques of Di Pietro et al. (2015).

References

Aghili, J., Boyaval, S., Di Pietro, D.A.: Hybridization of mixed high-order methods on general meshes and application to the Stokes equations. Comput. Methods Appl. Math. **15**(2), 111–134 (2015). https://doi.org/10.1515/cmam-2015-0004

Alboin, C., Jaffré, J., Roberts, J.E., Serres, C.: Modeling fractures as interfaces for flow and transport in porous media. In: American Mathematical Society Contemporary Mathematics (ed.) Fluid Flow and Transport in Porous Media: Mathematical and Numerical Treatment, vol. 295, pp. 13–24. American Mathematical Society, Providence (2002). https://doi.org/10.1090/conm/295/04999

Anderson, D., Droniou, J.: An arbitrary-order scheme on generic meshes for miscible displacements in porous media. SIAM J. Sci. Comput. **40**(4), B1020–B1054 (2018). https://doi.org/10.1137/17M1138807

Angot, P., Gallouët, T., Herbin, R.: Convergence of finite volume methods on general meshes for non smooth solution of elliptic problems with cracks. Finite Vol. Complex Appl. **II**, 215–222 (1999)

Angot, P., Boyer, F., Hubert, F.: Asymptotic and numerical modelling of flows in fractured porous media. ESAIM Math. Model. Numer. Anal. **43**(2), 239–275 (2009). https://doi.org/10.1051/m2an/2008052

Antonietti, P.F., Facciola, C., Russo, A., Verani, M.: Discontinuous Galerkin approximation of flows in fractured porous media on polytopic grids. MOX Report No. 55/2016 (2016a). https://www.mate.polimi.it/biblioteca/add/qmox/55-2016.pdf. Accessed 14 Jan 2019

Antonietti, P.F., Formaggia, L., Scotti, A., Verani, M., Verzotti, N.: Mimetic finite difference approximation of flows in fractured porous media. ESAIM Math. Model. Numer. Anal. **50**(3), 809–832 (2016b). https://doi.org/10.1051/m2an/2015087

Bastian, P., Chen, Z., Ewing, R.E., Helmig, R., Jakobs, H., Reichenberger, V.: Numerical simulation of multiphase flow in fractured porous media. Numer. Treat. Multiph. Flows Porous Media **52**, 50–68 (1999). https://doi.org/10.1007/3-540-45467-5_4

Benedetto, M.F., Berrone, S., Pieraccini, S., Scialò, S.: The virtual element method for discrete fracture network simulations. Comput. Methods Appl. Mech. Eng. **280**, 135–156 (2014). https://doi.org/10.1016/j.cma.2014.07.016

Berrone, S., Pieraccini, S., Scialò, S.: Non-stationary transport phenomena in networks of fractures: effective simulations and stochastic analysis. Comput. Methods Appl. Mech. Eng. **315**, 1098–1112 (2016). https://doi.org/10.1016/j.cma.2016.12.006

Berrone, S., Pieraccini, S., Scialò, S.: Flowsimulations in porous media with immersed intersecting fractures. J. Comput. Phys. **345**, 768–791 (2017). https://doi.org/10.1016/j.jcp.2017.05.049

Boffi, D., Di Pietro, D.A.: Unified formulation and analysis of mixed and primal discontinuous skeletal methods on polytopal meshes. ESAIM Math. Model. Numer. Anal. **52**(1), 1–28 (2018). https://doi.org/10.1051/m2an/2017036

Boon, W.M., Nordbotten, J.M., Yotov, I.: Robust discretization of flow in fractured porous media. SIAM J. Numer. Anal. (2018). https://doi.org/10.1137/17M1139102

Brenner, K., Hennicker, J., Masson, R., Samier, P.: Hybrid-dimensional modelling of twophase flow through fractured porous media with enhanced matrix fracture transmission conditions. J. Comput. Phys. **357**, 100–124 (2018). https://doi.org/10.1016/j.jcp.2017.12.003

Chave, F., Di Pietro, D.A., Formaggia, L.: A Hybrid High-Order method for darcy flows in fractured porous media. SIAM J. Sci. Comput. **40**(2), A1063–A1094 (2018). https://doi.org/10.1137/17M1119500

Chernyshenko, A., Olshanskii, M., Vassilevski, Y.: A hybrid finite volume-finite element method for bulk-surface coupled problems. J. Comput. Phys. **352**, 516–533 (2016). https://doi.org/10.1016/j.jcp.2017.09.064

Cockburn, B., Di Pietro, D.A., Ern, A.: Bridging the Hybrid High-Order and hybridizable discontinuous Galerkin methods. ESAIM Math. Model. Numer. Anal. **50**(3), 635–650 (2016). https://doi.org/10.1051/m2an/2015051

D'Angelo, C., Scotti, A.: A mixed finite element method for Darcy flow in fractured porous media with non-matching grids. ESAIM Math. Model. Numer. Anal. **46**(2), 465–489 (2012). https://doi.org/10.1051/m2an/2011148

Del Pra, M., Fumagalli, A., Scotti, A.: Well-posedness of fully coupled fracture/bulk Darcy flow with XFEM. SIAM J. Numer. Anal. **55**(2), 785–811 (2017). https://doi.org/10.1137/15M1022574

Di Pietro, D.A., Droniou, J.: A Hybrid High-Order method for Leray–Lions elliptic equations on general meshes. Math. Comput. **86**(307), 2159–2191 (2017). https://doi.org/10.1090/mcom/3180

Di Pietro, D.A., Ern, A.: Equilibrated tractions for the Hybrid High-Order method. C. R. Acad. Sci. Paris Ser. I **353**, 279–282 (2015). https://doi.org/10.1016/j.crma.2014.12.009

Di Pietro, D.A., Tittarelli, R.: Numerical methods for PDEs. State of the art techniques. In: Formaggia, L., Di Pietro, D.A., Ern, A. (eds.) SEMA-SIMAI 15. Springer (2018). Chap. An introduction to Hybrid High-Order methods. ISBN: 978-3-319-94675-7 (Print) 978-3-319-94676-4 (eBook). http://arxiv.org/abs/1703.05136

Di Pietro, D.A., Ern, A., Lemaire, S.: An arbitrary-order and compact-stencil discretization of diffusion on general meshes based on local reconstruction operators. Comput. Methods Appl. Math. **14**(4), 461–472 (2014). https://doi.org/10.1515/cmam-2014-0018. Open access

Di Pietro, D.A., Droniou, J., Ern, A.: A discontinuous-skeletal method for advection–diffusion–reaction on general meshes. SIAM J. Numer. Anal. **53**(5), 2135–2157 (2015). https://doi.org/10.1137/140993971

Faille, I., Flauraud, E., Nataf, F., Pégaz-Fiornet, S., Schneider, F., Willien, F.: A new fault model in geological basin modelling. Application of finite volume scheme and domain decomposition methods. In: Herbin, R., Kroner, D. (eds.) Finite Volumes for Complex Applications III, pp. 543–550. Kogan Page Science (2002)

Flemisch, B., Fumagalli, A., Scotti, A.: A review of the XFEM-basedapproximation of flowin fractured porous media. In: Ventura, G., Benvenuti, E. (eds.) Advances in Discretization Methods. SEMA-SIMAI, vol. 12. Springer, Cham (2016). https://doi.org/10.1007/978-3-319-41246-7_3

Formaggia, L., Quarteroni, A., Vergara, C.: On the physical consistency between threedimensional and one-dimensional models in haemodynamics. J. Comput. Phys. **244**, 97–112 (2013). https://doi.org/10.1016/j.jcp.2012.08.001

Fumagalli, A., Keilegavlen, E.: Dual virtual element method for discrete fractures networks. SIAM J. Sci. Comput. **40**(1), B228–B258 (2018). https://doi.org/10.1137/16M1098231

Fumagalli, A., Scotti, A.: A reduced model for flow and transport in fractured porous media with non-matching grids. Numer. Math. Adv. Appl. **2013**, 499–507 (2011). https://doi.org/10.1007/978-3-642-33134-3_53

Gross, S., Olshanskii, M.A., Reusken, A.: A trace finite element method for a class of coupled bulk-interface transport problems. ESAIM Math. Model. Numer. Anal. **49**(5), 1303–1330 (2015). https://doi.org/10.1051/m2an/2015013

Martin, V., Jaffré, J., Roberts, J.E.: Modeling fractures and barriers as interfaces for flow in porous media. SIAM J. Matrix Anal. Appl. **26**(5), 1667–1691 (2005). https://doi.org/10.1137/S1064827503429363

Peaceman, D.W.: Improved treatment of dispersion in numerical calculation of multidimensional miscible displacement. Soc. Pet. Eng. J. **6**, 213–216 (1966). https://doi.org/10.2118/1362-PA

Scotti, A., Formaggia, L., Sottocasa, F.: Analysis of a mimetic finite difference approximation of flows in fractured porous media. ESAIM Math. Model. Numer. Anal. (2017). https://doi.org/10.1051/m2an/2017028

Affiliations

Florent Chave[1,2] · Daniele A. Di Pietro[1] · Luca Formaggia[2]

Daniele A. Di Pietro
daniele.di-pietro@umontpellier.fr

Luca Formaggia
luca.formaggia@polimi.it

[1] Institut Montpelliérain Alexander Grothendieck, University of Montpellier, 34095 Montpellier, France

[2] Politecnico di Milano, MOX, 20133 Milan, Italy

GEM - International Journal on Geomathematics
https://doi.org/10.1007/s13137-018-0105-3

ORIGINAL PAPER

Advanced computation of steady-state fluid flow in Discrete Fracture-Matrix models: FEM–BEM and VEM–VEM fracture-block coupling

S. Berrone[1] · A. Borio[1] · C. Fidelibus[2] · S. Pieraccini[3] · S. Scialò[1] · F. Vicini[1]

Received: 31 March 2018 / Accepted: 18 July 2018

Abstract

In this note the issue of fluid flow computation in a Discrete Fracture-Matrix (DFM) model is addressed. In such a model, a network of percolative fractures delimits porous matrix blocks. Two frameworks are proposed for the coupling between the two media. First, a FEM–BEM technique is considered, in which finite elements on non-conforming grids are used on the fractures, whereas a boundary element method is used on the blocks; the coupling is pursued by a PDE-constrained optimization formulation of the problem. Second, a VEM–VEM technique is considered, in which a 2D and a 3D virtual element method are used on the fractures and on the blocks, respectively, taking advantage of the flexibility of VEM in using arbitrary meshes in order to ease the meshing process and the consequent enforcement of the matching conditions on fractures and blocks.

Keywords Coupling 3D BEM–2D FEM · Optimization procedure for non-matching grids · Coupling 3D VEM–2D VEM · Conforming polygonal-polyhedral meshes · Steady-state flows in porofractured media · Dual-porosity media

Mathematics Subject Classification 65N30 · 65N50 · 68U20 · 86-08 · 86A05 · 86A60

1 Introduction

Two main approaches are available for the simulation of the fluid flow regime in poro-fractured media: the Dual-Porosity Continuum (DPC) and the Discrete Fracture-

This research has been partially supported by INdAM-GNCS Project 2018, and by the MIUR project "Dipartimenti di Eccellenza 2018-2022". Computational resources were partially provided by HPC@POLITO (http://hpc.polito.it) and by CINECA Project IsC58 HP10CDFLWH.
Authors S.B., A.B., S.P., S.S., F.V. are members of the INdAM research group GNCS.

Extended author information available on the last page of the article

Matrix (DFM) model. The first model essentially consists of the overlap of two continua, hydraulically equivalent to the fracture network and to the matrix blocks, respectively, and locally exchanging fluid through a transfer function. In the second model, fractures at all scales are explicitly included and delimit matrix blocks.

Since the original works of Warren and Root (1963) and the advent of the first computers, the development of DPC models has been underway. Kazemi and Gilman (1993) provided a first significant advancement by extending the use to the reservoir scale. Subsequently, a blooming of numerical techniques for the solution of DPC models has been recorded (see among the others Aldejain 1999; Bai et al. 1994; Huyakorn and Pinder 1983).

The applicability of a DPC model depends on the consistency of the equivalence of the fracture network to a continuum. As a matter of fact, such an equivalence is practicable only when a Representative Elementary Volume (REV) exists with reference to all the hydraulic properties. Natural fracture networks generally show different intensities with the scale and significant variation in connectivity. In addition, highly persistent and strongly connected fractures, typically located near bedding planes and fault zones, may strongly affect the flow regime. Under these circumstances, the quest of a REV for the entire network is unsuccessful, thus DPC should not be utilized.

The DFM models are alternative to the DPC models. In such models, as previously mentioned, a Discrete Fracture Network (DFN) is combined with matrix blocks, therefore scale dependency and heterogeneity, peculiarities of the fracture networks, are preserved. Usually, the Darcy law is considered for the porous matrix blocks, whereas for the fractures the law is applied after averaging along the thickness, with interface conditions imposed for the match with the blocks (see e.g. Ahmed et al. 2015; Alboin et al. 1999; Angot et al. 2009; Antonietti et al. 2016; Brenner et al. 2016a, b; D'Angelo and Scotti 2012; Martin et al. 2005).

Even at relatively small scales, the discretization of both fractures and blocks leads to "monster" systems of equations and to a high computational demand. This circumstance is detrimental to the use of DFM models and hampers its applicability, therefore, the need emerges for specific techniques aiming at the reduction of this demand. In recent literature, several contributions addressing this issue are available. D'Angelo and Scotti (2012) and Fumagalli and Scotti (2013) used the eXtended Finite Element Method (XFEM) (see Fries and Belytschko 2010) in a mixed formulation for the reduced Darcy problem by arbitrarily placing fractures into the mesh of the porous medium. The Virtual Element Method (VEM), recently introduced by Beirão da Veiga et al. (2013a, b), was applied by Benedetto et al. (2016a, b, c); Berrone and Borio (2017a, b); Fumagalli and Keilegavlen (2018) in order to obtain flexible meshes for flow simulations in DFNs. Other approaches were proposed by Al-Hinai et al. (2015); Formaggia et al. (2016), in which mimetic finite differences (see Beirão da Veiga et al. 2014) are used, by Frih et al. (2012), who resorted to mixed finite elements and by Chave et al. (2018), where an Hybrid High Order discretization is proposed for Darcy problems in fractured media. Furthermore, several authors applied a Two-Point Flux Approximation (TPFA) or a Multi-Point Flux Approximation (MPFA) of Aavatsmark (2002) (see, e.g., Ahmed et al. 2015; Angot et al. 2009; Faille et al. 2016; Hajibeygi et al. 2011; Hyman et al. 2015; Makedonska et al. 2015; Reichenberger et al. 2006;

Sandve et al. 2012), recently investigated also in the general framework of gradient schemes (Brenner et al. 2016b). Benchmarks on different strategies for DFM model simulations are available in (Flemisch et al. 2018), and in (Fumagalli et al. 2018) for simulations in fracture networks with complex geometries.

Berrone et al. (2013a, b, 2014b), for the solution of steady-state fluid flow problems in DFNs, introduced an optimization approach, in which any conformity requirement between meshes of intersecting fractures is removed, resorting to the minimization of continuity mismatch and flux unbalance at traces (i.e., the line intersections among fractures). This minimization problem is constrained by the state equation of the hydraulic head on the fractures. The discretization of the fractures can be performed by several methods indifferently, like standard FEM (Berrone et al. 2015, 2016a), XFEM (Berrone et al. 2014a, 2016b) and VEM (Benedetto et al. 2014). The technique was proven effective, therefore the extension to a DFM model has been performed by (Berrone et al. 2017), where a FEM discretization of the fractures is combined with a Boundary Element Method (BEM) discretization of the matrix blocks boundaries. The optimization method has also shown very good approximation capabilities compared to a large variety of numerical schemes, both in primal and mixed formulation (Fumagalli et al. 2018).

In the present note the approach by (Berrone et al. 2017) is briefly recalled and enriched with a novel analysis on the behaviour of the method when the hydraulic conductivity of the matrix blocks varies. Furthermore, a new approach is presented, consisting of a new globally conforming technique based on the VEM on both fractures and matrix blocks. A comparison between the two techniques is reported and discussed.

The following issues are herein described: the mathematical model of the problem of the steady-state fluid flow in a DFM model (Sect. 3), with its variational formulation (Sect. 4); the PDE-constrained optimization formulation of the problem (Sect. 5); the FEM discretization of the fractures (Sect. 6) and the BEM discretization of the matrix blocks (Sect. 7) for the first technique; the VEM construction (Sect. 8) and the VEM discretization (Sect. 9) for the second technique; numerical tests on relatively small problems with remarks on the robustness and quality of the offered solutions in view of their use at larger scales (Sect. 10). Finally, conclusions are given in Sect. 11.

2 Notation

Throughout the text, the symbol $\mathcal{D}$ denotes matrix blocks, represented as polyhedral subset of $\mathbb{R}^3$, Γ denotes instead the block faces and, finally, F is used for the fractures.

Let $\mathcal{D}$ be a domain crossed by fractures F_i, for $i \in \mathcal{I}$, splitting $\mathcal{D}$ in sub-blocks $\mathcal{D}_k$, for $k \in \mathcal{K}$ and set $\mathcal{F} = \cup_{i \in \mathcal{I}} F_i$. It is assumed a fracture is enough persistent to entirely cross the blocks, namely, it does not end inside $\mathcal{D}$, rather it terminates against another fracture or the boundary. All the results presented can be extended to the case of non-persistent fractures, requiring a more complex notation and the introduction of some additional terms in the equations, similarly to what done by Berrone et al. (2017).

Each block face Γ_ℓ, for $\ell \in \mathcal{L}$, may either belong to $\partial\mathcal{D}$ or be given by the intersection of the closure of two blocks. In the last case, namely if $\Gamma_\ell = \bar{\mathcal{D}}_p \cap \bar{\mathcal{D}}_r$, for some

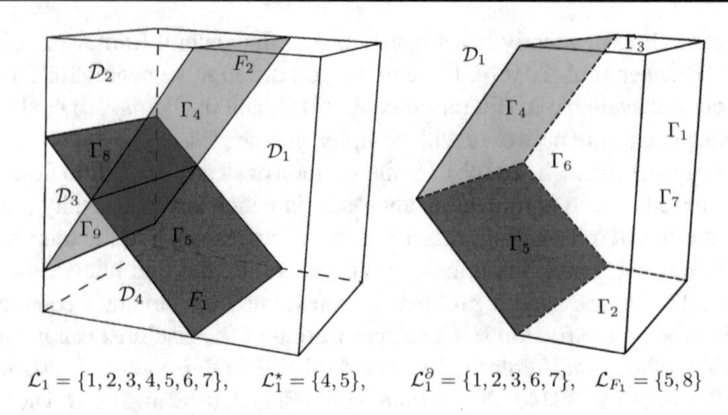

Fig. 1 Example of used notation

p and r, $\mathcal{K}_\ell := \{p, r\}$ is set. For each $k \in \mathcal{K}$, let $\mathcal{L}_k$ denote the set of indices of faces of $\mathcal{D}_k$, namely $\partial\mathcal{D}_k = \cup_{\ell\in\mathcal{L}_k}\Gamma_\ell$. A further subset of $\mathcal{L}_k$, denoted by $\mathcal{L}_k^\star$, is introduced, containing indexes of the faces lying on any fracture. Let also define $\mathcal{L}_k^\partial := \mathcal{L}_k \backslash \mathcal{L}_k^\star$ as the set of $\mathcal{D}_k$ faces lying on $\partial\mathcal{D}$. By collecting these subsets for all the blocks, the global sets $\mathcal{L}^\star = \cup_{k\in\mathcal{K}}\mathcal{L}_k^\star$ and $\mathcal{L}^\partial = \cup_{k\in\mathcal{K}}\mathcal{L}_k^\partial$ are obtained. Figure 1 reports an example of the used notation for a simple geometry.

Fractures are planar objects and mutually intersect giving rise to linear segments (traces), denoted by S. For each $i \in \mathcal{I}$, the set $\mathcal{M}_i$ indexing traces belonging to F_i is introduced. Let $\mathcal{M}$ be the global set of trace indexes, namely $\mathcal{M} = \cup_{i\in\mathcal{I}}\mathcal{M}_i$. For each trace $S_m = F_i \cap F_j$, the set $\mathcal{I}_m = \{i, j\}$ is introduced. Note that each fracture is divided by traces in sub-fractures, geometrically coincident with block faces, thus the set $\mathcal{L}_{F_i}$ of block faces belonging to F_i, namely $\bar{F}_i = \cup_{\ell\in\mathcal{L}_{F_i}}\bar{\Gamma}_\ell$ is also introduced. Correspondingly, a function $\varrho : \mathcal{L}^\star \mapsto \mathcal{I}$, such that $i = \varrho(\ell)$, is defined.

In the sequel, symbols F_i and Γ_ℓ are used to denote also the 2D objects described by means of a local tangential coordinate system.

For any sufficiently regular function v defined in $\mathcal{D}$, let $v^{(k)}$ be the restriction of v to $\mathcal{D}_k$ and be $v^{(\ell,k)}$ the trace of $v^{(k)}$ on Γ_ℓ. Similarly, $v^{((i))}$ denotes the trace of v on F_i and $v^{((\ell,i))}$ the restriction of $v^{((i))}$ to Γ_ℓ for $\ell \in \mathcal{L}_{F_i}$. For a function v defined on a fracture F_i, we will use the notation $v^{(\ell,i)}$ to denote its restriction to a face $\Gamma_\ell \subseteq F_i$.

3 Mathematical framework for a DFM model

In a DFM model, a basic assumption is that all the fractures are planar and of no thickness. For each fracture F_i, $i \in \mathcal{I}$ and for each face $\Gamma_\ell \subseteq F_i$, and with reference to a local coordinate system $\mathbf{x} = (x_1, x_2)$, the Darcy law applies:

$$\mathbf{q}^{(\ell,i)}(\mathbf{x}) = -T_i(\mathbf{x})\nabla h^{((\ell,i))}(\mathbf{x}), \tag{1}$$

where $\mathbf{q}$ is the fluid discharge across a fracture void of unit width and $h^{((\ell,i))}$ is the restriction to Γ_ℓ of the hydraulic head on fracture F_i, endowed with (isotropic) transmissivity T_i.

Then, the following equation can be written for the steady-state fluid flow on each face $\Gamma_\ell \subseteq F_i$ (Shapiro and Andersson 1983; Fidelibus et al. 1997):

$$\mathrm{div}_{\pi_i}\mathbf{q}^{(\ell,i)} = Q^{(\ell,p)} + Q^{(\ell,r)}, \tag{2}$$

where div_{π_i} is the tangetial divergence operator on the tangential plane π_i of F_i, $Q^{(\ell,p)}$, $Q^{(\ell,r)}$ are the fluid discharges from the blocks $\mathcal{D}_p$ and $\mathcal{D}_r$, respectively, with $p, r \in \mathcal{K}_\ell$. In Eq. (2), being the fracture of no thickness, the walls collapse to coincide with the mean plane of the fracture, thus the hydraulic head at the mean plane coincides with the hydraulic heads $h^{(\ell,p)}$, $h^{(\ell,r)}$ at the corresponding points of the adjacent blocks. However, in what follows, $h^{((\ell,i))}$, $h^{(\ell,p)}$, $h^{(\ell,r)}$ are kept distinct and continuity of the hydraulic head is explicitly imposed.

The traces affect the flow regime in a fracture, acting as linear sources or sinks. For each trace S_m with $m \in \mathcal{M}_i$, by introducing a tangential coordinate system $\mathbf{x} = x_1$, $U^{(m,i)}(\mathbf{x})$ is denoted as the fluid discharge per unit trace length exiting F_i from the intersecting fracture through trace S_m. Let Γ_ℓ and $\Gamma_{\ell'}$ be the two faces on fracture F_i sharing trace S_m. The quantities

$$q_{\mathrm{n}_\ell} = \mathbf{q}^{(\ell,i)} \cdot \mathbf{n}_{\pi_i,\ell}, \qquad q_{\mathrm{n}_{\ell'}} = \mathbf{q}^{(\ell',i)} \cdot \mathbf{n}_{\pi_i,\ell'}$$

are introduced, being $\mathbf{n}_{\pi_i,\ell}$, $\mathbf{n}_{\pi_i,\ell'}$ the unit normals to the trace pointing outwards the sub-fractures Γ_ℓ and $\Gamma_{\ell'}$, respectively. The discharge $U^{(m,i)}$ is then given by:

$$U^{(m,i)} = q_{\mathrm{n}_\ell} + q_{\mathrm{n}_{\ell'}} = -T_i\left(\frac{\partial h^{((\ell,i))}}{\partial \mathbf{n}_{\pi_i,\ell}} + \frac{\partial h^{((\ell',i))}}{\partial \mathbf{n}_{\pi_i,\ell'}}\right). \tag{3}$$

By continuously matching $h^{((\ell,i))}$ and $h^{((\ell',i))}$ on the two sides of each trace of F_i, the hydraulic head $h^{((i))}$ on F_i may be obtained.

For i and j indices of two fractures intersecting along a trace S_m, at each point $\mathbf{x}$ of the local coordinate system on S_m it is assumed that:

$$h^{((i))}_{|S_m}(\mathbf{x}) = h^{((j))}_{|S_m}(\mathbf{x}), \tag{4}$$

$$U^{(m,i)}(\mathbf{x}) + U^{(m,j)}(\mathbf{x}) = 0, \tag{5}$$

representing the continuity of the hydraulic head and the conservation of fluid volume across the trace, respectively. In the case of n fractures intersecting at the same trace, $n(n-1)/2$ coupling conditions of the type (4) should be considered, one for each different couple of intersecting fractures, whereas a single condition as equation (5) need to be enforced to balance all the fluxes.

With reference to a coordinate system $\mathbf{x} = (x_1, x_2, x_3)$, by assuming also that each block k of the rock mass is homogeneous and isotropic, the equation governing the fluid flow in k is the Laplace equation:

$$\mathrm{div}\mathbf{Q}^{(k)} = \mathrm{div}(-K^{(k)}\nabla h^{(k)}) = f^{(k)}, \tag{6}$$

being $\mathbf{Q}^{(k)}$ the specific discharge, $K^{(k)}$ the hydraulic conductivity of $\mathcal{D}_k$ and $f^{(k)}$ a volumetric source term. The normal component $Q^{(\ell,k)}$ of the specific discharge at the block face Γ_ℓ is equal to $-K^{(k)}\nabla h^{(k)} \cdot \mathbf{n}_{(\ell,k)}$, being $\mathbf{n}_{(\ell,k)}$ an outward normal to the block boundary $\mathcal{D}_k$ at face Γ_ℓ.

4 Variational formulation

The following functional spaces are defined on the fractures in $\mathcal{F}$:

$$\mathrm{H}_0^1(F_i) = \left\{v \in \mathrm{H}^1(F_i) : v_{|\partial F_i} = 0\right\},$$

whereas the functional space $\mathrm{H}^1_{\mathcal{F}}(\mathcal{D})$ is defined on the whole domain as:

$$\mathrm{H}^1_{\mathcal{F}}(\mathcal{D}) = \left\{v \in \mathrm{H}_0^1(\mathcal{D}) : v^{((i))} \in \mathrm{H}_0^1(F_i),\ i \in \mathcal{I},\right.$$
$$\left. v^{((i))}{}_{|S_m} = v^{((j))}{}_{|S_m},\ m \in \mathcal{M},\ i, j \in \mathcal{I}_{S_m}\right\}.$$

The space $\mathrm{H}^1_{\mathcal{F}}$ can be easily seen to be a Hilbert space endowed with the scalar product $(v, w)_{\mathcal{D},\mathcal{F}} = (\nabla v, \nabla w)_{\mathcal{D}} + \sum_{i\in\mathcal{I}} \left(\nabla_{\pi_i} v^{((i))}, \nabla_{\pi_i} w^{((i))}\right)_{F_i}, \forall v, w \in \mathrm{H}^1_{\mathcal{F}}$, being ∇_{π_i} the gradient on the tangential plane π_i of F_i.

The variational formulation of the problem given by (2) and (6) reads as follows: find $h \in \mathrm{H}^1_{\mathcal{F}}(\mathcal{D})$ such that, $\forall v \in \mathrm{H}^1_{\mathcal{F}}(\mathcal{D})$:

$$(K\nabla h, \nabla v)_{\mathcal{D}} = (f, v)_{\mathcal{D}} - \sum_{i\in\mathcal{I}} \sum_{\ell\in\mathcal{L}_{F_i}} \left\langle Q^{(\ell,p)} + Q^{(\ell,r)}, v^{(\ell,i)}\right\rangle_{\Gamma_\ell}, \tag{7}$$

$$\sum_{i\in\mathcal{I}} \left(T_i \nabla_{\pi_i} h^{((i))}, \nabla_{\pi_i} v^{((i))}\right)_{F_i} = \sum_{i\in\mathcal{I}} \sum_{\ell\in\mathcal{L}_{F_i}} \left\langle Q^{(\ell,p)} + Q^{(\ell,r)}, v^{((\ell,i))}\right\rangle_{\Gamma_\ell}, \tag{8}$$

where $p, r \in \mathcal{K}_\ell$. Well-posedness of the problem is shown in Berrone et al. (2017).

5 Optimization formulation

The solution of problem (7)–(8) can be seen as the constrained minimum of a properly designed cost functional. The following functional spaces are introduced:

- $\forall k \in \mathcal{K}$ the space $V_k = \mathrm{H}^1_{\mathcal{F}}(\mathcal{D})_{|\mathcal{D}_k}$,
  $\mathrm{H}^1_{\mathcal{F}}(\mathcal{D})_{|\mathcal{D}_k} = \left\{v \in \mathrm{H}^1(\mathcal{D}_k) : v_{|\partial\mathcal{D}\cap\partial\mathcal{D}_k} = 0, \exists w \in \mathrm{H}^1_{\mathcal{F}}(\mathcal{D}) : v = w_{|\mathcal{D}_k}\right\};$
- $\forall i \in \mathcal{I}$ the space $W_i = \mathrm{H}_0^1(F_i)$;

– $\forall \ell \in \mathcal{L}$ the space $\mathcal{H}_\ell^\Gamma = \mathrm{H}^1(\Gamma_\ell)$ and its dual $\mathcal{Q}_\ell = (\mathcal{H}_\ell^\Gamma)'$;
– $\forall m \in \mathcal{M}$ the space $\mathcal{H}_m^S = \mathrm{H}^{\frac{1}{2}}(S_m)$ and its dual $\mathcal{U}_m = (\mathcal{H}_m^S)'$.

Starting from the above definitions, the following product spaces are also introduced:

$$V = \prod_{k \in \mathcal{K}} V_k, \quad W = \prod_{i \in \mathcal{I}} W_i, \quad \mathcal{U}_{(i)} = \prod_{m \in \mathcal{M}_i} \mathcal{U}_m, \quad \mathcal{U} = \prod_{i \in \mathcal{I}} \mathcal{U}_{(i)}, \quad \mathcal{Q} = \prod_{\ell \in \mathcal{L}_\mathcal{D}} \mathcal{Q}_\ell.$$

Let $U^{(m,i)} \in \mathcal{U}_m$ be the jump of the co-normal derivative across S_m of a given function $h_F^{(i)} \in W_i$, we denote by $h_F^{(\ell,i)}$ its restriction to Γ_ℓ, $\ell \in \mathcal{L}_{F_i}$ and:

$$U^{(m,i)} = \left[\!\left[\left(-T_i \nabla_{\pi_i} h_F^{(i)} \right) \cdot \mathbf{n}_{\pi_i,m} \right]\!\right], \tag{9}$$

where $\mathbf{n}_{\pi_i,m}$ is a fixed normal to the trace S_m in the plane π_i, and then define

$$U^{(i)} = \prod_{m \in \mathcal{M}_i} U^{(m,i)} \in \mathcal{U}_{(i)}, \quad U = \prod_{i \in \mathcal{I}} U^{(i)} \in \mathcal{U}.$$

It is now possible to introduce a cost functional J defined for any function $h \in V$, $h_F \in W$, $U \in \mathcal{U}$, $Q \in \mathcal{Q}$ as:

$$\begin{aligned} J(h, h_F, U, Q) = & \sum_{\ell \in \mathcal{L}^\star} \left\| h^{(\ell,k)}(Q) - h_F^{(\ell,\varrho(\ell))}(Q) \right\|^2_{\mathcal{H}_\ell^\Gamma} \\ & + \sum_{m \in \mathcal{M}} \left(\left\| h_F^{(i)}{}_{|S_m}(U, Q) - h_F^{(j)}{}_{|S_m}(U, Q) \right\|^2_{\mathcal{H}_m^S} \right. \\ & \left. + \left\| U^{(m,i)} + U^{(m,j)} \right\|^2_{\mathcal{U}_m} \right) \end{aligned} \tag{10}$$

in which, for each $m \in \mathcal{M}$, $i, j \in \mathcal{I}_{S_m}$, $i \neq j$. In the case of multiple fractures intersecting in a trace, the second and third term in J should be expanded according to what done for equations (4)–(5).

In Eq. (10), h and h_F depend on U and/or Q via the following problems defined on each matrix block $\mathcal{D}_k$: find $h^{(k)} \in V_k$ such that for all $v \in V_k$:

$$\left(K^{(k)} \nabla h^{(k)}, \nabla v \right)_{\mathcal{D}_k} = (f_k, v)_{\mathcal{D}_k} - \sum_{\ell \in \mathcal{L}_k} \left\langle Q^{(\ell,k)}, v^{(\ell)} \right\rangle_{\Gamma_\ell}, \tag{11}$$

and on each fracture plane: find $h_F^{(i)} \in W_i$ such that for all $v \in W_i$:

$$\left(T_i \nabla h_F^{(i)}, \nabla v \right)_{F_i} = \sum_{\ell \in \mathcal{L}_{F_i}} \left(Q^{(\ell,p)} + Q^{(\ell,r)}, v \right)_{F_i} - \sum_{m \in \mathcal{M}_i} \left\langle U^{(m,i)}, v \right\rangle_{S_m}, \tag{12}$$

being $p, r \in \mathcal{K}_{\Gamma_\ell}$, $p \neq r$. It is to remark that problems (11)–(12) are well posed.

The DFM model flow problem, formulated as a PDE-constrained optimization problem, then reads as follows:

$$\min J(h, h_F, U, Q),$$
$$\text{constrained by (11)}, \forall k \in \mathcal{K}, \text{ and by (12)}, \forall i \in \mathcal{I}. \tag{13}$$

The solution of Eq. (13) is equivalent to the solution of Eqs. (7)–(8).

6 FEM discretization of the fractures

In the following, overloading notation, $h_F^{(l)}$, $h^{(l)}$, $Q^{(\ell,p)}$ and $U^{(m,i)}$ denotes also the finite dimensional approximation of the hydraulic head and flux on the corresponding object, either fracture, block, face or trace. The following finite dimensional subspaces are introduced:

- for each fracture F_i, W_i^δ of W_i,

$$W_i^\delta = \text{span}\left\{\varphi_l,\ l = 1, \ldots, n_{\text{dof}}^{(i)}\right\}$$

- for each trace S_m of fracture F_i, $\mathcal{U}_{m,i}^\delta$ of $\mathcal{U}_m$,

$$\mathcal{U}_{m,i}^\delta = \text{span}\left\{\psi_l,\ l = 1, \ldots, n_{\text{dof}}^{((m,i))}\right\}$$

- for each face Γ_ℓ of block $\mathcal{D}_k$, $\mathcal{Q}_{\ell,k}^\delta$ of $\mathcal{Q}_\ell$,

$$\mathcal{Q}_{\ell,k}^\delta = \text{span}\left\{\phi_l,\ l = 1, \ldots, n_{\text{dof}}^{(\ell,k)}\right\}$$

where n_{dof} is the number of degrees of freedom (dofs). The discrete weak formulation of Eq. (12) is as follows:

$$\int_{F_i} T_i \nabla h_F^{(i)} \nabla \varphi_s \mathrm{d}F_i = \sum_{\ell \in \mathcal{L}_{F_i}} \int_{\Gamma_\ell} (Q^{(\ell,p)} + Q^{(\ell,r)}) \varphi_s \mathrm{d}\Gamma_\ell - \sum_{m \in \mathcal{M}_i} \int_{S_m} U^{(m,i)} \varphi_s \mathrm{d}S_m. \tag{14}$$

where φ_s are basis functions.

Equation (14) can be re-written in algebraic form as:

$$\mathbf{A}\mathbf{h}_F = -\sum_{\ell \in \mathcal{L}_{F_i}} \left(\mathbf{B}^{(\ell,p)}\mathbf{Q}^{(\ell,p)} + \mathbf{B}^{(\ell,r)}\mathbf{Q}^{(\ell,r)}\right) - \sum_{m \in \mathcal{M}_i} \mathbf{C}^{(m)}\mathbf{U}^{(m)}, \tag{15}$$

where $\mathbf{A}, \mathbf{B}^{(\ell,p)}, \mathbf{B}^{(\ell,r)}, \mathbf{C}^{(m)}$ are matrices of coefficients resulting from the computation of the integrals involving the basis functions φ_s of Eq. (14). Vectors $\mathbf{h}_F$, $\mathbf{Q}^{(\ell,p)}$, $\mathbf{Q}^{(\ell,r)}$, $\mathbf{U}^{(m)}$ collect the nodal values of the corresponding functions. Note that, in order to ease the notation, the superscript (i) has been dropped.

The discrete counterpart of functional (10) can also be written by replacing the norm in $\mathcal{H}_\ell^\Gamma$ with the norm in $\mathrm{L}^2(\Gamma_\ell)$, and the norms in $\mathcal{H}_m^S$ and $\mathcal{U}_m$ with the norms in $\mathrm{L}^2(S_m)$. While referring to Berrone et al. (2017) for the details, it is outlined here that in compact form the structure of the discrete counterpart of (10) is:

$$J(\mathbf{h}, \mathbf{h}_F, \mathbf{U}, \mathbf{Q}) = \mathbf{h}^T \mathbf{G}^{\mathcal{D}} \mathbf{h} + 2\mathbf{h}^T \mathbf{G}^{\mathcal{D},F} \mathbf{h}_F + \mathbf{h}_F^T \mathbf{G}^F \mathbf{h}_F + \mathbf{U}^T \mathbf{G}^S \mathbf{U}, \tag{16}$$

being $\mathbf{h}$, $\mathbf{h}_F$, $\mathbf{U}$, $\mathbf{Q}$ the global vectors collecting all the dofs of the corresponding functions, and being matrices $\mathbf{G}^{\mathcal{D}}$, $\mathbf{G}^{\mathcal{D},F}$, $\mathbf{G}^F$, $\mathbf{G}^S$ suitably defined.

7 BEM local solution for the blocks

The relevant part of the fluid flow in a fractured rock mass takes places in the fracture network and in the flow equations only quantities at the boundaries of the blocks are involved. Therefore, a natural choice for the solution of a DFM model is to apply the Boundary Element Method for the discretization of the governing equations in the blocks. In fact, the BEM-derived algebraic equations include boundary quantities only. Application of BEM entails the division of the boundary faces of each block into elements. Let the block boundary $\partial\mathcal{D}_k$ be partitioned in $n_\mathrm{b}^{(k)}$ triangular elements, and consider a piecewise constant approximation in each element for the hydraulic head and the normal flux $\partial h/\partial n$, with collocation points corresponding to the centroids of the elements. After application of the second Green's theorem (see Brebbia et al. 1984), Eq. (6), with forcing functions $f^{(k)} = 0$, can be written at each collocation node $\mathbf{x}_s$, $s = 1, \ldots, n_\mathrm{b}^{(k)}$ as follows:

$$c_s h_s^{(k)} + \sum_{m=1}^{n_\mathrm{b}^{(k)}} h_m^{(k)} \int_{\Gamma_m} \frac{\partial \hat{h}^{(k)}}{\partial n}(\mathbf{x}, \mathbf{x}_s) \mathrm{d}\Gamma_m = \sum_{m=1}^{n_\mathrm{b}^{(k)}} \left(\frac{\partial h^{(k)}}{\partial n} \right)_m \int_{\Gamma_m} \hat{h}^{(k)}(\mathbf{x}, \mathbf{x}_s) \mathrm{d}\Gamma_m, \tag{17}$$

where c_s is a suitable coefficient (depending on the smoothness of the boundary in $\mathbf{x}_s$) and $h_m^{(k)}$ is the hydraulic head value at $\mathbf{x}_m$. The symbol $(\partial h/\partial n)_m$ represents the inward normal flux at point m; Γ_m is the m-th element; $\hat{h}^{(k)}(\mathbf{x}, \mathbf{x}_s)$ is the Green's function at point $\mathbf{x}_s$, solution of the following equation:

$$\Delta \hat{h}^{(k)}(\mathbf{x}, \mathbf{x}_s) + \delta_s = 0 \tag{18}$$

being δ_s the Dirac Delta function at point $\mathbf{x}_s$ on the surface of the k-th block. From Eq. (18) the Green's function is:

$$\hat{h}^{(k)}(\mathbf{x}, \mathbf{x}_s) = \frac{1}{4\pi r}, \tag{19}$$

being r the geodetic distance, i.e. the distance between $\mathbf{x}$ and the collocation node $\mathbf{x}_s$.

The integrals in (17) can be computed by means of Gaussian quadrature rules for $m \neq s$; for $m = s$, suitable quadrature rules are necessary in order to tackle the singularities of functions $\hat{h}^{(k)}$ and $(\partial \hat{h}^{(k)}/\partial n)$ at the collocation node.

After numerical integration, Eq. (17) is rewritten as follows:

$$\sum_{m=1}^{n_\mathrm{b}^{(k)}} H_{s,m}^{(k)} h_m^{(k)} = -\frac{1}{K^{(k)}} \sum_{m=1}^{n_\mathrm{b}^{(k)}} G_{s,m}^{(k)} Q_m^{(k,\ell)} \tag{20}$$

where the terms $H_{s,m}^{(k)}$, $G_{s,m}^{(k)}$ accounts for the integrals of $\hat{h}^{(k)}$ and $(\partial \hat{h}^{(k)}/\partial n)$ on the element Γ_m and with collocation node $\mathbf{x}_s$. Introducing the (dense) matrices $\mathbf{G}$ and $\mathbf{H}$, Eq. (17) is compactly written as:

$$\mathbf{Hh} = -\frac{1}{K^{(k)}} \mathbf{GQ}, \tag{21}$$

where the superscript (k) to the symbols of the matrices and vectors have been removed for simplicity.

8 The 2D and 3D Virtual Element Method

The Virtual Element Method (VEM) was first introduced in a seminal paper by Beirão da Veiga et al. (2013a) and further developed by other authors (see for example Ahmad et al. 2013; Beirão da Veiga et al. 2013b; Beirão Da Veiga et al. 2014; Beirão da Veiga et al. 2015; Brezzi et al. 2014). It is a generalization of the Finite Element Method and allows one to use general polytopal meshes. In this section, the functional spaces for the VEM on bidimensional domains are introduced first, then the spaces for the three-dimensional domains are defined. In the following, the order of the method is fixed to $n \in \mathbb{N}$.

Let $\Omega \subset \mathbb{R}^2$ be a bounded open domain, and let $\mathcal{T}_\delta(\Omega)$ be a polygonal tessellation of Ω, satisfying the following regularity assumption: exists $\gamma > 0$ independent of δ such that, for any given element $E \in \mathcal{T}_\delta(\Omega)$:

1. E is star-shaped with respect to a ball of radius γh_E, h_E being the diameter of E;
2. any given edge $e \subset \partial E$ has length $|e| \geq \gamma h_E$.

Let $\mathbb{P}_n(\mathcal{T}_\delta(\Omega))$ be the space of piecewise polynomials of order $\leq n$ on the elements of the tessellation. The operator $\Pi_n^\nabla : \mathrm{H}^1(\Omega) \to \mathbb{P}_n(\mathcal{T}_\delta(\Omega))$ is defined such that, for any $v \in \mathrm{H}^1(\Omega)$, $\left(\nabla \left(v - \Pi_n^\nabla v\right), \nabla p\right)_E = 0$, $\forall E \in \mathcal{T}_\delta(\Omega)$ and $\forall p \in \mathbb{P}_n(E)$. Next, given $E \in \mathcal{T}_\delta(\Omega)$, the local VEM space of order n is defined as follows:

$$W_\delta^E = \Big\{ v \in \mathrm{H}^1(E) \colon v|_{\partial E} \in C^0(\partial E),\ v|_e \in \mathbb{P}_n(e) \quad \forall e \subset \partial E,$$
$$\Delta v \in \mathbb{P}_n(E),\ \left(v - \Pi_n^\nabla v, p\right)_E = 0 \quad \forall p \in \mathbb{P}_n(E) \backslash \mathbb{P}_{n-2}(E) \Big\}. \tag{22}$$

Using (22), the bidimensional VEM space of order n is defined as follows:

$$W_\delta(\mathcal{T}_\delta(\Omega)) = \left\{ v \in \mathrm{H}^1(\Omega) \colon v \in C^0\left(\mathcal{S}\left(\mathcal{T}_\delta(\Omega)\right)\right),\ v \in W_\delta^E\ \forall E \in \mathcal{T}_\delta(\Omega) \right\}, \quad (23)$$

being $\mathcal{S}(\mathcal{T}_\delta(\Omega))$ the skeleton of $\mathcal{T}_\delta(\Omega)$. A function $v \in W_\delta(\mathcal{T}_\delta(\Omega))$ can be described entirely by the following set of dofs:

- the values at the mesh vertices;
- if $n \geq 2$, the values at $n-1$ points internal to each edge of the mesh;
- if $n \geq 2$, for each $E \in \mathcal{T}_\delta(\Omega)$, the moments $\frac{1}{|E|}(v, m)_E$, for each $m \in \mathcal{M}_{n-2}(E)$, $\mathcal{M}_{n-2}(E)$ being a basis of $\mathbb{P}_{n-2}(E)$.

Concerning the 3D VEM, consider a bounded open set $\Omega \subset \mathbb{R}^3$ and let $\mathcal{T}_\delta(\Omega)$ be a polyhedral tessellation of Ω with polyhedra satisfying the same regularity assumptions stated above for $\mathcal{T}_\delta(\Omega)$. Let $\Pi_n^\nabla : \mathrm{H}^1(\Omega) \to \mathbb{P}_n(\mathcal{T}_\delta(\Omega))$ be defined formally as above and define, for each $E \in \mathcal{T}_\delta(\Omega)$, the following local space:

$$\begin{aligned} V_\delta^E = \Big\{ & v \in \mathrm{H}^1(E) \colon v_{|f} \in W_\delta^f \quad \forall f \subset \partial E,\ \Delta v \in \mathbb{P}_n(E), \\ & \left(v - \Pi_n^\nabla v, p\right)_E = 0 \quad \forall p \in \mathbb{P}_n(E) \backslash \mathbb{P}_{n-2}(E) \Big\}, \end{aligned} \quad (24)$$

where W_δ^f is the 2D VEM space on a face of E, defined by (22). Then, the three-dimensional VEM space of order n can be defined as follows:

$$V_\delta(\mathcal{T}_\delta(\Omega)) = \left\{ v \in \mathrm{H}^1(\Omega) \colon v \in C^0\left(\mathcal{S}\left(\mathcal{T}_\delta(\Omega)\right)\right),\ v \in V_\delta^E \quad \forall E \in \mathcal{T}_\delta(\Omega) \right\}. \quad (25)$$

A possible set of dofs for a function $v \in V_\delta(\mathcal{T}_\delta(\Omega))$ is the following:

- the values at the mesh vertices;
- if $n \geq 2$, the values at $n-1$ points internal to each edge of the mesh;
- if $n \geq 2$, for each $E \in \mathcal{T}_\delta(\Omega)$ and for each face f of E, the moments $\frac{1}{|f|}\left(v, m_f\right)_f$, for each $m_f \in \mathcal{M}_{n-2}(f)$, $\mathcal{M}_{n-2}(f)$ being a basis of $\mathbb{P}_{n-2}(f)$;
- if $n \geq 2$, for each $E \in \mathcal{T}_\delta(\Omega)$, the moments $\frac{1}{|E|}(v, m)_E$, for each $m \in \mathcal{M}_{n-2}(E)$, $\mathcal{M}_{n-2}(E)$ being a basis of $\mathbb{P}_{n-2}(E)$.

The 2D VEM has already be applied in the framework of DFN simulations in Benedetto et al. (2016a, b, c). In the next section the application of the technique to the fracture-block coupling (VEM–VEM coupling) is shown.

9 VEM discretization of fractures and matrix blocks

As previously mentioned, the DFN $\mathcal{F}$ partitions the domain $\mathcal{D}$ in polyhedral blocks $\mathcal{D}_k$; moreover, the intersections among fractures generate a system of polygonal sub-fractures on each fracture F_i; these sub-fractures coincide with the faces Γ_ℓ of the surrounding blocks $\mathcal{D}_k$, with $k \in \mathcal{K}_\ell$.

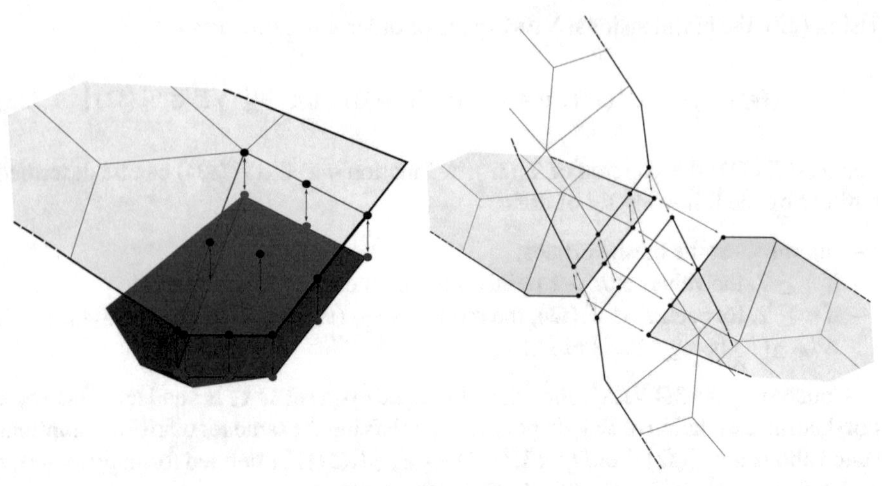

Fig. 2 Example of dofs matching for the VEM-based approach

By exploiting the possibility of the Virtual Element Method of using generic polyhedral and polygonal elements, a local VEM space on each block $\mathcal{D}_k$ and on each sub-fracture Γ_ℓ can be defined. A VEM of the same order n, both on 2D and 3D objects, can be considered. Two neighbouring 2D elements built on the two sub-fractures of a same fracture, sharing a trace S_m, have the same dofs on the trace itself; the same applies to the two sub-fractures of fracture F_j sharing with F_i the trace S_m (see Fig. 2). The same applies to 3D elements: two blocks sharing a face Γ_ℓ have the same dofs on Γ_ℓ; these dofs are also in common with the VEM element coincident with Γ_ℓ introduced for the discretization of the hydraulic head on the fracture.

Focusing on Eqs. (7) and (8), the global continuity of the discrete spaces W_i^δ and W_j^δ, corresponding to intersecting fractures, can be imposed by forcing the equality of VEM dofs lying on the traces. Similarly, the global continuity of the discrete spaces V_p^δ and V_r^δ for blocks $\mathcal{D}_p$, $\mathcal{D}_r$ and intersecting at face Γ_ℓ is obtained by imposing the equality of the dofs on the block faces. Furthermore, by imposing the equality also for the dofs on the sub-fractures built on $\Gamma_\ell \subseteq F_i$, one also obtains the global continuity with W_i^δ.

Within this framework, the formulation of the VEM–VEM coupling is: find $h^{(k)} \in V_k$, $k \in \mathcal{K}$; $h_F^{(i)} \in W_i$, $i \in \mathcal{I}$; $Q^{(\ell,k)} \in \mathcal{Q}_\ell$; $\ell \in \mathcal{L}_k^\star$, $k \in \mathcal{K}$, such that for all $\forall v \in \mathrm{H}^1_{\mathcal{F}}(\mathcal{D})$:

$$\sum_{k\in\mathcal{K}} \left(K^{(k)}\nabla h^{(k)}, \nabla v\right)_{\mathcal{D}_k} + \sum_{i\in\mathcal{I}} \left(T_i \nabla h_F^{(i)}, \nabla v\right)_{F_i} = \sum_{k\in\mathcal{K}} \left(f^{(k)}, v\right)_{\mathcal{D}_k}, \tag{26}$$

with the matching conditions

$$h^{(i)}_{F\,|S_m} = h^{(j)}_{F\,|S_m}, \quad \text{on } S_m, \forall m \in \mathcal{M}, \tag{27}$$

$$h^{(\ell,p)} = h_F^{(\ell,\varrho(\ell))}, \quad \text{in } \Gamma_\ell, \forall \ell \in \mathcal{L}^\star, \forall p \in \mathcal{K}_\ell. \tag{28}$$

The following local VEM finite dimensional sub-spaces are introduced:

- for each block $\mathcal{D}_k$, V_k^δ of V_k,

$$V_k^\delta = \text{span}\left\{\Phi_l,\ l = 1, \ldots, n_{\text{dof}}^{(k)}\right\},$$

- for each fracture F_i, W_i^δ of W_i,

$$W_i^\delta = \text{span}\left\{\varphi_l,\ l = 1, \ldots, n_{\text{dof}}^{(i)}\right\}.$$

By using standard techniques for saddle-point problems arising from domain decomposition methods, the discrete version of problem (26), coupled with (27), (28) can be easily obtained:

$$\begin{pmatrix} \mathbf{A} & \mathbf{L}^T \\ \mathbf{L} & 0 \end{pmatrix} \begin{pmatrix} \mathbf{h} \\ \lambda \end{pmatrix} = \begin{pmatrix} \mathbf{b} \\ 0 \end{pmatrix},$$

$$\mathbf{A} = \begin{pmatrix} \text{diag}\{\mathbf{A}_k\}_{k \in \mathcal{K}} & 0 \\ 0 & \text{diag}\{\mathbf{A}_i\}_{i \in \mathcal{I}} \end{pmatrix}, \quad \mathbf{L} = \begin{pmatrix} \cdots & \cdots & \cdots & \cdots & \cdots \\ \cdots & 1 & \cdots & -1 & \cdots \\ \cdots & \cdots & \cdots & \cdots & \cdots \end{pmatrix},$$

where $\mathbf{h}$ is the vector containing all the dofs for the hydraulic head in blocks and fractures, λ is the vector of the Lagrange multipliers imposing the matching conditions at the corresponding block-fracture (Fig. 2, left) or fracture-fracture (Fig. 2, right) dofs and $\mathbf{A}_k$ and $\mathbf{A}_i$ are the standard local stiffness matrices for block $\mathcal{D}_k$ and fracture F_i, respectively.

The matching conditions (27), (28) are imposed through matrix $\mathbf{L}$, in which each row includes the coupling between two dofs belonging to matching sub-fractures and block-faces (see Fig. 2, left) or to matching fractures (see Fig. 2, right). In detail, with reference to the same figure, in the case of matching between block-face dofs and sub-fracture dofs, the generic row of $\mathbf{L}$ has an element with value $+1$ in the column corresponding to the black bullet (representing a dof on the sub-fracture element), and a -1 in the column corresponding to the blue bullet (representing a dof on the block face). In the case of matching between fracture dofs at traces (right of Fig. 2), two different kinds of matching are considered, highlighted by arrows with different colors: matching between dofs on sub-fractures belonging to the same original fracture (black and red arrows) and matching between dofs on intersecting fractures (blue arrows). While the first kind of matching is already automatically imposed in VEM spaces W_i^δ on each fracture, the second kind of matching is imposed through matrix $\mathbf{L}$, with a row for each couple of dofs representing the same pointwise value of the hydraulic head on the two intersecting fractures.

The choice of using Lagrange multipliers to enforce continuity is in view of a possible parallel resolution of the problem. Clearly, this can be replaced by a simple renumbering of the degrees of freedom, thus avoiding a saddle-point formulation.

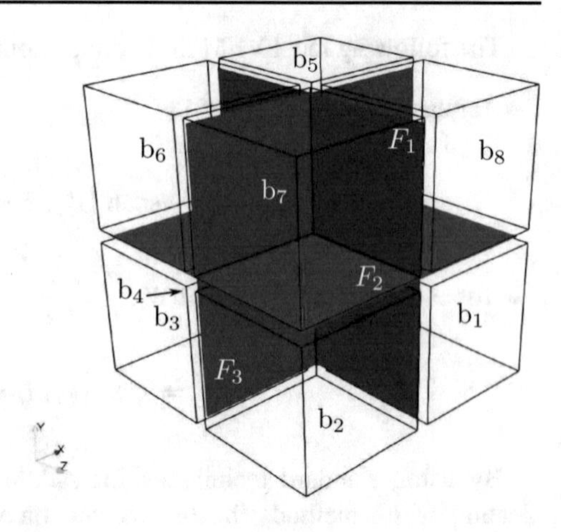

Fig. 3 Example 1: exploded view of sub-blocks and fractures

Table 1 Example 1: values of block hydraulic conductivity and fracture transmissivity for the considered test cases

Case	$K^{(k)}$	T
c_1	$K^{(1-8)} = 10^{-5}$	$T_{1-3} = 1$
c_2	$K^{(1-8)} = 1$	$T_{1-3} = 10^{-5}$
c_3	$K^{(1,4,6,7)} = 1$; $K^{(2,3,5,8)} = 10^{-5}$	$T_{1-3} = 1$
c_4	$K^{(1,4,6,7)} = 1$; $K^{(2,3,5,8)} = 10^{-5}$	$T_{1,3} = 1$; $T_2 = 10^{-5}$
c_5	$K^{(1-7)} = 1$; $K^{(8)} = 10^{-5}$	$T_{1-3} = 1$
c_6	$K^{(1-4,6,7)} = 1$; $K^{(5,8)} = 10^{-5}$	$T_{1-3} = 1$
c_7	$K^{(1-8)} = 0.3$	$T_{1-3} = 1$

10 Illustrative examples

Two examples are reported to illustrate the quality of the proposed techniques. For both the examples, a $1 \times 1 \times 1\,\mathrm{m}^3$ nucleus of a porous medium, sectioned by a DFN, is considered. In the first example, the DFN consists of three persistent mutually-orthogonal fractures, whereas in the second example 10 random 9-fracture DFNs are analyzed.

With reference to a x, y, z coordinate system, no-flow conditions are applied to the external boundaries parallel to $x-y$, $x-z$ planes, and a uniform directional gradient $\mathrm{d}h/\mathrm{d}x = -1$ is imposed by fixing the hydraulic head h equal to 1 and 0 at the $x = 0$, $x = 1$ boundaries, respectively. Single blocks and fractures are homogeneous and isotropic.

With reference to Fig. 3, for the first example seven different configurations (c_1–c_7) are considered, each of them corresponding to a given set of values for the block hydraulic conductivities $K^{(k)}$ and for the fracture transmissivities T_i, as detailed in Table 1.

The target of the simulations is the computation of the discharge from the matrix block faces ($Q_{\mathrm{b},x}$) and fracture edges ($Q_{\mathrm{f},x}$) on the plane $x = 1$ in all the configurations, by using both the FEM–BEM optimization technique and the VEM-based

Table 2 Example 1: $Q_{b,x}$ for test cases c_1-c_7

Config.	FEM–BEM		VEM	
	Coarse	Fine	Order 1	Order 2
c_1	1.12e−5	1.07e−5	1e−5*	1e−5*
c_2	1.05	1.02	1*	1*
c_3	0.54	0.50	0.47	0.47
c_4	0.49	0.47	0.46	0.46
c_5	0.87	0.82	0.78	0.78
c_6	0.61	0.57	0.54	0.54
c_7	0.33	0.32	0.3*	0.3*

Values with a * correspond to results correct up to machine precision

Table 3 Example 1: $Q_{f,x}$ for test cases c_1-c_7

Config.	FEM–BEM		VEM	
	Coarse	Fine	Order 1	Order 2
c_1	2.00	2.00	2*	2*
c_2	2.17e−5	2.0e−5	2e−5*	2e−5*
c_3	1.98	1.98	2.00	1.98
c_4	0.98	0.98	1.00	0.98
c_5	2.07	2.07	2.09	2.08
c_6	2.17	2.17	2.18	2.17
c_7	2.00	2.00	2*	2*

Values with a * correspond to results correct up to machine precision

technique. Two different meshes are used for the FEM–BEM technique, with a maximum element size of 10^{-2} m^2 for the coarse mesh and 2.5×10^{-3} m^2 for the fine mesh. We recall that linear finite elements are used in this approach, coupled to piecewise constant boundary elements. Virtual elements of order 1 and 2 are used instead for the VEM-based approach, with 3D elements entirely coinciding with the blocks $\mathcal{D}_k$ and 2D elements coinciding with the sub-fractures Γ_ℓ.

For the configurations with uniform $K^{(k)}$ and T_i (c_1, c_2 and c_7), the value of the discharge relative to blocks, $Q_{b,x}$, is equal to $K^{(k)}$, and the value of the discharge relative to fractures, $Q_{f,x}$, is equal to $2T_i$.

Due to the boundary conditions imposed and to the symmetry, blocks and fractures do not mutually exchange fluid, thus all the $Q^{(\ell,p)} + Q^{(\ell,r)}$ and $U^{(m,i)}$ values are nil. For the other configurations, no analytical solutions are available.

The obtained results are reported in Table 2 for $Q_{b,x}$, and in Table 3 for $Q_{f,x}$. In this and in all following tables results are reported with two significant digits; values with a * superscript correspond, instead, to results correct up to machine precision. It can be seen that both techniques provides similar results, in particular for the quantity $Q_{f,x}$, whereas larger differences are observed for $Q_{b,x}$. This is probably due to the poor quality of the BEM discretization with constant elements, here used, if compared to the VEM discretizations. This choice, however, increases the flexibility in mesh generation, as mesh elements on block faces are not required to match at block edges.

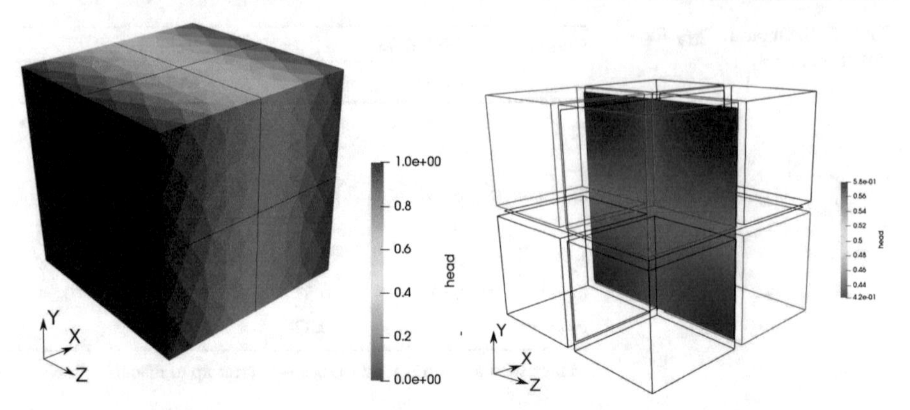

Fig. 4 Example 1: solution on block boundaries (left) and on fracture F_1 (right) for configuration c_3. Blocks with red boundaries have conductivity equal to 1, the others equal to 10^{-5} (colour figure online)

The differences in the approximation of $Q_{b,x}$ obtained with FEM–BEM and VEM–VEM, about 10% with the coarse FEM–BEM mesh, are reduced to about 5% with the fine mesh. Negligible differences are instead observed when increasing the VEM order from 1 to 2. Furthermore, the exact solution for configurations c_1, c_2 and c_7 belongs to the discrete functional subspace for the VEM-based technique only; indeed, in these cases, the computed solution reproduces the analytical solution within the machine precision.

Figure 4 reports the solution obtained with the FEM–BEM approach for the configuration c_3, on the outer skin of the matrix domain (left) and on fracture F_1 (right). The matrix blocks drawn in black have conductivity five orders of magnitude smaller than the others, and, thus, as expected, the flux along fracture F_1 establishes a connection among the highly conductive blocks. In Fig. 5 top-left, the FEM–BEM solution on fracture F_1 for c_3 is also shown, along with a representation of the solution obtained with the second order VEM, in solid red line. The representation of the VEM solution is the piece-wise linear interpolant of the VEM solution on the edges of each polygonal element. In Fig. 5 top-right to bottom-right, the solution on F_1 with the FEM–BEM approach and the VEM is shown for c_4 to c_6: in all these cases it can be seen that there is a non-zero flux along this fracture, as a consequence of the different conductivity of the surrounding blocks. It is also shown that the solutions given by the two numerical techniques are always in good agreement.

As previously mentioned, in the second example the porous nucleus is sectioned by 9 fractures. Ten different DFNs are obtained by randomly perturbing the orientation of three sets of three originally orthogonal and equally spaced fractures, and then the resulting geometrical settings are analyzed by applying the same boundary conditions of the first example. We remark that these random networks are not intended to reproduce any particular distribution, nor to have any statistical relevance. The random geometry is only targeted at showing the capabilities of the methods in handling complex geometrical configurations. The resulting geometrical settings g_1–g_{10} are shown in Fig. 6. The number of blocks ranges between 64 and 74. The block hydraulic conductivity values are stochastically generated from

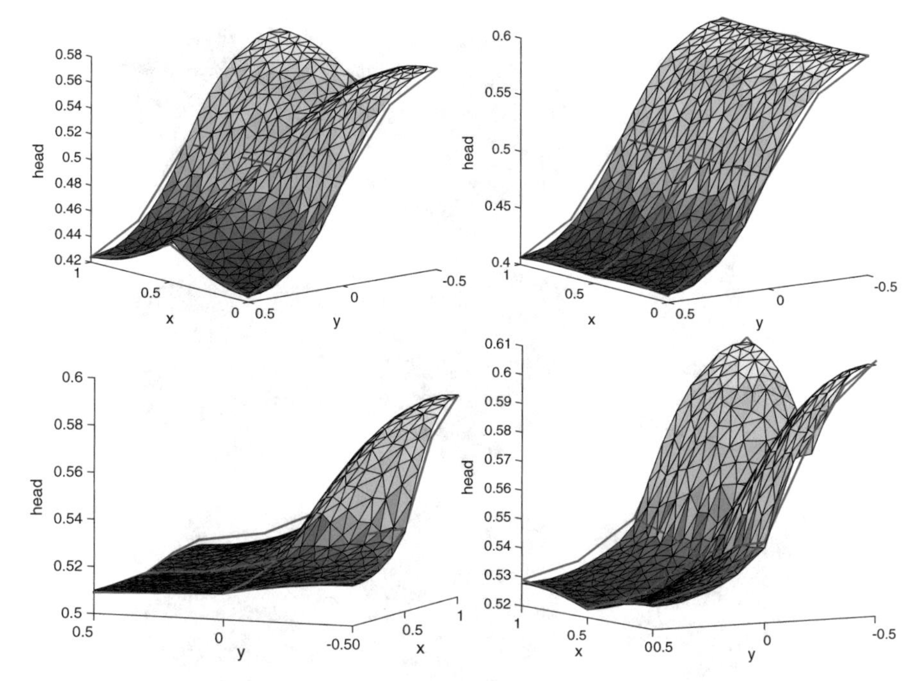

Fig. 5 Example 1: solution on fracture F_1. Configurations from c_3 (top-left) to c_6 (bottom-right)

a uniform distribution and reported in Fig. 7, whereas fracture transmissivity values are $T_{1-9} = 0.1, 0.35, 0.6, 0.7, 0.2, 0.5, 0.65, 0.43, 0.87$. The mesh size for the FEM–BEM technique coincides with the fine mesh of the previous example, and first order accuracy is used for the VEM approach.

In Fig. 6 the solutions in terms of h obtained with first order VEM are reported for the 10 configurations. Small blocks are present, delimited by fractures intersecting at small angles, like for example in g_5. The solution for this configuration, obtained with the FEM–BEM technique, is shown in Fig. 8. A further view of this solution on the fractures is reported in Fig. 9. In this last image, the mesh on one of the fractures and on one of its neighbouring blocks is reproduced, in order to highlight the non-conformity of the elements on the fractures and on the block faces; the non-conformity of the mesh elements on fractures at the traces is also shown. Figure 10 shows the solution, still limited to the fractures, for g_5, this time obtained with first order VEM. It is worth to remark that the proposed VEM technique easily allows dealing with complex geometries, making the generation of a mesh a rather simple task.

The computed values of $Q_{b,x}$ and $Q_{f,x}$ are reported in Table 4. Again, for $Q_{f,x}$ the results provided by the two techniques are in good agreement, whereas some discrepancies are again observed for $Q_{b,x}$; the explanation of these discrepancies advanced for the first example still holds.

As a whole, these numerical tests show the viability and effectiveness of the proposed techniques when dealing with the simulation of the fluid flow in DFMs.

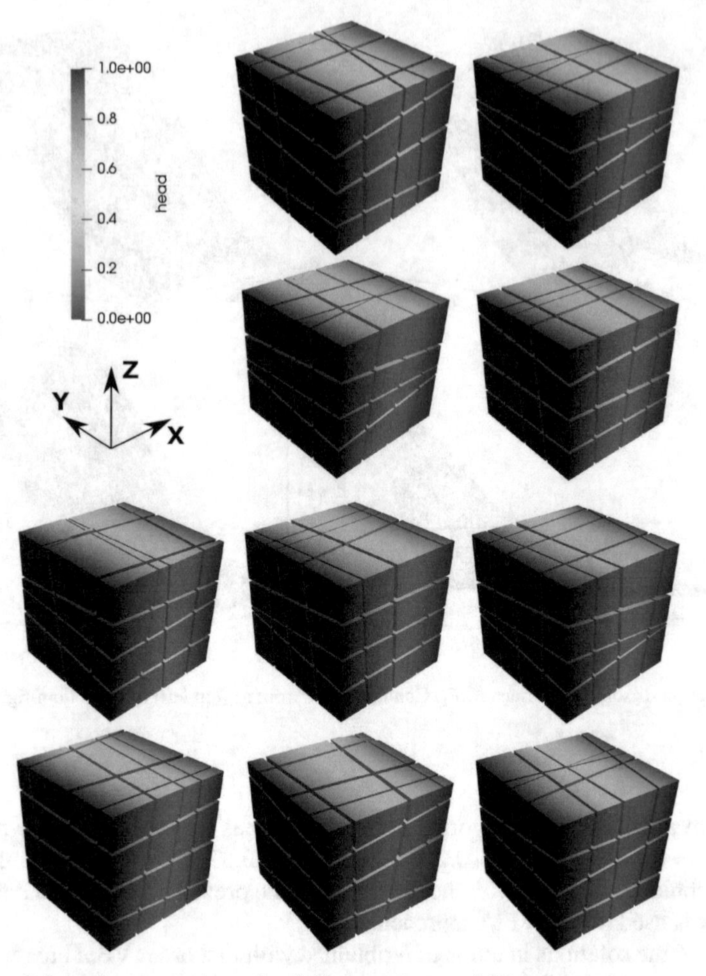

Fig. 6 Example 2: random geometries g_1 (top-left) to g_{10} (bottom-right)

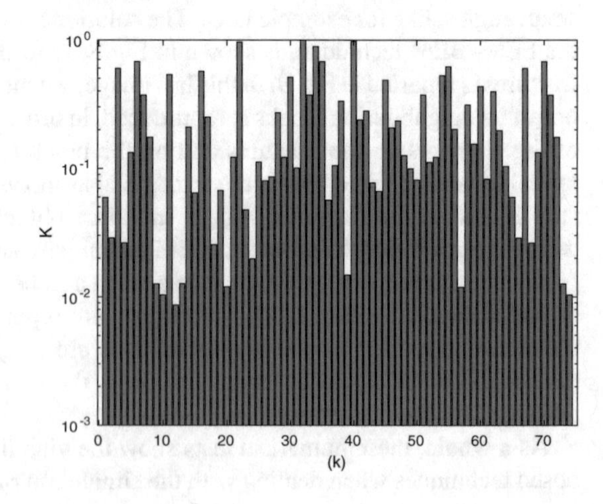

Fig. 7 Example 2: values of block hydraulic conductivity $K^{(k)}$ versus block index k

Fig. 8 Example 2: solution obtained with FEM–BEM coupling on geometry g_5

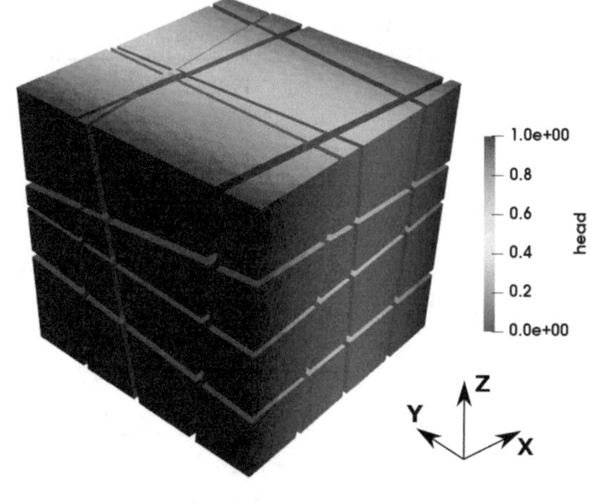

Fig. 9 Example 2, geometry g_5. Solution obtained with FEM–BEM coupling and mesh details

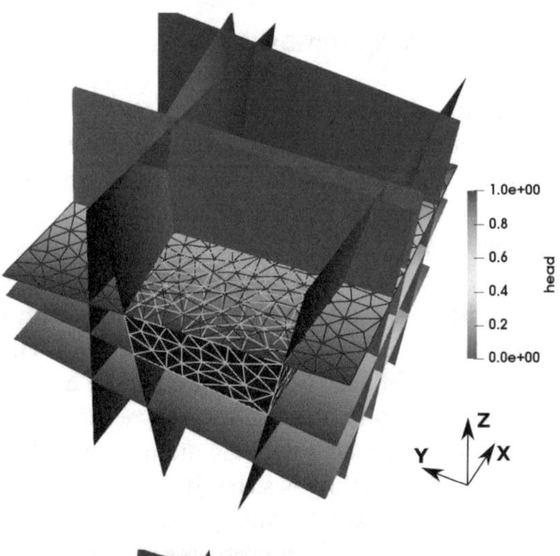

Fig. 10 Example 2, geometry g_5. Solution obtained with the VEM–VEM coupling

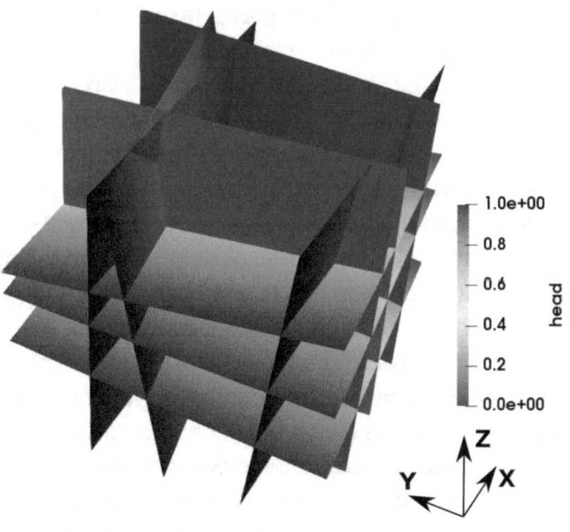

Table 4 Example 2: $Q_{b,x}$ and $Q_{f,x}$ for g_1 to g_{10} and for the two considered approaches

Method		g_1	g_2	g_3	g_4	g_5
FEM–BEM	$Q_{b,x}$	0.35	0.34	0.31	0.26	0.23
VEM		0.32	0.31	0.27	0.24	0.20
FEM–BEM	$Q_{f,x}$	3.31	3.30	3.31	3.39	3.44
VEM		3.32	3.31	3.32	3.41	3.44
Method		g_6	g_7	g_8	g_9	g_{10}
FEM–BEM	$Q_{b,x}$	0.27	0.34	0.41	0.42	0.18
VEM		0.24	0.30	0.36	0.39	0.16
FEM–BEM	$Q_{f,x}$	3.34	3.30	3.24	3.16	3.44
VEM		3.36	3.32	3.25	3.19	3.44

11 Concluding remarks

In this note the coupling of Darcy flows in a Discrete Fracture-Matrix (DFM) model is addressed. Two different techniques are proposed: 1) an optimization approach coupling FEM elements on the fractures and BEM elements on the blocks and 2) a fully-conforming VEM technique.

The optimization approach have been proven to be very effective in dealing with very complex Discrete Fracture Networks (DFNs). Its applicability does not depend on the specific discretization approach used on both the fracture and the blocks. It is used herein for the coupling of a Galerkin FEM used for the fractures with a collocation BEM for the blocks, however it could be extended to other discretization methods, even by using fully non-conforming meshes, i.e. the mesh used for the porous matrix blocks does not match with the mesh of the delimiting fractures.

The VEM coupling is strongly based on the conformity of the mesh of the blocks with the mesh of the delimiting fractures. However, when this conformity is ensured, the coupling of the flow problems of fractures and blocks is trivial when the same order is used in all the elements.

The optimization method is particularly indicated for very complex geometries, where the blocks generated after cutting the original domain according to the fracture network might be extremely badly shaped and of very different sizes. In this case the non-conformity allowed by the optimization approach allows to prevent the problem. Furthermore the optimization method naturally yields a scalable parallel approach, thus being suitable for large simulations. On the other hand the VEM-based approach is more standard and can take advantage of a large literature on solvers and preconditioners borrowed from domain decomposition methods.

The geometries considered in the numerical tests are quite simple if compared to the complex situations in practical applications, nevertheless these geometries include badly-shaped and sharp blocks and small blocks close to larger blocks. It is shown that problems do not emerge when using the two techniques, even in case of strong variation of both fracture trasmissivity and block hydraulic conductivity among close objects. In addition, the performance of the VEM technique looks noticeable, even by using low order elements. However, also the FEM–BEM technique performs rather well with a slight difference with respect to the VEM-based technique, mainly due to the low quality of the selected BEM representation.

References

Aavatsmark, I.: An introduction to multipoint flux approximations for quadrilateral grids. Comput. Geosci. **6**(3), 405–432 (2002). https://doi.org/10.1023/A:1021291114475

Ahmad, B., Alsaedi, A., Brezzi, F., Marini, L.D., Russo, A.: Equivalent projectors for virtual element methods. Comput. Math. Appl. **66**, 376–391 (2013)

Ahmed, R., Edwards, M., Lamine, S., Huisman, B., Pal, M.: Control-volume distributed multi-point flux approximation coupled with a lower-dimensional fracture model. J. Comput. Phys. **284**, 462–489 (2015). https://doi.org/10.1016/j.jcp.2014.12.047

Al-Hinai, O., Srinivasan, S., Wheeler, M.F.: Domain decomposition for flow in porous media with fractures. In: SPE Reservoir Simulation Symposium 23–25 February 2013, Houston, Texas, USA, Society of Petroleum Engineers (2015). https://doi.org/10.2118/173319-MS

Alboin, C., Jaffré, J., Roberts, J., Serres, C.: Domain decomposition for flow in porous media with fractures. In: Lai, C.H., Bjorstad, P.E., Cross, M., Widlund, O.B. (eds.) Proceedings of the 11th International Conference on Domain Decomposition Methods in Greenwich, pp. 371–379 (1999)

Aldejain, A.: Implementation of dual porosity model in a chemical flooding simulator. Ph.D. thesis, The University of Texas at Austin, Texas (1999)

Angot, P., Boyer, F., Hubert, F.: Asymptotic and numerical modelling of flows in fractured porous media. ESAIM: M2AN **43**(2), 239–275 (2009). https://doi.org/10.1051/m2an/2008052

Antonietti, P., Formaggia, L., Scotti, A., Verani, M., Verzott, N.: Mimetic finite difference approximation of flows in fractured porous media. ESAIM: M2AN **50**(3), 809–832 (2016). https://doi.org/10.1051/m2an/2015087

Bai, M., Ma, Q., Roegiers, J.: A nonlinear dual-porosity model. Appl. Math. Model. **18**(11), 602–610 (1994). https://doi.org/10.1016/0307-904X(94)90318-2

Beirão da Veiga, L., Brezzi, F., Cangiani, A., Manzini, G., Marini, L.D., Russo, A.: Basic principles of virtual element methods. Math. Models Methods Appl. Sci. **23**(01), 199–214 (2013a). https://doi.org/10.1142/S0218202512500492

Beirão da Veiga, L., Brezzi, F., Marini, L.D.: Virtual elements for linear elasticity problems. SIAM J. Numer. Anal. **51**(2), 794–812 (2013b). https://doi.org/10.1137/120874746

Beirão Da Veiga, L., Brezzi, F., Marini, L.D., Russo, A.: The hitchhiker's guide to the virtual element method. Math Models Methods Appl Sci **24**(8), 1541–1573 (2014)

Beirão da Veiga, L., Lipnikov, K., Manzini, G.: The Mimetic Finite Difference Method for Elliptic Problems, Modeling, Simulation and Applications, vol 11. Springer, Berlin (2014).

Beirão da Veiga, L., Brezzi, F., Marini, L.D., Russo, A.: Virtual element methods for general second order elliptic problems on polygonal meshes. Math. Models Methods Appl. Sci. **26**(04), 729–750 (2015). https://doi.org/10.1142/S0218202516500160

Benedetto, M., Berrone, S., Pieraccini, S., Scialò, S.: The virtual element method for discrete fracture network simulations. Comput. Methods Appl. Mech. Eng. **280**, 135–156 (2014). https://doi.org/10.1016/j.cma.2014.07.016

Benedetto, M., Berrone, S., Borio, A., Pieraccini, S., Scialò, S.: A hybrid mortar virtual element method for discrete fracture network simulations. J. Comput. Phys. **306**, 148–166 (2016a). https://doi.org/10.1016/j.jcp.2015.11.034

Benedetto, M., Berrone, S., Scialò, S.: A globally conforming method for solving flow in discrete fracture networks using the virtual element method. Finite Elem. Anal. Des. **109**, 23–36 (2016b). https://doi.org/10.1016/j.finel.2015.10.003

Benedetto, M.F., Berrone, S., Borio, A.: The Virtual Element Method for underground flow simulations in fractured media. Advances in Discretization Methods, SEMA SIMAI Springer Series, vol. 12, pp. 167–186. Springer International Publishing, Basel (2016c)

Berrone, S., Borio, A.: Orthogonal polynomials in badly shaped polygonal elements for the Virtual Element Method. Finite Elements Anal. Des. **129**, 14–31 (2017a). https://doi.org/10.1016/j.finel.2017.01.006

Berrone, S., Borio, A.: A residual a posteriori error estimate for the virtual element method. Math. Models Methods Appl. Sci. **27**(08), 1423–1458 (2017b). https://doi.org/10.1142/S0218202517500233

Berrone, S., Pieraccini, S., Scialò, S.: On simulations of discrete fracture network flows with an optimization-based extended finite element method. SIAM J. Sci. Comput. **35**(2), A908–A935 (2013a). https://doi.org/10.1137/120882883

Berrone, S., Pieraccini, S., Scialò, S.: A PDE-constrained optimization formulation for discrete fracture network flows. SIAM J. Sci. Comput. **35**(2), B487–B510 (2013b). https://doi.org/10.1137/120865884

Berrone, S., Fidelibus, C., Pieraccini, S., Scialò, S.: Simulation of the steady-state flow in discrete fracture networks with non-conforming meshes and extended finite elements. Rock Mech. Rock Eng. **47**(6), 2171–2182 (2014a). https://doi.org/10.1007/s00603-013-0513-5

Berrone, S., Pieraccini, S., Scialò, S.: An optimization approach for large scale simulations of discrete fracture network flows. J. Comput. Phys. **256**, 838–853 (2014b). https://doi.org/10.1016/j.jcp.2013.09.028

Berrone, S., Pieraccini, S., Scialò, S., Vicini, F.: A parallel solver for large scale DFN flow simulations. SIAM J. Sci. Comput. **37**(3), C285–C306 (2015). https://doi.org/10.1137/140984014

Berrone, S., Borio, A., Scialò, S.: A posteriori error estimate for a PDE-constrained optimization formulation for the flow in DFNs. SIAM J. Numer. Anal. **54**(1), 242–261 (2016a). https://doi.org/10.1137/15M1014760

Berrone, S., Pieraccini, S., Scialò, S.: Towards effective flow simulations in realistic discrete fracture networks. J. Comput. Phys. **310**, 181–201 (2016b). https://doi.org/10.1016/j.jcp.2016.01.009

Berrone, S., Pieraccini, S., Scialò, S.: Flow simulations in porous media with immersed intersecting fractures. J. Comput. Phys. **345**, 768–791 (2017). https://doi.org/10.1016/j.jcp.2017.05.049

Brebbia, C., Telles, J., Wrobel, L.: Boundary Element Techniques, Theory and Apllications in Engineering. Springer, Berlin (1984)

Brenner, K., Groza, M., Guichard, C., Lebeau, G., Masson, R.: Gradient discretization of hybrid dimensional darcy flows in fractured porous media. Numerische Mathematik **134**(3), 569–609 (2016a). https://doi.org/10.1007/s00211-015-0782-x

Brenner, K., Hennicker, J., Masson, R., Samier, P.: Gradient discretization of hybrid-dimensional darcy flow in fractured porous media with discontinuous pressures at matrix-fracture interfaces. IMA J. Numer. Anal. (2016b). https://doi.org/10.1093/imanum/drw044

Brezzi, F., Falk, R.S., Marini, L.D.: Basic principles of mixed virtual element methods. ESAIM Math. Model. Numer. Anal. **48**(4), 1227–1240 (2014). https://doi.org/10.1051/m2an/2013138

Chave, F., Di Pietro, D., Formaggia, L.: A hybrid high-order method for darcy flows in fractured porous media. SIAM J. Sci. Comput. **40**(2), A1063–A1094 (2018). https://doi.org/10.1137/17M1119500

D'Angelo, C., Scotti, A.: A mixed finite element method for darcy flow in fractured porous media with non-matching grids. ESAIM: M2AN **46**(2), 465–489 (2012). https://doi.org/10.1051/m2an/2011148

Faille, I., Fumagalli, A., Jaffré, J., Roberts, J.E.: Model reduction and discretization using hybrid finite volumes for flow in porous media containing faults. Comput. Geosci. **20**(2), 317–339 (2016). https://doi.org/10.1007/s10596-016-9558-3

Fidelibus, C., Barla, G., Cravero, M.: A mixed solution for two-dimensional unsteady flow in fractured porous media. Int. J. Numer. Anal. Methods Geomech. **21**(9), 619–633 (1997)

Flemisch, B., Berre, I., Boon, W., Fumagalli, A., Schwenck, N., Scotti, A., Stefansson, I., Tatomir, A.: Benchmarks for single-phase flow in fractured porous media. Adv. Water Resour. **111**, 239–258 (2018). https://doi.org/10.1016/j.advwatres.2017.10.036

Formaggia, L., Scotti, A., Sottocasa, F.: Analysis of a Mimetic Finite Difference approximation of flows in fractured media. Technical Report 49/2016, MOX, Mathematical Department, Politecnico di Milano (2016)

Fries, T.P., Belytschko, T.: The extended/generalized finite element method: an overview of the method and its applications. Int. J. Numer. Methods Eng. **84**(3), 253–304 (2010). https://doi.org/10.1002/nme.2914

Frih, N., Martin, V., Roberts, J.E., Saâda, A.: Modeling fractures as interfaces with nonmatching grids. Comput. Geosci. **16**(4), 1043–1060 (2012). https://doi.org/10.1007/s10596-012-9302-6

Fumagalli, A., Keilegavlen, E.: Dual virtual element method for discrete fractures networks. SIAM J. Sci. Comput. **40**, B228–B258 (2018). https://doi.org/10.1137/16M1098231

Fumagalli, A., Scotti, A.: A numerical method for two-phase flow in fractured porous media with non-matching grids. Adv. Water Resour. **62**, 454–464 (2013). https://doi.org/10.1016/j.advwatres.2013.04.001

Fumagalli, A., Keilegavlen, E., Scialò, S. (2018) Conforming, non-conforming and non-matching discretization couplings in discrete fracture network simulations. arXiv:1803.01732
Hajibeygi, H., Karvounis, D., Jenny, P.: A hierarchical fracture model for the iterative multiscale finite volume method. J. Comput. Phys. **230**(24), 8729–8743 (2011). https://doi.org/10.1016/j.jcp.2011.08.021
Huyakorn, P., Pinder, G.: The Computational Methods in Subsurface Flow. Academic Press, Cambridge (1983). doi: 10.1016/B978-0-12-363480-1.50001-4.
Hyman, J.D., Karra, S., Makedonska, N., Gable, C.W., Painter, S.L., Viswanathan, H.S.: dfnworks: a discrete fracture network framework for modeling subsurface flow and transport. Comput. Geosci. **84**, 10–19 (2015). https://doi.org/10.1016/j.cageo.2015.08.001
Kazemi, H., Gilman, J.: Multiphase flow in fractured petroleum reservoirs. In: Bear, J., Tsang, C., de Marsily, G. (eds.) Flow and Contaminant Transport in Fractured Rock, pp. 267–323. AcademicPress, San Diego (1993)
Makedonska, N., Painter, S.L., Bui, Q.M., Gable, C.W., Karra, S.: Particle tracking approach for transport in three-dimensional discrete fracture networks. Comput. Geosci. **19**(5), 1123–1137 (2015). https://doi.org/10.1007/s10596-015-9525-4
Martin, V., Jaffré, J., Roberts, J.E.: Modeling fractures and barriers as interfaces for flow in porous media. SIAM J. Sci. Comput. **26**(5), 1667–1691 (2005). https://doi.org/10.1137/S1064827503429363
Reichenberger, V., Jakobs, H., Bastian, P., Helmig, R.: A mixed-dimensional finite volume method for two-phase flow in fractured porous media. Adv. Water Resour. **29**(7), 1020–1036 (2006). https://doi.org/10.1016/j.advwatres.2005.09.001
Sandve, T., Berre, I., Nordbotten, J.: An efficient multi-point flux approximation method for discrete fracture-matrix simulations. J. Comput. Phys. **231**(9), 3784–3800 (2012). https://doi.org/10.1016/j.jcp.2012.01.023
Shapiro, A.M., Andersson, J.: Steady state fluid response in fractured rock: a boundary element solution for a coupled, discrete fracture continuum model. Water Resour. Res. **19**(4), 959–969 (1983). https://doi.org/10.1029/WR019i004p00959
Warren, M.A., Root, P.J.: The behavior of naturally fractured reservoirs. Soc. Petrol. Eng. J. **3**(3), 245–279 (1963)

Affiliations

S. Berrone[1] · A. Borio[1] · C. Fidelibus[2] · S. Pieraccini[3] · S. Scialò[1] · F. Vicini[1]

✉ S. Berrone
stefano.berrone@polito.it

A. Borio
andrea.borio@polito.it

C. Fidelibus
corrado.fidelibus@unisalento.it

S. Pieraccini
sandra.pieraccini@polito.it

S. Scialò
stefano.scialo@polito.it

F. Vicini
fabio.vicini@polito.it

[1] Dipartimento di Scienze Matematiche, Politecnico di Torino, Turin, Italy

[2] Dipartimento di Ingegneria dell'Innovazione, Università del Salento, Lecce, Italy

[3] Dipartimento di Ingegneria Meccanica e Aerospaziale, Politecnico di Torino, Turin, Italy

GEM - International Journal on Geomathematics
https://doi.org/10.1007/s13137-019-0118-6

ORIGINAL PAPER

Two-phase Discrete Fracture Matrix models with linear and nonlinear transmission conditions

Joubine Aghili[1,2] · Konstantin Brenner[1,2] · Julian Hennicker[3] · Roland Masson[1,2] · Laurent Trenty[4]

Received: 10 April 2018 / Accepted: 17 December 2018

Abstract

This work deals with two-phase Discrete Fracture Matrix models coupling the two-phase Darcy flow in the matrix domain to the two-phase Darcy flow in the network of fractures represented as co-dimension one surfaces. Two classes of such hybrid-dimensional models are investigated either based on nonlinear or linear transmission conditions at the matrix–fracture interfaces. The linear transmission conditions include the cell-centred upwind approximation of the phase mobilities classically used in the porous media flow community as well as a basic extension of the continuous phase pressure model accounting for fractures acting as drains. The nonlinear transmission conditions at the matrix–fracture interfaces are based on the normal flux continuity equation for each phase using additional interface phase pressure unknowns. They are compared both in terms of accuracy and numerical efficiency to a reference equi-dimensional model for which the fractures are represented as full-dimensional subdomains. The discretization focuses on Finite Volume cell-centred Two-Point Flux Approximation which is combined with a local nonlinear solver allowing to eliminate efficiently the additional matrix–fracture interfacial unknowns together with the nonlinear transmission conditions. 2D numerical experiments illustrate the better accuracy provided by the nonlinear transmission conditions compared to their linear approximations with a moderate computational overhead obtained thanks to the local nonlinear elimination at the matrix–fracture interfaces. The numerical section is complemented by a comparison of the reduced models on a 3D test case using the Vertex Approximate Gradient scheme.

Keywords Two-phase Darcy flow · Discrete fracture network · Discrete fracture matrix model · Nonlinear transmission condition · Finite volume discretization

Mathematics Subject Classification 65M08 · 65N08 · 76S05

✉ Konstantin Brenner
konstantin.brenner@unice.fr

Extended author information available on the last page of the article

1 Introduction

This work deals with numerical modeling of two-phase flow in fractured porous media, for which the fracture network is represented as a manifold of co-dimension one with respect to matrix domain. This approach gives rise to so-called Discrete Fracture Matrix (DFM) models also termed hybrid-dimensional models, which differ among themselves mostly by the interface condition imposed at matrix–fracture (*mf*) interfaces. Various interface conditions have been proposed in the literature, both for single-phase and multi-phase flows. For single-phase flows there are two major approaches—the first, designed for modeling highly conductive fractures and referred to as continuous pressure model (Alboin et al. 2002; Brenner et al. 2016b), assumes the continuity of the fluid pressure at the *mf* interface; the second approach, referred to as discontinuous pressure model (Flauraud et al. 2003; Jaffré et al. 2005; Angot et al. 2009; Brenner et al. 2016a), allows to represent fractures acting as permeability barriers by imposing Robin-type condition at *mf* interface.

When the modeling of multi-phase flow is concerned, three major types of models may be distinguished. The first and most common type is based on the straightforward adaptation of single-phase continuous pressure model to the multi-phase setting (see Bogdanov et al. 2003; Reichenberger et al. 2006; Monteagudo and Firoozabadi 2007; Hoteit et al. 2008; Brenner et al. 2015, 2017), it assumes the continuity of each phase pressure at *mf* interfaces. As for single-phase flow this approach can not account for fractures acting as barriers.

Another existing type of models, accounting for both drains or barriers, amounts to eliminate the interfacial phase pressures using, for each phase, the linear single-phase Darcy flux conservation equation at the *mf* interfaces. It is usually combined with Two-Point (Karimi-Fard et al. 2004) or MultiPoint (Tunc et al. 2012; Sandve et al. 2012; Ahmed et al. 2015) cell-centred finite volume schemes for which the interfacial unknowns can be easily eliminated when building the single phase Darcy flux transmissibilities. Both this type of model and the previous continuous pressure model are termed in the following hybrid-dimensional *mf* linear models in the sense that they are based on linear transmission conditions for the phase pressures.

Finally a few works have considered nonlinear interface conditions, namely Brenner et al. (2018), Droniou et al. (2018) using a formulation based on phase pressure and upwinding of the mobilities, and Jaffré et al. (2011), Ahmed et al. (2017) using a global pressure formulation. Such nonlinear interface conditions account for the permeability jump as well as for the discontinuity of phase mobility at *mf* interface. This type of model is termed in the following hybrid-dimensional *mf* nonlinear model since it is based on the nonlinear flux continuity equations at the *mf* interface.

This article is a follow-up of the work presented in Brenner et al. (2018) on the comparison of the hybrid-dimensional *mf* nonlinear and linear models using a reference equi-dimensional solution for which the fractures are represented as full-dimensional subdomains. The work presented in Brenner et al. (2018) is limited to rather large apertures due to the difficulty to obtain reference equi-dimensional solutions using nodal based discretizations like the Vertex Approximate Gradient (VAG) scheme. The present work focuses on cell-centred discretizations with Two-Point Flux Approximations (TPFA) on orthogonal meshes in order to obtain reference equi-dimensional solutions for small apertures. We will show in particular that, in contrast with single-phase flow (see Flemisch et al. 2018), the continuous pressure model may fail even when fractures are highly conductive. This typically happens when the fracture network is dry, and hence, due to the low relative permeability the wetting phase pressure is no longer continuous across the fracture. We also investigate the effect of gravitational phase segregation within the fracture network, which can only be captured by the nonlinear *mf* model. We also compare the efficiency of the models in terms of linear and nonlinear convergence as well as in terms of overall CPU time. The TPFA discretization of the hybrid-dimensional *mf* nonlinear model is combined with a new nonlinear interface solver in order to eliminate the *mf* interface unknowns. This strategy allows to obtain a better approximation at a moderate additional cost compared with the *mf* linear models.

The remainder of the present article is organized as follows. In Sect. 2 we recall the hybrid-dimensional *mf* nonlinear two-phase flow model from Brenner et al. (2018). In Sect. 3, the cell-centred TPFA discretization of this model is introduced as well as the TPFA discretizations of the *mf* linear two-phase flow models based on linear *mf* interface conditions. In Sect. 4 the local nonlinear interface problem is studied and an algorithm for finding its solutions is presented. In Sect. 5, numerical comparisons on 2D test cases of the hybrid dimensional *mf* nonlinear and linear models are considered using equi-dimensional reference solutions. Section 6 presents the comparison of the models for a 3D test case using the Vertex Approximate Gradient scheme introduced in Brenner et al. (2018) for hybrid-dimensional two phase Darcy flow models.

2 Continuous model

Let Ω be an open bounded domain of $\mathbb{R}^d$, $d = 2, 3$ assumed to be polyhedral for $d = 3$ (and polygonal for $d = 2$). To fix ideas the dimension will be fixed to $d = 3$ when it needs to be specified, for instance in the naming of the geometrical objects or for the space discretization in the next section. The adaptations to the case $d = 2$ are straightforward. Let $\overline{\Gamma} = \bigcup_{i \in I} \overline{\Gamma}_i$ denotes the network of fractures $\Gamma_i \subset \Omega$, $i \in I$, such that each Γ_i is a planar polygonal simply connected open domain included in some plane of $\mathbb{R}^d$. Without restriction of generality, we will assume that the fractures may intersect exclusively at their boundaries (see Fig. 1), that is for any $i, j \in I, i \neq j$ one has $\Gamma_i \cap \Gamma_j = \emptyset$, but not necessarily $\overline{\Gamma}_i \cap \overline{\Gamma}_j = \emptyset$.

In the matrix domain Ω (resp. in the fracture network Γ), we denote by $\phi_m(\mathbf{x})$ (resp. $\phi_f(\mathbf{x})$) the porosity, by $\Lambda_m(\mathbf{x})$ (resp. $\Lambda_f(\mathbf{x})$) the permeability (resp. tangential permeability) tensor. The permeability tensor in the fracture network is assumed to have the fracture normal vector as principal direction and $\lambda_{f,n}(\mathbf{x})$ denotes the corresponding fracture normal permeability. The thickness of the fractures is denoted by $d_f(\mathbf{x})$ for $\mathbf{x} \in \Gamma$.

For each phase $\alpha = w, nw$ (where w stands for "wetting" and nw for "non-wetting" phases) we denote by $k^{\alpha}_{r,m}(s)$ (resp. $k^{\alpha}_{r,f}(s)$), the phase relative permeabilities and by $S^{\alpha}_m(\pi)$ (resp. $S^{\alpha}_f(\pi)$) the capillary pressure–saturation curves. For simplicity and since we focus on matrix–fracture network interfaces, its is assumed that both the matrix and fracture domains are homogeneous with respect to capillary pressure and relative

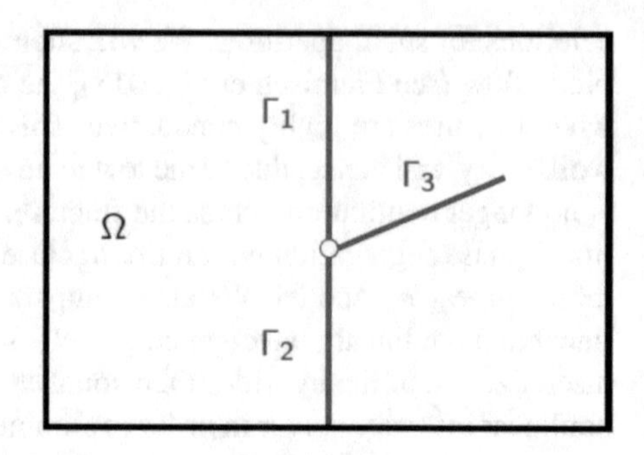

Fig. 1 Example of a 2D domain Ω with 3 intersecting fractures $\Gamma_i, i = 1, 2, 3$

permeability laws. We will assume that for each $j = m, f$ the capillary pressure–saturation and relative permeability curves satisfy

- S_j^{nw} is a non-decreasing continuous function from $\mathbb{R}$ onto [0, 1) such that there exist the so-called entry pressure $\pi_{j,e}$ satisfying $S_j^{nw}(\pi \leq \pi_{j,e}) = 0$. The function S_j^w satisfy $S_j^w(\pi) = 1 - S_j^{nw}(\pi)$ for all $\pi \in \mathbb{R}$.
- $k_{r,j}^\alpha$, for each $\alpha = w, nw$, is a non-decreasing continuous function on [0, 1] such that $k_{r,j}^\alpha(0) = 0$ and $k_{r,j}^\alpha(1) > 0$.

Typical examples of such relations are given by Brooks and Corey (1964), Bentsen and Anli (1977) or Van Genuchten–Mualem (Mualem 1976; Van Genuchten 1980) laws.

For $\alpha = w, nw$, we will denote by ρ^α the phase densities and by μ^α the phase dynamic viscosities, which for the sake of clarity, are assumed constant. The mobilities of the phases (ratio of the phase relative permeability to the phase viscosity) will be denoted by k_m^α and k_f^α. We denote by p_m^α (resp. p_f^α) the pressure of phase $\alpha = w, nw$ and by s_m^α (resp. s_f^α) the saturation of phase $\alpha = w, nw$ in the matrix (resp. the fracture network) domain.

2.1 Matrix equations

The Darcy flux of phase $\alpha = w, nw$ in the matrix domain is defined by

$$\mathbf{q}_m^\alpha = -k_m^\alpha(s_m^\alpha)\Lambda_m(\nabla p_m^\alpha - \rho^\alpha \mathbf{g}), \tag{1}$$

where $\mathbf{g} = -g\nabla z$ stands for the gravity vector. The flow in the matrix domain is described for both phases $\alpha = w, nw$ by the volume balance equation

$$\phi_m \partial_t s_m^\alpha + \mathrm{div}(\mathbf{q}_m^\alpha) = 0, \tag{2}$$

and the macroscopic capillary pressure law

$$s_m^\alpha = S_m^\alpha(p_m^{nw} - p_m^w). \tag{3}$$

2.2 Fracture network equations

For each fracture Γ_i we identify the two sides $+$ and $-$ of Γ_i in $\Omega \setminus \overline{\Gamma}$. We formally denote by γ_i^+ and γ_i^- the corresponding trace operators. The unit normal vectors at Γ_i outward to the side $+$ (resp. $-$) are denoted by $\mathbf{n}_i^+$ (resp. $\mathbf{n}_i^-$), and $\mathbf{q}_m^{\alpha,+} \cdot \mathbf{n}_i^+$ (resp. $\mathbf{q}_m^{\alpha,-} \cdot \mathbf{n}_i^-$) formally denotes the normal trace of the matrix fluxes at the $+$ side (resp. $-$ side) of the fracture Γ_i oriented outward to the matrix. In addition, for $i \in I$, we denote by ∇_{τ_i} the tangential gradient, and by div_{τ_i} the tangential divergence operators.

The Darcy flux of phase $\alpha = w, nw$ in the fracture Γ_i integrated over the width of the fracture is defined by

$$\mathbf{q}_{f,i}^{\alpha} = -d_f k_f^{\alpha}(s_f^{\alpha}) \Lambda_f (\nabla_{\tau_i} p_f^{\alpha} - \rho^{\alpha} \mathbf{g}_{\tau_i}), \tag{4}$$

with $\mathbf{g}_{\tau_i} = \mathbf{g} - (\mathbf{g} \cdot \mathbf{n}_i^+)\mathbf{n}_i^+$. The flow in each fracture Γ_i is described for both phases $\alpha = w, nw$ by

$$d_f \phi_f \partial_t s_f^{\alpha} + \mathrm{div}_{\tau_i}(\mathbf{q}_{f,i}^{\alpha}) - \mathbf{q}_m^{\alpha,+} \cdot \mathbf{n}_i^+ - \mathbf{q}_m^{\alpha,-} \cdot \mathbf{n}_i^- = 0, \tag{5}$$

and

$$s_f^{\alpha} = S_f^{\alpha}(p_f^{nw} - p_f^{w}). \tag{6}$$

2.3 Matrix–fracture interface conditions

The matrix and the fracture equations are coupled by Robin boundary conditions which make use of two-point approximations of the normal fluxes within the fracture. Let us first define for $i \in I$ and for both phases $\alpha = w, nw$ the single phase normal fluxes in the fracture

$$V_{f,i,n}^{\alpha,\pm} = \lambda_{f,n} \left(\frac{\gamma_i^{\pm} p_m^{\alpha} - p_f^{\alpha}}{d_f/2} - \rho^{\alpha} \mathbf{g} \cdot \mathbf{n}_i^{\pm} \right). \tag{7}$$

For any $a \in \mathbb{R}$, let us set $a^+ = \max\{0, a\}$ and $a^- = -(-a)^+$, the condition coupling matrix and fracture unknowns then reads

$$\mathbf{q}_m^{\alpha,\pm} \cdot \mathbf{n}_i^{\pm} = k_f^{\alpha}(S_f^{\alpha}(\gamma_i^{\pm} p_m^{nw} - \gamma_i^{\pm} p_m^{w}))(V_{f,i,n}^{\alpha,\pm})^+ + k_f^{\alpha}(s_f^{\alpha})(V_{f,i,n}^{\alpha,\pm})^-. \tag{8}$$

The hybrid dimensional two-phase flow model looks for $(p_m^{\alpha}, p_f^{\alpha})_{\alpha=w,nw}$ satisfying (1)–(8). In addition to (1)–(8) we prescribe no-flux boundary conditions at the tips of the immersed fractures, that is to say on $\partial\Gamma \setminus \partial\Omega$, and the flux conservation and pressure continuity conditions at the fracture intersections. We refer to Brenner et al. (2015) for a more detailed presentation of those conditions. Finally, one should provide some appropriate initial and boundary data.

3 Discrete models

The approximation of the diffusive fluxes relies on *Two-Point Flux Approximation*. This approximation is consistent if the permeability tensor is isotropic and if the mesh satisfies the so-called orthogonality condition (see e.g. Eymard et al. 2014). Below we introduce some notations related to the space discretization, for the sake of clarity we present the discretization for $d = 3$, the extension to $d = 2$ is straightforward.

3.1 Space discretization

We will denote by $\mathcal{M}$ the set of disjoint open polyhedral cells, by $\mathcal{F}$ the set of faces and by $\mathcal{E}$ the set of edges of the mesh. For each cell $K \in \mathcal{M}$, we denote by $\mathcal{F}_K \subset \mathcal{F}$ the set of its faces and for each face $\sigma \in \mathcal{F}$, we denote by $\mathcal{E}_\sigma$ the set of its edges. In addition, the set of cells sharing a given face $\sigma \in \mathcal{F}$ is denoted by $\mathcal{M}_\sigma$, and the set of faces sharing a given edge $e \in \mathcal{E}$ is denoted by $\mathcal{F}_e$.

The mesh is supposed to be conforming w.r.t. the fracture network $\overline{\Gamma}$ in the sense that, for each $i \in I$, there exists a subset $\mathcal{F}_{\Gamma_i}$ of $\mathcal{F}$ such that $\overline{\Gamma}_i = \bigcup_{\sigma \in \mathcal{F}_{\Gamma_i}} \overline{\sigma}$. We will denote by $\mathcal{F}_\Gamma$ the subset $\bigcup_{i \in I} \mathcal{F}_{\Gamma_i}$ of fracture faces.

For all $K \in \mathcal{M}$ and $\sigma \in \mathcal{F}_K$, we denote by $\mathbf{n}_{K\sigma}$ the unit outward normal vector to σ w.r.t K, and similarly, for all $\sigma \in \mathcal{F}_\Gamma$ and $e \in \mathcal{E}$, we denote by $\mathbf{n}_{\sigma e}$ the unit outward normal vector to e w.r.t σ. Finally, we denote by $|\sigma|$, $\sigma \in \mathcal{F}$ the $d-1$ dimensional measure of σ and by $|e|$, $e \in \mathcal{E}$ the $d-2$ dimensional measure of e.

The degrees of freedom of the discretization scheme are associated with the "cell centres" $\mathbf{x}_K \in K$, the "centres" of the fracture faces $\mathbf{x}_\sigma \in \sigma$, and the "centres" $\mathbf{x}_e \in e$ of certain edges of the mesh. More precisely, in order to deal with the intersection of multiple fractures at a given edge $e \in \mathcal{E}$ we associate to it a degree of freedom and a certain porous volume. We will denote by $\mathcal{E}_\star$ the subset of $\mathcal{E}$ such that for all $e \in \mathcal{E}_\star$ the set $\mathcal{F}_\Gamma \cap \mathcal{F}_e$ has at least 3 elements. The space of degrees of freedom is defined by

$$X_\mathcal{D} = \{p_K, p_\sigma, p_e \in \mathbb{R}, K \in \mathcal{M}, \sigma \in \mathcal{F}_\Gamma, e \in \mathcal{E}_\star\}.$$

3.1.1 Orthogonality condition

Let us denote by $\mathbf{x}_{K\sigma}$ the projection of $\mathbf{x}_K$ on the face $\sigma \in \mathcal{F}_K$ and by $\mathbf{x}_{\sigma e}$ the projection of $\mathbf{x}_\sigma$ on the edge $e \in \mathcal{E}_\sigma$. We assume that the mesh satisfies the following conditions

- for any interior face σ and $K, L \in \mathcal{M}$ such that $\sigma = \mathcal{F}_K \cap \mathcal{F}_L$ one has $\mathbf{x}_{K\sigma} = \mathbf{x}_{L\sigma}$,
- for any fracture face $\sigma \in \mathcal{F}_\Gamma$ and K such that $\sigma \in \mathcal{F}_K$ one has $\mathbf{x}_{K\sigma} = \mathbf{x}_\sigma$,

and

- for any interior fracture edge e and $\sigma, \sigma' \in \mathcal{F}_\Gamma$ such that $e = \mathcal{E}_\sigma \cap \mathcal{E}_{\sigma'}$ one has $\mathbf{x}_{\sigma e} = \mathbf{x}_{\sigma' e}$,
- for any intersection edge $e \in \mathcal{E}_\star$ and $\sigma \in \mathcal{F}_\Gamma$ such that $e \in \mathcal{E}_\sigma$ one has $\mathbf{x}_{\sigma e} = \mathbf{x}_e$.

In addition we assume that the porosity ϕ_m (resp. ϕ_f) is cell-wise (resp. face-wise) constant, and that the permeability tensor Λ_m (resp. Λ_f) is cell-wise (resp. face-wise) constant and isotropic.

3.2 Discrete equations

For $N \in \mathbb{N}^*$, let us consider the discretization $t^0 = 0 < t^1 < \cdots < t^{n-1} < t^n < \cdots < t^N = T$ of the time interval $[0, T]$. We denote the time steps by $\Delta t^n = t^n - t^{n-1}$ for all $n = 1, \ldots, N$. For $K \in \mathcal{M}$, $\sigma \in \mathcal{F}_K$ and $\alpha = w, nw$, we denote by $F_{K\sigma}^{\alpha,n}$ the approximation of the Darcy flux through the face σ of the cell K

$$F_{K\sigma}^{\alpha,n} \approx \frac{1}{\Delta t^n} \int_{t_{n-1}}^{t_n} \int_\sigma \Lambda_m(\mathbf{x}_K) k_m^\alpha(s_m^\alpha) \left(\nabla p_m^\alpha - \rho^\alpha \mathbf{g}\right) \cdot \mathbf{n}_{K\sigma} d\sigma(\mathbf{x}) dt.$$

Similarly, for $\sigma \in \mathcal{F}_{\Gamma_i}$, $i \in I$, $e \in \mathcal{E}_\sigma$ and $\alpha = w, nw$, we denote by $F_{\sigma e}^{\alpha,n}$ the approximation of the Darcy flux through the edge e of the fracture face σ

$$F_{\sigma e}^{\alpha,n} \approx \frac{1}{\Delta t^n} \int_{t_{n-1}}^{t_n} \int_e d_f(\mathbf{x}) \Lambda_f(\mathbf{x}_\sigma) k_f^\alpha(s_f^\alpha) \left(\nabla_{\tau_i} p_f^\alpha - \rho^\alpha \mathbf{g}_{\tau_i}\right) \cdot \mathbf{n}_{\sigma e} de(\mathbf{x}) dt.$$

Those discrete fluxes are going to be defined later in this section. Let us first present the general layout of the numerical scheme, for this purpose we define the porous volume associated with the degrees of freedom. For all $K \in \mathcal{M}$ we simply set

$$\phi_K = \int_K \phi_m(\mathbf{x}) d\mathbf{x},$$

while the fracture porous volume has to be distributed between the degrees of freedom associated with fracture faces $\sigma \in \mathcal{F}_\Gamma$ and intersection edges $e \in \mathcal{E}_\star$. For each $\sigma \in \mathcal{F}_\Gamma$ such that $\mathcal{E}_\sigma \cap \mathcal{E}_\star \neq \emptyset$ we define a set of non-negative volume fractions $(\alpha_{\sigma e})_{e \in \mathcal{E}_\sigma \cap \mathcal{E}_\star}$ satisfying $\sum_{e \in \mathcal{E}_\sigma \cap \mathcal{E}_\star} \alpha_{\sigma e} < 1$, and we set

$$\phi_\sigma = \left(1 - \sum_{e \in \mathcal{E}_\sigma \cap \mathcal{E}_\star} \alpha_{\sigma e}\right) \int_\sigma d_f(\mathbf{x}) \phi_f(\mathbf{x}) d\sigma(\mathbf{x}),$$

and

$$\phi_e = \sum_{\sigma \in \mathcal{F}_e \cap \mathcal{F}_\Gamma} \alpha_{\sigma e} \int_\sigma d_f(\mathbf{x}) \phi_f(\mathbf{x}) d\sigma(\mathbf{x}).$$

For each $\sigma \in \mathcal{F}_\Gamma$ such that $\mathcal{E}_\sigma \cap \mathcal{E}_\star = \emptyset$, we simply set $\phi_\sigma = \int_\sigma d_f(\mathbf{x}) \phi_f(\mathbf{x}) d\sigma(\mathbf{x})$. Note that the above conservative distribution of a porous volume from adjacent fracture faces to their intersecting edge $e \in \mathcal{E}_\star$ is a simple and efficient way to circumvent the classical bad conditioning issue induced by fracture intersections. Alternative approaches based on the Star Delta (Karimi-Fard et al. 2004) transformation or a Schur complement eliminating the edge $e \in \mathcal{E}_\star$ unknowns (Sandve et al. 2012) are basically limited to linear problems.

The finite volume scheme looks for $p^{\alpha,n} \in X_\mathcal{D}$, $\alpha = w, nw$, $n = 1, \ldots, N$ such that, for $\alpha = w, nw$, the following set of equations is satisfied:

$$\frac{\phi_K}{\Delta t^n}\left(s_K^{\alpha,n} - s_K^{\alpha,n-1}\right) + \sum_{\sigma\in\mathcal{F}_K\setminus\mathcal{F}_\Gamma} F_{K\sigma}^{\alpha,n} + \sum_{\sigma\in\mathcal{F}_K\cap\mathcal{F}_\Gamma} F_{K\sigma}^{\alpha,n} = 0, \tag{9}$$

for all $K \in \mathcal{M}$, and

$$\frac{\phi_\sigma}{\Delta t^n}\left(s_\sigma^{\alpha,n} - s_\sigma^{\alpha,n-1}\right) + \sum_{e\in\mathcal{E}_\sigma\setminus\mathcal{E}_\star} F_{\sigma e}^{\alpha,n} + \sum_{e\in\mathcal{E}_\sigma\cap\mathcal{E}_\star} F_{\sigma e}^{\alpha,n} - \sum_{K\in\mathcal{M}_\sigma} F_{K\sigma}^{\alpha,n} = 0, \tag{10}$$

for all $\sigma \in \mathcal{F}_\Gamma$, and

$$\frac{\phi_e}{\Delta t^n}\left(s_e^{\alpha,n} - s_e^{\alpha,n-1}\right) - \sum_{\sigma\in\mathcal{F}_e\cap\mathcal{F}_\Gamma} F_{\sigma e}^{\alpha,n} = 0, \tag{11}$$

for all $e \in \mathcal{E}_\star$, with

$$\begin{aligned} s_K^{\alpha,n} &= S_m^\alpha(p_K^{nw,n} - p_K^{w,n}) \quad \text{for all } K \in \mathcal{M}, \\ s_\nu^{\alpha,n} &= S_f^\alpha(p_\nu^{nw,n} - p_\nu^{w,n}) \quad \text{for all } \nu \in \mathcal{F}_\Gamma \cup \mathcal{E}_\star, \end{aligned}$$

where $p_\nu^{\alpha,n}$ denotes the approximation of the matrix (resp. fracture) phase pressure at point $\mathbf{x}_\nu$ for $\nu \in \mathcal{M}$ (resp. $\nu \in \mathcal{F}_\Gamma \cup \mathcal{E}_\star$).

3.3 Discrete fluxes

In this subsection, we give the definition of the discrete fluxes using the following additional notations. For all degrees of freedom $\nu \in \mathcal{M} \cup \mathcal{F}_\Gamma \cup \mathcal{E}_\star$ and $n = 1, \ldots, N$ we denote by $w_\nu^{\alpha,n}$ the approximation of the phase potential at point $\mathbf{x}_\nu$ defined by

$$w_\nu^{\alpha,n} = p_\nu^{\alpha,n} + \rho^\alpha g z_\nu.$$

3.3.1 Matrix–matrix fluxes

Let us consider an interface σ between two neighboring cells $K, L \in \mathcal{M}$ such that $\sigma = \mathcal{F}_K \cap \mathcal{F}_L \neq \emptyset$ and $\sigma \notin \mathcal{F}_\Gamma$. Let us define the half-transmissibilities defined for all $K \in \mathcal{M}$ and $\sigma \in \mathcal{F}_K$ by

$$T_{K\sigma} = \Lambda_m(\mathbf{x}_K)\frac{|\sigma|}{|\mathbf{x}_K - \mathbf{x}_\sigma|}.$$

Let $k_{K\sigma}^{\alpha,n}$ and $w_\sigma^{\alpha,n}$ be some approximate values of the phase mobility and of the phase potential at the interface σ. Then, we set

$$V_{K\sigma}^{\alpha,n} = T_{K\sigma}(w_K^{\alpha,n} - w_\sigma^{\alpha,n}) \quad \text{and} \quad F_{K\sigma}^{\alpha,n} = k_{K\sigma}^{\alpha,n} V_{K\sigma}^{\alpha,n},$$

and flux continuity is imposed at the interface

$$F_{K\sigma}^{\alpha,n} + F_{L\sigma}^{\alpha,n} = 0. \tag{12}$$

Since, except for the *mf* interfaces, the capillary pressure–saturation and relative permeability curves do not depend on the space variable, the saturation jump at the interface σ can be neglected and it is natural to assume that $k_{K\sigma}^{\alpha,n} = k_{L\sigma}^{\alpha,n}$. Hence equation (12) reduces to

$$V_{K\sigma}^{\alpha,n} + V_{L\sigma}^{\alpha,n} = 0, \tag{13}$$

and the interface potential $w_{\sigma}^{\alpha,n}$ can be linearly eliminated from (13) yielding

$$V_{K\sigma}^{\alpha,n} = \frac{T_{K\sigma} T_{L\sigma}}{T_{K\sigma} + T_{L\sigma}} \left(w_K^{\alpha,n} - w_L^{\alpha,n} \right). \tag{14}$$

Using the upstream approximation of the relative mobility, we come up with the following definition of the discrete flux

$$F_{K\sigma}^{\alpha,n} = k_m^{\alpha}(s_K^{\alpha,n}) \left(V_{K\sigma}^{\alpha,n} \right)^+ + k_m^{\alpha}(s_L^{\alpha,n}) \left(V_{K\sigma}^{\alpha,n} \right)^- . \tag{15}$$

3.3.2 Fracture–fracture fluxes

Similarly to the previous matrix–matrix fluxes, we define for all $\sigma \in \mathcal{F}_\Gamma$ and $e \in \mathcal{E}_\sigma$ the half-transmissibilities

$$T_{\sigma e} = \Lambda_f(\mathbf{x}_\sigma) \frac{d_f(\mathbf{x}_\sigma)|e|}{|\mathbf{x}_\sigma - \mathbf{x}_e|}.$$

Let σ and σ' be a couple of intersecting fracture faces and let $e = \overline{\sigma} \cap \overline{\sigma}' \notin \mathcal{E}_\star$, the flux between σ and σ' is defined by

$$F_{\sigma e}^{\alpha,n} = k_f^{\alpha}(s_\sigma^{\alpha,n}) \left(V_{\sigma e}^{\alpha,n} \right)^+ + k_f^{\alpha}(s_{\sigma'}^{\alpha,n}) \left(V_{\sigma e}^{\alpha,n} \right)^- ,$$

with

$$V_{\sigma e}^{\alpha,n} = \frac{T_{\sigma e} T_{\sigma' e}}{T_{\sigma e} + T_{\sigma e}} \left(w_\sigma^{\alpha,n} - w_{\sigma'}^{\alpha,n} \right).$$

Let $\sigma \in \mathcal{F}_\Gamma$ and $e \in \mathcal{E}_\sigma \cap \mathcal{E}_\star$, we set

$$F_{\sigma e}^{\alpha,n} = k_f^{\alpha}(s_\sigma^{\alpha,n}) \left(V_{\sigma e}^{\alpha,n} \right)^+ + k_f^{\alpha}(s_e^{\alpha,n}) \left(V_{\sigma e}^{\alpha,n} \right)^- , \tag{16}$$

with

$$V_{\sigma e}^{\alpha,n} = T_{\sigma e} \left(w_\sigma^{\alpha,n} - w_e^{\alpha,n} \right).$$

3.3.3 Matrix–fracture fluxes with nonlinear transmission conditions

For $K \in \mathcal{M}$ and $\sigma \in \mathcal{F}_K \cap \mathcal{F}_\Gamma$ let us denote by $w_{K\sigma}^{\alpha,n}$ an approximation of the trace of $w_m^\alpha(\cdot, t_n)|_K$ at the interface σ on the cell K side (interface $K\sigma$) using

$$z_{K\sigma} = z_\sigma + \frac{(z_K - z_\sigma)d_f(\mathbf{x}_\sigma)}{2|\mathbf{x}_K - \mathbf{x}_\sigma|}.$$

Let us define the phase saturation $s_{K,K\sigma}^{\alpha,n}$ at the interface $K\sigma$ on the matrix side by

$$s_{K,K\sigma}^{\alpha,n} = S_m^\alpha \left(w_{K\sigma}^{nw,n} - w_{K\sigma}^{w,n} - (\rho^{nw} - \rho^w)gz_{K\sigma}\right),$$

and the phase saturation $s_{\sigma,K\sigma}^{\alpha,n}$ at the interface $K\sigma$ on the fracture side by

$$s_{\sigma,K\sigma}^{\alpha,n} = S_f^\alpha \left(w_{K\sigma}^{nw,n} - w_{K\sigma}^{w,n} - (\rho^{nw} - \rho^w)gz_{K\sigma}\right).$$

The left-hand side of (8) is discretized by

$$F_{K,K\sigma}^{\alpha,n} = k_m^\alpha(s_K^{\alpha,n})T_{K\sigma}(w_K^{\alpha,n} - w_{K\sigma}^{\alpha,n})^+ + k_m^\alpha(s_{K,K\sigma}^{\alpha,n})T_{K\sigma}(w_K^{\alpha,n} - w_{K\sigma}^{\alpha,n})^-, \quad (17)$$

while the discretization of the right-hand side of (8) is given by $-F_{\sigma,K\sigma}^{\alpha,n}$ with

$$F_{\sigma,K\sigma}^{\alpha,n} = k_f^\alpha(s_\sigma^{\alpha,n})T_\sigma(w_\sigma^{\alpha,n} - w_{K\sigma}^{\alpha,n})^+ + k_f^\alpha(s_{\sigma,K\sigma}^{\alpha,n})T_\sigma(w_\sigma^{\alpha,n} - w_{K\sigma}^{\alpha,n})^-, \quad (18)$$

with

$$T_\sigma = \lambda_{f,n}(\mathbf{x}_\sigma)\frac{|\sigma|}{d_f(\mathbf{x}_\sigma)/2}.$$

Gathering (17) and (18) in the flux continuity equation (8) for each phase, we find

$$F_{K,K\sigma}^{\alpha,n} + F_{\sigma,K\sigma}^{\alpha,n} = 0, \quad \alpha = w, nw. \quad (19)$$

The interface potentials $\left(w_{K\sigma}^{\alpha,n}\right)_{\alpha=w,nw}$ are eliminated by solving the two equations (19) using the algorithm detailed in the next section. Then, the matrix–fracture flux $F_{K\sigma}^{\alpha,n}$ depending only on p_ν^β, $\beta = w, nw$, $\nu = K, \sigma$ is defined by

$$F_{K\sigma}^{\alpha,n} = F_{K,K\sigma}^{\alpha,n} = -F_{\sigma,K\sigma}^{\alpha,n}. \quad (20)$$

3.3.4 Matrix–fracture fluxes with linear transmission conditions

The evaluation of the matrix–fracture fluxes defined by (17), (18) and (19) requires the solution of local nonlinear problems which may produce a computational overhead compared to some simpler approaches. Following Brenner et al. (2018), we consider two alternative definitions of *mf* discrete fluxes. The first one corresponds to the classical flux approximation proposed in Karimi-Fard et al. (2004). It assumes, as for matrix–matrix (or fracture–fracture) fluxes, the continuity of the phase mobility at the matrix–fracture interface defined by its upstream value, precisely we define

$$F_{K\sigma}^{\alpha,n} = k_m^\alpha(s_K^{\alpha,n}) \left(V_{K\sigma}^{\alpha,n}\right)^+ + k_f^\alpha(s_\sigma^{\alpha,n}) \left(V_{K\sigma}^{\alpha,n}\right)^-, \tag{21}$$

with

$$V_{K\sigma}^{\alpha,n} = \frac{T_{K\sigma} T_\sigma}{T_{K\sigma} + T_\sigma} \left(w_K^{\alpha,n} - w_\sigma^{\alpha,n}\right). \tag{22}$$

The second alternative is inspired by the continuous pressure model proposed in Brenner et al. (2015), which assumes that the phase pressure is the same within the fracture and at the *mf* interfaces. This model can be extended to account also for low permeable fractures as follows, although it does not account for capillary barriers as will be checked in the numerical experiments. Precisely, it is defined by the flux

$$F_{K\sigma}^{\alpha,n} = k_m^\alpha(s_K^{\alpha,n}) \left(V_{K\sigma}^{\alpha,n}\right)^+ + k_m^\alpha(s_{K\sigma}^{\alpha,n}) \left(V_{K\sigma}^{\alpha,n}\right)^-, \tag{23}$$

with $V_{K\sigma}^{\alpha,n}$ given again by (22) and

$$s_{K\sigma}^{\alpha,n} = S_m^\alpha\left(w_\sigma^{nw,n} - w_\sigma^{w,n} - (\rho^{nw} - \rho^w) g z_\sigma\right). \tag{24}$$

For permeable fractures the flux designed by (23) is essentially equivalent to the one defined in Brenner et al. (2015).

The model provided by the matrix–fracture fluxes (21) is termed in the following the hybrid-dimensional *mf* linear *f* upwind model, while the model provided by the matrix–fracture fluxes (23), (24) is termed in the following the hybrid-dimensional *mf* linear *m* upwind model.

Remark 3.1 (Non homogeneous matrix and fracture network) We remark that the discrete fluxes defined by (15) and (16) make use of elimination of interface potential from the linear flux continuity equations (13). This is indeed possible because we have neglected the jump of the saturation across the interface. In order to extend the numerical scheme for the case of cells (and fracture faces) having different capillary pressure–saturation and relative permeability curves one may use the same methodology that has been applied at matrix fracture interfaces. That is instead of (14) and (15), one defines

$$V_{K\sigma}^{\alpha,n} = T_{K\sigma}(w_K^{\alpha,n} - w_\sigma^{\alpha,n}), \tag{25}$$

and

$$F_{K\sigma}^{\alpha,n} = k_m^\alpha(s_K^{\alpha,n}, \mathbf{x}_K) \left(V_{K\sigma}^{\alpha,n}\right)^+ + k_m^\alpha(s_{K\sigma}^{\alpha,n}, \mathbf{x}_K) \left(V_{K\sigma}^{\alpha,n}\right)^-, \tag{26}$$

with

$$s_{K\sigma}^{\alpha,n} = S_m^\alpha(w_\sigma^{nw,n} - w_\sigma^{w,n} - (\rho^{nw} - \rho^w) g z_\sigma, \mathbf{x}_K).$$

The flux is computed by solving (12) for $\alpha = w, nw$ with respect to $w_\sigma^{nw,n}$ and $w_\sigma^{w,n}$, which results in a couple of nonlinear equations. The same approach can be applied at fracture–fracture interfaces.

Remark 3.2 (Continuous representation of the linear transmission conditions) The equations (17)–(19) reproduce at the discrete level the coupling condition (8). Let us recover, in a formal way, the continuous coupling condition, which would correspond to the discrete *mf* flux defined by (21) or (23) instead of (20). Let $\sigma \in \mathcal{F}_{\Gamma_i}$, $i \in I$ and let $K \in \mathcal{M}_\sigma$ be such K lies on the "+" side of Γ_i (the argumentation is similar for the "−" side). Using (21) or (23) we write the discrete flux as

$$F_{K\sigma}^{\alpha,n} = k_{K\sigma}^{\alpha,n} \frac{T_{K\sigma} T_\sigma}{T_{K\sigma} + T_\sigma} \left(w_K^{\alpha,n} - w_\sigma^{\alpha,n} \right)$$

where $k_{K\sigma}^{\alpha,n}$ is an appropriate upstream mobility defined for the linear f upwind model as

$$k_{K\sigma}^{\alpha,n} = \begin{cases} k_m^\alpha(s_K^{\alpha,n}) & \text{if } V_{K\sigma}^{\alpha,n} \geq 0 \\ k_f^\alpha(s_\sigma^{\alpha,n}) & \text{else,} \end{cases}$$

and for the linear m upwind model by

$$k_{K\sigma}^{\alpha,n} = \begin{cases} k_m^\alpha(s_K^{\alpha,n}) & \text{if } V_{K\sigma}^{\alpha,n} \geq 0 \\ k_m^\alpha(s_{K\sigma}^{\alpha,n}) & \text{else} \end{cases}$$

with $s_{K\sigma}^{\alpha,n}$ given by (24). The transmissibility may be written as

$$\frac{T_{K\sigma} T_\sigma}{T_{K\sigma} + T_\sigma} = \frac{|\sigma| \lambda_{f,n}(\mathbf{x}_\sigma)}{d_f(\mathbf{x}_\sigma)/2 + \frac{\lambda_{f,n}}{\Lambda_m(\mathbf{x}_K)}(\mathbf{x}_\sigma)|\mathbf{x}_K - \mathbf{x}_\sigma|}$$

so that $\frac{T_{K\sigma} T_\sigma}{T_{K\sigma} + T_\sigma}$ tends to $\frac{|\sigma| \lambda_{f,n}(\mathbf{x}_\sigma)}{d_f(\mathbf{x}_\sigma)/2}$ when $\mathbf{x}_K \to \mathbf{x}_\sigma$. When the mesh size and the time step are small one would expect the discrete flux $F_{K\sigma}^{\alpha,n}$ to be an approximation of the quantity

$$\frac{1}{\Delta t_n} \int_{t_{n-1}}^{t_n} \int_\sigma k_{mf}^{\alpha,+} V_{f,i,n}^{\alpha,+} d\sigma(\mathbf{x}) dt$$

where

$$k_{mf}^{\alpha,+} = \begin{cases} k_m^\alpha(S_m^\alpha(\gamma_i^+ p_m^{nw} - \gamma_i^+ p_m^w)) & \text{if } V_{f,i,n}^{\alpha,+} \geq 0 \\ k_f^\alpha(S_f^\alpha(p_f^{nw} - p_f^w))) & \text{else} \end{cases}$$

for the linear f upwind model, and

$$k_{mf}^{\alpha,+} = \begin{cases} k_m^\alpha(S_m^\alpha(\gamma_i^+ p_m^{nw} - \gamma_i^+ p_m^w)) & \text{if } V_{f,i,n}^{\alpha,+} \geq 0 \\ k_m^\alpha(S_m^\alpha(p_f^{nw} - p_f^w))) & \text{else} \end{cases}$$

for the linear m upwind model. We then deduce that, for $i \in I$, the continuous coupling condition corresponding to f and m linear upwind models would be respectively

$$\mathbf{q}_m^{\alpha,\pm} \cdot \mathbf{n}_i^{\pm} = k_m^{\alpha}(S_m^{\alpha}(\gamma_i^{\pm} p_m^{nw} - \gamma_i^{\pm} p_m^{w}))(V_{f,i,n}^{\alpha,\pm})^+ + k_f^{\alpha}(S_f^{\alpha}(p_f^{nw} - p_f^{w})))(V_{f,i,n}^{\alpha,\pm})^-$$

and

$$\mathbf{q}_m^{\alpha,\pm} \cdot \mathbf{n}_i^{\pm} = k_m^{\alpha}(S_m^{\alpha}(\gamma_i^{\pm} p_m^{nw} - \gamma_i^{\pm} p_m^{w}))(V_{f,i,n}^{\alpha,\pm})^+ + k_m^{\alpha}(S_m^{\alpha}(p_f^{nw} - p_f^{w})))(V_{f,i,n}^{\alpha,\pm})^-.$$

4 Nonlinear interface problem

In this section, we present the algorithm for solving the local nonlinear interface problem (19) with respect to the unknowns $\left(w_{K\sigma}^{\alpha,n}\right)_{\alpha=w,nw}$. Let $T_m, T_f > 0$ and let us define for all $\pi \in \mathbb{R}$

$$M_m^{\alpha}(\pi) = T_m k_m^{\alpha}\left(S_m^{\alpha}(\pi)\right) \quad \text{and} \quad M_f^{\alpha}(\pi) = T_f k_f^{\alpha}\left(S_f^{\alpha}(\pi)\right).$$

In an abstract form, the local interface problem may be reformulated as follows:
Let $(w_j^{\alpha})_{j=m,f}^{\alpha=w,nw}$, z_m, z_f and z_{mf} be given and let us denote by π_j, for $j = m, f$, the capillary pressure

$$\pi_j = w_j^{nw} - w_j^{w} - (\rho^{nw} - \rho^{w})g z_j.$$

We look for a triplet (w^{nw}, w^{w}, π) satisfying

$$M_{m,mf}^{\alpha}(w_m^{\alpha} - w^{\alpha}) + M_{f,mf}^{\alpha}(w_f^{\alpha} - w^{\alpha}) = 0 \qquad \text{for } \alpha = w, nw \tag{27}$$

and

$$\pi = w^{nw} - w^{w} - (\rho^{nw} - \rho^{w})g z_{mf}, \tag{28}$$

where, for $j = m, f$, we set

$$M_{j,mf}^{\alpha} = \begin{cases} M_j^{\alpha}(\pi_j) & \text{if } w_j^{\alpha} - w^{\alpha} > 0, \\ M_j^{\alpha}(\pi) & \text{if } w_j^{\alpha} - w^{\alpha} \le 0. \end{cases} \tag{29}$$

Let us remark that capillary pressures π smaller then the minimal entry pressure $\pi_{min} = \min(\pi_{m,e}, \pi_{f,e})$ are not considered since for $\pi \le \pi_{min}$ it holds that $\partial_{\pi} M_m^{\alpha}(\pi) = \partial_{\pi} M_f^{\alpha}(\pi) = 0$, $\alpha = w, nw$.

We present below an algorithm for finding a solution to (27), (28). Let us briefly explain the idea behind it. It is assumed first that, for both phases, each of the Eq. (27) can be uniquely solved with respect to the corresponding w^{α}. This means that there exists a pair of functions $W^{\alpha}(\pi)$, $\alpha = w, nw$, such that, for a given $\pi \ge \pi_{min}$, the couple $(W^{\alpha}(\pi), \pi)$ constitute the unique solution of (27). Moreover, for practical purposes one has to be able to compute $W^\alpha(\pi)$ (and probably even $(W^\alpha)'(\pi)$). The system (27), (28) is then reduced to the equation

$$W^{nw}(\pi) = \widehat{W}^{nw}(\pi), \tag{30}$$

where $\widehat{W}^{nw}(\pi)$ denotes the non-wetting phase potential based on $W^w(\pi)$ and the Eq. (28), that is

$$\widehat{W}^{nw}(\pi) = W^w(\pi) + \pi - (\rho^w - \rho^{nw})gz_{mf}.$$

The Eq. (30) has to be solved numerically using some iterative method.

As it is going to be shown below, the function W^w is well defined and is continuous. In contrast, the values of $W^{nw}(\pi)$ are not in general uniquely defined. It turns out that in certain situations the set of w^{nw} satisfying (27) with $\pi = \pi_{min}$ is a half-line. This is due to the fact that the non-wetting phase may vanish. The structure of the non-wetting phase balance equation (27) is dictated by the presence of the concerned phase on the adjacent matrix and fracture sides of the interface. We will distinguish three following cases:

- Single phase flow characterized by an absence of the non-wetting phase on both sides of the interface.
- Degenerate two-phase flow taking place when the non-wetting phase is missing exclusively on the upwind side.
- Non-degenerate two-phase flow taking place when the non-wetting phase is present on the upwind side.

Now let us give a detailed description of the algorithm. Using the notations

$$(\mathrm{up}(\alpha), \mathrm{do}(\alpha)) = \begin{cases} (m, f) & \text{if } w_m^\alpha \geq w_f^\alpha, \\ (f, m) & \text{if } w_f^\alpha > w_m^\alpha. \end{cases}$$

we may rewrite (27) as

$$M^\alpha_{\mathrm{up}(\alpha),mf}(w^\alpha_{\mathrm{up}(\alpha)} - w^\alpha) + M^\alpha_{\mathrm{do}(\alpha),mf}(w^\alpha_{\mathrm{do}(\alpha)} - w^\alpha) = 0 \quad \text{for } \alpha = w, nw. \tag{31}$$

Since the wetting phase mobility does not vanish, the solution of the Eq. (31) for $\alpha = w$ at given π is defined by

$$w^w = W^w(\pi)$$

with

$$W^w(\pi) = w^w_{\mathrm{do}(w)} + M^w_{\mathrm{up}(w)}(\pi_{\mathrm{up}(w)}) \frac{w^w_{\mathrm{up}(w)} - w^w_{\mathrm{do}(w)}}{M^w_{\mathrm{up}(w)}(\pi_{\mathrm{up}(w)}) + M^w_{\mathrm{do}(w)}(\pi)}.$$

Let us remark that the function W^w is non-decreasing and continuous on $\mathbb{R}$, so that

$$\widehat{W}^{nw}(\pi) = W^w(\pi) + \pi - (\rho^w - \rho^{nw})gz_{mf}$$

is a strictly increasing and continuous function on $\mathbb{R}$. Now, we look for w^{nw} and $\pi \in [\pi_{min}, +\infty)$ satisfying

$$w^{nw} = \widehat{W}^{nw}(\pi) \tag{32}$$

and the Eq. (31) for the non-wetting phase $\alpha = nw$. Since the non-wetting phase mobility may vanish, we consider three following cases.

4.1 Single-phase case

If the non-wetting phase is missing at both sides of the interface, i.e $M_m^{nw}(\pi_m) = M_f^{nw}(\pi_f) = 0$, we set

$$\pi = \pi_{min}, \; w^w = W^w(\pi_{min}) \text{ and } w^{nw} = \widehat{W}^{nw}(\pi_{min}).$$

4.2 Degenerate two-phase case

If the non-wetting phase is missing only on the upwind side, that is $M_{\mathrm{up}(nw)}^{nw}(\pi_{\mathrm{up}(nw)}) = 0$ and $M_{\mathrm{do}(nw)}^{nw}(\pi_{\mathrm{do}(nw)}) > 0$, the Eq. (31) for $\alpha = nw$ is equivalent to either

$$\pi = \pi_{min} \text{ and } w^{nw} > w_{\mathrm{up}(nw)}^{nw}, \tag{33}$$

or

$$M_{\mathrm{do}(nw),mf}^{nw}(w_{\mathrm{do}(nw)}^{nw} - w^{nw}) = 0 \text{ and } w^{nw} \le w_{\mathrm{up}(nw)}^{nw}. \tag{34}$$

Moreover since $M_{\mathrm{do}(nw)}^{nw}(\pi_{\mathrm{do}(nw)}) > 0$, the Eq. (34) implies that $w^{nw} \ge w_{\mathrm{do}(nw)}^{nw}$ and $M_{\mathrm{do}(nw),mf}^{nw} = M_{\mathrm{do}(nw)}^{nw}(\pi)$. Hence, the Eq. (34) is equivalent to

$$M_{\mathrm{do}(nw)}^{nw}(\pi)(w_{\mathrm{do}(nw)}^{nw} - w^{nw}) = 0 \text{ and } w^{nw} \le w_{\mathrm{up}(nw)}^{nw}. \tag{35}$$

Let us consider the following complementary problem

$$\begin{cases} (\pi - \pi_{min})(w^{nw} - w_{\mathrm{do}(nw)}^{nw}) = 0, \\ \pi \ge \pi_{min}, \quad w^{nw} \ge w_{\mathrm{do}(nw)}^{nw}. \end{cases} \tag{36}$$

One can see that any solution of (36) satisfies (33) or (35). Let us notice that the system formed by (32) and (36) has a unique solution. In order to find it we proceed as follows (see also Fig. 2):

- If $\widehat{W}^{nw}(\pi_{min}) > w_{\mathrm{do}(nw)}^{nw}$ we set $\pi = \pi_{min}$, $w^w = W^w(\pi_{min})$ and $w^{nw} = \widehat{W}^{nw}(\pi_{min})$;
- If not, π is the unique solution of $\widehat{W}^{nw}(\pi) = w_{\mathrm{do}(nw)}^{nw}$, and we set $w^w = W^w(\pi)$ and $w^{nw} = w_{\mathrm{do}(nw)}^{nw}$.

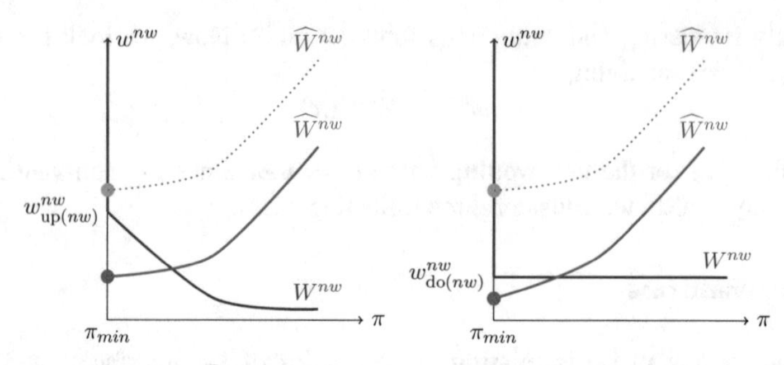

Fig. 2 The values of W^{nw} and $\widehat{W}^{nw}$ versus capillary pressure π in non-degenerate (left) and degenerate (right) two phase cases

4.3 Non-degenerate two-phase case

Finally, if $M^{nw}_{\text{up}(nw)}(\pi_{\text{up}}(nw)) > 0$ we define

$$W^{nw}(\pi) = w^{nw}_{\text{do}(nw)} + M^{nw}_{\text{up}(nw)}(\pi_{\text{up}(nw)})\frac{w^{nw}_{\text{up}(nw)} - w^{nw}_{\text{do}(nw)}}{M^{nw}_{\text{up}(nw)}(\pi_{\text{up}(nw)}) + M^{nw}_{\text{do}(nw)}(\pi)},$$

where $W^{nw}(\pi)$ is a non-increasing function on $[\pi_{min}, +\infty)$ such that $W^{nw}(\pi_{min}) = w^{nw}_{\text{up}(nw)}$. One can verify that the pair (w^{nw}, π) is the solution of (31) for $\alpha = nw$ (excluding $\pi < \pi_{min}$) if and only if it satisfies

$$\begin{cases} (\pi - \pi_{min})(w^{nw} - W^{nw}(\pi)) = 0, \\ \pi \geq \pi_{min}, \quad w^{nw} \geq W^{nw}(\pi). \end{cases} \tag{37}$$

Again the system formed by (32) and (37) has a unique solution and this solution is found by the following procedure (see Fig. 2):

- If $\widehat{W}^{nw}(\pi_{min}) > w^{nw}_{\text{up}(nw)}$ we set $\pi = \pi_{min}$, $w^w = W^w(\pi_{min})$ and $w^{nw} = \widehat{W}^{nw}(\pi_{min})$;
- If not, π is the unique solution of $W^{nw}(\pi) = \widehat{W}^{nw}(\pi)$, and we set $w^w = W^w(\pi)$ and $w^{nw} = W^{nw}(\pi)$.

5 2D numerical experiments

The objective of this section is to compare on 2D test cases, both in terms of accuracy and CPU time, the numerical solutions obtained with the hybrid-dimensional *mf* nonlinear and *mf* linear *f* upwind and *m* upwind models. The reference solution is defined by the equi-dimensional numerical solution obtained with 4 cells in the width of the fractures. The test cases presented in the following subsections consider various physical configurations including gravity, viscous or capillary dominant test cases with homogeneous or heterogeneous capillary pressures, and with highly or low permeable fractures.

The discretization is based on the Two-Point Flux Approximation defined in Sect. 3 using uniform Cartesian meshes. The fluxes at the matrix–fracture interfaces for the equi-dimensional model are discretized using (25), (26) rather than (14), (15) as mentioned in Remark 3.1 in order to account for the saturation jump accurately. This discretization is combined with the nonlinear interface solver introduced in Sect. 4 both for the hybrid-dimensional *mf* nonlinear model at matrice fracture interfaces and for the equi-dimensional model at interfaces between the matrix and fracture rocktypes.

The following notations and numerical parameters are used in this section for all the simulations. Let t_f denote the final time of the simulation, and let be given the initial time step Δt^1, a maximum time step Δt^{max}, and an objective variation of the saturation between two successive time steps $\delta s^{obj} \in]0, 1]$. The time stepping is adaptive and defined by

$$\Delta t^{n+1} = \max(\beta \Delta t^n, \Delta t^{max})$$

with $\beta = \min(1.2, \frac{\delta s^{obj}}{\delta s^n})$ where δs^n is the maximum variation of the saturation at time step n. At each time step, the set of nonlinear equations (9), (10), (11) is solved by a Newton–Raphson algorithm using the non-wetting phase pressure and non-wetting phase saturation as primary unknowns. This choice is known to provide a better nonlinear convergence than the one obtained with both phase pressures as primary unknowns. It can also allow vanishing capillary pressure curves. At each Newton–Raphson iteration, the Jacobian is computed analytically and the linear system is solved using a GMRes iterative solver combined with the Constrained Pressure Residual Algebraic Multigrid (CPR-AMG) right preconditioner (Lacroix et al. 2001; Scheichl et al. 2003). The time step is chopped by a factor 2 in case the Newton–Raphson algorithm has not converged in 50 iterations with the Newton–Raphson stopping criteria defined by the relative norm of the residual lower than 10^{-6} or a maximum normalized variation of the primary unknowns lower than 10^{-5}.

In the tables below, the following notations are used:

- **Nb Cells** is the number of cells of the mesh,
- **Nb dof** is the number of degrees of freedom (d.o.f.) (with two physical primary unknowns per d.o.f.) including in addition to the cells, the faces $\sigma \in \mathcal{F}_\Gamma$ and the edges $e \in \mathcal{E}_\star$ for the hybrid-dimensional models,
- $\mathbf{N}_{\Delta t}$ is the number of successful time steps and $\mathbf{N}_{Chop}$ is the number of time step chops,
- $\mathbf{N}_{Newton}$ is the average number of Newton–Raphson iterations per successful time step,
- $\mathbf{N}_{GMRes}$ is the average number of GMRes iterations per Newton–Raphson step using the stopping criteria 10^{-7} on the relative residual.

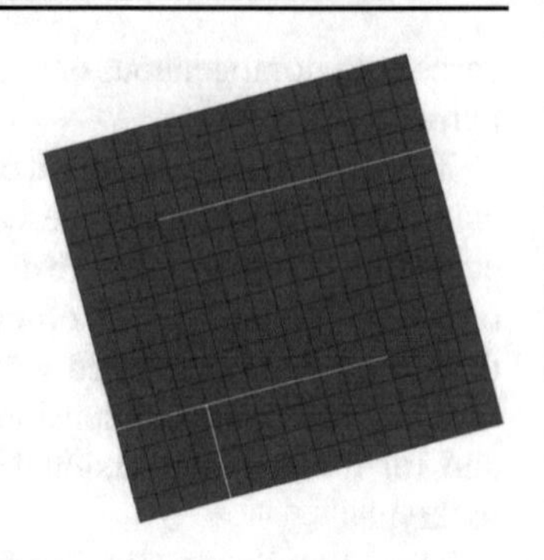

Fig. 3 Discrete Fracture Model with the four fractures Γ_i, $i = 1, \dots, 4$, and its uniform Cartesian mesh of size 16×16

5.1 Gravity dominant flow with homogeneous capillary pressures

We consider in this test case an oil (o) water (w) two-phase Darcy flow in the DFM model exhibited in Fig. 3 and defined by the domain

$$\Omega = \{(x = x'\cos(\theta) - y'\sin(\theta), y = x'\sin(\theta) + y'\cos(\theta)) \mid (x', y') \in (0, 100\text{ m})^2\}, \tag{38}$$

and the fracture network $\Gamma = \bigcup_{i=1}^{4} \Gamma_i$ with

$$\begin{aligned}
\Gamma_1 &= \{(x = x'\cos(\theta) - y'\sin(\theta), y = x'\sin(\theta) + y'\cos(\theta)) \mid x' \in (0, 25), y' = 25\},\\
\Gamma_2 &= \{(x = x'\cos(\theta) - y'\sin(\theta), y = x'\sin(\theta) + y'\cos(\theta)) \mid x' \in (25, 75), y' = 25\},\\
\Gamma_3 &= \{(x = x'\cos(\theta) - y'\sin(\theta), y = x'\sin(\theta) + y'\cos(\theta)) \mid x' \in (25, 100), y' = 75\},\\
\Gamma_4 &= \{(x = x'\cos(\theta) - y'\sin(\theta), y = x'\sin(\theta) + y'\cos(\theta)) \mid y' \in (0, 25), x' = 25\},
\end{aligned} \tag{39}$$

setting $\theta = \frac{\pi}{12}$.

The matrix and the fracture network have the same relative permeabilities and capillary pressures given by $k^o_{r,f}(s^o) = k^o_{r,m}(s^o) = (s^o)^2$, $k^w_{r,f}(s^w) = k^w_{r,m}(s^w) = (s^w)^2$ and by $P_{c,m}(s^o) = P_{c,f}(s^o) = -b\log(1 - s^o)$ with $b = 10$ Pa. The matrix is homogeneous and charaterized by the isotropic permeability $\Lambda_m = 0.1$ Darcy, and the porosity $\phi_m = 0.2$. The fracture properties are set to $d_f = 1$ cm, $\Lambda_f = \lambda_{f,n} = 10^3\Lambda_m$, $\phi_f = 0.35$. The fluid properties are defined by their dynamic viscosities $\mu^o = 5 \times 10^{-3}$, $\mu^w = 10^{-3}$ Pa s and their mass densities $\rho^w = 1000$ and $\rho^o = 800$ Kg.m^{-3}.

The matrix and fracture domains are initially saturated by the water phase. The top boundary conditions at $y' = 100$ are defined by the water saturation $s^w = 1$ and the water pressure

$$p^w(x, y) = (100(\cos(\theta) + \sin(\theta)) - y)g\rho^w \text{ Pa}.$$

The bottom boundary conditions at $y' = 0$ are impervious in the matrix and defined in the bottom fracture by the water pressure

$$p^w = (100(\cos(\theta) + \sin(\theta)) - 25\sin(\theta))g\rho^w \text{ Pa},$$

and the oil saturation $s^o = 0.9$. The lateral boundaries are assumed impervious.

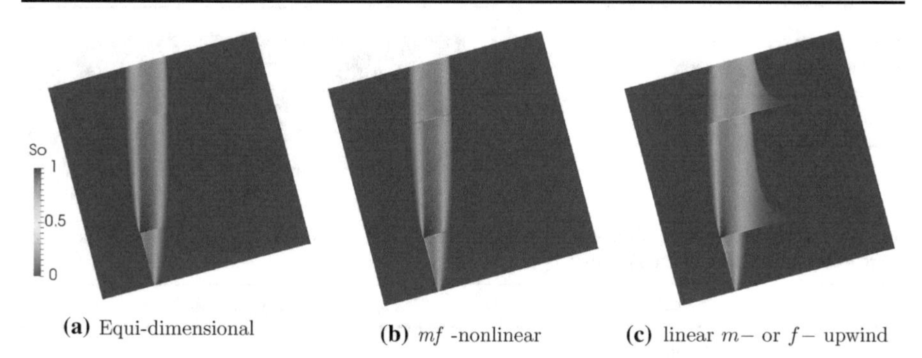

(a) Equi-dimensional (b) mf -nonlinear (c) linear $m-$ or $f-$ upwind

Fig. 4 Oil saturation at final time computed by the equi-dimensional and hybrid-dimensional models for the gravity dominant test case with homogeneous capillary pressures. The mesh is Cartesian of size 260 × 264 for the equi-dimensional model and of size 256 × 256 for the hybrid-dimensional models

The simulation time is set to $t_f = 35$ years. The time stepping is defined by the parameters $\Delta t^{max} = 30$ days, $\Delta t^1 = 0.01$ day, and $\delta s^{obj} = 0.5$.

Figure 4 compares the solutions obtained with the hybrid-dimensional *mf* nonlinear and *mf* linear models to a reference solution obtained using the equi-dimensional model meshed with 4 cells in the fracture width and a Cartesian grid of size 260 × 264 uniform in the matrix domain. The solutions of the hybrid-dimensional *mf* nonlinear and *mf* linear models are obtained using a uniform Cartesian grid of size 256 × 256. Note that the hybrid-dimensional *mf* linear *m* and *f* upwind models provide the same solution since the relative permeabilities and the capillary pressures are the same in the matrix and fracture domains for this test case.

Figure 4 shows that the hybrid-dimensional *mf* nonlinear model reproduces accurately the reference solution provided by the equi-dimensional model while the hybrid-dimensional *mf* linear model exhibits large differences with the reference solution. This discrepancy does not reduce when the mesh is refined and is due to the fact that the hybrid-dimensional *mf* linear model cannot account for gravity segregation in the fracture width since it uses a single saturation unknown in the fracture width. As a consequence, the top mobility of oil rising out of the fractures is averaged and reduced for the *mf* linear model. On the other hand, the hybrid-dimensional *mf* nonlinear model accounts for gravity segregation thanks to its additional saturation unknowns at the *mf* interfaces and provides a good approximation of the top *mf* interface oil mobility. It can be checked that this discrepancy is considerably reduced for a larger capillary pressure in the fractures (say for $b = 1000$ Pa) which smoothes out the gravity segregation in the fracture width as exhibited in Fig. 5. Table 1 exhibits a gain of a factor 4 in CPU time between the equi-dimensional and the hybrid-dimensional *mf* nonlinear model. This is mainly due to a higher number of GMRes iterations for the equi-dimensional model which can be explained by the smaller volumes in the fracture width and at the fracture intersection. The hybrid-dimensional *mf* nonlinear model is only 50% more expensive than the hybrid-dimensional *mf* linear model due to a higher number of nonlinear iterations.

Fig. 5 Zoom on the oil saturation in the Γ_2 fracture width for the equi-dimensional model with $b = 10$ Pa (left) and $b = 1000$ Pa (right)

Table 1 Numerical behavior of the equi-dimensional and hybrid-dimensional models for the gravity dominant test case with homogeneous capillary pressures

Model	Nb cells	Nb dof	$N_{\Delta t}$	N_{Newton}	N_{GMRes}	N_{Chop}	CPU[s]
Equi-dimensional	68640	68640	471	4.5	55	4	2574
mf nonlinear	65536	65985	463	4.0	11	2	612
mf linear	65536	65985	459	3.0	10.3	0	436

The mesh is Cartesian of size 260 × 264 for the equi-dimensional model and of size 256 × 256 for the hybrid-dimensional models

5.2 Gravity dominant flow with heterogeneous capillary pressures

We consider the same test case as in the previous subsection with different relative permeabilities and capillary pressures in the matrix and in the fracture network defined by

$$k^o_{r,f}(s^o) = s^o, \quad k^o_{r,m}(s^o) = (s^o)^2, \quad k^w_{r,f}(s^w) = s^w, \quad k^w_{r,m}(s^w) = (s^w)^2,$$

and

$$P_{c,m}(s^o) = 10^3(1 - \log(1 - s^o)) \text{ Pa}, \quad P_{c,f}(s^o) = -10 \log(1 - s^o) \text{ Pa}.$$

The simulation time is reduced to $t_f = 20$ years for this test case, and the time stepping is defined by the parameters $\Delta t^{max} = 30$ days, $\Delta t^1 = 0.01$ day, and $\delta s^{obj} = 0.5$. Other physical and numerical parameters are unchanged compared with the previous test case.

Figure 6 compares the solutions obtained using the hybrid-dimensional *mf* nonlinear and *mf* linear *f* and *m* upwind models to the reference solution obtained using the equi-dimensional model with 4 cells in the fracture width. Two different Cartesian mesh sizes are used, 64 × 64 (68 × 72 for the equi-dimensional model) and 256 × 256 (260 × 264 for the equi-dimensional model). The hybrid-dimensional *mf* linear *f* upwind model overestimates the oil flux rising out of the fractures for the coarsest mesh. This is due to the absence of an *mf* interface capillary pressure unknown for this model and to a cell size above the fracture not small enough compared with the entry pressure divided by the gravity constant g and by $\rho^w - \rho^o$. To be more specific, this model overestimates the buoyancy force at the top matrix fracture interface by the quantity $g(\rho^w - \rho^o)\frac{\Delta z}{2}$ where Δz is the cell height above the fracture. If this quantity is not small compared with the matrix entry pressure, this model

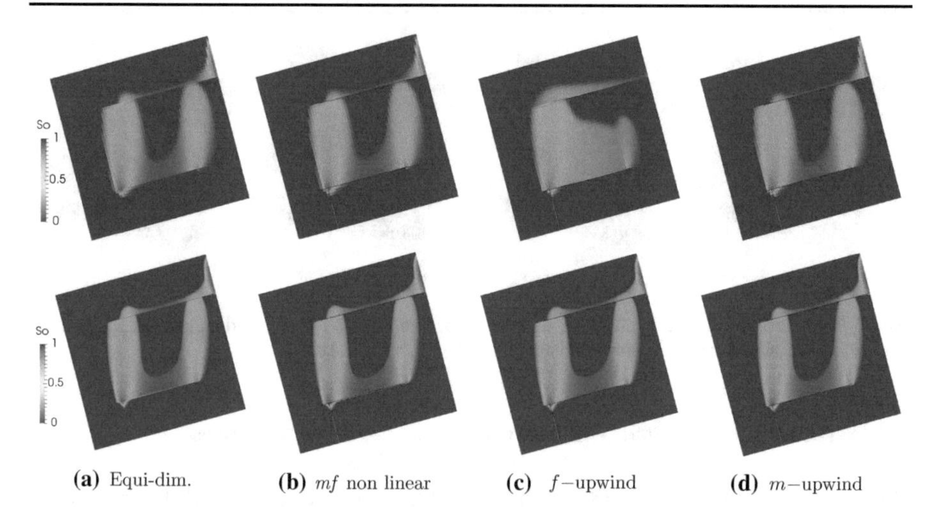

(**a**) Equi-dim. (**b**) mf non linear (**c**) f−upwind (**d**) m−upwind

Fig. 6 From left to right: oil saturation at final time for the equi-dimensional and hybrid-dimensional models. Top row: Cartesian mesh of size 68 × 72 for the equi-dimensional model and of size 64 × 64 for the hybrid-dimensional models. Bottom row: Cartesian mesh of size 260 × 264 for the equi-dimensional model and of size 256 × 256 for the hybrid-dimensional models

Table 2 Numerical behavior of the equi-dimensional and hybrid-dimensional models on both Cartesian meshes for the gravity dominant test case with heterogeneous capillary pressures

Model	Nb cells	Nb dof	$N_{\Delta t}$	N_{Newton}	N_{GMRes}	N_{Chop}	CPU[s]
Equi-dimensional	4896	4896	322	9.6	17.5	6	121
mf nonlinear	4096	4209	288	6.8	8.5	0	47.4
f upwind	4096	4209	280	3.1	8.8	0	22.7
m upwind	4096	4209	289	3.5	12	0	28.8
Equi-dimensional	68640	68640	425	13.7	42	5	5040
mf nonlinear	65536	65985	337	11.7	13.3	2	1296
f upwind	65536	65985	326	4.9	15.6	0	607
m upwind	65536	65985	336	4.4	28.8	0	908

will overestimate the oil flux rising out of the fractures. This is clearly observed on the coarsest mesh for which $g(\rho^w - \rho^o)\frac{\Delta z}{2} \simeq 780$ Pa which is of the same order of magnitude than the matrix entry pressure of 10^3 Pa. When the mesh is refined by a factor 4 in each direction the discrepancy is much smaller. The *mf* linear *m* upwind model is much better for this test case since it better captures the saturation jump at the *mf* interface. The *mf* nonlinear model cannot be distinguished from the reference model. Regarding the numerical behavior exhibited in Table 2, on the finest meshes, the *mf* linear *f* upwind model is twice cheaper than the *mf* nonlinear model which is four times cheaper than the equi-dimensional model.

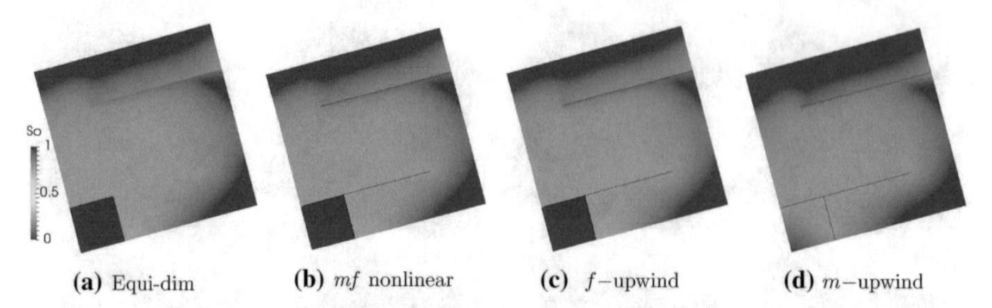

(a) Equi-dim **(b)** *mf* nonlinear **(c)** f−upwind **(d)** m−upwind

Fig. 7 From left to right: oil saturation at final time computed by the equi-dimensional and hybrid-dimensional models for the viscous dominant test case with overpressure of 2×10^5 Pa

5.3 Viscous dominant flow with heterogeneous capillary pressures

We consider the same test case as in the previous subsection with different relative permeabilities and capillary pressures now defined by

$$k^o_{r,f}(s^o) = k^o_{r,m}(s^o) = (s^o)^2, \quad k^w_{r,f}(s^w) = k^w_{r,m}(s^w) = (s^w)^2,$$

and

$$P_{c,m}(s^o) = 10^5(1 - \log(1 - s^o)) \text{ Pa}, \quad P_{c,f}(s^o) = -100 \log(1 - s^o) \text{ Pa}.$$

The water pressure in the bottom fracture at the bottom boundary $y' = 0$ now includes an overpressure and is set to

$$p^w = 2 \times 10^5 + (100(\cos(\theta) + \sin(\theta)) - 25\sin(\theta))g\rho^w \text{ Pa}.$$

The simulation time is set to $t_f = 19$ years for this test case. The time stepping is defined by the parameters $\Delta t^{max} = 30$ days, $\Delta t^1 = 0.01$ day, and $\delta s^{obj} = 1.2$ such that $\beta = 1.2$. The other physical and numerical parameters are the same as in the previous test case.

We compare in Fig. 7 the solutions obtained with the hybrid-dimensional models to the reference solution obtained with the equi-dimensional model with 4 cells in the fracture width. The meshes are Cartesian of size 68×72 for the equi-dimensional model and of size 64×64 for the hybrid-dimensional models. For this test case, the *mf* nonlinear model cannot be distinguished from the reference solution while the *mf* linear f upwind model slightly overestimates the oil flux rising out of the fractures. This small discrepancy is reduced on a finer mesh. The *mf* linear m upwind model considerably smoothes out the solution independently of the mesh size as soon as the fractures are filled with oil. As exhibited in Fig. 8, this is due to the fact that the capillary pressure inside the filled fractures is too high for the m upwind model which reduces the capillary barrier effect of the matrix. This discrepancy is a modeling error that occurs for filled fractures and increases with the overpressure imposed at the bottom fracture. The numerical behavior of the different models exhibited in Table 3 shows only small differences for this small mesh size.

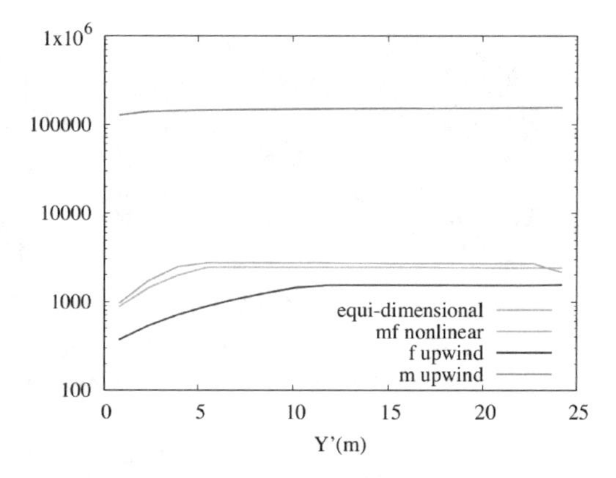

Fig. 8 Capillary pressure in Pa at final time in the bottom fracture Γ_4 as a function of y' for the equi-dimensional model (average in the fracture width), the *mf* nonlinear model, the *mf* linear *f* upwind model and the *mf* linear *m* upwind model

Table 3 Numerical behavior of the equi-dimensional and the hybrid-dimensional models for the viscous dominant test case with heterogeneous capillary pressures and overpressure of 2×10^5 Pa

Model	Nb cells	Nb dof	$N_{\Delta t}$	N_{Newton}	N_{GMRes}	N_{Chop}	CPU[s]
Equi-dimensional	4896	4896	269	6.0	12	1	56
mf nonlinear	4096	4209	269	5.6	10	1	36
f upwind	4096	4209	267	3.8	16	0	31
m upwind	4096	4209	267	4.0	22	0	40

The mesh is Cartesian of size 68 × 72 for the equi-dimensional model and of size 64 × 64 for the hybrid-dimensional models

We consider a second viscous dominant test case with one additional fracture

$$\Gamma_5 = \{(x = x' \cos(\theta) - y' \sin(\theta), y = x' \sin(\theta) + y' \cos(\theta)) \mid y' \in (25, 50), x' = 25\}.$$

This test case uses the fracture porosity $\phi_f = 0.4$, the matrix porosity $\phi_m = 0.2$, a higher fracture permeability $\Lambda_f = \lambda_{f,n} = 10^4$ Darcy, a lower matrix permeability $\Lambda_m = 0.01$ Darcy, and a higher overpressure at the bottom boundary of the bottom fracture where the following water pressure is imposed

$$p^w = 2 \times 10^6 + (100(\cos(\theta) + \sin(\theta)) - 25 \sin(\theta))g\rho^w \text{ Pa}.$$

The simulation time is set to $t_f = 2$ years for this test case. The time stepping is defined by the parameters $\Delta t^{max} = 30$ days, $\Delta t^1 = 0.001$ day, and $\delta s^{obj} = 0.5$. The other physical and numerical parameters are the same as in the previous test case.

We compare in Figs. 9 and 10 the solutions obtained with the hybrid-dimensional models to the reference solution obtained with the equi-dimensional model with 4 cells in the fracture width. The meshes are Cartesian of size 260 × 264 for the equi-dimensional model and of size 256 × 256 for the hybrid-dimensional models. For this test case with a high overpressure compared with the matrix entry pressure, the *mf* nonlinear model cannot be distinguished from the reference solution while the *mf* linear *f* upwind model clearly overestimates the oil flux rising out of the fractures

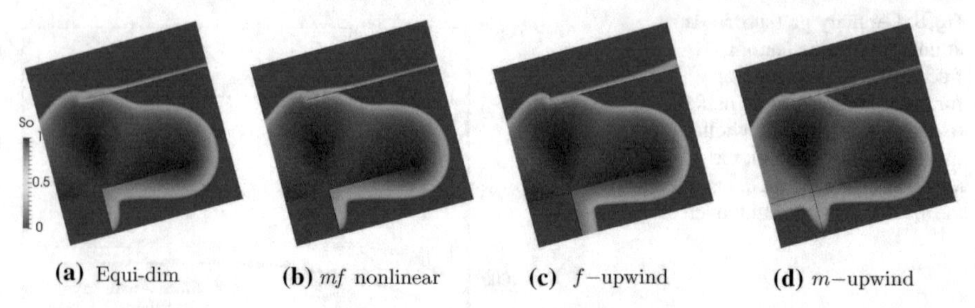

(a) Equi-dim **(b)** *mf* nonlinear **(c)** f−upwind **(d)** m−upwind

Fig. 9 From left to right: oil saturation at final time computed by the equi-dimensional and hybrid-dimensional models for the viscous dominant test case with overpressure of 2×10^6 Pa

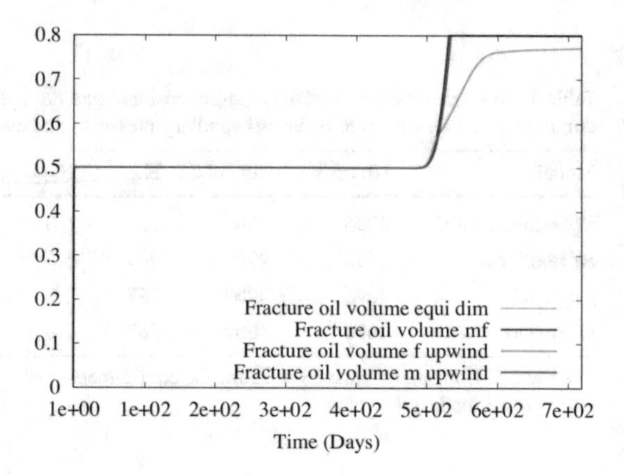

Fig. 10 Oil volume in the fracture network as a function of time for the equi-dimensional model, the *mf* nonlinear and the *mf* linear *f* upwind and *m* upwind models

Table 4 Numerical behavior of the equi-dimensional and hybrid-dimensional models for the viscous dominant test case with heterogeneous capillary pressures and overpressure of 2×10^6 Pa

Model	Nb cells	Nb dof	$N_{\Delta t}$	N_{Newton}	N_{GMRes}	N_{Chop}	CPU[s]
Equi-dimensional	68640	68640	338	14.8	23	16	2868
mf nonlinear	65536	66049	315	12.7	8.2	11	1141
f upwind	65536	66049	249	14.3	8.3	27	1060
m upwind	65536	66049	319	13.2	12.8	9	1726

The mesh is Cartesian of size 260×264 for the equi-dimensional model and of size 256×256 for the hybrid-dimensional models

even on this fine mesh. This can be explained by the definition of the matrix–fracture flux for this model which does not take into account the saturation jump at the *mf* interfaces in contrast to the *mf* nonlinear and *mf* linear *m* upwind models. The *mf* linear *m* upwind model provides a very good solution during the fracture filling as can be observed in Fig. 10 but again it smoothes out the solution as soon as the fractures are filled with oil. The numerical behavior of the different models exhibited in Table 4 shows that the *mf* nonlinear model is very competitive compared with the other hybrid-dimensional models with a CPU time almost 3 times lower than the CPU time of the equi-dimensional model.

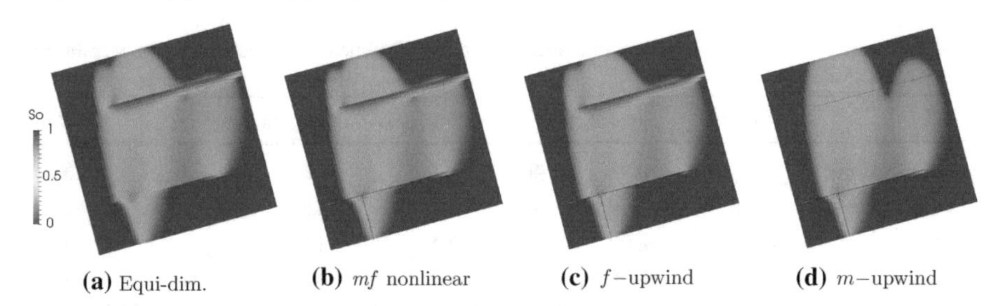

(a) Equi-dim. **(b)** *mf* nonlinear **(c)** f−upwind **(d)** m−upwind

Fig. 11 From left to right, oil saturation at final time computed by the equi-dimensional and hybrid-dimensional models for the three drains and one barrier test case

5.4 Three drains and one barrier test case

In this test case, the matrix properties are set to $\Lambda_m = 1$ Darcy, $\phi_m = 0.2$ and $P_{c,m}(s^o) = -5 \times 10^3(1 - \log(1 - s^o))$ Pa. The top fracture Γ_3 is a barrier with properties $\Lambda_b = \lambda_{b,n} = 10^{-1}\Lambda_m$, $\phi_b = 0.15$, and $P_{c,b}(s^o) = -10^4(1 - 1.5\log(1 - s^o))$ Pa. The bottom fractures Γ_1,Γ_2, Γ_4 are drains with properties $\Lambda_d = \lambda_{d,n} = 10^3\Lambda_m$, $\phi_d = 0.35$, and $P_{c,d}(s^o) = -10^3\log(1 - s^o)$ Pa. The fracture width is fixed to $d_f = 1$ cm and the relative permeabilities are set to $k_r^o(s^o) = (s^o)^2$ and $k_r^w(s^w) = (s^w)^2$ for all rocktypes.

The water pressure at the bottom boundary of the bottom fracture is set to

$$p^w = 10^4 + (100(\cos(\theta) + \sin(\theta)) - 25\sin(\theta))g\rho^w \text{ Pa.}$$

The final simulation time is set to $t_f = 30$ years. The time stepping is defined by the parameters $\Delta t^{max} = 30$ days, $\Delta t^1 = 0.01$ day, and $\delta s^{obj} = 0.5$. Other physical and numerical parameters are the same as in the previous test cases.

We compare in Fig. 11 the solutions obtained with the hybrid-dimensional models to a reference solution obtained with the equi-dimensional model with 4 cells in the fracture width. The mesh is Cartesian of size 68×72 for the equi-dimensional model and of size 64×64 for the hybrid-dimensional models. As before, the *mf* nonlinear model cannot be distinguished from the reference model and *mf* linear f upwind provide a good match with the reference solution for this test case with overpressure. On the other hand, the *mf* linear m upwind does not capture as expected the capillary barrier effect of the top fracture. The numerical behavior exhibited in Table 5 shows small differences between the hybrid-dimensional models and a factor roughly 2 with the equi-dimensional model.

5.5 Desaturation by suction of a fractured porous medium

We consider in this test case a water gas two-phase Darcy flow in the DFM model defined by the vertical domain $\Omega = (0, 10\,\text{m})^2$ and the fracture network exhibited in Fig. 12. The matrix is homogeneous and charaterized by the isotropic permeability $\Lambda_m = 1$ Darcy, the porosity $\phi_m = 0.2$, the relative permeabilities $k_{r,m}^g(s^g) = (s^g)^2$,

Table 5 Numerical behavior of the equi-dimensional and hybrid-dimensional models for the 3 drains and 1 barrier test case

Model	Nb cells	Nb dof	$N_{\Delta t}$	N_{Newton}	N_{GMRes}	N_{Chop}	CPU[s]
Equi-dimensional	4896	4896	406	3.8	22	0	73
mf nonlinear	4096	4209	405	3.4	12	0	42
f upwind	4096	4209	403	3.5	16.4	0	47
m upwind	4096	4209	404	3.2	13.7	0	37

The mesh is Cartesian of size 68 × 72 for the equi-dimensional model and of size 64 × 64 for the hybrid-dimensional models

$k_{r,m}^w(s^w) = (s^w)^3$, and the capillary pressure $P_{c,m}(s^g) = -b_m \log(1 - s^g)$ with $b_m = 10^6$ Pa. The fracture properties are set to $d_f = 1$ mm, $\Lambda_f = \lambda_{f,n} = 10^4 \Lambda_m$, $\phi_f = 0.35, k_{r,f}^g = k_{r,m}^g, k_{r,f}^w = k_{r,m}^w$, and $P_{c,f}(s^g) = -b_f \log(1-s^g)$ with $b_f = 10^4$ Pa. The fluid properties are defined by their dynamic viscosities $\mu^g = 2 \times 10^{-4}$, $\mu^w = 10^{-3}$ Pa s and their mass densities $\rho^w = 1000$ and $\rho^g = 1$ Kg.m^{-3}.

The matrix and fracture domains are initially saturated by the water phase. The top boundary conditions are defined by the water saturation $s^w = 1$ and the water pressure $p^w = 10^5$ Pa. The bottom boundary conditions are defined by the gas pressure $p^g = 10^5$ Pa and the gas saturation fixed to $s^g = 0.1$ in the matrix and to $s^g = P_{c,f}^{-1}(P_{c,m}(0.1))$ in the fractures. The lateral boundaries are assumed impervious.

The simulation time is set to $t_f = 10^7$ s which suffices to reach the stationary state for the three models. The time stepping is defined by the parameters $\Delta t^1 = 10$ s, $\Delta t^{max} = 10^5$ s, and $\delta s^{obj} = 0.3$. The mesh is a uniform Cartesian grid of size 160 × 160.

Figures 12, 13, 14 exhibit the large discrepancies between the reference solution which is here given by the hybrid dimensional *mf* nonlinear model and both *mf* linear models. The *mf* linear *f* upwind model considerably overestimates the capillary barrier effect at the *mf* interfaces due to bad approximation of the water mobility at the matrix–fracture interfaces for dry fractures. The *mf* linear *m* upwind model hardly sees the fractures in the sense that it does not capture the jumps of the water pressure and saturation at the matrix–fracture interfaces. The numerical behavior of the three models is similar as exhibited in Table 6.

6 3D numerical experiments

In this section, the numerical solutions of the hybrid-dimensional *mf* nonlinear and *mf* linear *f* and *m* upwind models are compared on a 3D numerical test case using a family of refined tetrahedral meshes. On such meshes, the Two-Point Flux Approximation of Sect. 3 is not consistent which motivates the use in this section of the alternative Vertex Approximate Gradient scheme described in Brenner et al. (2018). This scheme is particularly well suited to tetrahedral meshes since it is based essentially on nodal unknowns. We refer to Brenner et al. (2018) for its detailed presentation for the two-phase hybrid-dimensional *mf* nonlinear model. Let us remark that with such nodal

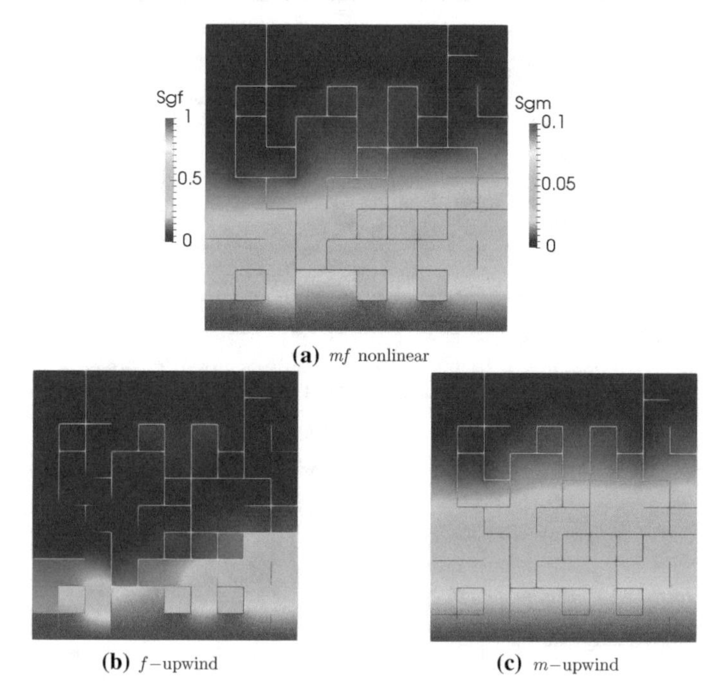

(a) mf nonlinear

(b) f−upwind

(c) m−upwind

Fig. 12 Gas saturation at final time computed by the hybrid-dimensional models for the desaturation by suction test case

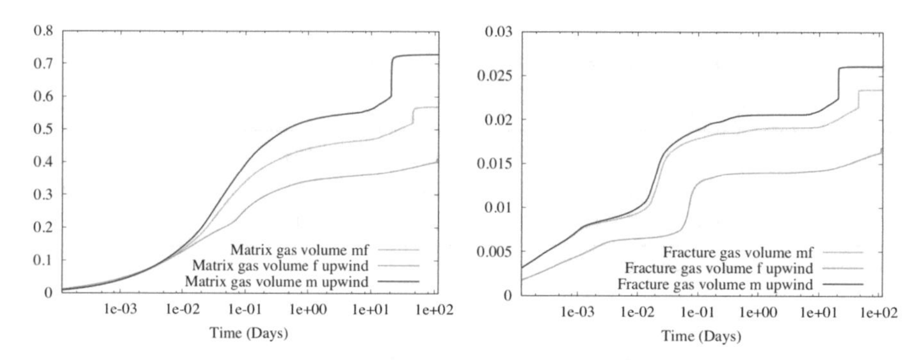

Fig. 13 Gas volume in the matrix and in the fracture network as a function of time for the hybrid-dimensional models

scheme, the *mf* interface unknowns are coupled along the connected fractures and hence cannot be eliminated. Since the test case considers highly permeable fractures, the *mf* linear models can be further approximated using a continuous pressure model such as the one introduced in Brenner et al. (2015, 2017). This continuous pressure model can be combined with either the *m* upwinding of the mobilities as in Brenner et al. (2015, 2017) or the more usual *f* upwinding of the mobilities. The advantage of this continuous pressure approximation is to eliminate readily the interface unknowns for the *mf* linear models with a negligible additional error for fractures acting as drains.

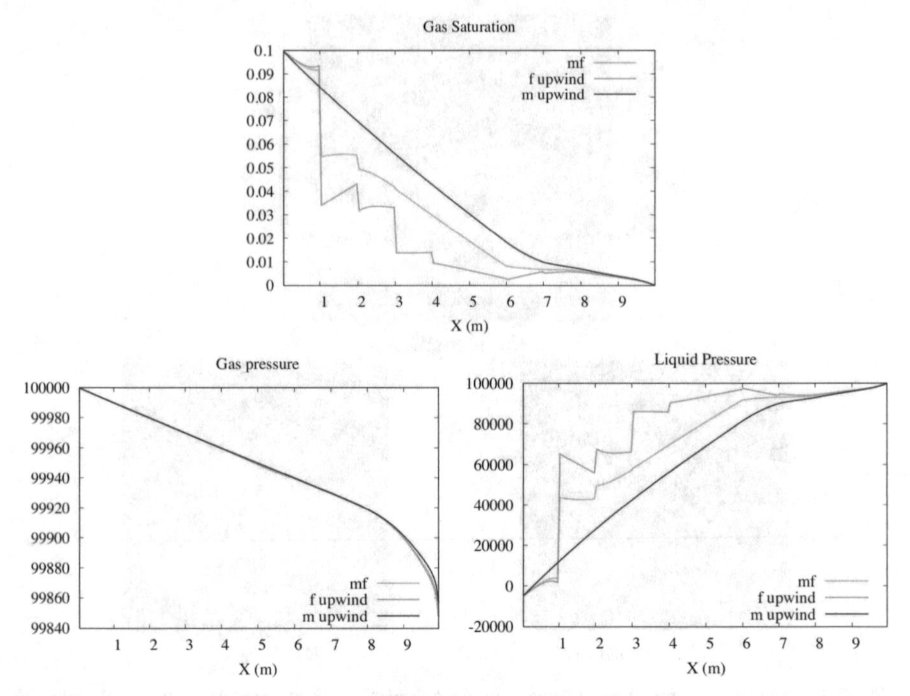

Fig. 14 Cuts at $x = 7.5$ m of the gas saturation (top), water pressure (left bottom) and gas pressure (right bottom) at final simulation time for the for the hybrid-dimensional models

Table 6 Numerical behavior of the hybrid-dimensional models for the desaturation test case

Model	Nb cells	Nb dof	$N_{\Delta t}$	N_{Newton}	N_{GMRes}	N_{Chop}	CPU[s]
mf nonlinear	25600	27117	227	11.2	105	20	2726
f upwind	25600	27117	199	10.9	111	15	2601
m upwind	25600	27117	245	10.8	92	24	2525

The mesh is Cartesian of size 160×160

The geometry of the domain and its coarsest tetrahedral mesh conforming to the fracture network are illustrated in Fig. 15. The domain consists of a matrix domain of extension $100\,\text{m}\times 100\,\text{m}\times 100\,\text{m}$ and of an homogeneous anisotropic permeability Λ_m, which is of 0.1 Darcy in x and in y directions and 0.01 Darcy in z direction. The fracture network is assumed to be of constant aperture $d_f = 1$ cm and of isotropic permeability $\Lambda_f = \lambda_{f,n} = 10$ Darcy. The matrix porosity ϕ_m is set to 0.2 and the fracture porosity ϕ_f to 0.4. The matrix capillary pressure is given by $P_{c,m}(s^o) = -b_m \log(1 - s^o)$, with $b_m = 10^4$ Pa, and the fracture capillary pressure by $P_{c,f}(s^o) = -b_f \log(1 - s^o)$, with $b_f = 100$ Pa. The matrix and the fracture network have the same relative permeabilities given by $k^o_{r,f}(s^o) = k^o_{r,m}(s^o) = (s^o)^2$ and $k^w_{r,f}(s^w) = k^w_{r,m}(s^w) = (s^w)^2$. The fluid properties are characterized by their dynamic viscosities $\mu^o = 5 \times 10^{-3}$, $\mu^w = 10^{-3}$ Pa s and by their mass densities $\rho^w = 1000$ and $\rho^o = 700$ Kg.m^{-3}.

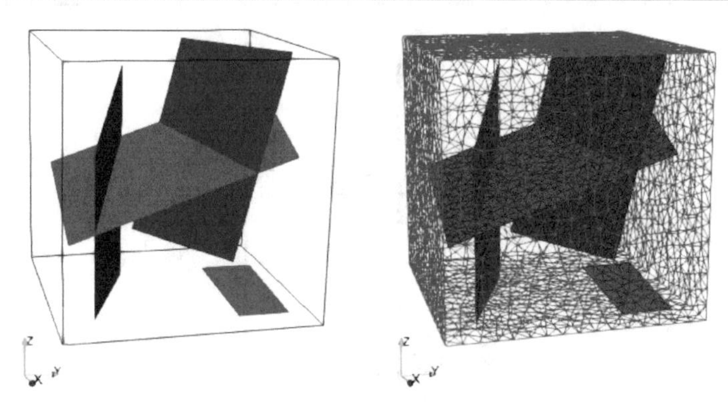

Fig. 15 Geometry of the domain $\Omega = 100\,\text{m} \times 100\,\text{m} \times 100\,\text{m}$ with the fracture network in red (left), coarsest tetrahedral mesh with 47670 cells (right) (color figure online)

At initial time, the reservoir is fully saturated with water. Then, oil is injected from below, which is managed by imposing Dirichlet conditions at the bottom and at the top of the reservoir. We also impose 20×10^5 Pa of overpressure w.r.t. the hydrostatic distribution of pressure. The oil then mounts by gravity, thanks to its lower density compared to water, and by the pressure gradient.

The sizes of the four tetrahedral meshes used for the simulations are reported in Table 7 together with the number of d.o.f. of the hybrid-dimensional *mf* nonlinear and linear models. Note that the cell unknowns used for the VAG discretization are eliminated without any fill-in by static condensation after assembly of the Jacobian system at each Newton iteration resulting in a much lower number of d.o.f. compared with the number of cells. The number of d.o.f. for both *mf* linear models is reduced compared with the number of d.o.f. of the *mf* nonlinear model since the *mf* interface d.o.f. are not included when using continuous pressure models. The parameters for the numerical resolution are as follows: $crit^{rel}_{Newton} = 10^{-6}$, $crit^{rel}_{GMRes} = 10^{-6}$, $\delta s^{obj} = 0.5$. The simulation of the flow covers the period of one year. The maximal time step size is $\Delta t_{max} = 0.1$ days, except for the *mf* non linear model solved on the finest mesh, for which we set $\Delta t_{max} = 0.03$ days.

This test shows the discrepancy between the *mf* linear *f* upwind model and the two other hybrid-dimensional models for a simulation period of fracture filling time scale (See Fig. 16). As one would expect, Figs. 17 and 18 reveal that the *mf* linear *f* upwind model overestimates the exchange of oil from the fracture network into the matrix, which leads on the one hand to a higher prediction of oil volume in the global domain and on the other hand to a slower filling of the fracture network. This is basically due to the fact that the flux going out of the fractures does not take into account the saturation jump at the *mf* interfaces for the *mf* linear *f* upwind model in contrast with the *mf* nonlinear and the *mf* linear *m* upwind models. Also, from Fig. 18 the solution of the *mf* linear *f* upwind model is strongly mesh dependent. For the *mf* linear *m* upwind model and the *mf* nonlinear model, we observe a much better convergence in space. Table 8 exhibits a large computational overhead for the *mf* nonlinear model compared with the *mf* linear models especially on fine meshes. Such a large computational overhead was not observed on the 2D test cases using

Table 7 **Nb cells** is the number of cells of the mesh; **Nb dof** is the number of d.o.f. (with two physical primary unknowns per d.o.f.); **Nb dof el.** is the number of d.o.f. after elimination of cell unknowns without fill-in

Mesh	Nb cells	Nb dof (*mf* non lin.)	Nb dof (*mf* lin.)	Nb dof el. (*mf* non lin.)	Nb dof el. (*mf* lin.)
1	47670	62763	57696	15093	9278
2	124245	159744	148505	35499	23187
3	253945	321670	301643	67725	46283
4	452401	566243	535085	113842	80965

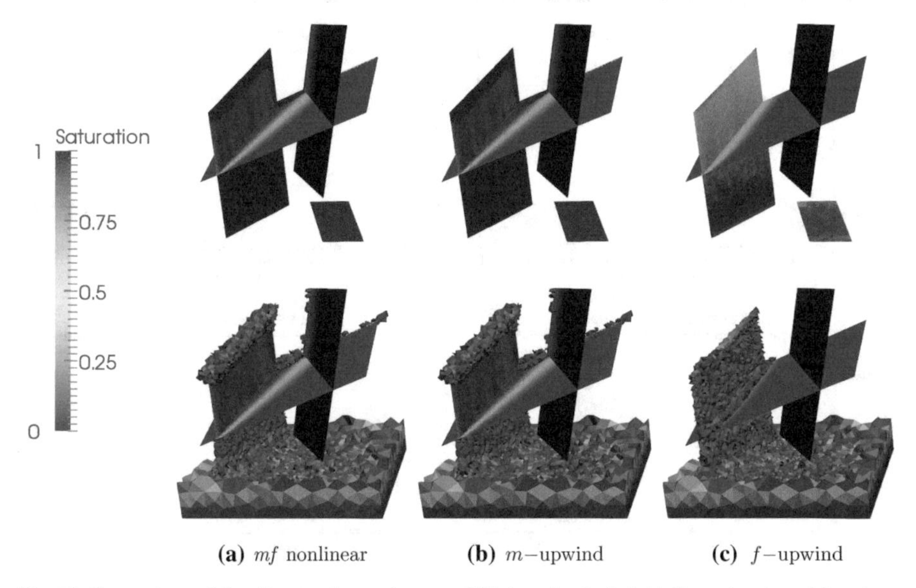

(a) mf nonlinear **(b)** $m-$upwind **(c)** $f-$upwind

Fig. 16 Comparison of the oil saturation at time $t = 295$ days for the hybrid-dimensional models using a matrix oil saturation threshold at 0.01. Upper line: fracture network only

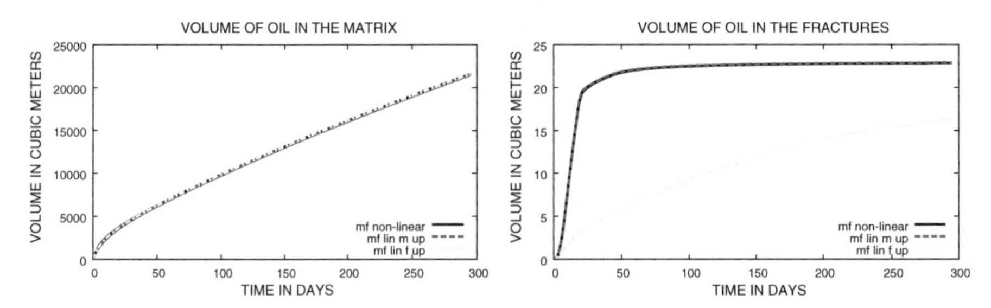

Fig. 17 Comparison of the matrix and fracture volumes occupied by oil as a function of time for the different models

the TPFA discretization, probably thanks to the nonlinear and linear eliminations of the *mf* interface unknowns. Such eliminations cannot be performed for the VAG discretization due to the coupling of the node unknowns along the connected fractures. On the other hand, for this test case, the *mf* nonlinear model on the coarsest mesh is more accurate than the *mf* linear f upwind model on the finest mesh.

7 Conclusion

The numerical experiments exhibit on various test cases a better accuracy of the hybrid-dimensional *mf* nonlinear model based on nonlinear transmission conditions compared with the hybrid-dimensional models based on linear transmission conditions.

The hybrid-dimensional *mf* linear m upwind model, which matches basically with the continuous phase pressures model (Bogdanov et al. 2003; Reichenberger et al.

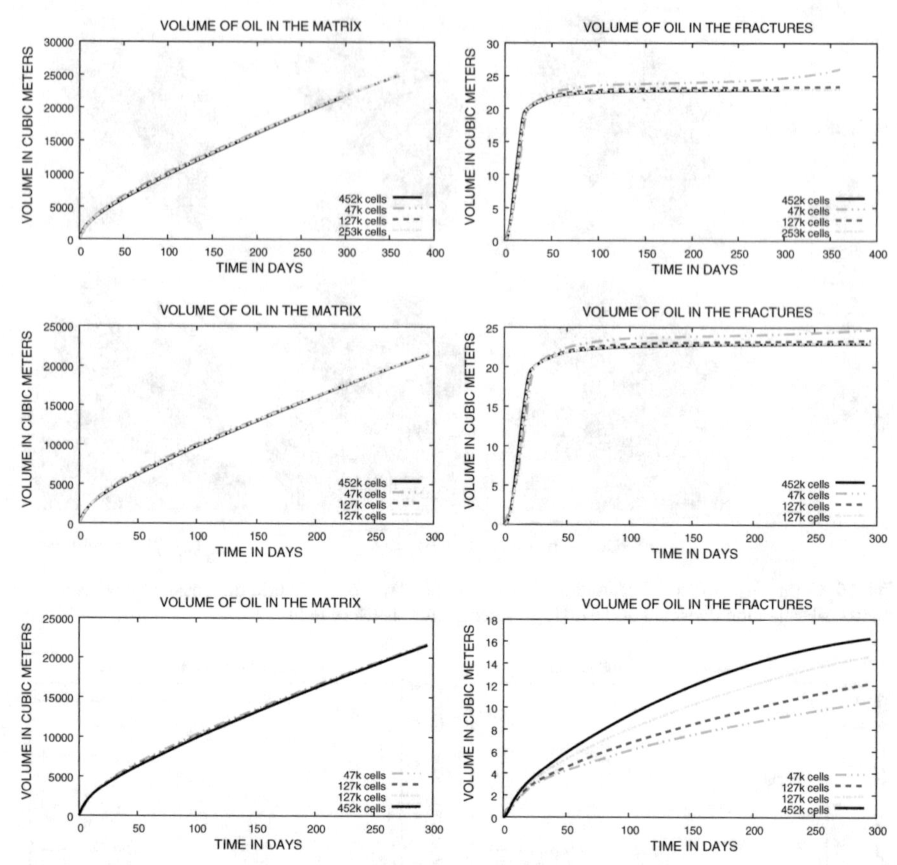

Fig. 18 Comparison of the convergence for the different models: *mf* non linear, *mf* linear *m* upwind and *mf* linear *f* upwind schemes (above to below)

2006; Monteagudo and Firoozabadi 2007; Hoteit et al. 2008; Brenner et al. 2015, 2017) in the case of fractures acting as drains, is shown to fail to produce a good approximation of the equi-dimensional model once the fractures are filled with the non-wetting phase. This is a modeling error due to an overestimation of the capillary pressure inside the filled fractures. This situation is quite different to what happens for single phase flows for which continuous and discontinuous pressure models basically match as soon as the ratio between the fracture normal permeability and the fracture width is large compared with the ratio between the matrix permeability and its characteristic length. As expected, for fractures acting as capillary barriers, the hybrid-dimensional *mf* linear *m* upwind model also fails to reproduce the capillary barrier effect and only accounts for permeability barriers.

The hybrid-dimensional *mf* linear *f* upwind model tends to overestimate the non-wetting phase flux going out of the permeable fractures since it does not capture the saturation jump at *mf* interfaces in contrast with the *mf* nonlinear and *mf* linear *m* upwind models. Both *mf* linear *f* and *m* upwind models fail to reproduce an accurate

Table 8 **Mesh** refers to the meshes defined in Table 7; $\mathbf{N}_{\Delta t}$ is the number of successful time steps; $\mathbf{N}_{Newton}$ is the total number of Newton iterations (for successful time steps); $\mathbf{N}_{Chop}$ is the number of time step chops

Mesh	$N_{\Delta t}$	N_{Newton}	N_{GMRes}	N_{Chop}	CPU[s]
mf non linear model					
1	3103	9283	445677	6	14113
2	3367	11846	869702	19	64326
3	3890	16413	1365365	38	207589
4	9947	32625	2792423	29	775567
mf linear m upwind model					
1	3075	3655	202178	0	3042
2	3075	3781	282025	0	10325
3	3075	4122	387035	0	27839
4	3075	4737	536641	0	63766
mf linear f upwind model					
1	3075	3621	199874	0	2879
2	3075	3782	283473	0	9943
3	3075	4125	388904	0	26917
4	3075	4703	538231	0	63700

solution for the desaturation by suction test case. The *f* upwind model overestimates the capillary barrier effect at *mf* interfaces while the *m* upwind model does not capture the saturation and water pressure jumps at dry fracture interfaces.

None of the *mf* linear models are able to take into account the gravity segregation in the fracture width which can have a strong impact for small capillary pressures or for large apertures such as in the case of faults.

A moderate computational overhead of the *mf* nonlinear model compared with the *mf* linear models is obtained in the 2D simulations using the TPFA scheme thanks to the elimination of the *mf* interface unknowns at both the linear and nonlinear levels. The VAG scheme does not allow such eliminations for the *mf* nonlinear model due to the coupling of the nodes along the connected fractures. It results that the *mf* nonlinear model combined with the VAG discretization exhibits a large computational overhead compared with the continuous pressure model. To extend the elimination of the *mf* interface unknowns to consistent discretizations on general meshes, a promising approach developped in Aghili et al. (2018) is to use face-based discretizations (Brenner et al. 2016a) rather than nodal-based discretizations.

References

Aghili, J., Brenner, K., Hennicker, J., Masson, R., Trenty, L.: Hybrid finite volume discretization of two-phase Discrete Fracture Matrix models with nonlinear interface solver. In: ECMOR XVI—16th European Conference on the Mathematics of Oil Recovery, Barcelona, France (2018) https://doi.org/10.3997/2214-4609.201802272

Alboin, C., Jaffré, J., Roberts, J., Serres, C.: Modeling fractures as interfaces for flow and transport in porous media. Fluid Flow Transport Porous Med. **295**, 13–24 (2002)

Ahmed, R., Edwards, M.G., Lamine, S., Huisman, B.A.H.: Control-volume distributed multi-point flux approximation coupled with a lower-dimensional fracture model. J. Comput. Phys. **284**, 462–489 (2015)

Ahmed, E., Jaffré, J., Roberts, J.E.: A reduced fracture model for two-phase flow with different rock types. Math. Comput. Simul. **7**, 49–70 (2017)

Angot, P., Boyer, F., Hubert, F.: Asymptotic and numerical modeling of flows in fractured porous media. Math. Model. Numer Anal **43**(2), 239–275 (2009)

Bentsen, R.G., Anli, J.: Using parameter estimation techniques to convert centrifuge data into a capillary-pressure curve. SPE J. **17**(1), 57–64 (1977)

Bogdanov, I., Mourzenko, V., Thovert, J.-F., Adler, P.M.: Two-phase flow through fractured porous media. Phys. Rev. E **68**, 026703 (2003)

Brenner, K., Groza, M., Guichard, C., Masson, R.: Vertex approximate gradient scheme for hybrid-dimensional two-phase Darcy flows in fractured porous media. ESAIM Math. Model. Numer. Anal. **49**, 303–330 (2015)

Brenner, K., Hennicker, J., Masson, R., Samier, P.: Gradient discretization of hybrid-dimensional Darcy flow in fractured porous media with discontinuous pressures at matrix–fracture interfaces. IMA J. Numer. Anal. **37**(3), 1551–1585 (2016a)

Brenner, K., Groza, M., Guichard, C., Lebeau, G., Masson, R.: Gradient discretization of hybrid-dimensional Darcy flows in fractured porous media. Numerische Mathematik **134**(3), 569–609 (2016b)

Brenner, K., Groza, M., Jeannin, L., Masson, R., Pellerin, J.: Immiscible two-phase Darcy flow model accounting for vanishing and discontinuous capillary pressures: application to the flow in fractured porous media. Comput. Geosci. **21**, 5–6 (2017)

Brenner, K., Hennicker, J., Masson, R., Samier, P.: Hybrid-dimensional modeling of two-phase flow through fractured porous media with enhanced matrix fracture transmission conditions. J. Comput. Phys. **357**, 100–124 (2018)

Brooks, R.H., Corey, A.T.: Hydraulic properties of porous media and their relation to drainage design. Trans. ASAE **7**(1), 0026–0028 (1964)

Droniou, J., Hennicker, J., Masson, R.: Numerical analysis of a two-phase flow discrete fracture model. Numerische Mathematik (2018). https://doi.org/10.1007/s00211-018-0994-y

Eymard, R., Gallouët, T., Guichard, C., Herbin, R., Masson, R.: TP or not TP, that is the question. Comput. Geosci. **18**, 285–296 (2014)

Flauraud, E., Nataf, F., Faille, I., Masson, R.: Domain decomposition for an asymptotic geological fault modeling. C. R. à l'Académie des Sciences, Mécanique **331**, 849–855 (2003)

Flemisch, B., Berre, I., Boon, W., Fumagalli, A., Schwenck, N., Scotti, A., Stefansson, I., Tatomir, A.: Benchmarks for single-phase flow in fractured porous media. Adv. Water Resour. **111**, 239–258 (2018)

Hoteit, H., Firoozabadi, A.: Numerical modeling of two-phase flow in heterogeneous permeable media with different capillarity pressures. Adv. Water Resour. **31**, 56–73 (2008)

Jaffré, J., Martin, V., Roberts, J.E.: Modeling fractures and barriers as interfaces for flow in porous media. SIAM J. Sci. Comput. **26**(5), 1667–1691 (2005)

Jaffré, J., Mnejja, M., Roberts, J.E.: A discrete fracture model for two-phase flow with matrix–fracture interaction. Procedia Comput. Sci. **4**, 967–973 (2011)

Karimi-Fard, M., Durlovski, L.J., Aziz, K.: An efficient discrete-fracture model applicable for general-purpose reservoir simulators. SPE J. **9**, 227–236 (2004)

Lacroix, S., Vassilevski, Y.V., Wheeler, M.F.: Decoupling preconditioners in the implicit parallel accurate reservoir simulator (IPARS). Numer. Linear Algebra Appl. **8**, 537–549 (2001)

Monteagudo, J.E.P., Firoozabadi, A.: Control-volume model for simulation of water injection in fractured media: incorporating matrix heterogeneity and reservoir wettability effects. SPE J. **12**, 3 (2007)

Mualem, Y.: A new model for predicting the hydraulic conductivity of unsaturated porous media. Water Resour. Res. **12**, 513–522 (1976)

Reichenberger, V., Jakobs, H., Bastian, P., Helmig, R.: A mixed-dimensional finite volume method for multiphase flow in fractured porous media. Adv. Water Resour. **29**(7), 1020–1036 (2006)

Scheichl, R., Masson, R., Wendebourg, J.: Decoupling and block preconditioning for sedimentary basin simulations. Comput. Geosci. **7**, 295–318 (2003)

Sandve, T.H., Berre, I., Nordbotten, J.M.: An efficient multi-point flux approximation method for Discrete Fracture–Matrix simulations. JCP **231**, 3784–3800 (2012)

Tunc, X., Faille, I., Gallouët, T., Cacas, M.C., Havé, P.: A model for conductive faults with non matching grids. Comput. Geosciences **16**, 277–296 (2012)

Van Genuchten, M.T.: A closed-form equation for predicting the hydraulic conductivity of unsaturated soils. Soil Sci. Soc. Am. J. **44**, 892–898 (1980)

Publisher's Note Springer Nature remains neutral with regard to jurisdictional claims in published maps and institutional affiliations.

Affiliations

Joubine Aghili[1,2] · Konstantin Brenner[1,2] · Julian Hennicker[3] · Roland Masson[1,2] · Laurent Trenty[4]

1 Team COFFEE, Inria Sophia Antipolis - Méditerranée, Valbonne, France

2 Laboratoire J.A. Dieudonné, UMR 7351 CNRS, Université Côte d'Azur, Nice, France

3 Université de Genève, Geneva, Switzerland

4 Andra, Chatenay-Malabry, France

GEM - International Journal on Geomathematics
https://doi.org/10.1007/s13137-019-0116-8

ORIGINAL PAPER

A hybrid-dimensional discrete fracture model for non-isothermal two-phase flow in fractured porous media

Dennis Gläser[1] · Bernd Flemisch[1] · Rainer Helmig[1] · Holger Class[1]

Received: 31 March 2018 / Accepted: 2 November 2018

Abstract

We present a hybrid-dimensional numerical model for non-isothermal two-phase flow in fractured porous media, in which the fractures are modeled as entities of codimension one embedded in a bulk domain. Potential fields of applications of the model could be radioactive waste disposal or geothermal energy production scenarios in which a two-phase flow regime develops or where CO_2 is used as working fluid. We test the method on synthetic test cases involving compressible fluids and strongly heterogeneous, full tensor permeability fields by comparison with a reference solution obtained from an equi-dimensional discretization of the domain. The results reveal that especially for the case of a highly conductive fracture, the results are in good agreement with the reference. While the model qualitatively captures the involved phenomena also for the case of a fracture acting as both hydraulic and capillary barrier, it introduces larger errors than in the highly-conductive fracture case, which can be attributed to the lower-dimensional treatment of the fracture. Finally, we apply the method to a three-dimensional showcase that resembles setups for the determination of upscaled parameters of fractured blocks.

Keywords Porous media · Fractures · Finite volumes · Multi-point flux approximation

Mathematics Subject Classification 76S05

1 Introduction

The presence of fractures may substantially alter the hydraulic behavior of porous media, which, due to the fact that almost all rocks exhibit fractures (see e.g. Jaeger

✉ Dennis Gläser
dennis.glaeser@iws.uni-stuttgart.de

1 Department of Hydromechanics and Modelling of Hydrosystems, University of Stuttgart, Pfaffenwaldring 61, 70569 Stuttgart, Germany

et al. 2007), manifests their importance particularly in geotechnical engineering applications. For instance, oil recovery applications or environmental issues as e.g. the integrity of radioactive waste disposal sites are strongly influenced by fractures, while geothermal energy production or unconventional gas production techniques even rely on them. Many of these applications involve complex physics, as not only two-phase flow processes but also non-isothermal effects often play an important role. For example, enhanced geothermal systems using supercritical CO_2 as working fluid implicate two-phase flow regimes driven by both the injection as well as buoyancy forces, together with temperature changes caused by heat exchange and by the compression and expansion of CO_2 (Pruess 2006).

However, the numerical simulation of fractured porous media is very challenging due to the complex geometries involved in arbitrary networks of fractures, and the typically very small apertures in comparison with the considered spatial scales. Many approaches to the numerical modeling of fractured porous media have been developed, which can be classified into continuum fracture models and discrete fracture–matrix (dfm) models. Dual-continuum models belong to the first class of models, in which the fractures are not resolved geometrically but are accounted for by a second overlapping continuum with transfer functions describing the mutual interaction (see e.g. Warren and Root 1963; Kazemi et al. 1976). An extension to multiple continua is also possible (see e.g. Pruess 1992; Tatomir 2013). In dfm models, the fracture geometries are discretely captured either by an equi-dimensional discretization of the fractures (see e.g. Matthai et al. 2007), or by lower-dimensional geometries with corresponding values for the apertures. The latter, i.e. hybrid-dimensional discrete fracture models, can be realized in a conforming way by constraining the element facets of the discretization used for the matrix domain to the fracture geometries. This type of models has been realized for single-phase flow e.g. using mixed finite elements (see e.g. Martin et al. 2005) or cell-centered finite volumes (see e.g. Karimi-Fard et al. 2004; Sandve et al. 2012; Ahmed et al. 2015), while for two-phase flow, Reichenberger et al. (2006) and Brenner et al. (2014) present models on the basis of a node-centered finite volume scheme and the vertex approximate gradient scheme, respectively. A big advantage of non-conforming approaches is the ability to use independent discretizations which substantially facilitates the grid creation, especially for highly fractured reservoirs. Examples for this type of models on the basis of an extended finite element formulation can be found e.g. in Schwenck (2015) for single-phase and in Fumagalli and Scotti (2013) for two-phase flow. An approach using finite volumes is presented in Tene et al. (2017). All of the three mentioned models are developed such that they are also applicable to fractures acting as barriers to flow, while this is generally difficult to handle in non-conforming approaches.

In this work we present a conforming hybrid-dimensional model based on a finite-volume formulation for non-isothermal two-phase flow, which has been implemented into the open source simulator DuMu^x (Flemisch et al. 2011). We state the governing set of equations in Sect. 2, before we outline its formulation in the hybrid-dimensional context in Sect. 3. Subsequently, we illustrate the finite-volume discretization in Sect. 4 and apply the method to two- and three-dimensional synthetic test cases in Sect. 5. A summary and conclusion is then given in Sect. 6.

2 Two-phase non-isothermal flow equations

We consider two immiscible fluid phases indexed by $\alpha \in \{\mathrm{w}, \mathrm{n}\}$, where w and n refer to the wetting and the non-wetting phase, respectively. Flow processes occur through a rigid solid matrix and we furthermore assume local thermal equilibrium between the three constituents, which allows us to formulate the governing set of equations as follows:

$$\phi \frac{\partial (\rho_\alpha S_\alpha)}{\partial t} + \nabla \cdot (\rho_\alpha \mathbf{v}_\alpha) = q_\alpha, \tag{1}$$

$$\phi \sum_\alpha \frac{\partial \rho_\alpha S_\alpha u_\alpha}{\partial t} + (1-\phi) \frac{\partial \rho_\mathrm{s} c_\mathrm{s} T}{\partial t} + \nabla \cdot \left(\sum_\alpha \rho_\alpha h_\alpha \mathbf{v}_\alpha - \lambda \nabla T \right) = q_\mathrm{h}, \tag{2}$$

$$\mathbf{v}_\alpha = -\frac{k_{\mathrm{r}\alpha}}{\mu_\alpha} \mathbf{K} \left(\nabla p_\alpha - \rho_\alpha \mathbf{g} \right). \tag{3}$$

The Eq. (1) is the mass balance equations of the fluid phases, (2) is the energy balance equation of the porous medium and Eq. (3) are the momentum balance equations of the fluid phases, expressed by means of the standard multi-phase extension of Darcy's law. Here, ρ_α, μ_α, p_α, S_α and $\mathbf{v}_\alpha$ denote the fluid phase densities, viscosities, pressures, saturations and velocities. Moreover, $k_{\mathrm{r}\alpha}$, u_α and h_α are the fluid phase relative permeabilities, specific internal energies and specific enthalpies. The quantities ρ_s and c_s denote the density and the heat capacity of the solid matrix, while ϕ, $\mathbf{K}$ and λ are the porosity, the absolute permeability and the thermal conductivity of the porous medium. The vector $\mathbf{g}$ refers to the gravitational acceleration, while q_α and q_h denote mass and heat sources, respectively. We close the system of equations by the constraint $S_\mathrm{w} + S_\mathrm{n} = 1$ and by introducing the capillary pressure, $p_\mathrm{c} = p_\mathrm{n} - p_\mathrm{w}$, for which we will use a constitutive relationship as a function of the wetting phase saturation. Note that all the involved parameters might exhibit a non-linear dependency on pressure, temperature and/or saturation, while $\mathbf{K}$ and λ are possibly also discontinuous in space. After insertion of (3) in (1) and (2) we are left with three equations, for which we use the wetting phase pressure p_w, the non-wetting phase saturation S_n and the temperature T of the porous medium as the primary variables. We use a fully-implicit solution strategy applying the implicit Euler scheme for the temporal discretization. Furthermore, all equations fully coupled and monolithically with the Newton–Raphson method as non-linear solver together with the direct solver UMFPack (Davis 2004) to solve the linear system of equations in each Newton iteration.

3 Flow in a domain with a fracture

As mentioned above, the parameters describing the porous medium might be highly heterogeneous within the domain as a consequence of the heterogeneous nature of geological systems. In particular, fractures can be seen as geological features of very small extent in one coordinate direction in comparison to the domain size considered. In spite of the small size, fractures can have a strong influence on the global flow field as

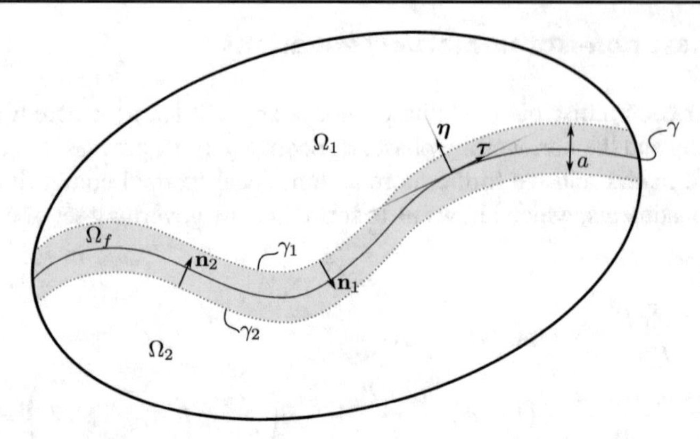

Fig. 1 Illustration of a domain intersected by a fracture

their hydraulic properties might be orders of magnitude different compared to those of the surrounding porous matrix. Conceptually, we treat fractures as porous media with a very different parameterization in comparison to the surrounding matrix. Geologically, this corresponds to situations where the fractures are filled with matter, e.g. due to abrasion during shear failure of the fracture plane, and might not be adequate when considering open fractures. In order to define the problem at hand [i.e. the Eqs. (1)–(2)] on such a system, let us consider an open bounded connected polygonal domain $\Omega \subset \mathbb{R}^d, d = \{2, 3\}$, which is split by a fracture, Ω_{f}, into two sub-domains Ω_1 and Ω_2 (see Fig. 1). We denote by $\Gamma = \partial\Omega$ the boundary of the domain and by $\Gamma_i = \partial\Omega_i \cap \Gamma$, $i \in \{1, 2, \mathrm{f}\}$ the part of the boundary associated with the sub-domains. We further split those into two parts $\Gamma_{i\mathrm{N}}$ and $\Gamma_{i\mathrm{D}}$, where we define Neumann and Dirichlet boundary conditions, respectively. Moreover, $\gamma_1 = \bar{\Omega}_{\mathrm{f}} \cap \bar{\Omega}_1$ and $\gamma_2 = \bar{\Omega}_{\mathrm{f}} \cap \bar{\Omega}_2$ denote the interfaces between the fracture and the two neighboring sub-domains. On this setting the system thus reads:

$$\begin{aligned}
&\phi_i \frac{\partial\left(\rho_\alpha S_{i\alpha}\right)}{\partial t} + \nabla\cdot\mathbf{F}_{i\alpha} = q_{i\alpha} && \text{on } \Omega_i,\\
&\phi_i \sum_\alpha \frac{\partial \rho_\alpha S_{i\alpha} u_\alpha}{\partial t} + (1-\phi_i)\frac{\partial \rho_{is} c_{is} T_i}{\partial t} + \nabla\cdot\mathbf{F}_{i\mathrm{h}} = q_{i\mathrm{h}} && \text{on } \Omega_i,\\
&\mathbf{F}_{i\alpha}\cdot\mathbf{n} = \bar{F}_{i\alpha},\ \ \mathbf{F}_{i\mathrm{h}}\cdot\mathbf{n} = \bar{F}_{i\mathrm{h}} && \text{on } \Gamma_{i\mathrm{N}},\\
&p_{i\mathrm{w}} = \bar{p}_{i\mathrm{w}},\ \ S_{i\mathrm{n}} = \bar{S}_{i\mathrm{n}},\ \ T_i = \bar{T}_i && \text{on } \Gamma_{i\mathrm{D}},
\end{aligned} \tag{4}$$

where $i \in \{1, 2, \mathrm{f}\}$ and where we have introduced the mass and heat fluxes

$$\mathbf{F}_{i\alpha} = -\rho_\alpha \frac{k_{\mathrm{ir}\alpha}}{\mu_\alpha}\mathbf{K}_i\left(\nabla p_{i\alpha} - \rho_\alpha \mathbf{g}\right) = \rho_\alpha \mathbf{v}_{i\alpha}, \tag{5}$$

$$\mathbf{F}_{i\mathrm{h}} = \sum_\alpha \rho_\alpha h_\alpha \mathbf{v}_{i\alpha} - \lambda_i \nabla T_i. \tag{6}$$

At the interfaces between material discontinuities, i.e. on interfaces between sub-domains, we enforce continuity of pressures, temperature and fluxes:

$$\left.\begin{aligned} p_{\mathrm{fw}} &= p_{j\mathrm{w}} \\ p_{\mathrm{fc}} &= p_{j\mathrm{c}} \\ T_{\mathrm{f}} &= T_j \\ \mathbf{F}_{\mathrm{f}\alpha} \cdot \mathbf{n}_{\mathrm{f}} + \mathbf{F}_{j\alpha} \cdot \mathbf{n}_j &= 0 \\ \mathbf{F}_{\mathrm{fh}} \cdot \mathbf{n}_{\mathrm{f}} + \mathbf{F}_{j\mathrm{h}} \cdot \mathbf{n}_j &= 0 \end{aligned}\right\} \quad \text{on } \gamma_j,\ j \in \{1, 2\}. \tag{7}$$

Note that this presentation of the equations corresponds to the case of the fracture intersecting the entire domain. For fractures ending inside the domain the equations can be derived analogously, with the difference being that the two interfaces γ_1 and γ_2 end in a single common point inside the domain. Instead of the equations being formulated on the three sub-domains $\Omega_i, i \in \{1, 2, \mathrm{f}\}$, the system would then be formulated on only two sub-domains representing the matrix and fracture, respectively. Even though the test cases presented in this work do consider fractures ending inside the domain, the authors think the presentation of the equations is more comprehensive for the case of fractures intersecting the entire domain.

In the following we will outline the derivation of the lower-dimensional formulation for the fracture, inspired by the works of Martin et al. (2005), Jaffré et al. (2011) and Fumagalli and Scotti (2013). In order to derive a $(d-1)$-dimensional formulation on γ, the primary variables are averaged over cross sections of Ω_{f}, i.e. along the η-coordinate of the fracture-local coordinate system (see Fig. 1). Please note that this procedure requires γ_1 and γ_2 to be parallel, i.e. we assume $\mathbf{n}_1 \approx -\mathbf{n}_2$. Consequently, these average values are denoted by $S_{\gamma\mathrm{n}}$, $p_{\gamma\mathrm{w}}$ and T_γ and are used for the constitutive relationships to evaluate fluid properties and the relative permeability. Accordingly, we introduce ϕ_γ, $\mathbf{K}_{\gamma\tau}$, λ_γ, $\rho_{\gamma s}$ and $c_{\gamma s}$ as the cross section-averaged fracture porous medium properties, where $\mathbf{K}_{\mathrm{f}}$ has been split in a tangential and a normal part with respect to the fracture, $\mathbf{K}_{\mathrm{f}\tau}$ and $k_{\mathrm{f}\eta}$. Similarly, we split the fluxes $\mathbf{F}_{\mathrm{f}\alpha} = \mathbf{F}_{\mathrm{f}\alpha\tau} + \mathbf{F}_{\mathrm{f}\alpha\eta}$ and $\mathbf{F}_{\mathrm{fh}} = \mathbf{F}_{\mathrm{fh}\tau} + \mathbf{F}_{\mathrm{fh}\eta}$ and approximate the tangential and normal parts of these fluxes by means of the averaged variables and porous medium properties:

$$\mathbf{F}_{\mathrm{f}\alpha\tau} \approx -\rho_\alpha \frac{k_{\mathrm{r}\alpha}}{\mu_\alpha} \mathbf{K}_{\gamma\tau} \left(\nabla_\tau p_{\gamma\alpha} - \rho_\alpha \mathbf{g}_\tau\right) = \rho_\alpha \mathbf{v}_{\gamma\alpha\tau}, \tag{8}$$

$$\mathrm{F}_{\mathrm{f}\alpha\eta} \approx -\rho_\alpha \frac{k_{\mathrm{r}\alpha}}{\mu_\alpha} k_{\mathrm{f}\eta} \left(\nabla_\eta p_{\mathrm{f}\alpha} - \rho_\alpha g_\eta\right) = \rho_\alpha \mathrm{v}_{\mathrm{f}\alpha\eta}, \tag{9}$$

$$\mathbf{F}_{\mathrm{fh}\tau} \approx \sum_\alpha \rho_\alpha h_\alpha \mathbf{v}_{\gamma\alpha\tau} - \lambda_\gamma \nabla_\tau T_\gamma, \tag{10}$$

$$\mathrm{F}_{\mathrm{h}\eta} \approx \sum_\alpha \rho_\alpha h_\alpha \mathrm{v}_{\mathrm{f}\alpha\eta} - \lambda_\gamma \nabla_\eta T_{\mathrm{f}}. \tag{11}$$

Here, we have introduced the tangential and normal gradient operators ∇_τ and ∇_η. Using the averaged variables and the above fluxes in (4) for the fracture domain, integration over a cross section of Ω_{f} yields the following equations on γ:

$$
\begin{aligned}
& a\phi_\gamma \frac{\partial\left(\rho_\alpha S_{\gamma\alpha}\right)}{\partial t} + a\nabla_\tau \cdot \mathbf{F}_{\mathrm{f}\alpha\tau} \\
& \quad = a q_{\gamma\alpha} + \mathbf{F}_{1\alpha} \cdot \mathbf{n}_1 + \mathbf{F}_{2\alpha} \cdot \mathbf{n}_2, && \text{on } \gamma \\
& a\phi_\gamma \sum_\alpha \frac{\partial \rho_\alpha S_{\gamma\alpha} u_\alpha}{\partial t} + a\left(1-\phi_\gamma\right) \frac{\partial \rho_{\gamma \mathrm{s}} c_{\gamma \mathrm{s}} T_\gamma}{\partial t} + a\nabla_\tau \cdot \mathbf{F}_{\mathrm{fh}\tau} \\
& \quad = a q_{\gamma \mathrm{h}} + \mathbf{F}_{1\mathrm{h}} \cdot \mathbf{n}_1 + \mathbf{F}_{2\mathrm{h}} \cdot \mathbf{n}_2, && \text{on } \gamma \\
& \mathbf{F}_{\mathrm{f}\alpha\tau} \cdot \mathbf{n} = \bar{F}_{\mathrm{f}\alpha}, \;\; \mathbf{F}_{\mathrm{fh}\tau} \cdot \mathbf{n} = \bar{F}_{\mathrm{fh}} && \text{on } \Gamma_{\gamma N}, \\
& \mathbf{F}_{\mathrm{f}\alpha\tau} \cdot \mathbf{n} = 0, \;\; \mathbf{F}_{\mathrm{fh}\tau} \cdot \mathbf{n} = 0 && \text{on } \Gamma_{\gamma I}, \\
& p_{\gamma \mathrm{w}} = \bar{p}_{\gamma \mathrm{w}}, \;\; S_{\gamma \mathrm{n}} = \bar{S}_{\gamma \mathrm{n}}, \;\; T_\gamma = \bar{T}_\gamma && \text{on } \Gamma_{\gamma D}.
\end{aligned}
\tag{12}
$$

Note that here we introduced the boundary segments $\Gamma_{\gamma D} \subset \Gamma_\gamma$ and $\Gamma_{\gamma N} \subset \Gamma_\gamma$, with $\Gamma_\gamma = \partial\gamma \cap \partial\Omega$, on which we apply Dirichlet and Neumann boundary conditions, respectively. We use no-flow boundary conditions on fracture tips inside the domain, i.e. on $\Gamma_{\gamma I} = \partial\gamma \backslash \Gamma_\gamma$. Furthermore, we have used the divergence theorem, have introduced the tangential and normal divergence operators $\nabla_\tau \cdot$ and $\nabla_\eta \cdot$ and have assumed, that the sources are independent of the η-coordinate. The two additional source contributions arising on the right side of the first two equations of (12) account for the exchange fluxes with the matrix. For these we have to formulate transmission conditions, which can be done on the basis of the interface conditions (7) and which now read for the interface γ:

$$
\left.
\begin{aligned}
\mathbf{F}_{j\alpha} \cdot \mathbf{n}_j &\overset{!}{=} -\mathrm{F}_{\mathrm{f}\alpha\eta} \approx \rho_\alpha \frac{k_{\mathrm{fr}\alpha}}{\mu_\alpha} k_{\mathrm{f}\eta} \left(\frac{p_{j\alpha} - p_{\gamma\alpha}}{a/2} - \rho_\alpha g_\eta \right) \\
\mathbf{F}_{j\mathrm{h}} \cdot \mathbf{n}_j &\overset{!}{=} -\mathrm{F}_{\mathrm{fh}\eta} \approx \textstyle\sum_\alpha h_\alpha \mathbf{F}_{j\alpha} \cdot \mathbf{n}_j + \lambda_\gamma \frac{T_j - T_\gamma}{a/2}
\end{aligned}
\right\}
\quad \text{on } \gamma_j, \; j \in \{1, 2\}. \tag{13}
$$

Note that we have used a simple finite-difference appoximation of the normal fluxes inside the fracture. Finally, we solve (12) in the lower-dimensional fracture domain γ and (4) in the resulting domain for the matrix, i.e. $\Omega^* = \Omega \backslash \gamma$, with the transmission conditions (13) on γ. The simplification of using averaged variables on the fracture will introduce errors depending on the actual distributions along the η-coordinate. Note that the determination of ranges of applicability for this model is an open question and is not addressed in this work. For fractures that are highly conductive in comparison to the surrounding materials, gradients in this direction will be relatively small. In that case, the simplifications introduced above are assumed to be acceptable, but they are expected to provoke more apparent errors for low-permeable fractures, where strong gradients across their width can occur. We want to study the impact of this effect qualitatively by numerical simulation of a synthetic test case and comparison with a reference solution.

4 Finite-volume discretization

The capillary pressure is often described by constitutive models that are functions of the wetting phase saturation S_w (see e.g. Matthai et al. 2007; Reichenberger et al. 2006).

Thus, as a result from the conditions (7), saturations are generally discontinuous across material interfaces with different $p_c - s_w$ relationships. We employ a cell-centered finite volume scheme to discretize the equations, for which the discrete solution space is the space of piecewise constants on the grid cells and is therefore naturally able to capture the resulting discontinuities. For the ease of presentation, we outline the finite-volume scheme on the model problem (14), representative for the balance equation of a quantity e, function of an unknown u:

$$\begin{aligned} \frac{\partial e(u)}{\partial t} - \nabla \cdot (\boldsymbol{\Lambda} \nabla u) &= g \quad \text{in } \Omega, \\ u &= \bar{u} \quad \text{on } \Gamma_D, \\ -(\boldsymbol{\Lambda} \nabla u) \cdot \mathbf{n} &= f \quad \text{on } \Gamma_N. \end{aligned} \tag{14}$$

Here, $\Omega \in \mathbb{R}^d$, $d \in \mathbb{N}^*$ is again an open bounded connected polygonal domain with boundary $\Gamma = \partial\Omega$ and Dirichlet and Neumann boundary segments $\Gamma_D \subset \Gamma$ and $\Gamma_N \subset \Gamma$. Λ is a symmetric and positive definite tensor. We now denote by $\mathcal{T}$ the set of cells of the discretization such that $\Omega = \cup_{K \in \mathcal{T}} K$. Integration of equation (14) over the control volume $K \in \mathcal{T}$ yields:

$$\int_K \frac{\partial e(u)}{\partial t} \, \mathrm{d}x - \int_{\partial K} (\boldsymbol{\Lambda} \nabla u) \cdot \mathbf{n} \, \mathrm{d}x = \int_K g \, \mathrm{d}x. \tag{15}$$

Let us now introduce the set of faces $\mathcal{E}$ of the discretization and for each cell $K \in \mathcal{T}$ the subset $\mathcal{E}_K$ such that $\partial K = \cup_{\sigma \in \mathcal{E}_K} \sigma$. The second integral of the left side of equation (15) can be split into the integrals over the faces of the cell, i.e.:

$$\int_K \frac{\partial e(u)}{\partial t} \, \mathrm{d}x - \sum_{\sigma \in \mathcal{E}_K} \int_\sigma (\boldsymbol{\Lambda} \nabla u) \cdot \mathbf{n} \, \mathrm{d}x = \int_K g \, \mathrm{d}x. \tag{16}$$

At the core of finite volume schemes lies the construction of the discrete flux approximation

$$F_{K,\sigma} \approx \int_\sigma (\boldsymbol{\Lambda} \nabla u) \cdot \mathbf{n} \, \mathrm{d}x, \tag{17}$$

which can be substantially different among the schemes. In cell-centered methods, the fluxes $F_{K,\sigma}$ are usually constructed introducing intermediate face unknowns $\bar{u}_\sigma$ (see Fig. 2), which are subsequently eliminated on the basis of interface conditions (see e.g. Aavatsmark 2002; Droniou 2014). These are equivalent to the conditions (7), which were introduced here for material discontinuities. For an interior face σ between the two cells K, L, i.e. $\sigma = \bar{K} \cap \bar{L}$, these conditions read (we denote by $\mathcal{E}_{int}$ the set of interior faces):

$$\begin{aligned} F_{K,\sigma} + F_{L,\sigma} &= 0 \\ \bar{u}_{K,\sigma} = \bar{u}_{L,\sigma} &= \bar{u}_\sigma \end{aligned} \quad \text{on } \sigma \in \mathcal{E}_{int}. \tag{18}$$

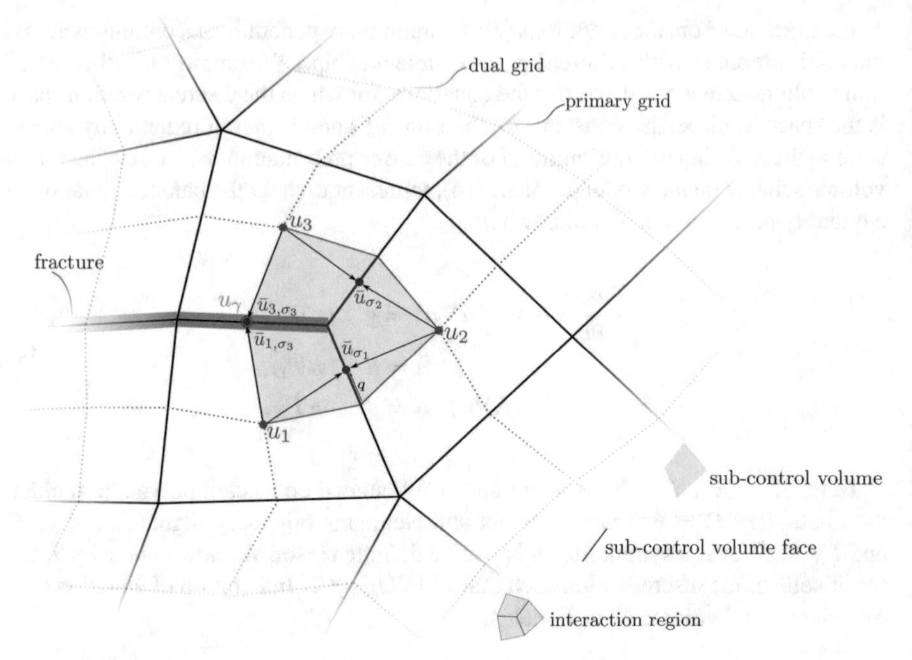

Fig. 2 Exemplary interaction volume of the mpfa–dfm scheme on a quadrilateral grid. The red dots represent the cell unknowns u_i, the blue dots illustrate the intermediate face unknowns $\bar{u}_{\sigma_i}$ and the fracture unknown u_γ within the interaction volume is shown. Note that geometrically, $\bar{u}_{1,\sigma_3}$, $\bar{u}_{3,\sigma_3}$ and u_γ lie on the same point (color figure online)

If we approximate the integral in (17) using only one integration point $\mathbf{x}_\sigma^{\text{ip}} \in \sigma$, the first condition of (18) results in

$$\left.\begin{aligned} |\sigma|\,(\boldsymbol{\Lambda}_K \nabla u)\cdot \mathbf{n}_K|_{\mathbf{x}_\sigma^{\text{ip}}} + |\sigma|\,(\boldsymbol{\Lambda}_L \nabla u)\cdot \mathbf{n}_L|_{\mathbf{x}_\sigma^{\text{ip}}} = 0 \\ \Rightarrow (\boldsymbol{\Lambda}_K \nabla u)|_{\mathbf{x}_\sigma^{\text{ip}}}\cdot \mathbf{n}_\sigma = (\boldsymbol{\Lambda}_L \nabla u)|_{\mathbf{x}_\sigma^{\text{ip}}}\cdot \mathbf{n}_\sigma \end{aligned}\right\} \quad \text{on } \sigma \in \mathcal{E}_{int}. \tag{19}$$

Note that we have introduced the averaged tensor $\boldsymbol{\Lambda}_K := \frac{1}{|K|}\int_K \boldsymbol{\Lambda}(\mathbf{x})\,\mathrm{d}x$ on a cell K, where the integral is meant component-wise, and we have defined $\mathbf{n}_\sigma := \mathbf{n}_K|_{\mathbf{x}_\sigma^{\text{ip}}} = -\mathbf{n}_L|_{\mathbf{x}_\sigma^{\text{ip}}}$.

In the classical two-point flux approximation scheme (tpfa), $\nabla u|_{\mathbf{x}^{\text{ip}}}$ is constructed within the cells using only the cell-centered value and the intermediate face unknown $\bar{u}_\sigma$. However, it is well known that this scheme does not converge to exact solutions on non-K-orthogonal grids (see e.g. Edwards and Rogers 1998). In order to be more flexible with respect to the discretization (i.e. unstructured grids) and choice for the tensors, we employ a multi-point flux approximation scheme (mpfa), which has been introduced as the mpfa-o scheme in Aavatsmark (2002). There is a wide range of different mpfa schemes available in the literature (see e.g. Aavatsmark et al. 2008; Agélas et al. 2010; Friis and Edwards 2011), some of which could also be extended to grid-conforming discrete fracture models. For example, the mpfa-o scheme with full pressure support (Friis and Edwards 2011) was used in the context of fractured porous media in Ahmed et al. (2017). The original mpfa-o scheme in conjunction with fractures was used in Sandve et al. (2012) and Ahmed et al. (2015) for single phase problems and explicit tracer transport simulations. A conforming discrete fracture model using tpfa can be found in Karimi-Fard et al. (2004). In the mpfa-o scheme, a dual grid is constructed around the grid vertices by connecting the barycenters of all geometrical entities adjacent to the vertices (i.e. cells and faces in 2d; cells, faces and edges in 3d). Doing so, the cells and faces are splitted into sub-control volumes and sub-control volume faces, respectively. The cluster of sub-control volumes and faces around a vertex is called interaction region, which is exemplarily depicted in Fig. 2. The gradients ∇u are then constructed within the sub-control volumes (and assumed to be constant within them), using the intermediate face unknowns of all embedded sub-control volume faces. The position of the intermediate unknown on a sub-control volume face is parameterized by a parameter q, $0 \leq q < 1$, where $q = 0$ corresponds to the primary grid face barycenter and $q = 1$ would correspond to the grid vertex position (see Fig. 2). The conditions (18) are imposed on all sub-control volume faces inside an interaction region, which leads to a local system of equations that can be solved to eliminate the intermediate unknowns $\bar{u}_\sigma$ and which results in flux expressions of the form

$$F_{K,\sigma} = \sum_{i \in \mathcal{S}_\sigma} t_{\sigma i} u_i \tag{20}$$

for each sub-control volume face σ embedded in cell K. The face stencil $\mathcal{S}_\sigma$ contains all the cells that contribute to the fluxes across σ with the product of the cell value u_i and associated transmissibility $t_{\sigma i}$. Generally, the face stencil comprises the cells that are contained in the interaction region.

As mentioned previously, we require the grid to be conforming with the fracture geometries. This enables us to treat the fracture as an interior Robin-type boundary condition for the matrix domain (see Fig. 2). In analogy to the previously introduced conditions (13) we can define them for the model problem (14) for all sub-control volume faces embedded in a cell K that lie on γ:

$$F_{K,\sigma} = (\boldsymbol{\Lambda}_K \nabla u)|_{\mathbf{x}_\sigma^{ip}} \cdot \mathbf{n}_\sigma = -|\sigma| \Lambda_{f,\eta} \frac{\bar{u}_{K,\sigma} - \bar{u}_\gamma}{a/2} \quad \text{on } \sigma \in \mathcal{E}_{int} \cap \gamma. \tag{21}$$

Note that this implies that $\bar{u}_{K,\sigma} \neq \bar{u}_{L,\sigma}$ and the local systems to solve within interaction regions intersected by one or more fractures become larger. The final flux expressions for faces within such interaction regions are now functions of both matrix and fracture unknowns:

$$F_{K,\sigma} = \sum_{i \in \mathcal{S}_\sigma^m} t_{\sigma i} u_i + \sum_{j \in \mathcal{S}_\sigma^\gamma} t_{\sigma j} u_{\gamma j}, \tag{22}$$

where the face stencils now consist of two parts containing the dependencies on matrix and fracture cell unknowns, $\mathcal{S}_\sigma^m$ and $\mathcal{S}_\sigma^f$. The increased size of the local systems to be solved comes with the benefit that this allows us to reproduce the jump in solution across the fracture associated with situations where $\mathbf{n}^T \boldsymbol{\Lambda}_K \mathbf{n} > \Lambda_{f,\eta}$. It should also be mentioned that the fracture is treated strictly lower-dimensional, i.e. the matrix grid is not modified in any way and the integration points for the enforcement of the conditions (21) still lie on the original grid face. The lower-dimensional fracture domain is modeled using standard tpfa or mpfa scheme with the source terms introduced in the previous section to account for the exchange fluxes with the matrix. It should be mentioned that on intersections of multiple fractures, we use the approach presented in a previous publication (Gläser et al. 2017), which is based on a scheme given in Sandve et al. (2012), to distribute the fluxes among the adjacent branches. The volume error associated with the lower-dimensional treatment of the fracture is illustrated by the gray area in Fig. 2, which depicts the occupied volume of the fracture by extrusion of the grid face with the aperture. Additional volume error contributions are introduced at intersections of fractures (not depicted in the figure), where multiple fracture geometries might overlap partly. In the subsequent chapters, we will refer to the method presented above as the mpfa–dfm scheme.

4.1 Flux approximations of mass and heat fluxes

The flux terms (5) and (6) appearing in the system of equations considered in this work differ from the one in (14), which has been chosen for ease of presentation. However, we will see that the interface conditions (7) and (13) can be reduced to the form given in (19) and (21), by introducing further simplifications. In this work, we apply a first order upwind scheme on the advectively transported quantities, i.e. for the terms $\rho_\alpha \frac{k_{ir\alpha}}{\mu_\alpha}$ and $\rho_\alpha h_\alpha \frac{k_{ir\alpha}}{\mu_\alpha}$. This means they are evaluated on the upstream side of a face σ, depending on the sign of the term $-\mathbf{K}_K \left(\nabla p_{K\alpha} - \rho_\alpha \mathbf{g}\right)$. Additionally, for the density appearing in the latter, i.e. in the term $\rho_\alpha \mathbf{g}$, we use the arithmetic average $\rho_{\sigma\alpha} = \left(\rho_{K\alpha} + \rho_{L\alpha}\right)/2$ on interior cell faces and we use the density inside the fracture on γ. Thus, the flux equality condition for the advective mass flux, i.e. the first equation in (7), reduces to the form given in (19), resulting in the following interface conditions on interior faces:

$$\left.\begin{aligned} p_{K\mathrm{w},\sigma} &= p_{L\mathrm{w},\sigma} \\ p_{K\mathrm{c},\sigma} &= p_{L\mathrm{c},\sigma} \\ T_{K,\sigma} &= T_{L,\sigma} \\ \left(\mathbf{K}_K \nabla p_{K\alpha}\right) \cdot \mathbf{n}_\sigma &= \left(\mathbf{K}_L \nabla p_{L\alpha}\right) \cdot \mathbf{n}_\sigma \\ \left(\lambda_K \nabla T_K\right) \cdot \mathbf{n}_\sigma &= \left(\lambda_L \nabla T_L\right) \cdot \mathbf{n}_\sigma \end{aligned}\right\} \quad \text{on } \sigma \in \mathcal{E}_{int} \setminus \left(\mathcal{E}_{int} \cap \gamma\right). \tag{23}$$

Analogously to (21), the transmission conditions (13) on γ then read:

$$\left.\begin{aligned} \left(\mathbf{K}_K \nabla p_{K\alpha}\right) \cdot \mathbf{n}_\sigma &= -|\sigma| k_{\mathrm{f}\eta} \frac{p_{K\alpha,\sigma} - p_{\gamma\alpha}}{a/2} \\ \left(\lambda_K \nabla T_K\right) \cdot \mathbf{n}_\sigma &= -|\sigma| \lambda_\gamma \frac{T_{K,\sigma} - T_\gamma}{a/2} \end{aligned}\right\} \quad \text{on } \sigma \in \mathcal{E}_{int} \cap \gamma. \tag{24}$$

5 Numerical experiments

The numerical scheme presented in this work has been studied in Ahmed et al. (2015) for incompressible single-phase flow and explicit tracer transport. The convergence tests shown therein have been extended in a previous publication of the authors (Gläser et al. 2017), where the individual error contributions of both the assumption that the fracture can be treated as lower-dimensional and the introduced volume error are elaborated. For that particular test case it was seen that the volume error is the dominant error contribution. In the same work, it was then studied the performance of the scheme, by comparison with an equi-dimensional discretization, for incompressible two-phase flow on a simple geometry and using K-orthogonal grids for both highly conductive and low-permeable fractures. The results revealed that the mpfa–dfm scheme could reproduce the equi-dimensional solution very well for the case of the highly permeable fracture, whereas, as expected, deviations were larger in the case of the low-permeable fracture. For that study, capillary pressure-saturation relationships after Brooks and Corey (1964) have been used together with very different entry pressures for the fracture and matrix domains as well as initial conditions that did not correspond to capillary pressure equilibrium between matrix and fractures. This situation is expected to be rather unfavorable for the presented scheme. For this reason, in this work we want to investigate whether or not the results improve when using the relationships after Van Genuchten (1980), in which the capillary pressures always vanish when the medium is fully wetting-phase saturated or at residual non-wetting phase saturation. Moreover, we want to increase the physical complexity of the test cases by considering compressible fluids, non-isothermal effects and full tensor permeabilities including strong heterogeneities within both the matrix and the fracture domain. We employ realistic descriptions of the fluid phase properties (see Table 1), for which we use water and liquid mesitylene as the wetting and non-wetting phase, respectively. The thermal conductivity of the porous medium is modeled using the relationship presented in Somerton et al. (1974), thus, it is a function of the saturations and the thermal conductivities of the fluid phases and the solid matrix.

5.1 Domain with a single fracture

In this synthetic test case we consider a domain of 50 m × 50 m with a single fracture connecting an aquifer, confined by a low-permeable caprock, to an overlying reservoir (see Fig. 3). The fracture extends up to the upper boundary. Except for the lower boundary, where Neumann no-flow boundary conditions are applied, Dirichlet boundary conditions are set everywhere. Pressure is fixed according to hydrostatic conditions and $T = 285\,\mathrm{K}$ and $S_n = 0$ are set everywhere except for the lower 10 m of the left boundary, where $S_n = 0.2$, $T = 335\,\mathrm{K}$ and pressure is increased by $\Delta p = 10\,\mathrm{bar}$. The parameters used for the solid matrix in the different layers are given in Table 2. Please note here that for the tensorial permeability of a layer i it is $\mathbf{K}_i = \mathbf{R}^{-1}(\delta_i)\begin{pmatrix} k_{\mathrm{h},i} & 0 \\ 0 & \kappa_i k_{\mathrm{h},i} \end{pmatrix}\mathbf{R}(\delta_i)$, where $\mathbf{R}(\delta)$ is the rotation matrix in counter-

Table 1 Fluid properties and relationships used in this work

Property	Value/function	Unit	References
Water density ρ_w	$f(p_w, T)$	kg/m^3	Wagner and Kretzschmar (2008)
Water viscosity μ_w	$f(p_w, T)$	Pa s	Wagner and Kretzschmar (2008)
Water heat capacity c_w	$f(T, p_w)$	J/(kgK)	Wagner and Kretzschmar (2008)
Water spec. enthalpy h_w	$f(p_w, T)$	J/kg	Wagner and Kretzschmar (2008)
Water spec. int. energy u_w	$h_w - p_w/\rho_w$	J/kg	–
Water thermal cond. λ_w	$f(p_w, T)$	W/(m K)	Wagner and Kretzschmar (2008)
Mesitylene density ρ_n	$f(T)$	kg/m^3	Reid et al. (1987)
Mesitylene viscosity μ_n	$f(T)$	Pa s	–
Mesitylene heat capacity c_n	$f(T)$	J/(kgK)	Reid et al. (1987)
Mesitylene spec. enthalpy h_n	$\int_{273.15}^{T} c_n \mathrm{d}T$	J/kg	–
Mesitylene spec. int. energy u_n	$h_n - p_n/\rho_n$	J/kg	–
Mesitylene thermal cond. λ_n	0.1351	W/(m K)	Kauffman and Jurs (2001)

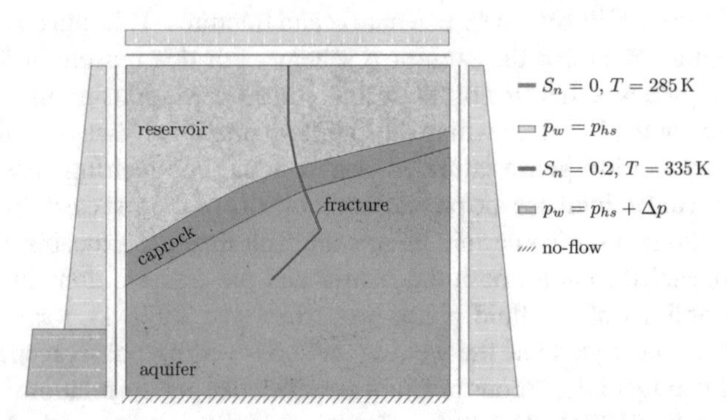

Fig. 3 Illustration of the domain and boundary conditions of the single fracture test case. p_{hs} refers to a pressure distribution according to hydrostatic conditions

clockwise direction through the angle δ. Besides, we set the non-wetting phase residual saturations to zero for all materials.

We evaluate the quality of the results by comparison with solutions obtained from an equi-dimensional discretization of the domain, where the grid is refined towards the fracture such that there are around 4–5 elements discretizing its width, with apertures lying within the range of $a = [0.14,\ 0.2]$ m. These are very high aperture values and were chosen in order for the equi-dimensional mesh to end up with a comparatively small number of elements on which reference solutions can be computed efficiently, and are not intended to be representative for real geological systems. Also, the case of high apertures compared to the domain dimensions presents an unfavorable situation for the lower-dimensional model and makes drawbacks of the scheme more likely to appear. In this setting, the total fracture area is 6.5 m^2 which corresponds to a volume error of 0.26% for the lower-dimensional discretization. Even though the mpfa–dfm model requires the mesh to be conforming with the fracture geometries, a much coarser

Table 2 Rock properties used in the different layers in the single fracture test cases

Property	Aquifer	Reservoir	Caprock	Unit
Hor. permeability $k_{h,i}$	1×10^{-13}	1×10^{-12}	1×10^{-15}	m^2
Anisotropy ratio κ_i	1.25	1.4	1.0	–
Permeability angle δ_i	25	−15	0	°
Porosity ϕ_i	0.15	0.2	0.05	–
Solid density $\rho_{s,i}$	2700	2400	1760	kg/m^3
Solid heat capacity c_i	790	1145	1704	J/kg
Solid therm. cond. λ_i	2.8	1.5	1.3	W/(m K)
Res. water saturation $S_{wr,i}$	0.05	0.05	0.1	–
Van Genuchten alpha $\alpha_{vg,i}$	2×10^{-4}	8×10^{-4}	8×10^{-5}	1/Pa
Van Genuchten N $N_{vg,i}$	3.3	2	3.6	–

The fracture properties differ between the different tests and are mentioned in the respective paragraphs

discretization of the domain can be chosen in comparison to an equi-dimensional approach. This being the main benefit of the scheme, we want to evaluate the quality of the results for coarse grids and the maximum possible time step sizes. Additionally, we perform simulations with a grid that is locally refined around the fracture and is more comparable to the one used for the equi-dimensional discretization, which decreases the discretization error and allows for a better visualization of the error introduced by the model assumptions.

The reference solution, however, does not qualify as a reference in the sense of a converged solution as the chosen grid is still much too coarse. The authors are aware of this fact and want to point out that only a qualitative comparison is sought at this point. We chose this discretization to be able to produce equi-dimensional solutions efficiently and we do not claim to reproduce all occurring effects accurately here. For the highly-conductive fracture case we performed an additional simulation with a grid consisting of 175,668 elements, for which the solution did not show differences in magnitudes that would affect the outcomes of the qualitative comparisons presented here. All grids used for the results presented in the subsequent sections are depicted in Fig. 4 and were generated using the open-source mesh generator Gmsh (Geuzaine and Remacle 2009).

With respect to the time discretization, we start with an initial time step size of $\Delta t = 10\,\mathrm{s}$ in all simulations shown here and adapt it according to the convergence behaviour of the non-linear solver. For this we set 10 iterations as the desired number of Newton iterations that should be necessary to perform a time step. If the number of iterations needed is above that, the time step size is decreased for the next time step and increased if it is below. Furthermore, we use 18 iterations as the maximum permissive number of iterations required per time step. If this number is reached, time integration is repeated with half the time step size. The authors are aware of the fact that this might lead to a very different temporal resolution and thus to less comparable solutions due to e.g. issues with numerical diffusion. However, we enforce the time

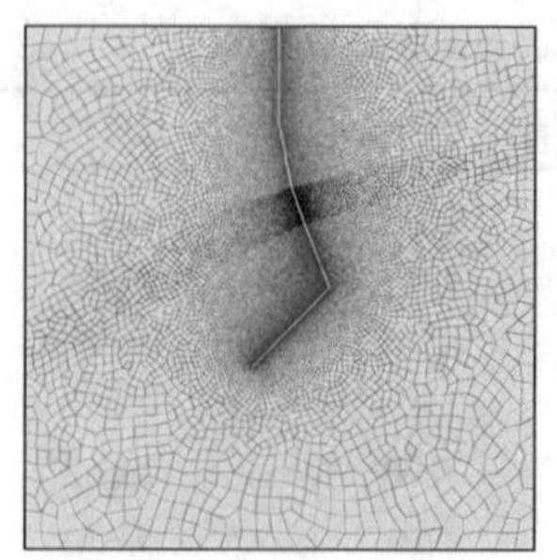

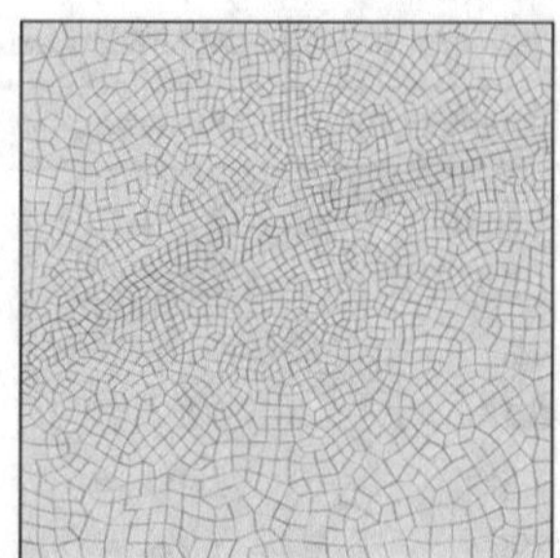

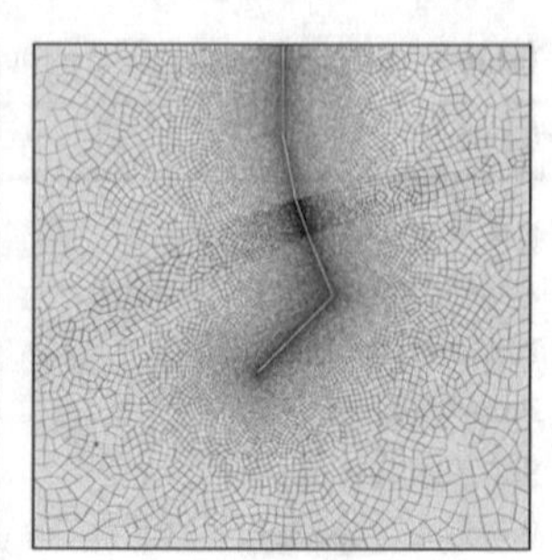

Fig. 4 Left: the grid used to produce the equi-dimensional reference solution with 37,938 cells. Center: the grid used for the mpfa–dfm model with 2217 two-dimensional and 40 one-dimensional cells. Right: the fine grid used in conjunction with the mpfa–dfm model comprising of 30,912 two-dimensional and 641 one-dimensional cells. The different colors used for the edges indicate the different material layers the elements are embedded in (color figure online)

discretization scheme to respect time intervals of 2 months such that we have solutions on identical time levels for all simulations shown here.

Highly conductive fracture

In this scenario, the fracture acts as a preferential pathway with $k_\mathrm{f} = 10^{-9}\,\mathrm{m}^2$ and $\phi_\mathrm{f} = 0.4$ chosen for the permeability and porosity. Additionally we set $\alpha_{vg,f} = 10^{-3}\,\mathrm{Pa}^{-1}$, $N_{vg,f} = 2.4$ and $S_{wr,f} = 0.05$, leading to lower capillary pressures than in the surrounding media which motivates the mesitylene to enter and to stay in the fracture. In contrast to that, the surrounding caprock acts as both hydraulic and capillary barrier (see Table 2). Driven by the higher pressures on the lower left boundary as well as slight buoyancy forces, the mesitylene is distributed in the domain, accumulates below the caprock and rapidly flows through the fracture towards the upper boundary. The mesitylene saturation distributions obtained with both the equi-dimensional discretization and the mpfa–dfm model at two different time steps are depicted in Fig. 5. Steep saturation gradients can be observed close to the fracture, where mesitylene is sucked in by capillary forces. In the overlying reservoir, where the capillary pressure differences between fracture and the surrounding porous medium are lower, mesitylene is slightly invading the matrix, however, saturations are generally low as most of the mesitylene leaves the domain through the fracture across the upper boundary. Figure 5 suggests that the mpfa–dfm model overestimates the saturations close to the fracture and underestimates the saturations at the interface between the aquifer and the caprock, but, one has to take into account the very different discretizations. The steep saturation gradients that occur here cannot be captured by the low resolution, but in an average sense the solution obtained with mpfa–dfm model is in good agreement with the reference solution. This can be seen in more detail in Fig. 6, where plots of the non-wetting phase saturation and the temperature along a part of the diagonal through the domain (see Fig. 5) are shown. These furthermore reveal that the saturation in the fracture is also reproduced well by the mpfa–dfm model, for which the values in the fracture represent quantities averaged over a cross section. The zooms provided

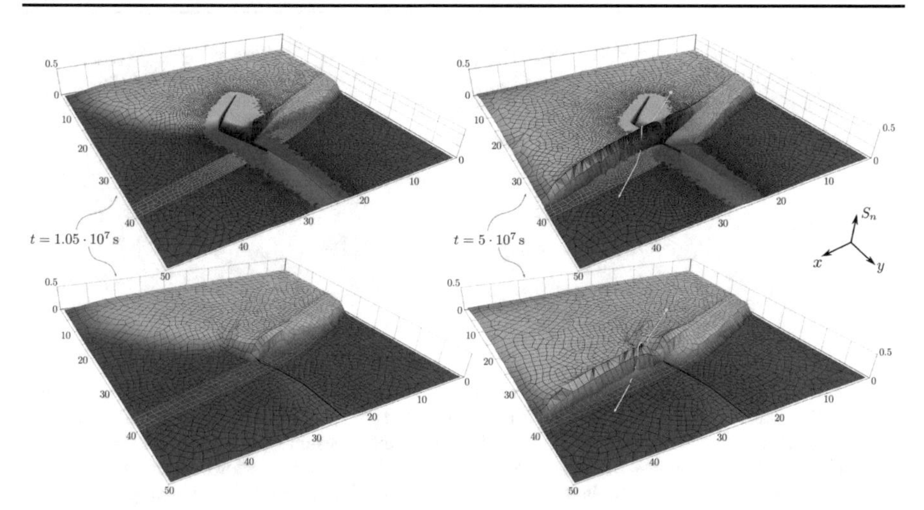

Fig. 5 S_n-distribution at 1.05×10^7 s (left column) and the final simulation time of 5×10^7 s (right column) of the reference (upper row) and the mpfa–dfm model (lower row) for the highly-conductive fracture case. The white diagonal lines illustrate the path along which the saturation and temperature profiles are plotted in Fig. 6. For the reference solution, the outlines of the grid elements are suppressed towards the fracture where the elements become very small and would impede the visibility of features in the solution. For better visibility of the different layers, the outlines of the elements within the caprock are depicted in gray (color figure online)

in the plots in Fig. 6 show that, when compared to the actual cross-section average values for temperature and saturation computed on the reference solution, only small deviations are observed. For the saturation this deviation is in the order of 5% and for the temperature less than 1%. On the fine grid, the mpfa–dfm model produces results that are very close to the reference, which indicates that most of the errors are introduced due to the coarser discretization. In particular, the numerical diffusion introduced by the first order upwind scheme used here (see previous section) is much higher on the coarse grid and leads to a more rapid propagation of the mesitylene front in the domain. Also, it should be mentioned that due to the chosen time stepping scheme (see above), 190 and 188 time steps were performed for the reference solution and the mpfa–dfm solution on the fine grid, respectively, with maximum time step sizes of $\Delta t_{max} \approx 1.5 \times 10^6$ s. In contrast to that, only 61 time steps were necessary on the coarse grid where the time steps reached maximum values of $\Delta t_{max} \approx 4 \times 10^6$ s. This introduces additional numerical diffusion in the case of the coarse grid. Nevertheless, the differences appear to be relatively small. A look at the mass distribution of mesitylene in the different materials (see Fig. 7) indicates that, even with the coarse discretization, the errors in mass contained in the layers stay below 10% except for the caprock where it is slightly above.

The temperature distribution in the domain is depicted in Fig. 8, which was seen to be in good agreement already in the plot in Fig. 6. With advection being the dominating process on this setup and the considered time scales, the temperature is highest in regions where high flow velocities occur and the rock matrix is steadily brought in contact with fluids of higher temperatures coming from the inflow boundary.

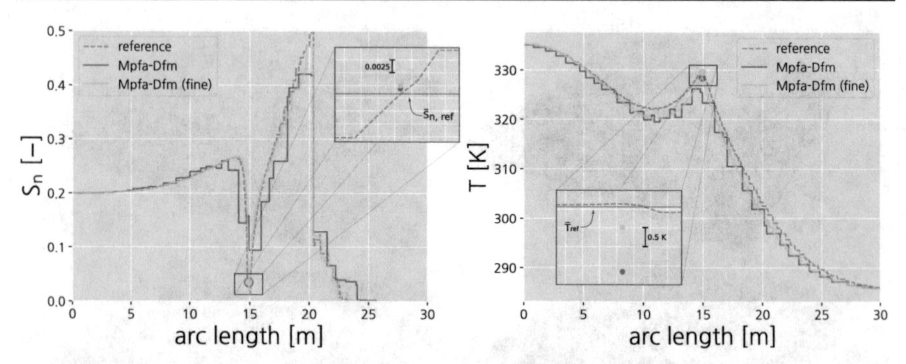

Fig. 6 Plots of S_n (left) and T (right) along a part of the diagonal from the lower left to the upper right corner of the domain (see Fig. 5) at the final simulation time of 5×10^7 s for the highly-conductive fracture case. The circles represent the solution inside the fracture in the lower-dimensional fracture domain of the mpfa–dfm model. In the zooms, the parts of the plots inside the fracture are shown, where the horizontal black lines depict the average values in the reference solution obtained over the corresponding cross section

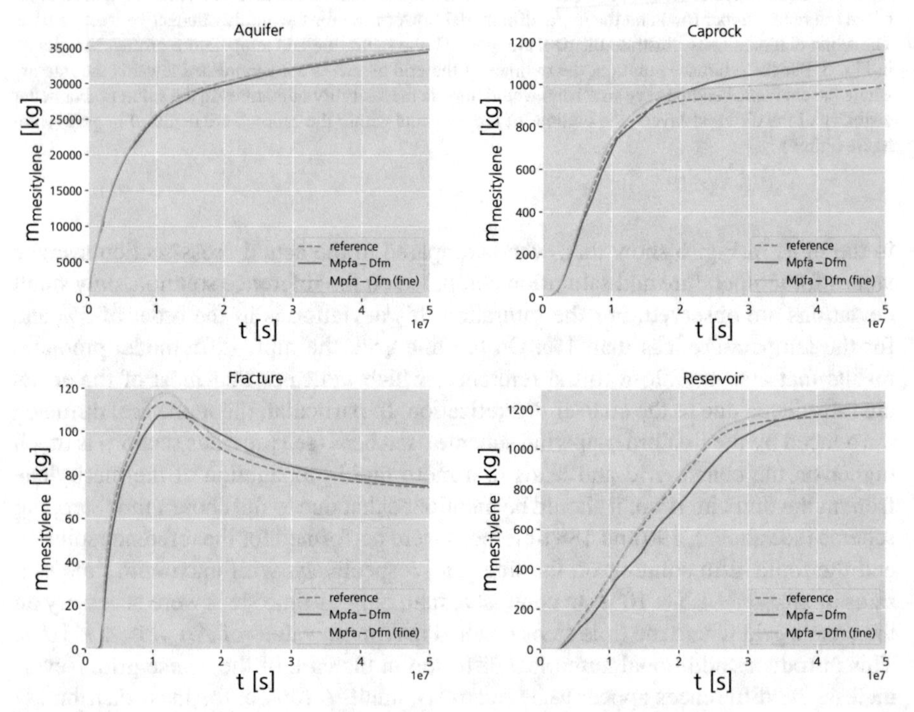

Fig. 7 Temporal evolution of the mesitylene mass contained within the different geological layers (see individual plot titles) for both the mpfa–dfm model and the reference solution for the highly-conductive fracture case

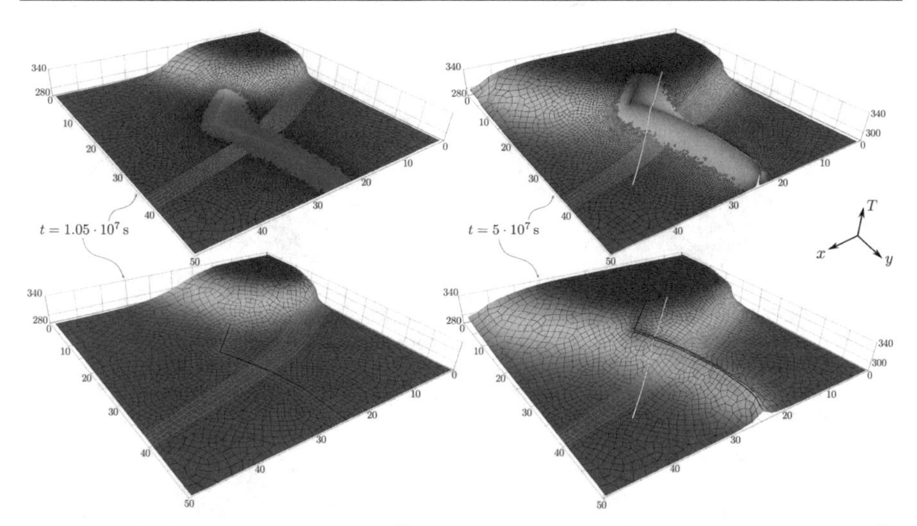

Fig. 8 Temperature distribution at 1.05×10^7 s (left column) and the final simulation time of 5×10^7 s (right column) of the reference (upper row) and the mpfa–dfm model (lower row) for the highly-conductive fracture case. The white diagonal lines illustrate the path along which the saturation and temperature profiles are plotted in Fig. 6. For the reference solution, the outlines of the grid elements are suppressed towards the fracture where the elements become very small and would impede the visibility of features in the solution. For better visibility of the different layers, the outlines of the elements within the caprock are depicted in gray (color figure online)

Heterogeneous, low-permeable fracture

In this section we use different geological parameters for the part of the fracture domain that lies in the caprock and the aquifer, such that it acts as both hydraulic and capillary barrier. In these parts we set $k_\mathrm{f} = k_\mathrm{c}/100, \phi_\mathrm{f} = 0.75\phi_\mathrm{c}, \alpha_{\mathrm{vg,f}} = \alpha_{\mathrm{vg,c}}/10$ and $S_{\mathrm{wr,f}} = 2 \cdot S_{\mathrm{wr,c}}$ while for the remaining parameters we use the same values as in the caprock. The upper part of the fracture within the reservoir is parameterized as in the highly-conductive test case. The high capillary pressures and low permeability in the lower part impede the mesitylene to enter the fracture such that most of it accumulates below and inside the caprock at the interface to the fracture (see Fig. 9), where steep saturation gradients towards and strong jumps in saturation across the fracture develop. The coarse resolution of the grid used in conjunction with the mpfa–dfm model cannot capture these effects and leads to a much more flattened out solution. However, Fig. 10 shows that on the locally refined grid the solution comes much closer to the reference while qualitatively capturing its distinct features. But, the accumulation of mesitylene at the interface between caprock and the low-permeable fracture is not reproduced to the full extent. There seems to be more mass transfer to the other side of the fracture, where saturations are higher than in the reference. This is an outcome that one expects from the dimension-reduction and the use of averaged values over cross sections of the fracture, as the saturation gradients that are present across the fracture lead to very different relative permeabilities at the two interfaces with the surrounding matrix. The averaged value in the lower-dimensional model thus overestimates the relative

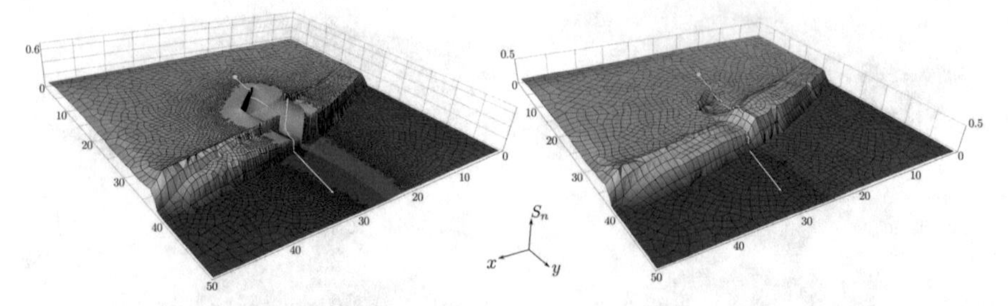

Fig. 9 S_n-distribution at the final simulation time of 5×10^7 s of the reference (left) and the mpfa–dfm model (right) for the low-permeable fracture case. The white lines illustrate the path along which the saturation profile is plotted in Fig. 10. For the reference solution, the outlines of the grid elements are suppressed towards the fracture where the elements become very small and would impede the visibility of features in the solution (color figure online)

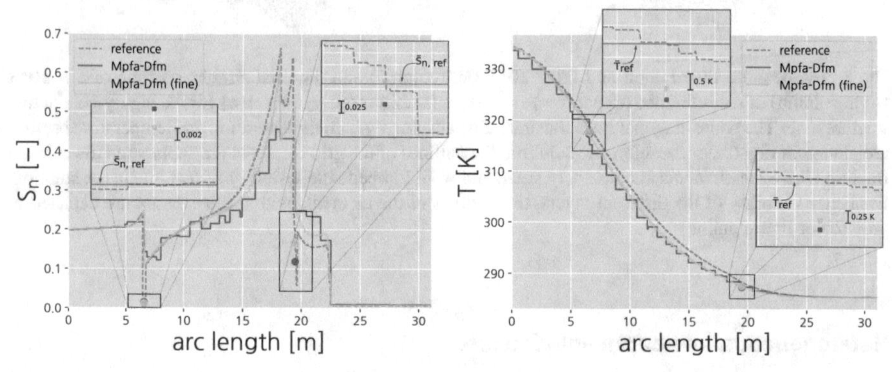

Fig. 10 Plots of S_n (left) and T (right) along the lines illustrated in Fig. 9 at the final simulation time of 5×10^7 s. The circles represent the solution inside the fracture in the lower-dimensional fracture domain of the mpfa–dfm model. In the zooms, the parts of the plots inside the fracture are shown, where the horizontal black lines depict the average values in the reference solution obtained over the corresponding cross section

permeability on the downstream interface which leads to more intrusion of mesitylene into the matrix.

The mesitylene mass distributions among the different layers (see Fig. 11) show that for the coarse grid, deviations up to $\approx$ 50% are visible in the fracture, $\approx$ 25% in the reservoir and coincide rather well for the aquifer. Unexpectedly, the deviations to the reference in the fracture are higher on the refined grid towards the end of the simulation time, while in the remaining layers the mass distributions are in very good agreement with the reference. But, since the mass contained in the fracture is very low in comparison to the other layers, this is very sensitive to small discrepancies within any of the latter and thus also very sensitive to e.g. small differences in the numerical grid, which makes it difficult to determine the origin of this observation. In any case and as expected beforehand, the lower-dimensional treatment of fractures acting as barriers to flow seems to introduce more artifacts than for highly conductive ones. However, the scheme is able to qualitatively capture the occurring phenomena and is furthermore expected to give much better results on fractures with lower apertures.

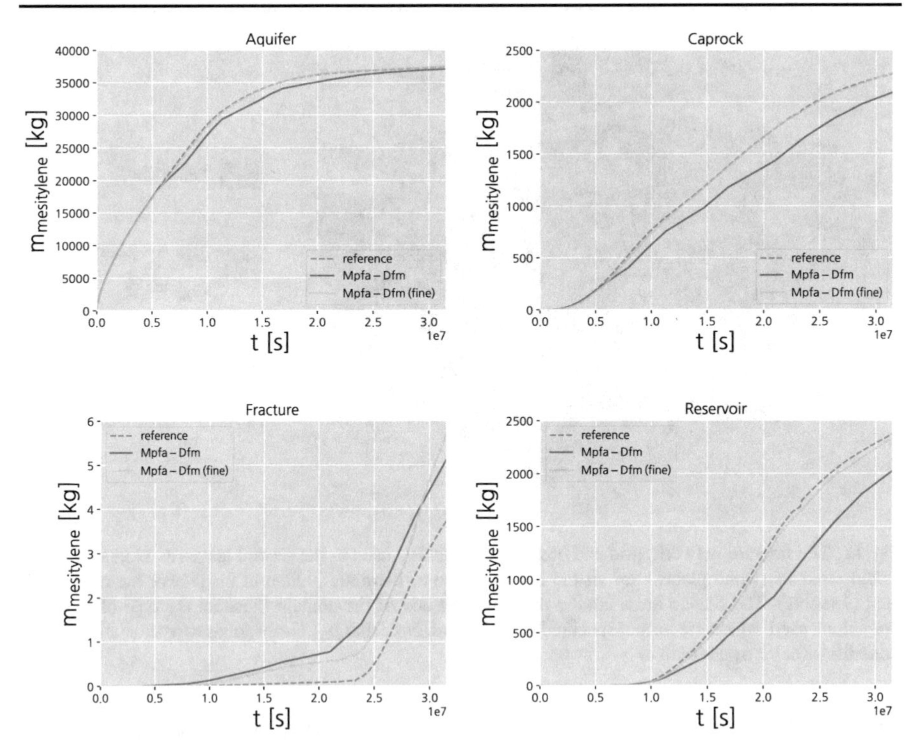

Fig. 11 Temporal evolution of the mesitylene mass contained within the different geological layers (see individual plot titles) for both the mpfa–dfm model and the reference solution for the low-permeable fracture case

As mentioned in Sect. 2, the determination of ranges of applicability of the scheme remains an open question.

With respect to the temporal resolution, for this test case the chosen time stepping scheme led to similar maximum time step sizes as in the previous case of a fully high-permeable fracture. However, these were reached more frequently throughout the simulation and resulted in 126 time steps for the equi-dimensional reference solution, while 69 and 133 time steps were performed with the mpfa–dfm model on the coarse and fine grid, respectively.

5.2 Stochastically generated fracture network

For this test case we created a three-dimensional fracture network geometry on a block of $8\,\text{m} \times 8\,\text{m} \times 5\,\text{m}$ using the stochastic fracture network generator Frac3d (Silberhorn-Hemminger 2003; Assteerawatt 2008). Subsequently, the geometry has been meshed using an algorithm for constructing almost regular triangulations (Fuchs 2001). This algorithm detects the intersection lines of fracture planes and the intersection points of intersection lines and ensures conformity of the grid with respect to these lines and points. The resulting geometry and the grid are visualized in Fig. 12 together with the boundary conditions applied here. Initially, the entire domain is half saturated with

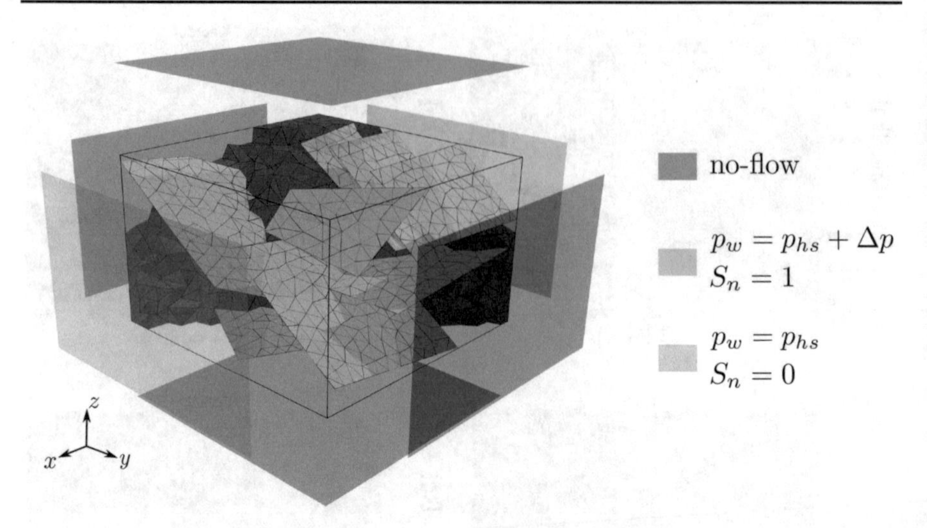

Fig. 12 The fracture network, grid and boundary conditions used in the second test case. Note that not all fracture planes (white planes) are visible here as some of them are hidden in the part of the matrix grid (red elements). The shaded areas around the bounding box of the domain illustrate the type of boundary conditions used, which are further specified in the legend. Note that p_{hs} corresponds to hydrostatic pressure conditions (color figure online)

Table 3 Rock properties used for fractures and matrix in the stochastic fracture network test case

Property	Matrix	Fractures	Unit
Permeability k	1×10^{-13}	1×10^{-7}	m^2
Porosity ϕ	0.25	0.75	–
Solid density $\rho_{s,i}$	2700	2700	kg/m^3
Solid heat capacity $c_{s,i}$	790	790	J/kg
Solid therm. cond. $\lambda_{s,i}$	2.8	2.8	W/(m K)
Van Genuchten alpha $\alpha_{vg,i}$	2×10^{-4}	1×10^{-3}	1/Pa
Van Genuchten N $N_{vg,i}$	3.3	2.4	–

mesitylene ($S_n = 0.5$) and a hydrostatic water pressure distribution prevails. Note that due buoyancy forces and due to the fact that we are using different capillary pressure-saturation relationships in matrix and fracture (see Table 3), these initial conditions do not represent an equilibrated state. We apply an increased pressure ($\Delta p = 2.64$ bar) and $S_n = 1$ as Dirichlet boundary conditions on one lateral side, while on the opposite side the initial pressure and $S_n = 0$ is set. The remaining boundaries are closed. Setups of this kind come into play in the determination of upscaled parameters for fractured blocks, however, most likely with different initial conditions and more realistic choices for the rock parameters and the residual saturations, which have all been set to zero here. The presentation of this test case should simply show the potential of the scheme to be used for such applications, although an actual evaluation would require a more rigorous approach as e.g. presented in Matthai et al. (2007).

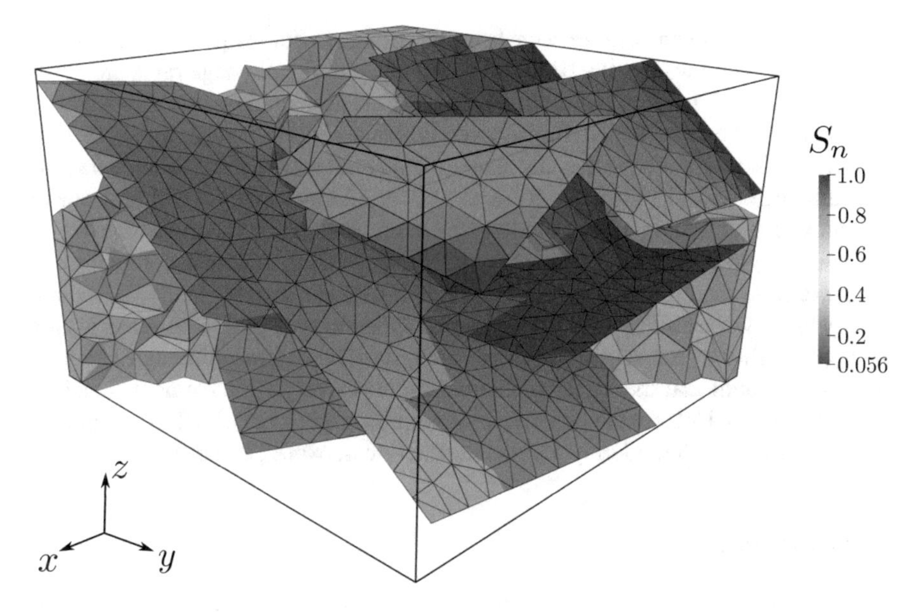

Fig. 13 Mesitylene saturation distribution in the stochastic fracture network at $t \approx 1$ d

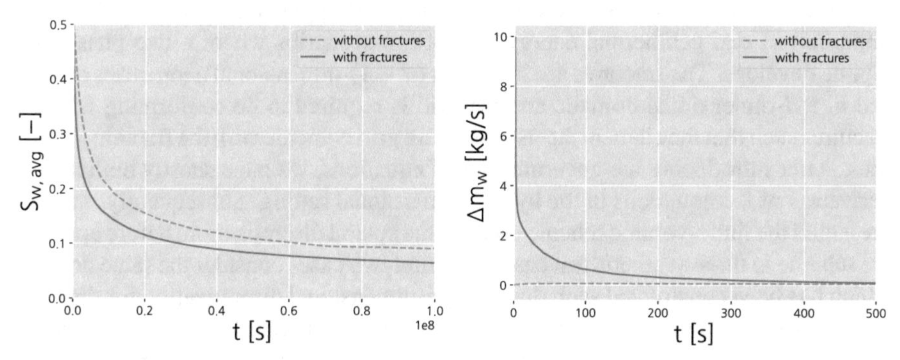

Fig. 14 Average water saturation (left) and water drainage rate (right) obtained with the mpfa–dfm model and a simulation neglecting the fractures. Note that the average water saturation has been computed by dividing the mass of water contained in the domain by the available pore volume multiplied with the average density in the domain

For the fracture apertures we set $a = 0.01$ m, which corresponds to a volume error of about 1%. We now want to generate the drainage curve for the considered fractured block under the given initial conditions, and, for simplicity, under isothermal conditions. This means we only solve for the Eq. (1) and assume a constant temperature of 10 °C. A snapshot of the mesitylene saturation in the domain after 3483.79 s is depicted in Fig. 13 and shows that already after this short time, all fractures that have a connection to the inflow boundary are almost entirely mesitylene-saturated. Thus, at this point most of the fractures are already drained, capillary pressures in the fractures become very high and the drainage of the remaining water is limited by the matrix conductivity. This can also be seen from the drainage curves provided in Fig. 14, where the average water saturation in the domain and the drainage rates are plotted over time for both the mpfa–dfm model and a simulation on the same grid where the fractures have been neglected. Initially, the drainage rates are much higher when the fractures are considered, but they decrease quickly until approaching the non-fractured rock drainage rates already after 500 s. On fractured blocks with a higher density of fractures or more complex connectivity patterns, the drainage curves for fractured and non-fractured rock would deviate much more, leading to very different upscaled parameters. As mentioned earlier, this is not done here and this test has been conducted mainly to show the ability of the model to be applied to such problems. Also, it shows the applicability of the scheme to three-dimensional problems, although one has to take into account that the mpfa-o scheme introduces very large element stencils on tetrahedral grids. Unfortunately, a transition to cubic grids is not trivial when enforcing grid conformity with randomly distributed fracture network geometries.

6 Conclusions

In this work we have presented a hybrid-dimensional numerical model for non-isothermal two-phase flow in fractured porous media, which is a problem that is particularly relevant for radioactive waste disposal applications, but could also be important in e.g. geothermal energy production scenarios where a two-phase flow regime develops. The fractures are treated as $(d-1)$-dimensional geometries embedded in a d-dimensional domain, and the grid is required to be conforming with the fractures such that the element facets of the bulk grid coincide with the fracture geometries. After introducing the governing set of equations, we have shortly outlined the derivation of the equations in the hybrid-dimensional setting. Subsequently, we have presented the finite-volume scheme used for the spatial discretization, before applying the scheme to three synthetic test cases. The first two cases consider the same domain, which has been constructed such that an equi-dimensional discretization for the computation of a reference solution could be carried out efficiently. While the first test case considers a fracture that is highly-conductive in comparison to the surrounding porous matrix, parts of the fracture have been defined as both hydraulic and capillary barriers in the second test case. In contrast to a previous publication where we have presented test cases for isothermal and incompressible single- and two-phase flow, compressible fluid phases using realistic descriptions of the fluid properties and both heterogeneous and full tensor permeability fields have been employed in this work. In addition, non-isothermal effects were taken into account. The presented model showed to be able to capture the relevant processes rather well, even for a coarse discretization, for the case of a highly conductive fracture. For a low-permeable fracture, larger errors became apparent in the saturation distributions and the mass transfer between matrix and fracture, which was expected beforehand as a result of the averaging procedure in the derivation process. Qualitatively, the model was able to reproduce the dominating effects, but, especially the accumulation of the non-wetting phase in front of the low-permeable fracture was underestimated. In the last section, the model is applied to a three-dimensional stochastically generated fracture network in a simple test case, which is representative for efforts towards the numerical determination of upscaled parameters of fractured rocks. However, an upscaling procedure is not performed here and the test case is to be seen solely as a showcase towards potential fields of application of the model as well as its applicability to three-dimensional problems. But, while the computational complexity is manageable in two dimensions, a major drawback of the scheme clearly is the computational complexity of the underlying mpfa-o scheme (Aavatsmark 2002) on tetrahedral grids. On such grids the element stencils can get very large with a big impact on both the assembly as well as the linear solver CPU times. The scheme performs much better on cubic grids, which unfortunately is unlikely to be realizable on random fracture network geometries such that the dependency on tetrahedral grids remains inevitable. Future investigations should include CPU time comparisons between the equi-dimensional model and the mpfa–dfm model, which was not included in this work because the latter is still in an early development stage and the implementation is not optimized yet. Additionally, it would be interesting to consider cases of intersecting fractures in comparisons with an equi-dimensional reference solution, as the treatment of the fracture intersections within the lower-dimensional model might introduce additional discrepancies. Further efforts could also be directed towards the incorporation of the fracture model to different, more recently developed finite volume schemes, as e.g. the one presented in Schneider et al. (2017), in order to decrease the computational complexity while maintaining the applicability to general grids and full tensor permeabilities. Moreover, especially for radioactive waste disposal applications, it would be beneficial to include geomechanical effects in the model.

Acknowledgements The research has received funding from the European Community Horizon 2020 Research and Innovation Programme under Grant Agreement No. 636811 in the scope of the FracRisk Project. Furthermore, the authors would like to thank the Cluster of Excellence in Simulation Technology (EXC 310/2) at the University of Stuttgart for the support.

References

Aavatsmark, I.: An introduction to multipoint flux approximations for quadrilateral grids. Comput. Geosci. **6**(3), 405–432 (2002). https://doi.org/10.1023/A:1021291114475

Aavatsmark, I., Eigestad, G., Mallison, B., Nordbotten, J.: A compact multipoint flux approximation method with improved robustness. Numer. Methods Partial Differ. Equ. **24**(5), 1329–1360 (2008). https://doi.org/10.1002/num.20320

Agélas, L., Di Pietro, D.A., Droniou, J.: The g method for heterogeneous anisotropic diffusion on general meshes. ESAIM Math. Model. Numer. Anal. **44**(4), 597–625 (2010). https://doi.org/10.1051/m2an/2010021

Ahmed, R., Edwards, M., Lamine, S., Huisman, B., Pal, M.: Control-volume distributed multi-point flux approximation coupled with a lower-dimensional fracture model. J. Comput. Phys. **284**, 462–489 (2015). https://doi.org/10.1016/j.jcp.2014.12.047

Ahmed, R., Edwards, M.G., Lamine, S., Huisman, B.A., Pal, M.: CVD-MPFA full pressure support, coupled unstructured discrete fracturematrix Darcy-flux approximations. J. Comput. Phys. **349**, 265–299 (2017). https://doi.org/10.1016/j.jcp.2017.07.041

Assteerawatt, A.: Flow and transport modelling of fractured aquifers based on a geostatistical approach. Ph.D. Thesis, Universitätsbibliothek der Universität Stuttgart, Stuttgart (2008). https://doi.org/10.18419/opus-289

Brenner, K., Groza, M., Guichard, C., Masson, R.: Vertex approximate gradient scheme forhybrid dimensional two-phase darcy flowsin fractured porous media. In: Fuhrmann, J., Ohlberger, M., Rohde, C. (eds.) Finite Volumes for Complex Applications VII-Elliptic, Parabolic and Hyperbolic Problems, pp. 507–515. Springer, Cham (2014)

Brooks, R.H., Corey, A.T.: Hydraulic properties of porous media and their relation to drainage design. Trans. ASAE **7**(1), 26–0028 (1964)

Davis, T.A.: Algorithm 832: Umfpack v4.3—an unsymmetric-pattern multifrontal method. ACM Trans. Math. Softw. **30**(2), 196–199 (2004). https://doi.org/10.1145/992200.992206

Droniou, J.: Finite volume schemes for diffusion equations: introduction to and review of modern methods. Math. Models Methods Appl. Sci. **24**(08), 1575–1619 (2014). https://doi.org/10.1142/S0218202514400041

Edwards, M.G., Rogers, C.F.: Finite volume discretization with imposed flux continuity for the general tensor pressure equation. Comput. Geosci. **2**(4), 259–290 (1998). https://doi.org/10.1023/A:1011510505406

Flemisch, B., Darcis, M., Erbertseder, K., Faigle, B., Lauser, A., Mosthaf, K., Müthing, S., Nuske, P., Tatomir, A., Wolff, M., Helmig, R.: DuMu^x: DUNE for multi-phase, component, scale, physics,.flow and transport in porous media. Adv. Water Resour. **34**, 1102–1112 (2011). https://doi.org/10.1016/j.advwatres.2011.03.007

Friis, H.A., Edwards, M.G.: A family of mpfa finite-volume schemes with full pressure support for the general tensor pressure equation on cell-centered triangular grids. J. Comput. Phys. **230**(1), 205–231 (2011). https://doi.org/10.1016/j.jcp.2010.09.012

Fuchs, A.: Almost regular triangulations of trimmend nurbs-solids. Eng. Comput. **17**(1), 55–65 (2001)

Fumagalli, A., Scotti, A.: A numerical method for two-phase flow in fractured porous media with non-matching grids. Adv. Water Resour. **62**, 454–464 (2013). https://doi.org/10.1016/j.advwatres.2013.04.001

Geuzaine, C., Remacle, J.F.: Gmsh: a 3-D finite element mesh generator with built-in pre- and post-processing facilities. Int. J. Numer. Methods Eng. **79**(11), 1309–1331 (2009). https://doi.org/10.1002/nme.2579

Gläser, D., Helmig, R., Flemisch, B., Class, H.: A discrete fracture model for two-phase flow in fractured porous media. Adv. Water Resour. **110**, 335–348 (2017). https://doi.org/10.1016/j.advwatres.2017.10.031

Jaeger, J., Cook, N., Zimmerman, R.: Fundamentals of Rock Mechanics. Wiley, Hoboken (2007)

Jaffré, J., Mnejja, M., Roberts, J.: A discrete fracture model for two-phase flow with matrix–fracture interaction. Procedia Comput. Sci. **4**, 967–973 (2011). https://doi.org/10.1016/j.procs.2011.04.102

Karimi-Fard, M., Durlofsky, L., Aziz, K.: An efficient discrete-fracture model applicable for general-purpose reservoir simulators. SPE J. **9**, 227–236 (2004)

Kauffman, G.W., Jurs, P.C.: Prediction of surface tension, viscosity, and thermal conductivity for common organic solvents using quantitative structure property relationships. J. Chem. Inf. Comput. Sci. **41**(2), 408–418 (2001). https://doi.org/10.1021/ci000139t. (pMID: 11277730)

Kazemi, H., Merrill Jr., L., Porterfield, K., Zeman, P., et al.: Numerical simulation of water-oil flow in naturally fractured reservoirs. Soci. Pet. Eng. J. **16**(06), 317–326 (1976). https://doi.org/10.2118/5719-PA

Martin, V., Jaffré, J., Roberts, J.E.: Modeling fractures and barriers as interfaces for flow in porous media. SIAM J. Sci. Comput. **26**(5), 1667–1691 (2005). https://doi.org/10.1137/S1064827503429363

Matthai, S.K., Mezentsev, A., Belayneh, M.: Finite element—node-centered finite-volume two-phase-flow experiments with fractured rock represented by unstructured hybrid-element meshes. Soc. Pet. Eng. (2007). https://doi.org/10.2118/93341-PA

Pruess, K.: Brief Guide to the MINC-Method for Modeling Flow and Transport in Fractured Media. United States, Department of Energy, Washington, DC (1992)

Pruess, K.: Enhanced geothermal systems (EGS) using CO_2 as working fluida novel approach for generating renewable energy with simultaneous sequestration of carbon. Geothermics **35**(4), 351–367 (2006). https://doi.org/10.1016/j.geothermics.2006.08.002

Reichenberger, V., Jakobs, H., Bastian, P., Helmig, R.: A mixed-dimensional finite volume method for two-phase flow in fractured porous media. Adv. Water Resour. **29**(7), 1020–1036 (2006)

Reid, R., Prausnitz, J., Poling, B.: The Properties of Gases and Liquids. McGraw-Hill Inc., New York City (1987)

Sandve, T., Berre, I., Nordbotten, J.: An efficient multi-point flux approximation method for discrete fracture–matrix simulations. J. Comput. Phys. **231**(9), 3784–3800 (2012). https://doi.org/10.1016/j.jcp.2012.01.023

Schneider, M., Agélas, L., Enchry, G., Flemisch, B.: Convergence of nonlinear finite volume schemes for heterogeneous anisotropic diffusion on general meshes. J. Comput. Phys. **351**, 80–107 (2017). https://doi.org/10.1016/j.jcp.2017.09.003

Schwenck, N.: An XFEM-based model for fluid flow in fractured porous media. Ph.D. Thesis, Universitätsbibliothek der Universitäat Stuttgart, Stuttgart (2015). https://doi.org/10.18419/opus-162

Silberhorn-Hemminger, A.: Modellierung von kluftaquifersystemen: Geostatistische analyse und deterministisch-stochastische kluftgenerierung. Ph.D. Thesis, Universitätsbibliothek der Universität Stuttgart, Stuttgart (2003)

Somerton, W., Keese, A., Chu, L.: Thermal behavior of unconsolidated oil sands. Soc. Pet. Eng. J. **14**, 513–521 (1974)

Tatomir, A.B.: From discrete to continuum concepts of flow in fractured porous media. Ph.D. Thesis, Universitätsbibliothek der Universität Stuttgart, Stuttgart (2013). https://doi.org/10.18419/opus-476

Tene, M., Bosma, S.B., Kobaisi, M.S.A., Hajibeygi, H.: Projection-based embedded discrete fracture model (pEDFM). Adv. Water Resour. **105**, 205–216 (2017). https://doi.org/10.1016/j.advwatres.2017.05.009

Van Genuchten, M.: A closed-form equation for predicting the hydraulic conductivity of unsaturated soils1. Soil Sci. Soc. Am. J. **44**, 892 (1980)

Wagner, W., Kretzschmar, H.J.: IAPWS industrial formulation 1997 for the thermodynamic properties of water and steam. In: International Steam Tables, pp. 7–150. Springer, Berlin (2008)

Warren, J., Root, P.: The behavior of naturally fractured reservoirs. Soc. Pet. Eng. J. **3**(03), 245–255 (1963)

GEM - International Journal on Geomathematics
https://doi.org/10.1007/s13137-019-0115-9

ORIGINAL PAPER

Mathematical analysis, finite element approximation and numerical solvers for the interaction of 3D reservoirs with 1D wells

Daniele Cerroni[1] · Federica Laurino[1,2] · Paolo Zunino[1]

Received: 20 March 2018 / Accepted: 6 December 2018

Abstract
We develop a mathematical model for the interaction of a three-dimensional reservoir with the flow through wells, namely narrow cylindrical channels cutting across the reservoir. Leak off or sink effects are taken into account. To enable the simulation of complex configurations featuring multiple wells, we apply a model reduction technique that represents the wells as one-dimensional channels. The challenge in this case is to account for the interaction of the reservoir with the embedded one-dimensional wells. The resulting problem consists of coupled partial differential equations defined on manifolds with heterogeneous dimensionality. The existence and regularity of weak solutions of such problem is thoroughly addressed. Afterwards, we focus on the numerical discretization of the problem in the framework of the finite element method. We notice that the numerical scheme does not require conformity between the computational mesh of the reservoir and the one of the wells. From the standpoint of the solvers, we discuss the application of multilevel algorithms, such as the algebraic multigrid method. Finally, the reduced mathematical model and the discretization method is applied to a few configurations of reservoir with wells, with the purpose of verifying the theoretical properties and to assess the generality of the approach.

Keywords Perforated reservoirs · Dimensional model reduction · Finite elements · Multigrid solvers

Mathematics Subject Classification 35J47 · 35Q35 · 35Q86 · 58J05 · 76S05 · 65N30

✉ Paolo Zunino
paolo.zunino@polimi.it

Daniele Cerroni
daniele.cerroni@polimi.it

Federica Laurino
federica.laurino@polimi.it

[1] MOX, Department of Mathematics, Politecnico di Milano, Milan, Italy

[2] Istituto Italiano di Tecnologia, Genoa, Italy

1 Introduction

The simulation of multiscale, multiphysics, multimodel systems is among the grand challenges in Computational Science and Engineering. In this context, the application of topological (or geometrical) model reduction techniques plays an essential role. For example, small inclusions of a continuum can be described as zero-dimensional (0D) or one-dimensional (1D) concentrated sources in order to reduce the computational cost of simulations. Many problems in this area are not well investigated yet, such as the coupling of three-dimensional (3D) continua with embedded (1D) networks, although it arises in applications of paramount importance such as flow through perforated media.

Despite the literature of computational models for describing fractured reservoirs is extremely rich and lively (it would be excessively reductive to make some examples here) the analogous problems of the interaction of reservoirs with wells are much less explored. Indeed, the seminal work by Peaceman (1978, 1983, 1987) is still widely used by the scientific community. Only recently (with respect to the previous works) some new approaches have been proposed (Bellout et al. 2012; Chen and Yue 2003; Jenny and Lunati 2009; Wolfsteiner et al. 2006; Wu 2000).

Our objective is contributing to the development of advanced computational models for the interaction of reservoirs with wells. We aim to develop an approach that is appealing for industrial applications, involving realistic geological models and real configurations of multiple wells. A simplified sketch of the applications we aim to address is reported in Fig. 1. For this reason, we adopt a geometrical model reduction technique that transforms the flow equations in the wells into a 1D model. The resulting problem consists of coupled partial differential equations (PDEs) on manifolds with heterogeneous dimensionality. This approach was originally proposed in D'Angelo (2007, 2012), D'Angelo and Quarteroni (2008). It has recently attracted the attention of several researchers from the perspective of theory and applications. On one hand, it requires particular attention to prove existence of a solution in the weak (or variational) sense (Köppl et al. 2016; Köppl and Wohlmuth 2014; Koeppl et al. 2018; Kuchta et al. 2016; Boon et al. 2017; Notaro et al. 2016). On the other hand, it is relevant for applications to microcirculation (Cattaneo and Zunino 2014a, b; Nabil et al. 2015; Nabil and Zunino 2016; Possenti et al. 2018, 2019).

After discussing the model, we discretize the equations using the finite element method (FEM) and we focus on the computational aspects of the problem. Even though FEM is a well established computational method and several open-source and commercial numerical solvers are available, the implementation of general three-dimensional FEM solvers able to efficiently handle one or two-dimensional inclusions is still a significant challenge for the scientific computing community.

The method proposed here facilitates this task, because it does not require conformity between the computational mesh of the reservoir and the one of the wells. However, because of the non-standard coupling of the flow problems in 3D and 1D, the algebraic structure of the discrete problem is modified with respect to the standard FEM case. For this reason, with a few test cases with increasing complexity, we present and discuss preliminary results on the application of high performance, multilevel algebraic solvers to this problem.

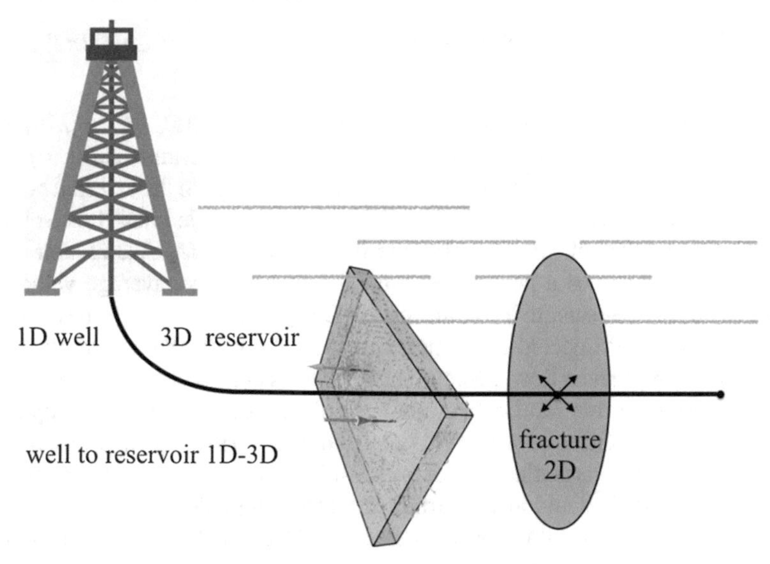

Fig. 1 A sketch of the problem involving the interaction of a reservoir with wells

This work is organized as follows. In Sect. 2, we address the full model formulation and we thoroughly describe the topological model reduction technique. We also derive the variational formulation of the reduced problem. Then, in Sect. 3 we study the well posedness and the regularity of the weak problem, which can be cast in the framework of the Lax-Milgram lemma. The numerical discretization is performed and analyzed in Sect. 4, with particular attention to the application of multigrid solvers. Finally, in Sect. 5 we discuss the numerical results in view of the available theory and we apply the computational model to some simplified configurations of 1D wells embedded into 3D reservoirs.

2 Problem setting for a reservoir with one well

The domain is denoted as Ω and composed by two parts, Ω_w and $\Omega_p = \Omega \backslash \overline{\Omega}_w$, where $\overline{\Omega}_w$ denotes the closure of Ω_w. We assume that Ω_w is the well and Ω_p the surrounding reservoir. Let Ω_w be a cylinder swept by a circle of radius ρ along a curve. More precisely, let $\boldsymbol{\lambda}(s) = [\xi(s), \nu(s), \zeta(s)]$, $s \in (0, S)$ be a $\mathcal{C}^2$-regular curve in the three-dimensional space. Let $\Lambda = \{\boldsymbol{\lambda}(s),\ s \in (0, S)\}$ be the centerline of the cylinder. For simplicity, let us assume that $\|\boldsymbol{\lambda}'(s)\| = 1$ such that the arc-length and the coordinate s coincide. Let $\boldsymbol{T}, \boldsymbol{N}, \boldsymbol{B}$ be the Frenet frame related to the curve. Let $\mathcal{D} = [r\cos\theta, r\sin\theta] : [0, \rho) \times [0, 2\pi) \rightarrow \mathbb{R}^2$ be a parametrization of the cross section. Let us also define the boundary of the cross section as $\partial\mathcal{D} = [\rho\sin\theta, \rho\cos\theta] : [0, 2\pi) \rightarrow \mathbb{R}^2$. Then, the cylinder Ω_w can be defined as follows

$$\Omega_w = \{\boldsymbol{\lambda}(s) + r\cos\theta\boldsymbol{N}(s) + r\sin\theta\boldsymbol{B}(s),\ r \in [0, \rho),\ s \in (0, S),\ \theta \in [0, 2\pi)\},$$

and the lateral boundary of it, denoted with Γ is,

$$\Gamma = \{\boldsymbol{\lambda}(s) + \rho \cos\theta \boldsymbol{N}(s) + \rho \sin\theta \boldsymbol{B}(s),\ s \in (0, S),\ \theta \in [0, 2\pi)\}.$$

We notice that Ω_w has *top* and *bottom* boundaries, which are $\partial\Omega_w \backslash \Gamma = \{\boldsymbol{\lambda}(0) + \mathcal{D}\} \cup \{\boldsymbol{\lambda}(S) + \mathcal{D}\}$. To model wells, without loss of generality, we assume that $\{\boldsymbol{\lambda}(0) + \mathcal{D}\}$ is the injection section of the well and for this reason it belongs to the external boundary, namely $\partial\Omega_p$. The other well tip, $\{\boldsymbol{\lambda}(S) + \mathcal{D}\}$ may be embedded into the reservoir.

Let $\boldsymbol{u}_p$, p_p be the velocity and pressure in the reservoir Ω_p. We assume that the reservoir is described as a porous medium. By consequence, average velocity and pressure within a representative volume obey to Darcy's law. Let $\boldsymbol{u}_w$, p_w be the velocity and pressure of fluid injected or extracted through the well. We model the well by means of pressure-driven flow. More precisely, we assume that the pressure gradient is the main driving force for the flow motion. This is a fairly general assumption that can be motivated in different ways. On one hand, it is verified by the Poiseuille flow, that can be seen as a particular case of Stokes-like flow applied to straight cylinders. On the other hand, it would also be true for a Darcy-type flow, where the motion is determined by the interaction between pressure and friction. For the sake of simplicity (homogeneity with the reservoir model) we adopt the second standpoint, but we remark that the mathematical structure of the final governing equations would be the same in the former case. However, the coefficients of the equations would differ in the two cases. Let $\boldsymbol{K}_p$ and $\boldsymbol{K}_w$ be the permeability tensors in the reservoir and in the well, respectively. We assume that these are symmetric positive definite tensors and that there exist constants $K_p,\ K_w > 0$ such that $\boldsymbol{v}^T \boldsymbol{K}_* \boldsymbol{v} \geq K_* \|\boldsymbol{v}\|^2$ with $* = p, w$. Finally, it is assumed that the interface Γ is permeable, namely it is crossed by a normal flux proportional to $K_\Gamma \left(p_p - p_w\right)$. The coefficient $K_\Gamma \geq 0$ denotes the permeability of the borehole lateral surface. It can take null values on part of the well.

For the boundary conditions, without loss of generality, we have prescribed that the pressure is fixed to values $p_{w,0}$, $p_{w,1}$ at the endpoints of the well. For the reservoir, we split $\partial\Omega_p \backslash \Gamma$ into two complementary parts, namely Σ_N, Σ_D such that $\Sigma_N \cup \Sigma_D = \partial\Omega_p \backslash \Gamma$. On Σ_N we set Robin-type boundary conditions, where the parameter $K_{\partial\Omega}$ may also vanish and $p_{p,0}$ stands for a reference pressure outside the reservoir, e.g. the hydrostatic pressure. On Σ_D we set Dirichlet type condition, for a given pressure value $p_{p,1}$.

As a result of these assumptions, we describe flow and pressure in the system by means of the following prototype problem,

$$\begin{cases} \nabla \cdot \boldsymbol{u}_p = 0, \quad \boldsymbol{u}_p + \boldsymbol{K}_p \nabla p_p = 0 & \text{in } \Omega_p, \\ \nabla \cdot \boldsymbol{u}_w = 0, \quad \boldsymbol{u}_w + \boldsymbol{K}_w \nabla p_w = 0 & \text{in } \Omega_w, \\ \boldsymbol{u}_p \cdot \boldsymbol{n}_p = K_\Gamma \left(p_p - p_w\right) & \text{on } \Gamma, \\ \boldsymbol{u}_w \cdot \boldsymbol{n}_w = K_\Gamma \left(p_w - p_p\right) & \text{on } \Gamma, \\ \boldsymbol{u}_p \cdot \boldsymbol{n}_p = K_{\partial\Omega} \left(p_p - p_{p,0}\right) & \text{on } \Sigma_N, \\ p_p = p_{p,1} & \text{on } \Sigma_D, \\ p_w = p_{w,0} & \text{on } \{\boldsymbol{\lambda}(0) + \mathcal{D}\}, \\ p_w = p_{w,1} & \text{on } \{\boldsymbol{\lambda}(S) + \mathcal{D}\}. \end{cases} \tag{1}$$

2.1 Topological model reduction and coupled problems with hybrid dimensionality

The objective of this work is to consider a simplified version of problem (1), where the domain Ω_w shrinks to its centerline Λ and the corresponding partial differential equation is averaged on the cylinder cross section $\mathcal{D}$. This new problem setting will be also called the *reduced problem*. Form the mathematical standpoint it is more challenging than (1), because it involves the coupling of 3D/1D elliptic equations.

Before proceeding, let us reformulate (1) as a problem for the pressure variables solely. We obtain the following equations,

$$\begin{cases} -\nabla \cdot (\boldsymbol{K}_p \nabla p_p) = 0 & \text{in } \Omega_p, \\ -\nabla \cdot (\boldsymbol{K}_w \nabla p_w) = 0 & \text{in } \Omega_w, \\ -\boldsymbol{K}_p \nabla p_p \cdot \boldsymbol{n}_p = K_\Gamma \left(p_p - p_w\right) & \text{on } \Gamma, \\ -\boldsymbol{K}_w \nabla p_w \cdot \boldsymbol{n}_w = K_\Gamma \left(p_w - p_p\right) & \text{on } \Gamma, \\ -\boldsymbol{K}_p \nabla p_p \cdot \boldsymbol{n}_p = K_{\partial\Omega} \left(p_p - p_{p,0}\right) & \text{on } \Sigma_N, \\ p_p = p_{p,1} & \text{on } \Sigma_D, \\ p_w = p_{w,0} & \text{on } \{\boldsymbol{\lambda}(0) + \mathcal{D}\}, \\ p_w = p_{w,1} & \text{on } \{\boldsymbol{\lambda}(S) + \mathcal{D}\}. \end{cases} \tag{2}$$

Then, let us scale the domains Ω_p and Ω_w and let us rewrite the equations in dimensionless form. More precisely, let $\chi_{\Omega_p}(\boldsymbol{x}) = \boldsymbol{x}/L$ where $L = \text{diam}(\Omega_p)$ be a scaling function and let be $\Omega_{p_\chi} = \chi_{\Omega_p}(\Omega_p)$, $\Omega_{w_\chi} = \chi_{\Omega_p}(\Omega_w)$ be the scaled domains. For simplicity of notation, and without loss of generality, form now on we will implicitly refer to the scaled domains dropping the sub-index χ. For the derivation of the dimensionless form of the equations, we take L as the characteristic length, $p_{w,0}$ as the characteristic pressure and K_p as the reference permeability. The dimensionless pressure is denoted as $u_* = p_*/p_{w,0}$, which represents the unknown of the dimensionless problem. Then, problem (2) is equivalent to find u_p, u_w such that,

$$\begin{cases} -\nabla \cdot (\boldsymbol{k}_p \nabla u_p) = 0 & \text{in } \Omega_p, \\ -\nabla \cdot (\boldsymbol{k}_w \nabla u_w) = 0 & \text{in } \Omega_w, \\ -\boldsymbol{k}_p \nabla u_p \cdot \boldsymbol{n}_p = \kappa \left(u_p - u_w\right) & \text{on } \Gamma, \\ -\boldsymbol{k}_w \nabla u_w \cdot \boldsymbol{n}_w = \kappa \left(u_w - u_p\right) & \text{on } \Gamma, \\ -\boldsymbol{k}_p \nabla u_p \cdot \boldsymbol{n}_p = \mu \left(u_p - \frac{p_{p,0}}{p_{w,0}}\right) & \text{on } \Sigma_N, \\ u_p = \frac{p_{p,1}}{p_{w,0}} & \text{on } \Sigma_D, \\ u_w = 1 & \text{on } \{\boldsymbol{\lambda}(0) + \mathcal{D}\}, \\ u_w = \frac{p_{w,1}}{p_{w,0}} & \text{on } \{\boldsymbol{\lambda}(S) + \mathcal{D}\}. \end{cases} \tag{3}$$

where we have denoted with $\boldsymbol{k}_* = \boldsymbol{K}_*/K_p$ the dimensionless permeability tensors, $\kappa = (K_\Gamma\, L)/K_p$ is the dimensionless permeability of the borehole lateral surface and $\mu = (K_{\partial\Omega}\, L)/K_p$ is the permeability at the external surface of the reservoir. For notational convenience, but without loss of generality, from now on we will assume that the external pressures $p_{p,0},\ p_{p,1},\ p_{w,1} = 0$. In this way, all the external forcing terms of the problem vanish, except from the unit injection pressure.

2.1.1 Topological model reduction of the well

The disadvantage of modelling a narrow borehole in three dimensions is that it requires the resolution of the geometry, which in many real applications can be difficult to handle in the context of a reservoir model. Therefore we apply a topological model reduction, namely we go from a 3D–3D to a 3D–1D formulation. The model reduction approach that we adopt is based on the following fundamental assumption:

(A0) *The diameter of the well is small compared to the diameter of the reservoir.*

The previous assumption implies that for the scaled domains the radius of the borehole $R = \rho/L$ is such that $0 < R \ll 1$. As a consequence of the previous fundamental assumption we formulate also the following:

(A1) the function u_w has a *uniform profile* on each cross section $\mathcal{D}$, namely in cylindrical coordinates $u_w(r, s, \theta) = U(s)$. We make a similar assumption on the coefficients of the problem. The permeability tensor in the borehole is isotropic, namely $\boldsymbol{k}_w = k_w \boldsymbol{I}$ and it is uniform on each cross section of the hole, that is $k_w(r, s, \theta) = k_w(s)$. The same restriction is enforced on the parameter κ on Γ, precisely $\kappa(\theta, s) = \kappa(s)$.

Since the derivation of the reduced model is based on averaging, we introduce the following notation,

$$\int_\Lambda \int_{\mathcal{D}} w d\sigma ds = \int_\Lambda \pi R^2 \overline{\overline{w}}(s) ds\,,$$
$$\int_\Lambda \int_{\partial\mathcal{D}} w d\gamma ds = \int_\Lambda 2\pi R \overline{w}(s) ds\,,$$

where $\overline{\overline{w}},\ \overline{w}$ denote the following mean values respectively,

$$\overline{\overline{w}}(s) = (\pi R^2)^{-1} \int_{\mathcal{D}} w d\sigma\,,$$
$$\overline{w}(s) = (2\pi R)^{-1} \int_{\partial\mathcal{D}} w d\gamma\,,$$

and $d\omega = r\, d\theta\, dr\, ds,\ d\sigma = r\, d\theta\, dr,\ d\gamma = R\, d\theta$ represent volume, surface and curvilinear measures.

We apply the averaging technique to Eq. (3). In particular, we consider an arbitrary portion $\mathcal{P}$ of the cylinder, bounded by two perpendicular sections to Λ with $s_1 < s_2$. We have,

$$
\begin{aligned}
\int_{\mathcal{P}} \nabla \cdot (k_w \nabla u_w) d\Omega &= \int_{\partial \mathcal{P}} k_w \nabla u_w \cdot \boldsymbol{n}_w \\
&= -\int_{\mathcal{D}(s_1)} k_w \partial_s u_w d\sigma + \int_{\mathcal{D}(s_2)} k_w \partial_s u_w d\sigma \\
&\quad + \int_{\Gamma} k_w \nabla u_w \cdot \boldsymbol{n}_w d\sigma
\end{aligned}
$$

By the fundamental theorem of integral calculus we have,

$$
\begin{aligned}
-\int_{\mathcal{D}(s_1)} k_w \partial_s u_w d\sigma + \int_{\mathcal{D}(s_2)} k_w \partial_s u_w d\sigma &= \int_{s_1}^{s_2} d_s \left(\int_{\mathcal{D}} (k_w \partial_s u_w) d\sigma \right) ds \\
&= \int_{s_1}^{s_2} \pi R^2 d_s (k_w \partial_s \overline{\overline{u}}_w) ds
\end{aligned}
$$

By means of the interface conditions between the well and the reservoir we obtain,

$$
\begin{aligned}
\int_{\Gamma} -k_w \nabla u_w \cdot \boldsymbol{n}_w d\sigma &= \int_{\Gamma} \kappa (u_w - u_p) d\sigma \\
&= \int_{s_1}^{s_2} \int_{\partial \mathcal{D}} \kappa (u_w - u_p) R d\theta ds \\
&= \int_{s_1}^{s_2} 2\pi R \kappa (\overline{u}_w - \overline{u}_p) ds \, .
\end{aligned}
$$

From the combination of all the above terms with the right hand side, we obtain that the solution u_w of (3) satisfies,

$$
\int_{s_1}^{s_2} \left[-\pi R^2 d_s (k_w \partial_s \overline{\overline{u}}_w) + 2\pi R \kappa (\overline{u}_w - \overline{u}_p) \right] ds = 0 \, .
$$

Since the choice of the points s_1, s_2 is arbitrary, we conclude that the following equation holds true,

$$
-\pi R^2 d_s (k_w \partial_s \overline{\overline{u}}_w) + 2\pi R \kappa (\overline{u}_w - \overline{u}_p) = 0 \quad \text{on } \Lambda \, . \tag{4}
$$

Using the boundary conditions we conclude that the function $u_w(r, s, \theta) = U(s)$ satisfies the following equations on Λ,

$$
\begin{aligned}
& -\pi R^2 d_s (k_w d_s U) + 2\pi R \kappa U = 2\pi R \kappa \overline{u}_p && \text{on } \Lambda \, , && \text{(5a)} \\
& U = 1 && \text{on } s = 0 && \text{(5b)} \\
& U = 0 && \text{on } s = S \, . && \text{(5c)}
\end{aligned}
$$

The weak form of the previous problem consists to find $U \in H_0^1(\Lambda)$ such that

$$
\pi R^2 (k_w d_s U, d_s V)_\Lambda + 2\pi R (\kappa U, V)_\Lambda = 2\pi R (\kappa (\overline{u}_p - W), V)_\Lambda, \quad \forall V \in H_0^1(\Lambda) \, , \tag{6}
$$

where $W = 1 - s$ denotes a suitable linear lifting of the Dirichlet boundary conditions of U on Λ.

2.1.2 Topological model reduction of the interface conditions

Let us now consider the weak formulation of Eq. (3) that govern the pressure on Ω_p. Integrating by parts and using boundary and interface conditions, we obtain:

$$\begin{aligned} 0 &= \int_{\Omega_p} \nabla \cdot (-\boldsymbol{k}_p \nabla u_p) v \, d\omega = \int_{\Omega_p} \boldsymbol{k}_p \nabla u_p \cdot \nabla v \, d\omega - \int_{\partial \Omega_p} \boldsymbol{k}_p \nabla u_p \cdot \boldsymbol{n}_p v \, d\sigma \\ &= \int_{\Omega_p} \boldsymbol{k}_p \nabla u_p \cdot \nabla v \, d\omega + \int_{\Sigma_N} \mu u_p v \, d\sigma + \int_{\Gamma} \kappa u_p v \, d\sigma \\ &\quad - \int_{\Gamma} \kappa u_w v \, d\sigma, \ \forall v \in H^1_{\Sigma_D}(\Omega). \end{aligned} \tag{7}$$

Let us split the solutions and the test functions on Γ, namely $u_*|_\Gamma$ for $* = p, w$ and $v|_\Gamma$, as their average plus some fluctuation, namely

$$u_* = \overline{u}_* + \tilde{u}_*, \qquad v = \overline{v} + \tilde{v},$$

where $\overline{\tilde{v}} = 0$ for any function. Therefore, using the cylindrical coordinates system (s, θ) on Γ, we have

$$\begin{aligned} \int_{\Gamma} \kappa u_* v \, d\sigma &= \int_{\Lambda} \kappa(s) \int_0^{2\pi} (\overline{u}_* + \tilde{u}_*)(\overline{v} + \tilde{v}) R \, d\theta ds \\ &= \int_{\Lambda} 2\pi R \kappa \overline{u}_* \overline{v} \, ds + \int_{\Lambda} \kappa \int_0^{2\pi} \tilde{u}_* \tilde{v} R \, d\theta ds, \end{aligned}$$

Then, we make the following modelling assumptions:

(A2) we identify the domain Ω_p with the entire Ω, and we correspondingly omit the subscript p to the functions defined on Ω, namely

$$\int_{\Omega_p} v_p d\omega \simeq \int_{\Omega} v d\omega \, .$$

(A3) we assume that the product of fluctuations is small, namely

$$\int_0^{2\pi} \tilde{u}_* \tilde{v} R \, d\theta \simeq 0 \, .$$

By means of the previous calculations, reminding that $u_w(r, s, \theta) = U(s)$, we obtain that $u \in H^1_{\Sigma_D}(\Omega)$ solves the following problem,

$$(\boldsymbol{k}_p \nabla u, \nabla v)_\Omega + (\mu u, v)_{\Sigma_N} + 2\pi R (\kappa \overline{u}, \overline{v})_\Lambda = 2\pi R (\kappa U, \overline{v})_\Lambda \, , \quad \forall v \in H^1_{\Sigma_D}(\Omega) \, , \tag{8}$$

where $H^1_{\Sigma_D}(\Omega)$ denotes the subspace of $H^1(\Omega)$ of functions with vanishing traces on Σ_D.

3 Mathematical analysis and numerical approximation of the problem

After applying the model reduction technique, the problem of finding the pressures u_p, u_w in the reservoir and the well, respectively, has transformed into solving a 3D problem for u in Ω and a 1D problem for U in Λ. In variational form, it consists of finding $u \in H^1_{\Sigma_D}(\Omega)$ and $U \in H^1_0(\Lambda)$ such that

$$\begin{cases} (\boldsymbol{k}_p \nabla u, \nabla v)_\Omega + (\mu u, v)_{\Sigma_N} + 2\pi R(\kappa \overline{u}, \overline{v})_\Lambda = 2\pi R(\kappa U, \overline{v})_\Lambda\,, & \forall v \in H^1_{\Sigma_D}(\Omega) \\ \pi R^2(k_w d_s U, d_s V)_\Lambda + 2\pi R(\kappa U, V)_\Lambda = 2\pi R(\kappa(\overline{u} - W), V)_\Lambda, & \forall V \in H^1_0(\Lambda)\,. \end{cases} \tag{9}$$

For what follows, it is convenient to define the bilinear forms:

$$\begin{aligned} a_\Omega(w, v) &= (\boldsymbol{k}_p \nabla w, \nabla v)_\Omega + (\mu w, v)_{\Sigma_N}, \\ a_\Lambda(w, v) &= \pi R^2(k_w d_s w, d_s v)_\Lambda \\ b_\Lambda(w, v) &= 2\pi R(\kappa w, v)_\Lambda. \end{aligned}$$

Let us now introduce a compact formulation for problem (9). In particular, we define $\mathcal{V} = [v, V]$ as a generic function of the space $\mathbb{V} = H^1_{\Sigma_D}(\Omega) \times H^1_0(\Lambda)$ and we name $\mathcal{U} = [u, U]$ the couple of unknowns of problem (9). Any function $\mathcal{V} \in \mathbb{V}$ is endowed with the norm $|||\mathcal{V}|||^2 = \|v\|^2_{H^1(\Omega)} + \|V\|^2_{H^1(\Lambda)}$. Then, we introduce the following bilinear form in $\mathbb{V} \times \mathbb{V}$,

$$\mathcal{A}(\mathcal{U}, \mathcal{V}) = a_\Omega(u, v) + a_\Lambda(U, V) + b_\Lambda(\overline{u} - U, \overline{v} - V)\,,$$

and the linear functional in $\mathbb{V}$,

$$\mathcal{F}(\mathcal{V}) = -b_\Lambda(W, V).$$

Then, the compact form of problem (9) consists of finding $\mathcal{U} \in \mathbb{V}$ such that

$$\mathcal{A}(\mathcal{U}, \mathcal{V}) = \mathcal{F}(\mathcal{V}), \quad \forall \mathcal{V} \in \mathbb{V}\,. \tag{10}$$

We name the previous problem as the *3D–1D coupled problem*. This problem is an extension to 3D of the one considered in Koeppl et al. (2018). The analysis can be pursued using the Lax-Milgram lemma. Before addressing the central result, that is Theorem 1, we present some auxiliary tools.

Lemma 1 *If $v \in H^1(\Omega)$ or alternatively $v \in L^2(\Gamma)$, then $\overline{v} \in L^2(\Lambda)$ and the following inequality holds*

$$\|\overline{v}\|^2_{L^2(\Lambda)} \leq \frac{1}{2\pi R}\|v\|^2_{L^2(\Gamma)} \leq \frac{C_T(\Gamma)}{2\pi R}\|v\|^2_{H^1(\Omega)}, \tag{11}$$

being $C_T(\Gamma)$ the (positive) constant of the trace inequality from $L^2(\Gamma)$ to $H^1(\Omega)$.

Using the results of Fernández Bonder and Rossi (2002) it can be proved that the constant $C_T(\Gamma)$ is uniformly upper bounded with respect to the radius of the inclusions, R.

Proof Let us consider

$$\int_\Lambda \overline{v}^2\, ds = \int_\Lambda \left(\frac{1}{2\pi R}\int_0^{2\pi} vR\, d\theta\right)^2 ds = \frac{1}{4\pi^2 R^2}\int_\Lambda \left(\int_0^{2\pi} vR\, d\theta\right)^2 ds. \tag{12}$$

Using Jensen's inequality, we obtain

$$\frac{1}{4\pi^2 R^2}\int_\Lambda \left(\int_0^{2\pi} vR\, d\theta\right)^2 ds \leq \frac{1}{2\pi R}\int_\Lambda\int_0^{2\pi} v^2 R\, d\theta\, ds \tag{13}$$

and consequently

$$\begin{aligned}\int_\Lambda \overline{v}^2\, d\gamma &\leq \frac{1}{2\pi R}\int_\Lambda\int_0^{2\pi} v^2 R\, d\theta\, ds \\ &= \frac{1}{2\pi R}\int_\Gamma v^2\, d\sigma = \frac{1}{2\pi R}\|v\|^2_{L^2(\Gamma)} \leq \frac{C_T(\Gamma)}{2\pi R}\|v\|^2_{H^1(\Omega)}.\end{aligned} \tag{14}$$

If the inequality (11) holds, it follows immediately that $\overline{v} \in L^2(\Lambda)$, since $v \in H^1(\Omega)$. □

Lemma 2 *Provided that Σ_D is a non-empty Lipshitz continuous subset of $\partial\Omega$, for any $v \in H^1_{\Sigma_D}(\Omega)$, there exists a positive constant, $C_P(\Omega)$, s.t.*

$$\|v\|^2_{L^2(\Omega)} \leq C_P(\Omega)\|\nabla v\|^2_{L^2(\Omega)}. \tag{15}$$

Inequality (15) is a Poincaré-type inequality, see for example Quarteroni and Valli (1994, Theorem 1.3.3) or Salsa (2016, Theorem 7.16). We will use (15) to control the H^1-norm of the solution by means of the energy of the problem (10). We can now address the coercivity and the continuity of the bilinear form $\mathcal{A}$.

Lemma 3 *Under the assumptions that $k_{p_{i,j}}, \in \left[L^\infty(\Omega)\right]^{3,3}$, $k_w \in L^\infty(\Lambda)$, $\kappa \in L^\infty(\Lambda)$, $\mu \in L^\infty(\Sigma_N)$, k_w is strictly positive with minimum k_{min} and κ, μ are nonnegative, the operator $\mathcal{A}$ is continuous and coercive. More precisely, there exist constants $M, m > 0$ such that,*

$$\mathcal{A}(\mathcal{U},\mathcal{V}) \leq M|||\mathcal{U}|||\,|||\mathcal{V}|||, \quad \mathcal{A}(\mathcal{V},\mathcal{V}) \geq m|||\mathcal{V}|||^2 \quad \forall \mathcal{U},\mathcal{V} \in \mathbb{V}.$$

Proof The coercivity of $\mathcal{A}$ follows from (15) and the ellipticity assumption. Indeed

$$\begin{aligned}\mathcal{A}(\mathcal{V}, \mathcal{V}) &= a_\Omega(v, v) + a_\Lambda(V, V) + b_\Lambda(\overline{v} - V, \overline{v} - V)\\ &\geq a_\Omega(v, v) + a_\Lambda(V, V),\end{aligned} \tag{16}$$

being $b_\Lambda(\overline{v} - V, \overline{v} - V)$ nonnegative. For the first term, using (15) and reminding that $\boldsymbol{k}_{p,ij} \geq 1$ we obtain

$$\begin{aligned}a_\Omega(v, v) &= \left(\boldsymbol{k}_p \nabla v, \nabla v\right)_\Omega + (\mu v, v)_{\Sigma_N}\\ &\geq \|\nabla v\|^2_{L^2(\Omega)}\\ &\geq \frac{1}{1 + C_P(\Omega)} \|v\|^2_{H^1(\Omega)}.\end{aligned}$$

For the second term,

$$\begin{aligned}a_\Lambda(V, V) &= \pi R^2 \,(k_w d_s V, d_s V)\\ &\geq \pi R^2 k_{min} \|d_s V\|^2_{L^2(\Lambda)}\\ &\geq \pi R^2 k_{min} \frac{1}{1 + C_p(\Lambda)} \|V\|^2_{H^1(\Lambda)},\end{aligned}$$

where $C_p(\Lambda)$ denotes the constant in the stantard Poincaré inequality for $H^1_0(\Lambda)$ functions. As a result, the coercivity constant is

$$m = \min\left(\frac{1}{1 + C_P(\Omega)}, \pi R^2 k_{min} \frac{1}{1 + C_p(\Lambda)}\right).$$

In order to prove the continuity of the bilinear form $\mathcal{A}$ we consider again each term of $\mathcal{A}(u, v)$ separately. For the first bilinear form $a_\Omega(u, v)$ we have

$$\begin{aligned}a_\Omega(u, v) &= \left(\boldsymbol{k}_p \nabla u, \nabla v\right)_\Omega + (\mu u, v)_{\Sigma_N}\\ &\leq \max_{i,j} \|k_{p_{i,j}}\|_{L^\infty(\Omega)} \|\nabla u\|_{L^2(\Omega)} \|\nabla v\|_{L^2(\Omega)}\\ &\quad + \|\mu\|_{L^\infty(\Sigma_N)} \|u\|_{L^2(\Sigma_N)} \|v\|_{L^2(\Sigma_N)}\\ &\leq \max_{i,j} \|k_{p_{i,j}}\|_{L^\infty(\Omega)} \|u\|_{H^1(\Omega)} \|v\|_{H^1(\Omega)}\\ &\quad + C_T\,(\Sigma_N)\, \|\mu\|_{L^\infty(\Sigma_N)} \|u\|_{H^1(\Omega)} \|v\|_{H^1(\Omega)}\\ &\leq \left(\max_{i,j} \|k_{p_{i,j}}\|_{L^\infty(\Omega)} + C_T\,(\Sigma_N)\, \|\mu\|_{L^\infty(\Sigma_N)}\right) |||\mathcal{U}|||\, |||\mathcal{V}|||,\end{aligned}$$

where $C_T(\Sigma_N)$ is the constant of the trace inequality from $L^2(\Sigma_N)$ to $H^1(\Omega)$. For the second term $a_\Lambda(U, V)$, we easily obtain

$$\begin{aligned} a_\Lambda(U,V) &= \pi R^2 (k_w d_s U, d_s V)_\Lambda \\ &\le \pi R^2 \|k_w\|_{L^\infty(\Lambda)} \|U\|_{H^1(\Lambda)} \|V\|_{H^1(\Lambda)} \\ &\le \pi R^2 \|k_w\|_{L^\infty(\Lambda)} |||\mathcal{U}||| \, |||\mathcal{V}|||. \end{aligned}$$

For the last term $b_\Lambda(\overline{u} - U, \overline{v} - V)$ using Lemma 1, we obtain

$$\begin{aligned} b_\Lambda(\overline{u} - U, \overline{v} - V) &\le 2\pi R \|\kappa\|_{L^\infty(\Lambda)} \|\overline{u} - U\|_{L^2(\Lambda)} \|\overline{v} - V\|_{L^2(\Lambda)} \\ &\le 2\pi R \|\kappa\|_{L^\infty(\Lambda)} \left(\|\overline{u}\|_{L^2(\Lambda)} + \|U\|_{L^2(\Lambda)}\right) \left(\|\overline{v}\|_{L^2(\Lambda)} + \|V\|_{L^2(\Lambda)}\right) \\ &\le 2\pi R \|\kappa\|_{L^\infty(\Lambda)} \left(1 + \sqrt{\frac{C_T(\Gamma)}{2\pi R}}\right)^2 |||\mathcal{U}||| \, |||V|||. \end{aligned}$$

Therefore,

$$\begin{aligned} M = \max_{i,j} \|k_{p_{i,j}}\|_{L^\infty(\Omega)} &+ C_T(\Sigma_N) \|\mu\|_{L^\infty(\Sigma_N)} + \pi R^2 \|k_w\|_{L^\infty(\Lambda)} \\ &+ 2\pi R \|\kappa\|_{L^\infty(\Lambda)} \left(1 + \sqrt{\frac{C_T(\Gamma)}{2\pi R}}\right)^2 . \end{aligned}$$

□

Lemma 3 shows that the coercivity and continuity constants depend of the boundary conditions and the parameters of the problem. In particular, they are not robust with respect to R because $m \to 0$ in the limit $R \to 0$. In other words, the subproblem on Λ looses coercivity in the limit case.

Lemma 4 *The functional $\mathcal{F}$ is continuous in $\mathbb{V}$.*

Proof Indeed,

$$\begin{aligned} \mathcal{F}(\mathcal{V}) &= -b_\Lambda(W, V) \\ &= -2\pi R (\kappa W, V)_\Lambda \\ &\le 2\pi R \|\kappa\|_{L^\infty(\Lambda)} \|W\|_{L^2(\Lambda)} \|V\|_{L^2(\Lambda)} \le 2\pi R \|\kappa\|_{L^\infty(\Lambda)} \|W\|_{L^2(\Lambda)} |||\mathcal{V}|||. \end{aligned}$$

□

Theorem 1 *Problem* (10) *has a unique solution $\mathcal{U} \in \mathbb{V}$. Furthermore, the following stability estimate holds true,*

$$|||\mathcal{U}||| \le \frac{1}{m} 2\pi R \|\kappa\|_{L^\infty(\Lambda)} \|W\|_{L^2(\Lambda)}. \tag{17}$$

Proof Owing to the Lax-Milgram Theorem, the result is a direct consequence of the previous lemmas. □

3.1 Finite element approximation

Let us consider a quasi-uniform partition $\mathcal{T}_\Omega^h$ of Ω and an admissible partition $\mathcal{T}_\Lambda^h$ of Λ with the same characteristic size h and let $\mathbb{V}_h = V_h^\Omega \times V_h^\Lambda \subset \mathbb{V}$ be continuous k_1, k_2-order Lagrangian finite element space defined on $\mathcal{T}_\Omega^h$, $\mathcal{T}_\Lambda^h$ respectively. The numerical approximation of the variational formulation (10) consists of finding $\mathcal{U}_h \in \mathbb{V}_h$ solution of

$$\mathcal{A}(\mathcal{U}_h, \mathcal{V}_h) = \mathcal{F}(\mathcal{V}_h) \qquad \forall \mathcal{V}_h \in \mathbb{V}_h. \tag{18}$$

We notice that in problem (18) it is implicitly assumed that numerical integration is performed exactly. In practice, the average operator $\overline{(\cdot)}$ is approximated by means of numerical quadrature. The effect of this further approximation shall be analyzed in a future development of this work.

We exploit the conformity of the finite element space combined with Lemma 3, in order to prove that $\mathcal{U}_h$ satisfies a Ceá-type inequality,

$$|||\mathcal{U} - \mathcal{U}_h||| \leq \frac{M}{m} \inf_{v_h \in V_h^\Omega, V_h \in V_h^\Lambda} \left(\|u - v_h\|_{H^1(\Omega)} + \|U - V_h\|_{H^1(\Lambda)} \right). \tag{19}$$

The convergence of the finite element method follows from (19) combined with approximation properties of the finite element space. For the latter property, additional regularity of the solution is required. The solution U on Λ is in $H^2(\Lambda)$. The regularity of U descends from the standard theory of elliptic operators in convex domains, Gilbarg and Trudinger (2001, Theorem 8.12), being W and $\overline{u}$ both in $L^2(\Lambda)$. Then, for the solution U on Λ, the standard finite element approximation estimate ensures that

$$\inf_{V_h \in V_h^\Lambda} \|U - V_h\|_{H^1(\Lambda)} \lesssim h^{r_2} \|U\|_{H^2(\Lambda)} \quad r_2 = \min(k_2, 1),$$

where $a \lesssim b$ is equivalent to the inequality $a \leq Cb$ being C is a generic constant, possibly dependent on Ω but independent of the parameters of the problem.

Conversely, the regularity of u can not be derived from standard results, because the right hand side in the first equation of (9) can be represented as a Dirac measure defined on Λ, the 1D manifold embedded into Ω. For this reason, we simply state the convergence result provided that u belongs to a suitable Sobolev space $W^{l,q}(\Omega)$. Let π^h be the Scott-Zhang interpolation operator from $W^{t,q}(\Omega) \cap H_0^1(\Omega)$ to V_h with $1 \leq q \leq \infty$ and $1 \leq t \leq l$ with the additional constraint $l > 1/q$ when $q > 1$. Then, the following interpolation estimate holds true [see for example Ern and Guermond (2004, Lemma 1.130)]

$$\|v - \pi^h v\|_{W^{t,q}(\Omega)} \lesssim h^{r_1 - t} \|v\|_{W^{l,q}(\Omega)}, \quad r_1 = \min(l, k_1 + 1).$$

Therefore, combining (19) and the previous inequalities for piecewise affine approximation, we obtain

$$|||\mathcal{U}-\mathcal{U}_h||| \lesssim h^{r_1-t}\|v\|_{W^{l,q}(\Omega)} + h^{r_2}\|U\|_{H^2(\Lambda)}.$$

4 Numerical solution strategies

In this section we analyze the properties of the matrix A arising from the discretization with linear finite elements of problem (10). We are particularly interested in investigating the performance of the Algebraic Multigrid Method (AMG) applied to this problem. For this purpose, we use the AMG library developed in Stüben (1983, 1999), Stüben et al. (2007).

Let us denote with $\varphi^*_{h,i} \in V^*_h$, $i = 1, \ldots, \dim(V^*_h)$ be the Lagrangian basis functions of V^*_h, where the symbol $*$ denotes either the domain Ω or Λ, and let $\boldsymbol{v}^* = \{v^*_i\}$ be the vector of the degrees of freedom relative to the generic finite element function v^*_h such that

$$v^*_h = \sum_{i=1}^{\dim(V^*_h)} v^*_i \varphi^*_{h,i}\,.$$

Let A be the stiffness matrix corresponding to the bilinear form $\mathcal{A}(\cdot,\cdot)$. More precisely, problem (10) is equivalent to the following algebraic problem,

$$A\boldsymbol{u} = \mathbf{f} \;\Leftrightarrow\; \begin{bmatrix} A^\Omega & -B \\ -B^T & A^\Lambda \end{bmatrix} \cdot \begin{bmatrix} \boldsymbol{u}^\Omega \\ \boldsymbol{u}^\Lambda \end{bmatrix} = \begin{bmatrix} \mathbf{0} \\ \mathbf{f}^\Lambda \end{bmatrix} \tag{20}$$

where the matrices and the right hand side have the following expressions

$$\begin{aligned}
A^\Omega_{ij} &= \left(\boldsymbol{k}_p \nabla\varphi^\Omega_{h,j}, \nabla\varphi^\Omega_{h,i}\right) + \left(\mu\varphi^\Omega_{h,j}, \varphi^\Omega_{h,i}\right)_{\Sigma_N} + 2\pi R(\kappa\overline{\varphi}^\Omega_{h,j}, \overline{\varphi}^\Omega_{h,i})_\Lambda\,,\\
A^\Lambda_{ij} &= \pi R^2(k_w d_s\varphi^\Lambda_{h,j}, d_s\varphi^\Lambda_{h,i})_\Lambda + 2\pi R(\kappa\varphi^\Lambda_{h,j}, \varphi^\Lambda_{h,i})_\Lambda\,,\\
B_{ij} &= 2\pi R(\kappa\varphi^\Lambda_{h,j}, \overline{\varphi}^\Omega_{h,i})_\Lambda\,,\\
\mathbf{f}^\Lambda_i &= -2\pi R(\kappa W, \varphi^\Lambda_{h,i})_\Lambda\,.
\end{aligned}$$

We aim to study system (20). We observe that, because of the different dimensionality, usually the discretization of Ω has many more degrees of freedom than the one of Λ. More precisely we have $\dim(V^\Omega_h) \gg \dim(V^\Lambda_h)$. Then, it is convenient to rewrite (20) as follows,

$$\begin{aligned}
&A^\Lambda\boldsymbol{u}^\Lambda = \mathbf{f}^\Lambda + B^T\boldsymbol{u}^\Omega\,,\\
&\left(A^\Omega - B\left(A^\Lambda\right)^{-1}B^T\right)\boldsymbol{u}^\Omega = B\left(A^\Lambda\right)^{-1}\mathbf{f}^\Lambda\,.
\end{aligned}$$

Since the matrix A^Λ can be easily solved and factorized because of the small size, this shows that the major cost for solving (20) is due to the subproblem on Ω. For this reason, in what follows we focus on the spectral properties and the solvers for matrix A^Ω solely. In this case, the discrete function identified by $\boldsymbol{u}^\Lambda$ is given a-priori. We name this problem as the *3D problem with 1D inclusions*.

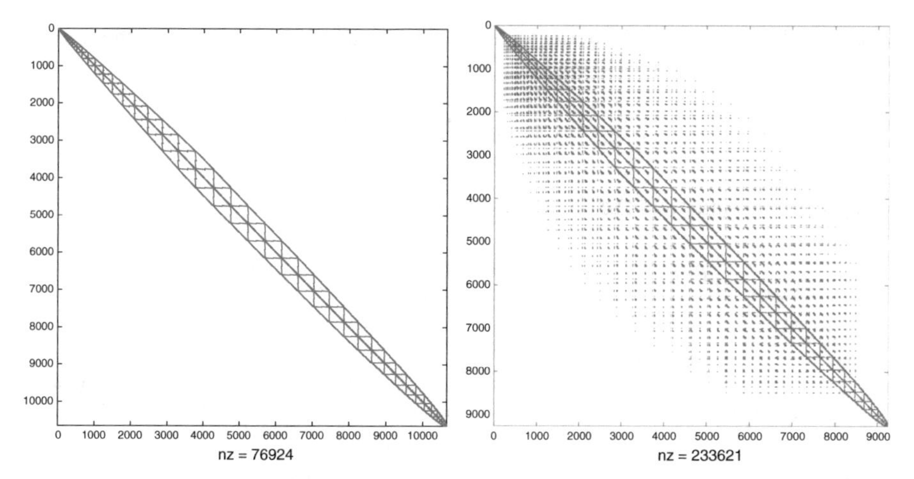

Fig. 2 The pattern of the standard stiffness matrix is reported on the left and compared with the pattern of the modified stiffness matrix A^{Ω}, on the right. The effect of the nonlocal coupling terms is clearly visible

4.1 Spectral properties and conditioning of the discrete 3D problem with 1D inclusions

We observe that A^{Ω} can be seen as the sum of a standard stiffness matrix (and possible boundary terms) with a matrix C^{Ω} relative to the coupling terms, such that $C^{\Omega}_{ij} = 2\pi R(\kappa\overline{\varphi}_{h,j}, \overline{\varphi}_{h,i})_{\Lambda}$. Since the coupling terms are non local, the sparsity pattern of the matrix will be significantly different than the one of a standard stiffness matrix. In this case, the rows of A^{Ω} may feature many non-zero entries. More precisely, Fig. 2 shows the pattern of standard finite element stiffness matrix on the left, namely $\left(\boldsymbol{k}_p\nabla\varphi^{\Omega}_{h,j}, \nabla\varphi^{\Omega}_{h,i}\right)$, and on the right we show the new matrix modified by the 3D–1D coupling terms, namely matrix A^{Ω} defined above. We immediately observe that the coupling terms affect significantly the sparsity pattern of the algebraic problem. This poses some questions about the applicability of state of art numerical solvers to the problem, which will be directly addressed in the next section.

We proceed here with the analysis of the spectrum of A^{Ω} that is related to the bilinear form

$$
\begin{aligned}
A^{\Omega}_{ij} &= \mathcal{A}^{\Omega}\left(\varphi^{\Omega}_{h,j}, \varphi^{\Omega}_{h,i}\right) \\
&= \left(\boldsymbol{k}_p\nabla\varphi^{\Omega}_{h,j}, \nabla\varphi^{\Omega}_{h,i}\right) + \left(\mu\varphi^{\Omega}_{h,j}, \varphi^{\Omega}_{h,i}\right)_{\Sigma_N} + 2\pi R(\kappa\overline{\varphi}^{\Omega}_{h,j}, \overline{\varphi}^{\Omega}_{h,i})_{\Lambda}.
\end{aligned}
$$

The eigenvalues of a self-adjoint operator are real and for the largest and smallest eigenvalues the following expressions are valid:

$$
\lambda_{min} = \min_{v^{\Omega}_h \neq 0 \in V^{\Omega}_h} \frac{\mathcal{A}^{\Omega}\left(v^{\Omega}_h, v^{\Omega}_h\right)}{\|\boldsymbol{v}^{\Omega}\|^2}, \qquad \lambda_{max} = \max_{v^{\Omega}_h \neq 0 \in V^{\Omega}_h} \frac{\mathcal{A}^{\Omega}\left(v^{\Omega}_h, v^{\Omega}_h\right)}{\|\boldsymbol{v}^{\Omega}\|^2},
$$

where $\|\cdot\|$ is the standard Euclidean norm. In the case of Lagrangian finite elements in a three dimensional space, we have

$$h^3\|\boldsymbol{v}^{\Omega}\|^2 \lesssim \|v_h^{\Omega}\|^2_{L^2(\Omega)} \lesssim h^3\|\boldsymbol{v}^{\Omega}\|^2.$$

Using the continuity and the coercivity of $\mathcal{A}^{\Omega}$, we obtain

$$\lambda_{min} = \min_{v_h^{\Omega}\in V_h^{\Omega}} \frac{\mathcal{A}^{\Omega}(v_h^{\Omega}, v_h^{\Omega})}{\|\boldsymbol{v}\|^2} \geq m^{\Omega}\frac{\|v_h^{\Omega}\|^2_{H^1(\Omega)}}{\|\boldsymbol{v}^{\Omega}\|} \geq m^{\Omega}\frac{\|v_h^{\Omega}\|^2_{L^2(\Omega)}}{\|\boldsymbol{v}^{\Omega}\|} \gtrsim m^{\Omega}h^3,$$

$$\lambda_{max} = \max_{v_h^{\Omega}\in V_h^{\Omega}} \frac{\mathcal{A}^{\Omega}\left(v_h^{\Omega}, v_h^{\Omega}\right)}{\|\boldsymbol{v}^{\Omega}\|^2} \lesssim M^{\Omega}\frac{\|v_h^{\Omega}\|^2_{H^1(\Omega)}}{\|\boldsymbol{v}^{\Omega}\|^2} \lesssim M^{\Omega}h^{-2}\frac{\|v_h^{\Omega}\|^2_{L^2(\Omega)}}{\|\boldsymbol{v}^{\Omega}\|^2} \lesssim M^{\Omega}h,$$

where, according to Lemma 3, constants m^{Ω}, M^{Ω} are the following,

$$m^{\Omega} = \frac{1}{1+C_P(\Omega)},$$
$$M^{\Omega} = \max_{i,j}\|k_{p_{i,j}}\|_{L^{\infty}(\Omega)} + C_T(\Sigma_N)\|\mu\|_{L^{\infty}(\Sigma_N)} + C_T(\Gamma)\|\kappa\|_{L^{\infty}(\Lambda)}.$$

For self-adjoint problems, upper and lower bounds of the spectrum give information about the spectral condition number $\mathcal{K}(\mathcal{A}^{\Omega})$. Precisely, in this case we have

$$\mathcal{K}(\mathcal{A}^{\Omega}) \lesssim \frac{M^{\Omega}}{m^{\Omega}}h^{-2}.$$

From the previous spectral analysis, we deduce that the discretization in Ω is robust with respect to small inclusion radii, being the continuity and coercivity constants bounded in the limit as R tends to zero, because the constants $C_P(\Omega)$ and $C_T(\Sigma_N)$ are independent of R and $C_T(\Gamma)$ is uniformly upper bounded with respect to R.

4.2 Application of Algebraic Multigrid to the solution of the 3D problem with 1D inclusions

The adequacy of the Algebraic Multigrid algorithm (AMG) to solve problems such as the one of Fig. 2 (right) is an interesting question. On one hand, matrix A^{Ω} is symmetric positive definite, which makes AMG to be a good solution method. On the other hand, there are concerns about the influence of the modified sparsity patterns on the algebraic coarsening process. For example, Fig. 3, shows how the sparsity pattern of matrix A^{Ω} changes after three levels of coarsening. The coarsest level (level three) is almost full and it is solved using a direct algorithm. Then, we aim to investigate how the coarsening algorithm transforms matrix A^{Ω} and how standard smoothing techniques behave when applied to these coarse problems. Finally, we will address the question of robustness of the solver with respect to reaction dominated problems with concentrated sources.

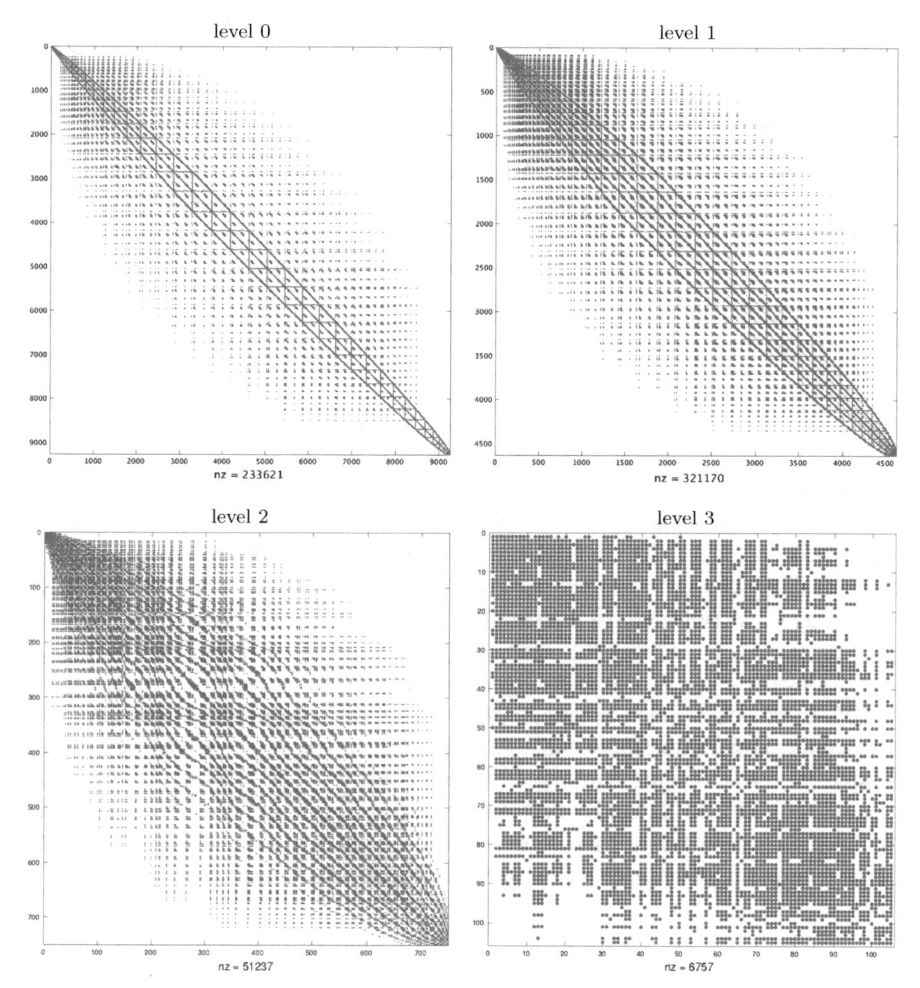

Fig. 3 Spartity patterns of the fine (level 0) and coarse levels (respectively 1, 2, 3) of the AMG solver

Keeping in mind the previous objectives, we review here some results about AMG. Since the recursive extension of any two level process to a multi-level process is formally straightforward, we describe the components of the AMG algorithm only on the basis of two level methods. Indices h and H mark the fine and the coarse level, respectively. We rewrite the sub-problem related to A^Ω as follows,

$$A^\Omega \boldsymbol{u}^\Omega = \boldsymbol{f}^\Omega \quad \Leftrightarrow \quad A_h \boldsymbol{u}_h = \boldsymbol{f}_h \,. \tag{21}$$

We remark that A_h in (21) stands for the matrix relative to the subproblem in Ω. It shall not be confused with the matrix of the coupled problem defined in (20). In this framework the coarse level system is

$$A_H \boldsymbol{u}_H = \boldsymbol{f}_H \,, \tag{22}$$

where the matrix A_H is defined as the Galerkin operator

$$A_H := I_h^H A_h I_H^h, \tag{23}$$

where I_H^h and I_h^H denote the interpolation and the restriction operators, respectively.

We now define a smoothing process with a corresponding smoothing operator S_h as

$$\hat{\boldsymbol{u}}^h = S_h \boldsymbol{u}^h + (I_h - S_h) A_h^{-1} \boldsymbol{f}^h, \tag{24}$$

where I_h denotes the identity operator. We focus our attention on the Gauss-Seidel relaxation scheme, which implies the following choice of the smother

$$S_h = \left(I_h - Q_h^{-1} A_h \right), \tag{25}$$

with Q_h being the lower triangular part of A_h.

We remark that A_h is a symmetric positive definite matrix for which we can define the restriction (I_h^H) as the transpose of interpolation (I_H^h). As a consequence A_H, defined in (23), is also symmetric and positive definite, *regardless* of the choice of I_H^h as long as it has full rank. In Stüben (1999) the authors show that for a symmetric positive definite matrix any full rank interpolation operator generate a coarse level correction operator, which is an orthogonal projector with respect to the scalar product of $\mathcal{A}^{\Omega}$. In this case Galerkin based coarse grid corrections minimize the Euclidean norm of the error. Furthermore, the smoother iterations converge, because the operator S_h satisfies the *smoothing property*. More precisely, let $\boldsymbol{e}_h$ be the difference between the exact solution of (21) and the current approximation before smoothing. In Stüben (1999) it is proven that for the smoother defined in (25), there exists σ, a strictly positive constant independent of $\boldsymbol{e}_h$, such that the following inequality holds true,

$$\|S_h \boldsymbol{e}_h\|_1^2 \leq \|\boldsymbol{e}_h\|_1^2 - \sigma \|\boldsymbol{e}_h\|_2^2,$$

where $\|\cdot\|_0$, $\|\cdot\|_1$ and $\|\cdot\|_2$ are norms induced by the following inner products,

$$(\boldsymbol{u}, \boldsymbol{v})_0 = (D_h \boldsymbol{u}, \boldsymbol{v}), \quad (\boldsymbol{u}, \boldsymbol{v})_1 = (A_h \boldsymbol{u}, \boldsymbol{v}),$$
$$(\boldsymbol{u}, \boldsymbol{v})_2 = \left(D_h^{-1} A_h \boldsymbol{u}, A_h \boldsymbol{v} \right), \quad D_h = diag(A_h).$$

This property implies that S_h is efficient in reducing the error $\boldsymbol{e}_h$ and more in general the performance of the AMG solver is good for symmetric positive definite problems such as (21). Moreover, the solver is not affected by the pattern of the matrix provided that the restriction operator has full rank and the coarsest problem can be appropriately handled (by a direct method).

We conclude this section with some considerations about the dependence of problem (10) from the physical parameters. In cases where $2\pi R\kappa \gg 1$, problem (10) is (locally) reaction dominated. As it will be discussed later, this is the case of applications to wells. The numerical experiments addressed in the next section suggest that the AMG solver is robust with respect to this case. This property can be justified following the lines of Schatz and Wahlbin (1983) and Olshanskii and Reusken (2000), where the convergence of the finite element method and of the geometric multigrid solver, respectively, are studied for reaction dominated problems.

The main reason why locally reaction-perturbed elliptic operators do not pose substantial problems at the level of finite element approximation, can be found in the analysis developed by Schatz and Wahlbin (1983). There, the authors prove the local uniform convergence of finite elements for singularly perturbed reaction–diffusion problems with Dirichlet boundary conditions, that is

$$\|u - u_h\|_{L^\infty(\Omega_0)} \leq C \ln^{\frac{3}{2}}(1/h) \min_{v_h \in V_h} \|u - v_h\|_{L^\infty(\Omega_1)} + Ch^2 \ln^{\frac{1}{2}}(1/h), \qquad (26)$$

where $\Omega_0 \subseteq \Omega_1 \subseteq \Omega$ and $\mathrm{dist}(\Omega_0, \partial\Omega_1) > 0$ and C denote generic constants uniformly independent of the parameters of the problem. This estimate confirms that uniform convergence takes place, away from the region where the singular behavior of the operator appears as a boundary layer. By exploiting this theory, Olshanskii and Reusken (2000) have shown that the geometric multigrid method applied to reaction diffusion problems converges with a rate that is not affected by the physical parameters, because the deterioration of the approximation properties in the boundary layer are compensated by improved smoothing properties.

Even though we can not exhibit a rigorous proof, we observe that the uniform convergence rate of the multigrid method with respect to parameters applies also to our case, where reaction dominates the elliptic operator on some very local portions of the domain only. Again, this may be qualitatively justified by (26), because the effect to the singular perturbation does not affect the approximation away from the singularity.

5 Numerical experiments and discussion

In this section we support with numerical evidence the claims stated about the properties of the AMG solver applied to the *3D problem with 1D inclusions*. Afterwards, we apply the *3D–1D coupled problem* to simulate some simple configurations of perforated reservoir.

5.1 Numerical experiments for the 3D problem with 1D inclusions

We consider a unit cubic domain $\Omega = (0, 1)^3$ with a centered cylindrical inclusion of radius $R = \{0.01, 0.2\}$. The parameter κ is set equal to 1 and the source U varies linearly between 0 and 1. Homogeneous Neumann conditions are imposed on the boundary $\partial\Omega$. We use a family of four uniform meshes. A graphical description of the test case is shown in Fig. 4.

To solve the linear system deriving from the discrete formulation (18) we use the SAMG library, which provides both direct and AMG solvers for linear systems. We use the AMG solver on multiple V-cycle levels as a stand-alone solver (named stand-alone AMG), the AMG solver on multiple levels as a preconditioner for BiCGstab

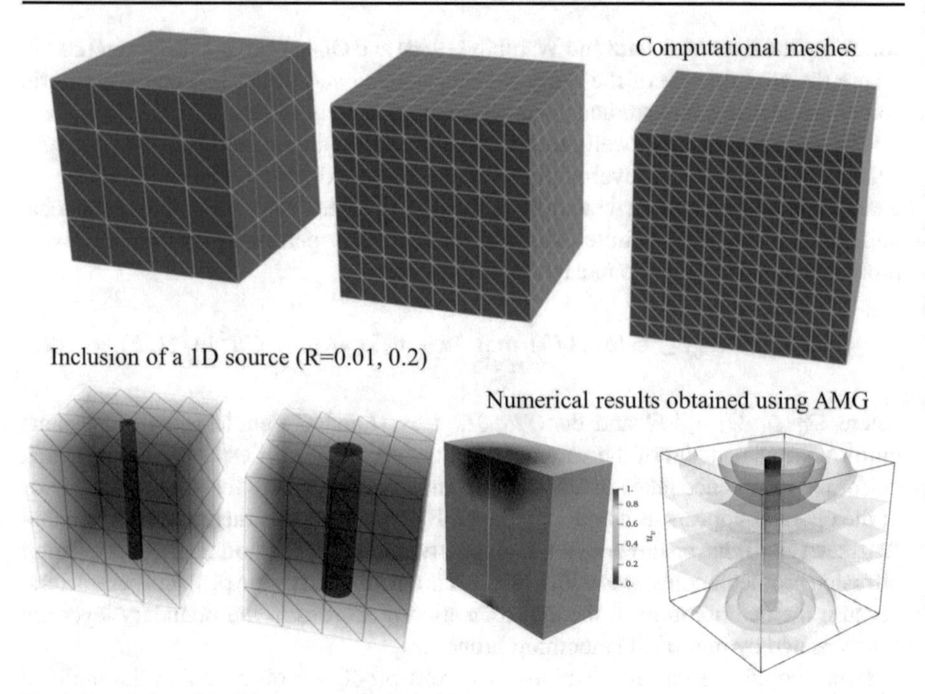

Fig. 4 Description of the 3D test case with 1D inclusions

(accelerated AMG) and Intel's Pardiso as direct solver (also part of the library). Despite the name, there isn't any guarantee that the accelerated AMG outperforms the standalone approach. Indeed, for rather small matrices, the additional cost of initializing the data structures for the BiCGstab algorithm may offset the computational time of accelerated approach.

The results are reported in Table 1 and in Fig. 5. We observe that in all the cases the numerical results confirm the theoretical expectations. More precisely, the computational cost scales with optimal complexity. Indeed, it is *linearly* dependent on the number of degrees of freedom (dof). On the contrary, as expected, the direct solver complexity is superlinear, in this case approximately *quadratic* with respect to the number of dof. This is a fundamental difference in performance, which makes the AMG solver 10 times faster than the direct solvers for systems with (only) 5×10^5 dof. We also observe that the solver is robust with respect to the main parameters of the problem, precisely R and κ, because the computational time is almost insensitive to variations of these.

Finally, the last dataset of Table 1 shows that the computational times for solving the standard stiffness matrix, namely $A^{\Omega} - C^{\Omega}$, or the modified matrix A^{Ω} are comparable (surprisingly less in the latter case). It means that the modified matrix pattern induced by the 3D–1D coupling terms does not affect the performance of AMG, even though it does modify the systems that are built and solved at the coarse levels. To sum up, the AMG solver is suitable for the problem at hand.

Table 1 CPU time (s) for the solution of the linear system using SAMG library

Nb. of sub.	Nb. of dof	Nb. nz entries	Stand-alone AMG	Accelerated AMG	Pardiso
Test with $R = 0.01, \kappa = 1$					
[20, 20, 20]	9261	7289	10.41299	10.62252	8.055254
[40, 40, 40]	68321	506289	16.61915	19.88221	18.01701
[60, 60, 60]	226981	1615081	25.33673	26.62287	123.8901
[80, 80, 80]	531441	3765281	44.91651	45.72543	680.3187
Test with $R = 0.2, \kappa = 1$					
[20, 20, 20]	9261	253997	12.9794	11.07768	8.129677
[40, 40, 40]	68921	1483169	22.4713	24.79993	22.00561
[60, 60, 60]	226981	3298703	40.4724	38.28844	139.3998
[80, 80, 80]	531441	5785529	71.54043	67.63766	722.6398
Test with $R = 0.01, \kappa = 10$					
[20, 20, 20]	9261	70837	10.71116	13.3264	8.448821
[40, 40, 40]	68921	1496781	16.13912	18.72223	21.52432
[60, 60, 60]	226981	1615081	24.45888	27.93736	117.4818
[80, 80, 80]	531441	3765281	47.71254	47.35248	694.8176
Test with $R = 0.01, \kappa = 100$					
[20, 20, 20]	9261	69581	10.9014	14.10448	8.489918
[40, 40, 40]	68921	496013	13.93065	18.26737	20.187
[60, 60, 60]	22698	161508	25.92969	26.54227	116.3273
[80, 80, 80]	531441	3765281	45.92995	46.85656	658.9241
Test with $R = 0.01, \kappa = 1$, standard stiffness matrix $A - C$.					
[20, 20, 20]	9261	66981	10.78995	12.177191	8.389091
[40, 40, 40]	68921	491561	16.55888	17.38915	19.30781
[60, 60, 60]	22698	1609741	27.79404	28.93468	125.8552
[80, 80, 80]	531441	3757521	52.23508	50.62627	667.1364

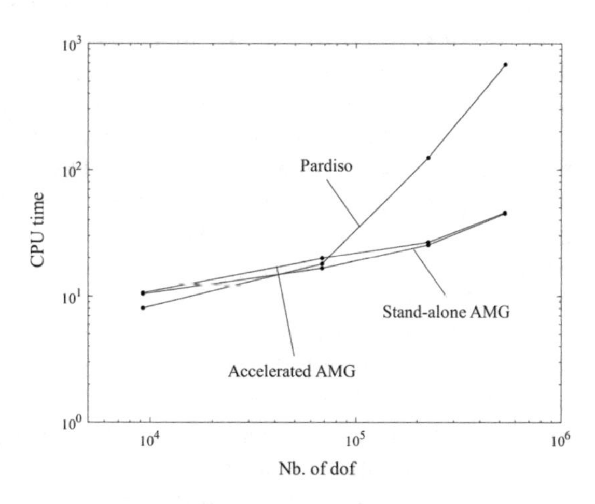

Fig. 5 CPU time (s) for the solution of the linear sistem using SAMG library

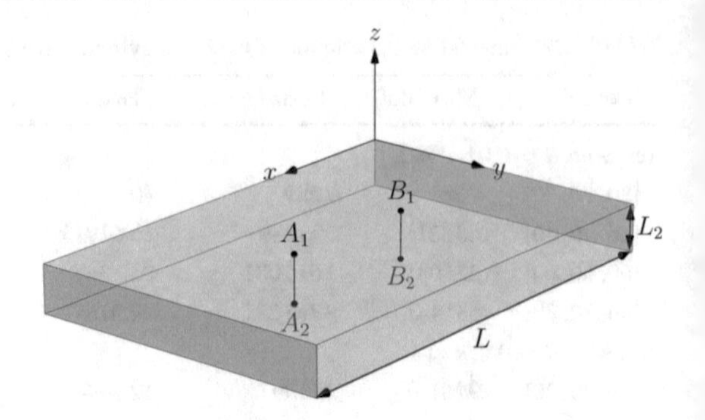

Fig. 6 Geometrical description of the problem

5.2 A 3D–1D coupled problem applied to reservoirs with two wells

We consider here data of a field experiment described in Lee (2012). In particular, the computational domain is a slab located at the surface, of dimension $\Omega = L \cdot ([0, 1] \times [0, 2/3] \times [-0.1, 0])$ where L=100 m. A sketch of the domain is shown in Fig. 6. In dimensionless coordinates, the wells are located at $x = 1/3,\ y = 1/3$ (labelled with B in Fig. 6) and in $x = 2/3,\ y = 1/3$ (labelled with A) and they extend to the entire thickness of the domain.

We assume that the material that surrounds the wells is isotropic. As a result, for a given value of K_p the tensor $\boldsymbol{k}_p$ is the identity in problem (3). The wells are much more permeable to flow than the reservoir. For this reason, we take $k_w = 1000$ in dimensionless form. Finally, we assume that the permeability of the well surface is larger than the intrinsic one of the rock. More precisely, we assume that it scales as $K = 100K_p/\rho$ that implies $\kappa = 100L/\rho = 100/R$.

We address *two test cases*, that differ for the boundary conditions enforced at the endpoints of the wells, while on the external boundary $\partial\Omega$ homogeneous Neumann conditions are imposed. In the first one (named *Test 1*), we assume that the injection well (A) is subject to an overpressure equal to the atmospheric one. At depth $z = -L_2 = -0.1$ (points A_2 and B_2) we enforce a pressure equal to $0.7p_{w,0}$ and at the top of the production well is open to the atmosphere. The overview of the numerical solution in both the slab and the wells is shown in Fig. 7. The Darcy velocity in the slab is visualized by the arrows. Also the injection and the extraction wells are marked in the same way. The pressure field is visualized by means of the color scale in Fig. 7. In particular, Fig. 7 shows the pressure field together with the isobaric curve in the middle plane of the domain ($z = -0.05$). From these results we conclude that all the expected behaviors of the problem are captured by the proposed scheme. More precisely, the injected fluid travels towards the extraction well and a moderate pressure gradient is established inside the slab. Finally, in Fig. 8 we plot the dimensionless pressure in the wells. We notice that the pressure varies linearly along the wells, matching the prescribed values at the endpoints. As a result, the velocity is almost constant and each well carries an almost steady flow. Only a small part of it leaks-off to the exterior from the lateral surface of the wells.

In the other test case (named *Test 2*), the bottom end of the wells is impermeable to flow. More precisely, an injection pressure equal to $p_{w,0}$ is set at the surface (point

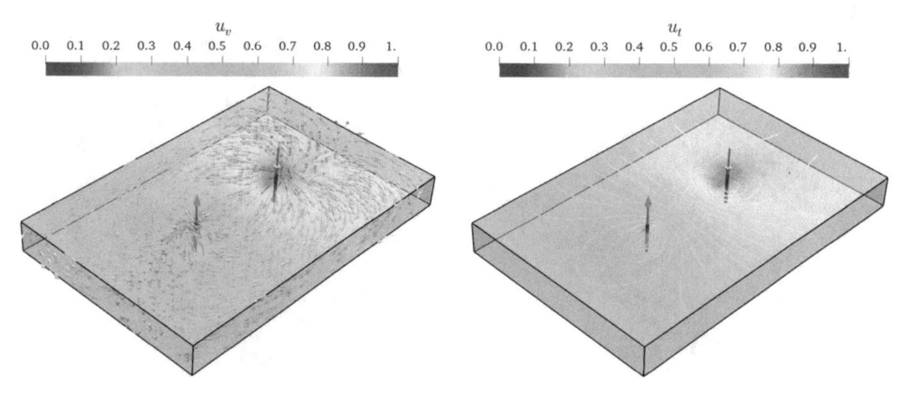

Fig. 7 Test 1: (left panel) Pressure field in the 3D/1D domain, the vectors point in the direction of the flow. (Right panel) Pressure field in the 3D/1D domain together with the isoline of the pressure field in the middle plane ($z = -0.05$)

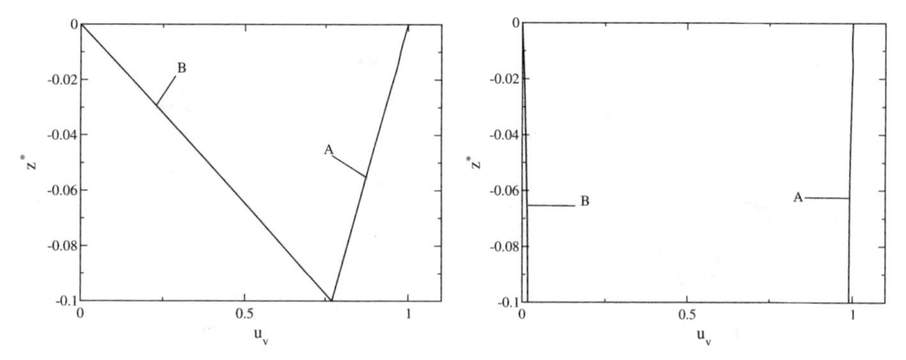

Fig. 8 Pressure field in the injection (A) and the extraction (B) well, left panel is Test 1 and right panel is Test 2, respectively

A_1), while no-flux condition is enforced at depth $z = -L_2 = -0.1$ (point A_2). For the production well we assume that the surface endpoint is exposed to the atmospheric pressure (point B_1) while no-flux condition is set at depth $z = -L_2 = -0.1$ (point B_2). As shown in Fig. 8, pressure is almost constant along each well. This setting resembles the case with prescribed sources analyzed in Sect. 3.

The maps of flow and pressure of *Test 2* are reported in Fig. 9. Comparing these results with the ones of the previous test, we observe that the pressure gradient in this case is more marked than before. It means that more flow propagates through the slab. This can be explained noticing that the wells are closed at the bottom. As a result, all the fluid that enters the injection well at the top must reach the extraction well.

5.3 A 3D–1D coupled problem applied to reservoirs with multiple wells

Here we test the ability of the model and of the corresponding implementation to handle multiple wells. In particular, thanks to the generality of the formulation, which does not require any conformity between the computational meshes and the discretization

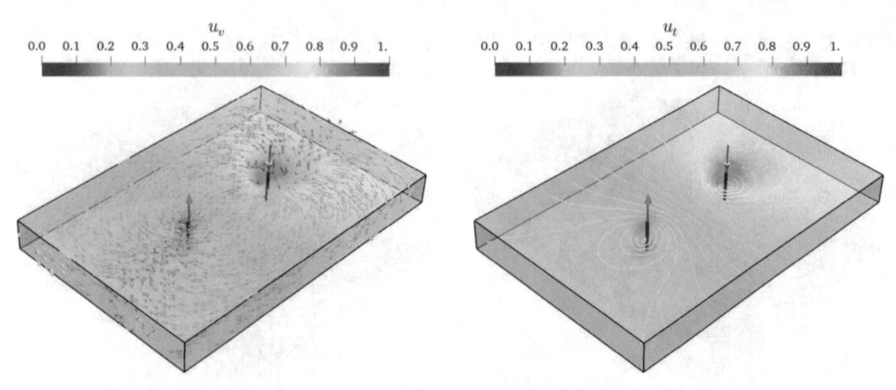

Fig. 9 Test 2: (left panel) Pressure field in the 3D/1D domain, the vectors point in the direction of the flow. (Right panel) Pressure field in the 3D/1D domain tohether with the isoline of the pressure field in the middle plane

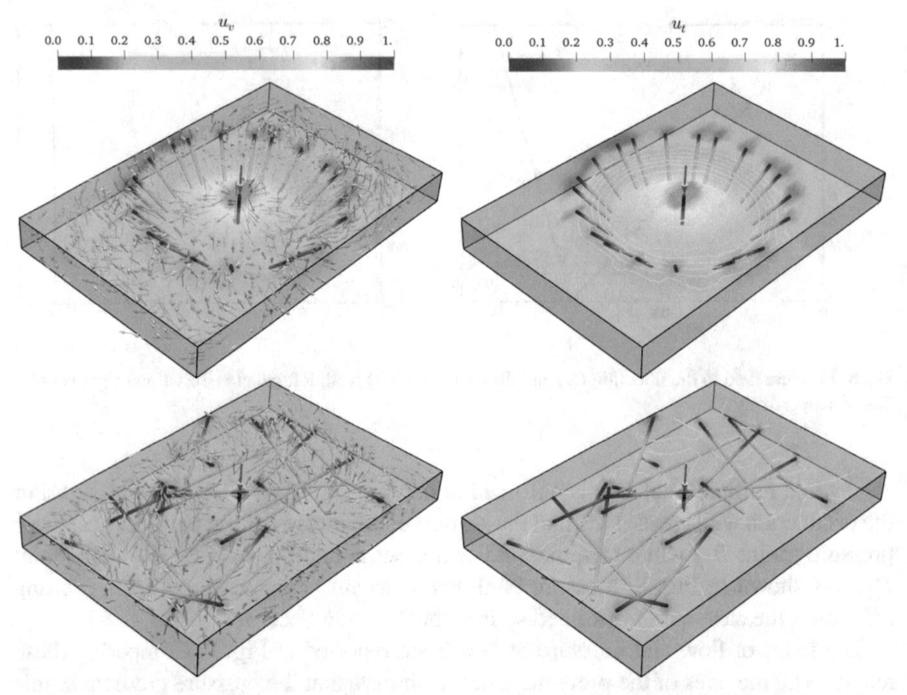

Fig. 10 Pressure (with isoline of the pressure field in the middle plane at $z = -0.05$) and velocity vector fields (arrows denote the dicection, not the magnitude) in the 3D/1D domain, for two idealized configurations with multiple wells. Parameters and boundary conditions are the same of Test 1

of the 3D and 1D subdomains, handling several wells does not poses any additional difficulty than modeling only two of them.

In Fig. 10, we show the simulations obtained with two idealized configurations. All the parameters and boundary conditions are the same than the ones of Test 1. On the top we have multiple production wells located on the surface of a cone with axis aligned with the injection well. This configuration forms a sort of pressure cushion around the production well. On the bottom, we have randomly located the endpoints of the wells on the top and bottom surfaces of the slab. We notice that the pressure field reflects the pattern of the wells while the velocity is rather chaotic.

6 Conclusions

We have developed a computational approach to model the interaction of wells with a reservoir. The problem shows non-standard features: (i) at the level of modeling for the application of topological model reduction techniques; (ii) at the level of analysis because it consists of coupled PDEs defined on manifolds of different dimensionality; and (iii) at the level of numerical discretization because it corresponds to a linear system with an unusual pattern.

A natural development of this work, already in progress, consists of extending the model and the analysis to the mixed formulation of the flow, where both pressure and velocity fields are modeled and approximated simultaneously. A preliminary report of this study is available in Notaro et al. (2016). More in general, this approach can be extended to model fractures, fracture tips and wells as two-dimensional (2D) and one-dimensional (1D) manifolds embedded into a three-dimensional (3D) reservoir. We believe that this extension would be particularly attractive for subsurface flow applications, because it provides a unified approach to model the fluid injection and borehole interaction with the surrounding reservoir (using the 3D–1D multiscale coupling) and the formation of the fracture network (using the 3D–2D–1D approach for the interaction of reservoir, fracture and fracture tip, respectively).

Acknowledgements The author names are listed in alphabetical order. All the authors are members of the INdAM Research group GNCS. The author Federica Laurino acknowledges the support of the Italian Institute of Technology with the fellowship: *Sviluppo di metodi computazionali di tipo multi-scala per la simulazione del trasporto vascolare ed extra vascolare di molecole, nano-costrutti e cellule in tessuti neoplastici*.

References

Bellout, M., Echeverra Ciaurri, D., Durlofsky, L., Foss, B., Kleppe, J.: Joint optimization of oil well placement and controls. Comput. Geosci. **16**(4), 1061–1079 (2012). https://doi.org/10.1007/s10596-012-9303-5

Boon, W., Nordbotten, J., Vatne, J.: Mixed-Dimensional Elliptic Partial Differential Equations. Tech. rep., arXiv, Cornell University Library (2017). arXiv:1710.00556v2

Cattaneo, L., Zunino, P.: A computational model of drug delivery through microcirculation to compare different tumor treatments. Int. J. Numer. Methods Biomed. Eng. **30**(11), 1347–1371 (2014a). https://doi.org/10.1002/cnm.2661

Cattaneo, L., Zunino, P.: Computational models for fluid exchange between microcirculation and tissue interstitium. Netw. Heterog. Media **9**(1), 135–159 (2014b). https://doi.org/10.3934/nhm.2014.9.135

Chen, Z., Yue, X.: Numerical homogenization of well singularities in the flow transport through heterogeneous porous media. Multiscale Model. Simul. **1**(2), 260–303 (2003). https://doi.org/10.1137/S1540345902413322

D'Angelo, C.: Multi Scale Modelling of Metabolism and Transport Phenomena in Living Tissues, PhD Thesis. EPFL, Lausanne (2007)

D'Angelo, C.: Finite element approximation of elliptic problems with dirac measure terms in weighted spaces: applications to one- and three-dimensional coupled problems. SIAM J. Numer. Anal. **50**(1), 194–215 (2012). https://doi.org/10.1137/100813853

D'Angelo, C., Quarteroni, A.: On the coupling of 1d and 3d diffusion–reaction equations. Application to tissue perfusion problems. Math. Models Methods Appl. Sci. **18**(8), 1481–1504 (2008). https://doi.org/10.1142/S0218202508003108

Ern, A., Guermond, J.L.: Theory and Practice of Finite Elements, *Applied Mathematical Sciences*, vol. 159. Springer, New York (2004). https://doi.org/10.1007/978-1-4757-4355-5

Fernández Bonder, J., Rossi, J.: Asymptotic behavior of the best sobolev trace constant in expanding and contracting domains. Comm. Pure Appl. Anal. **1**, 359–378 (2002). https://doi.org/10.3934/cpaa.2002.1.359

Gilbarg, D., Trudinger, N.: Elliptic Partial Differential Equations of Second Order, vol. 224. Springer, Berlin (2001)

Jenny, P., Lunati, I.: Modeling complex wells with the multi-scale finite-volume method. J. Comput. Phys. **228**(3), 687–702 (2009). https://doi.org/10.1016/j.jcp.2008.09.026

Koeppl, T., Vidotto, E., Wohlmuth, B., Zunino, P.: Mathematical modelling, analysis and numerical approximation of second order elliptic problems with inclusions. Math. Models Methods Appl. Sci. **28**(5), 953–978 (2018)

Köppl, T., Wohlmuth, B.: Optimal a priori error estimates for an elliptic problem with dirac right-hand side. SIAM J. Numer. Anal. **52**(4), 1753–1769 (2014)

Köppl, T., Vidotto, E., Wohlmuth, B.: A local error estimate for the Poisson equation with a line source term. In: Numerical Mathematics and Advanced Applications ENUMATH 2015, pp. 421–429. Springer, Berlin (2016)

Kuchta, M., Nordaas, M., Verschaeve, J., Mortensen, M., Mardal, K.A.: Preconditioners for saddle point systems with trace constraints coupling 2d and 1d domains. SIAM J. Sci. Comput. **38**(6), B962–B987 (2016). https://doi.org/10.1137/15M1052822

Lee, K.S.: Efficiency of horizontal and vertical well patterns on the performance of micellar-polymer flooding. Energy Procedia **16**, 889–894 (2012)

Nabil, M., Zunino, P.: A computational study of cancer hyperthermia based on vascular magnetic nanoconstructs. R. Soc. Open Sci. (2016). https://doi.org/10.1098/rsos.160287

Nabil, M., Decuzzi, P., Zunino, P.: Modelling mass and heat transfer in nano-based cancer hyperthermia. R. Soc. Open Sci. (2015). https://doi.org/10.1098/rsos.150447

Notaro, D., Cattaneo, L., Formaggia, L., Scotti, A., Zunino, P.: A Mixed Finite Element Method for Modeling the Fluid Exchange Between Microcirculation and Tissue Interstitium, pp. 3–25. Springer, Berlin (2016). https://doi.org/10.1007/978-3-319-41246-7-1

Olshanskii, M., Reusken, A.: On the convergence of a multigrid method for linear reaction–diffusion problems. Computing **65**(3), 193–202 (2000)

Peaceman, D.: Interpretation of well-block pressures in numerical reservoir simulation. Soc. Pet. Eng. AIME J. **18**(3), 183–194 (1978)

Peaceman, D.W.: Interpretation of well-block pressures in numerical reservoir simulation with nonsquare grid blocks and anisotropic permeability. Soc. Pet. Eng. J. **23**(3), 531–543 (1983). https://doi.org/10.2118/10528-PA

Peaceman, D.: Interpretation of Well-Block Pressures in Numerical Reservoir Simulation—Part 3: Some Additional Well Geometries, pp. 457–471 (1987)

Possenti, L., di Gregorio, S., Gerosa, F., et al.: A computational model for microcirculation including Fahraeus–Lindqvist effect, plasma skimming and fluid exchange with the tissue interstitium. Int J Numer Meth Biomed Engng. (2018). https://doi.org/10.1002/cnm.3165

Possenti, L., Casagrande, G., Di Gregorio, S., Zunino, P., Costantino, M.L.: Numerical simulations of the microvascular fluid balance with a non-linear model of the lymphatic system. Microvasc. Res. **122**, 101–110 (2019). https://doi.org/10.1016/j.mvr.2018.11.003

Quarteroni, A., Valli, A.: Numerical Approximation of Partial Differential Equations, Springer Series in Computational Mathematics, vol. 23. Springer, Berlin (1994)

Salsa, S.: Partial Differential Equations in Action, Unitext, vol. 99, 3rd edn. Springer, Cham (2016). https://doi.org/10.1007/978-3-319-31238-5. From modelling to theory, La Matematica per il 3+2

Schatz, A.H., Wahlbin, L.B.: On the finite element method for singularly perturbed reaction–diffusion problems in two and one dimensions. Math. Comp. **40**(161), 47–89 (1983)

Stüben, K.: Algebraic multigrid (AMG): experiences and comparisons. Appl. Math. Comput. **13**(3–4), 419–451 (1983). https://doi.org/10.1016/0096-3003(83)90023-1

Stüben, K.: Algebraic Multigrid (AMG): An Introduction with Applications; Updated Version of GMD Report No 53. GMD-Report. GMD-Forschungszentrum Informationstechnik (1999)

Stüben, K., Clees, T., Klie, H., Lu, B., Wheeler, M.: Algebraic Multigrid Methods (AMG) for the Efficient Solution of Fully Implicit Formulations in Reservoir Simulation, pp. 59–69 (2007)

Wolfsteiner, C., Lee, S., Tchelepi, H.: Well modeling in the multiscale finite volume method for subsurface flow simulation. Multiscale Model. Simul. **5**(3), 900–917 (2006). https://doi.org/10.1137/050640771

Wu, Y.S.: A virtual node method for handling well bore boundary conditions in modeling multiphase flow in porous and fractured media. Water Resour. Res. **36**(3), 807–814 (2000). https://doi.org/10.1029/1999WR900336

GEM - International Journal on Geomathematics
https://doi.org/10.1007/s13137-019-0121-y

ORIGINAL PAPER

Fast and robust flow simulations in discrete fracture networks with GPGPUs

S. Berrone[1] · A. D'Auria[1] · F. Vicini[1]

Received: 28 July 2018 / Accepted: 23 November 2018

Abstract
A new approach for flow simulation in very complex discrete fracture networks based on PDE-constrained optimization has been recently proposed in Berrone et al. (SIAM J Sci Comput 35(2):B487–B510, 2013b; J Comput Phys 256:838–853, 2014) with the aim of improving robustness with respect to geometrical complexities. This is an essential issue, in particular for applications requiring simulations on geometries automatically generated like the ones used for uncertainty quantification analyses and hydro-mechanical simulations. In this paper, implementation of this approach in order to exploit Nvidia Compute Unified Device Architecture is discussed with the main focus to speed up the linear algebra operations required by the approach, being this task the most computational demanding part of the approach. Furthermore, two different approaches for linear algebra operations and two storage formats for sparse matrices are compared in terms of computational efficiency and memory constraints.

Keywords Discrete fracture network flow simulations · Simulations in complex geometries · GPGPU · CUDA

Mathematics Subject Classification 65N30 · 68U20 · 68W10 · 68W40 · 86-08 · 86A05

This research has been partially supported by INdAM-GNCS Projects 2017 and 2018, and by the MIUR Project "Dipartimenti di Eccellenza 2018–2022". Computational resources were partially provided by HPC@POLITO (http://hpc.polito.it) and by CINECA Project IsC58 HP10CDFLWH.
Authors are members of the INdAM research group GNCS.

✉ S. Berrone
stefano.berrone@polito.it

A. D'Auria
alessandro.dauria@polito.it

F. Vicini
fabio.vicini@polito.it

[1] Dipartimento di Scienze Matematiche, Politecnico di Torino, Turin, Italy

1 Introduction

The growing interest in fluid flow simulations in fractured porous media is related to a wide range of applications such as oil/gas recovery, carbon dioxide geological storage, water resource management, nuclear waste, geological storage, geothermal applications and many others.

In the present work we focus on underground fractured media in which fractures create a preferential path for the flow in the rock, and the surrounding rock matrix has a moderate contribution to the flow. In the discrete fracture network (DFN) model the fractures are represented as planar polygons in the space with straight intersections. This model is commonly used to describe fractured media resorting to a set of 2D flow models (Fidelibus et al. 2009; Cacas et al. 1990; Dershowitz and Einstein 1988; Jaffré and Roberts 2012). Among the many well-known difficulties, e.g. the correct flow model or the huge uncertainty on the geometrical and hydro-geological parameters, the computational costs and the mesh generation problems have a relevant impact. To deal with the uncertainty of the problem, several DFNs are randomly generated starting from probabilistic distribution of position, orientation, dimension of the fractures and for the hydrogeological properties (see for example de Dreuzy et al. 2001). Then a relevant number of simulations is performed to compute statistical moments for the quantities of interest. In order to perform such simulations is advisable to resort to a very fast and robust code. The method on which the present implementation relies has proven to be very robust in dealing with complex geometrical configurations (Berrone et al. 2016b, 2018) circumventing all the mesh generation problems, and can be extended to deal with the coupling with the surrounding rock matrix (Berrone et al. 2017a; Garipov et al. 2016), as well.

Within the DFN model many approaches can be used to discretize the governing equations, such as Boundary Element Methods (Fidelibus 2007; Lenti and Fidelibus 2003), Extended and Mixed Finite Element Methods (Al-Hinai et al. 2017; Fumagalli and Scotti 2014; Flemisch et al. 2018; Pichot et al. 2010, 2012, 2014; de Dreuzy et al. 2013; Berrone et al. 2013a), Finite Volume Methods (Manzoor et al. 2018; Ahmed et al. 2017; Parramore et al. 2016; Franc et al. 2017), Gradient schemes (Brenner et al. 2017; Xing et al. 2017), Hybrid High Order methods (Chave et al. 2018), and Virtual Element Method (Beirão da Veiga et al. 2013) (see for example Benedetto et al. 2014, 2016, 2017; Fumagalli and Keilegavlen 2018; Berrone and Borio 2017). In addition, an approach resorting to Mimetic Finite Difference, Beirão da Veiga et al. (2014), was used in Antonietti et al. (2016). The conforming mesh generation on DFNs is a well-know issue (Mustapha and Mustapha 2007). Applications based on conforming meshes can be found in Hyman et al. (2014), Hadgu et al. (2017), Sentís and Gable (2017), Ngo et al. (2017), and other interesting approaches in Noetinger and Jarrige (2012), Noetinger (2015). In Dershowitz and Fidelibus (1999) the network of fractures is reduced to a network of pipes in order to get a simplified model of the flow. Multi-phase flows are also considered in Al-Hinai et al. (2013).

In order to avoid all the mesh generation difficulties, here, an optimization based approach is used to decouple the flow problem in the fractures and to allow the use of independent meshes on each fracture, generated disregarding intersections with other fractures (Berrone et al. 2013a) (see Figs. 5, 6 for two mesh examples). This optimization approach is based on a reformulation of the problem as a PDE constrained optimization problem in which the functional to be minimized is formed by norms of the hydraulic head continuity mismatches and the flux unbalance conservation at the fracture intersections. In Berrone et al. (2015) a parallel C++ implementation of the method, using message passing interface (MPI), has been proposed. The parallelization is based on the decoupling of the single fracture problems followed by a suitable imposition of matching conditions. This makes possible to perform flow simulations in realistic DFNs with a very large number of fractures. Accuracy can be improved by a posteriori error estimates (Berrone et al. 2016a) and mesh adaptivity. The method has been largely tested in the solution of steady and unsteady transport problems, providing very good performances in terms of accuracy and robustness (Berrone et al. 2013a, 2016b, 2017b; Fumagalli et al. 2018; Berrone et al. 2018). In this work we discuss issues related to an implementation of the method that can take advantage of the large computational resources of recent GPUs. Since 2007, Nvidia has equipped its GPUs with a parallel architecture called Compute Unified Device Architecture (CUDA), which allows to use the GPUs as a co-processor. A development kit has been distributed in order to use GPUs for general purpose applications. In particular, thanks to the hardware structure and optimized linear algebra libraries, GPUs can reach very high performances with dense matrices. On the other hand, as discussed by Merrill and Garland (2016), sparse matrix format strongly bounds the peak FLOPs performances. This issue is also largely tackled in this paper. Nevertheless the GPUs are widely spread and can be considered a fundamental device to improve computational power. In June 2018 SUMMIT HPC system became the fastest HPC system in the world. It is based on a IBM Power9 architecture with six GPU per node, with a peak of 122 petaFLOPs with LinPack test in double precision.

The aim of this paper is to show how it is possible to reduce the elapsed time for DFN flow simulations in a workstation equipped with a single common, generally relatively cheap, Nvidia GPU. For this reason, we choose a low cost Nvidia GPU belonging to the GeForce family designed for gaming stations, to strongly reduce the purchase and running workstation costs with respect to a more expensive Nvidia Tesla GPU, requiring suitable not common and expensive cooling and power supply systems.

We develop an interface called *CUDASolver* which makes use of two standard libraries of the Nvidia development CUDA Toolkit (Nvidia 2008b, c): CuBlas (Nvidia 2008a) and CuSparse (Nvidia 2008d) and we discuss its implementation.

The paper is organized as follows. In Sect. 2 there is a summary about the continuum model and its discrete counterpart. In Sect. 3 we describe CUDASolver and its implementation. In Sect. 4 we report results on the tests performed and the obtained performance improvements. Section 5 contains conclusions gathered by the tests, that also apply to a wider range of simulations based on Finite Elements like discretizations, yielding linear systems with similar patterns.

2 Continuum and discrete model

2.1 Continuum model description

In this subsection we introduce the appropriate mathematical formulation of the problem and its reformulation as a PDE constrained optimization, its discretization is discussed in the next subsection.

Let us consider a network of connected planar fractures F_i, for $i = 1, \ldots, I$, in which a single phase flow is confined. The intersection between two fractures is called *trace* and denoted by S_m, with $m = 1, \ldots, M$. The set I_m collects the fracture indices of the two fractures whose intersection generates the trace $S_m = \bar{F}_i \cap \bar{F}_j$ with $i \neq j$, and $\mathcal{S}_i$ is the set of the traces on the fracture F_i.

The flow model within the fracture network is the Darcy model with the imposition of flux balance and hydraulic head continuity at the fracture intersections. For more details on the formulation here applied we refer the reader to Berrone et al. (2013b), Jaffré and Roberts (2012), Pichot et al. (2010), Vohralík et al. (2007).

On each fracture the following functional spaces are required for well-posedness: $V_i = \mathrm{H}^1_0(F_i) = \left\{v \in \mathrm{H}^1(F_i) : v_{|\Gamma_{iD}} = 0\right\}$ and $V_i^D = \mathrm{H}^1_D(F_i) = \left\{v \in \mathrm{H}^1(F_i) : v_{|\Gamma_{iD}} = G_i^D\right\}$; we have $H_i \in V_i^D$. The weak formulation of Darcy law in each fracture with non homogeneous boundary conditions can be expressed by the following system of equations:

$$\int_{F_i} \mathbb{K}_i \nabla H_i \nabla v \mathrm{d}\Omega = \int_{F_i} q_i v \mathrm{d}\Omega + \int_{\Gamma_{iN}} G_i^N v_{|\Gamma_{iN}} \mathrm{d}\gamma$$
$$+ \sum_{S_m \in \mathcal{S}_i} \int_{S_m} \left[\!\left[\frac{\partial H_i}{\partial \hat{\nu}^i_{S_m}} \right]\!\right] v_{|S_m} \mathrm{d}\gamma, \quad \forall v \in V_i, \quad \forall i = 1, \ldots, I, \tag{1}$$

$$H_{i\,|S_m} - H_{j\,|S_m} = 0, \quad \text{for } i, j \in I_{S_m}, \quad \forall m = 1, \ldots, M, \tag{2}$$

$$\left[\!\left[\frac{\partial H_i}{\partial \hat{\nu}^i_{S_m}} \right]\!\right] + \left[\!\left[\frac{\partial H_j}{\partial \hat{\nu}^j_{S_m}} \right]\!\right] = 0, \quad \text{for } i, j \in I_{S_m}, \quad \forall m = 1, \ldots, M, \tag{3}$$

where $\Omega = \cup_{i=1,\ldots,I} F_i$ denotes the DFN in the 3D space. $\mathbb{K}_i$ is a symmetric positive definite transmissivity tensor on F_i, H denotes the global hydraulic head in the full DFN, and H_i its restriction to fracture F_i. For each trace S_m and on each fracture F_i we fix a normal unit vector $\hat{\nu}^i_m$; $\left[\!\left[\frac{\partial H_i}{\partial \hat{\nu}^i_m} \right]\!\right]$ denotes the jump of the co-normal derivative $\mathbb{K}_i \nabla H_i \cdot \hat{\nu}^i_m$ of H_i along $\hat{\nu}^i_m$ and it represents the incoming flux in the fracture F_i trough trace S_m.

Γ_D and Γ_N are the Dirichlet and Neumann boundary, respectively, such that $\Gamma_D \cap \Gamma_N = \emptyset$ and $\Gamma_D \cup \Gamma_N = \partial\Omega$. We introduce Γ_{iD}, Γ_{iN} the local boundary on each fracture F_i, such that $\Gamma_{iD} = \partial F_i \cap \Gamma_D$ and $\Gamma_{iN} = \partial F_i \cap \Gamma_N$. The Dirichlet and Neumann boundary conditions are described by the functions G^D and G^N respectively, and their restriction to fracture F_i by G_i^D and G_i^N. To ensure the uniqueness of the solution, Γ_D has to be non-empty. Nonetheless Γ_{iD} can be empty for some i (see Berrone et al. 2014 for more details).

Quantity $q_i \in L^2(F_i)$ is the source term for fracture F_i. Equation (2) ensures the continuity of the hydraulic head at the fracture intersections and (3) imposes the balance of the flux across the traces.

This problem has been reformulated in Berrone et al. (2013a, b, 2014) as a PDE constrained optimization. With this approach, Eqs. (2), (3) are replaced by a cost functional $J(H, U)$ to be minimized constrained by Eq. (1). In order to have a coercive constraint equation on each fracture F_i, a control variable $U_i^m \in \mathrm{H}^{-\frac{1}{2}}(S_m)$ for the constrained minimization problem is defined on each trace S_m as

$$U_i^m = \left[\!\left[\frac{\partial H_i}{\partial \hat{\nu}^i_{S_m}} \right]\!\right] + \alpha H_{i|S_m},$$

with a strictly positive parameter α. The new problem is:

$$\min J(H, U) := \sum_{m=1}^{M} \Bigg(\| H_{i|S_m} - H_{j|S_m} \|^2_{\mathrm{H}^{\frac{1}{2}}(S_m)} + \left\| U_i^m + U_j^m - \alpha \left(H_{i|S_m} + H_{j|S_m} \right) \right\|^2_{\mathrm{H}^{-\frac{1}{2}}(S_m)} \Bigg), \tag{4}$$

with $i, j \in I_m$ and such that

$$\int_{F_i} \mathbb{K}_i \nabla H_i \nabla v \mathrm{d}\Omega + \alpha \sum_{S_m \in \mathcal{S}_i} \int_{S_m} H_{i|S_m} v_{|S_m} \mathrm{d}\gamma = \int_{F_i} q_i v \mathrm{d}\Omega + \int_{\Gamma_{iN}} G_i^N v_{|\Gamma_{iN}} \mathrm{d}\gamma + \sum_{S_m \in \mathcal{S}_i} \int_{S_m} U_i^m v_{|S_m} \mathrm{d}\gamma, \quad \forall v \in V_i, \quad \forall i = 1, \ldots, I. \tag{5}$$

In Berrone et al. (2013a, b) it is shown that (4), (5) are equivalent to (1)–(3).

2.2 Discrete problem

In this section we briefly recall and reformulate the presentation of the method provided in Berrone et al. (2015). The reformulation is related to a suitable matrices memorization collecting contributions from all the fractures; this approach is performed in order to improve efficiency in performing matrix vector computations, both in the GPU and in the CPU. The improved efficiency is related to a better use of the cache memory on the CPU and a similar better use of the GPU. We remark that this storage solution does not affect the local nature of the resulting algorithm which is the same described in Berrone et al. (2015). In this presentation a single computational process is dealing with the whole DFN, so we can introduce a simplified notation avoiding the definition of communicating and non-communicating fractures.

We introduce a Finite Element discretization on fractures and traces. We note that the approach can be applied to other discretizations. In order to catch the advantages of the approach we stress that each fracture is meshed independently and we choose to consider an induced discretization for the control variable on traces in each fracture.

We use small case letters for the discrete counterpart of the variables introduced in the previous section. On each fracture $F_i, i = 1, \dots, I$, we introduce the finite element basis $\{\varphi_{i,k}\}_{k=1,\dots,N_i}$ and we define the discrete hydraulic head $h_i = \sum_{k=1}^{N_i} h_{i,k}\varphi_{i,k}$. Similarly, for each trace S_m of the fracture F_i, we set $\{\psi^i_{m,k}\}_{k=1,\dots,N_i^m}$ the finite element basis and we define the discrete control variable $u^i_m = \sum_{k=1}^{N^i_m} u^i_{m,k}\psi^i_{m,k}$. The quantities N_i and N^i_m represent the number of degrees of freedom (DOFs) on the fracture F_i and on the trace S_m in the fracture F_i, respectively. For each fracture we define the vector of DOFs for the hydraulic head h_i, $i = 1, \dots, I$, which, with an abuse of notation, we denote with the same symbol used for the hydraulic head function on the fracture. Moreover, with a similar abuse of notation, for each trace we define two vectors u^i_m, and u^j_m, $m = 1, \dots, M$ with $i, j \in I_m$, i.e. we have two different control variables per each trace, one for each fracture sharing the trace.

In this way it is possible to define $h = (h_1, \dots, h_I)$ as the vector that concatenates h_i for all $i = 1, \dots, I$, and $u = (u_1^-, u_1^+, \dots, u_M^-, u_M^+)$ as the vector that collects the control variables on the traces, where u_m^- denotes the trace on F_i and u_m^+ denotes the trace on F_j, with $i, j \in I_m$, with the convention $i < j$. The dimensions of vectors h and u are denoted by $N^F = \sum_{i=1,\dots,I} N_i$, as the total number of DOFs for the hydraulic head on the fractures, and by $N^T = \sum_{i=1,\dots,I} N_{\mathcal{S}_i}$, as the total number of DOFs for the control variables, and $N_{\mathcal{S}_i} = \sum_{S_m \in \mathcal{S}_i} N^i_m$, is the number of DOFs for the control variable on the fracture F_i.

Upon discretization, Darcy equation (5) reads

$$A_i h_i = q_i + \mathcal{B}_i u_i, \tag{6}$$

where

- $A_i \in \mathbb{R}^{N_i \times N_i}$ is the stiffness matrix;
- $\mathcal{B}_i \in \mathbb{R}^{N_i \times N_{\mathcal{S}_i}}$ is the matrix which collects the integrals of the basis functions $\{\varphi_{i,k|S_m}\}$, $k = 1, \dots, N_i$, against the basis functions $\{\psi^i_{m,k}\}$, $k = 1, \dots, N^i_m$, $S_m \in \mathcal{S}_i$;
- $q_i \in \mathbb{R}^{N_i}$ is the forcing term including terms provided by non homogeneous boundary conditions.

In order to minimize the communication through the Bus PCI-Express and improve the bandwidth use it is more convenient to deal with a unique matrix suitably collecting all the local matrices. Furthermore, the data continuity in the GPU memory facilitates the compiler to perform optimizations and improve the cache hits reducing the delay time to collect data form the memory. Collecting all the matrices A_i diagonal-wise, we get the matrix $A = diag(A_i) \in \mathbb{R}^{N^F \times N^F}$, moreover, collecting matrices $\mathcal{B}_i$ in a single matrix $\mathcal{B} \in \mathbb{R}^{N^F \times N^T}$, with a mapping from the u-DOFs numeration on the fractures to the global indexing used in vector u, the set of systems (6) can be rewritten as

$$Ah = q + \mathcal{B}u, \tag{7}$$

where q is the concatenation of q_i column-wise.

Replacing $\mathrm{H}^{\frac{1}{2}}(S_m)$ and $\mathrm{H}^{-\frac{1}{2}}(S_m)$ with the $\mathrm{L}^2(S_m)$ norm in the Eq. (4), we can write the following discrete functional

$$\begin{aligned} J(h,u) = \frac{1}{2}\sum_{i=1}^{I}\sum_{S_m\in\mathcal{S}_i}\Bigg(\int_{S_m}\left(\sum_{k=1}^{N_i} h_{i,k}\varphi_{i,k|S_m} - \sum_{k=1}^{N_j} h_{j,k}\varphi_{j,k|S_m}\right)^2 \mathrm{d}\gamma \\ + \int_{S_m}\Bigg(\sum_{k=1}^{N_m^i} u_{m,k}^i\psi_{m,k}^i + \sum_{k=1}^{N_m^j} u_{m,k}^j\psi_{m,k}^j \\ - \alpha\sum_{k=1}^{N_i} h_{i,k}\varphi_{i,k|S_m} - \alpha\sum_{k=1}^{N_j} h_{j,k}\varphi_{j,k|S_m}\Bigg)^2 \mathrm{d}\gamma\Bigg). \end{aligned} \tag{8}$$

The functional (8) can be written in a matrix form as:

$$J(h,u) := \frac{1}{2}\left(h^T G^h h - \alpha h^T B^h u - \alpha u^T B^u h + u^T G^u u\right) \tag{9}$$

where

- $G^h \in \mathbb{R}^{N^F\times N^F}$ collects the integrals of products of hydraulic head basis functions restricted on the traces;
- $G^u \in \mathbb{R}^{N^T\times N^T}$ collects the integrals of products of control variable basis functions;
- $B^h \in \mathbb{R}^{N^F\times N^T}$ and $B^u = \left(B^h\right)^T \in \mathbb{R}^{N^T\times N^F}$ contain the mixed terms given by the integral on traces of $\varphi_{i,k|S_m}$ against $\psi_{m,l}^j$, with $i, j \in S_m$, and $m = 1\ldots M$.

Using (7) and (9), the DFN problem (4), (5) can be rewritten as a finite dimensional equality constrained quadratic problem:

$$\begin{aligned} &\min\ J(h,u) := \frac{1}{2}\left(h^T G^h h - \alpha h^T B^h u - \alpha u^T B^u h + u^T G^u u\right) \\ &\text{s.t. } Ah - \mathcal{B}u = q. \end{aligned} \tag{10}$$

Well poselness of problem 10 is shown in Berrone et al. (2013a). In Berrone et al. (2013b) it is proven the uniqueness of the solution of problem (10).

If we formally eliminate h from J using the constraint (7), namely we replace $h = A^{-1}(\mathcal{B}u + q)$ in J, we obtain an equivalent unconstrained minimization problem

$$\begin{aligned} \min \hat{J}(u) := \frac{1}{2}u^T(\mathcal{B}^T A^{-T} G^h A^{-1}\mathcal{B} + G^u - \alpha\mathcal{B}^T A^{-T} B^h - \alpha B^u A^{-1}\mathcal{B})u \\ + q^T A^{-T}(G^h A^{-1}\mathcal{B} - \alpha B^h)u, \end{aligned} \tag{11}$$

that is a quadratic functional in the control variable u. As shown in Berrone et al. (2015), the Hessian matrix

$$\hat{H} = \mathcal{B}^T A^{-T} G^h A^{-1} \mathcal{B} + G^u - \alpha \mathcal{B}^T A^{-T} B^h - \alpha B^u A^{-1} \mathcal{B}$$

is symmetric positive definite. The minimum of the functional can be computed by a conjugate gradient method solving

$$\hat{H} u = -(G^h A^{-1} \mathcal{B} - \alpha B^h)^T A^{-1} q.$$

The computation of the matrix $\hat{H}$ can be avoided resorting to the dual variable $p := A^{-T}(G^h h - \alpha B^h u) \in \mathbb{R}^{N^F}$, i.e. the gradient of functional $\hat{J}$ is computed as follows:

$$\nabla \hat{J}(u) = \mathcal{B}^T p + G^u u - \alpha B^u h. \tag{12}$$

As a consequence (see Berrone et al. 2015 for full details), the minimization problem (10) can be solved with a conjugate gradient algorithm, where the gradient direction is obtained by the following system of equations:

$$Ah = q + \mathcal{B} u, \tag{13}$$

$$A^T p = G^h h - \alpha B^h u, \tag{14}$$

$$\nabla \hat{J}(u) = \mathcal{B}^T p + G^u u - \alpha B^u h. \tag{15}$$

This system of equation can be easily solved in parallel on a GPU, with very few communications between CPU and GPU, as described in the following section.

3 CUDASolver

CUDASolver is a code developed with the help of two external libraries used for linear algebra operations on GPU: *CuSparse*, to perform sparse matrix operations, and *CuBlas*, to perform dense matrix operations. The two libraries are developed by Nvidia and are suitably optimized for Nvidia GPUs. Simulations here presented are performed with the CUDA Toolkit version 8.0. Furthermore, to perform matrix and linear algebra operations on the CPU we resort to the library Eigen v3.3 (Guennebaud et al. 2010). Sparse matrices on the CPU are stored in CSR (Compressed Sparse Row) format, as it is the format supported both by CuSparse and Eigen functions. In this paper we are considering pure diffusive phenomena, so the full block diagonal matrix A is symmetric positive definite. To solve the linear systems (13), (14) in h and p we apply a classical Cholesky factorization to the matrix A. Possible extensions to convective problems, Berrone et al. (2017b), with non-symmetric matrices can be considered with few changes applying a LU factorization to solve the linear systems (13), (14).

The Cholesky Factorization (LL^T) of the stiffness matrix A is performed once at the beginning of the code by the CPU using an Eigen function and then it is reused at each gradient iteration. To reduce the fill-in of matrix L, we resort to the AMD (Approximate Minimum Degree) algorithm in order to perform a renumbering of the DOFs of the hydraulic head (and of the dual variable p) to minimize the bandwidth of the matrix A; thus $PAP^{-1} = LL^T$ is computed. Only the block diagonal factor L and a vector representing the permutation matrix P are sent to the GPU, to minimize the device memory occupation.

In the following we report the algorithm of the conjugate gradient method as implemented by CUDASolver:

1. Choose an initial guess u^0
   (a) $\rightarrow$ CPU to GPU communication of u^0, $\mathcal{B}$, G^h, G^u, B^h, B^u, q and $\mathbb{M}$;
2. Factorize A computing $PAP^{-1} = LL^T$;
   (a) $\rightarrow$ CPU to GPU communication of L and P;
   (b) Cusparse analysis of the factor L;
3. Initialize the algorithm
   (a) Compute h^0,p^0 and g^0 by:
      i. Solve $(LL^T)\ Ph^0 = P\left(\mathcal{B}u^0 + q\right)$
      ii. Solve $(LL^T)\ Pp^0 = P\left(G^h h^0 - \alpha B^h u^0\right)$
      iii. Compute $g^0 = \mathcal{B}^T p^0 + G^u u^0 - \alpha B^u h^0$
      iv. Apply preconditioner $\tilde{g}^0 = \mathbb{M}^{-1} g^0$
   (b) Compute $\beta_N^0 = g^0 \cdot \tilde{g}^0$ and set $d^0 = -\tilde{g}^0$
4. Set $k = 0$
5. While $\tilde{g}^k \neq 0$
   (a) Compute δg^k:
      i. Solve $(LL^T)\ P\delta h^k = P\mathcal{B}d^k$
      ii. Solve $(LL^T)\ P\delta p^k = P\left(G^h \delta h^k - \alpha B^h d^k\right)$
      iii. Compute $\delta g^k = \mathcal{B}^T \delta p^k + G^u d^k - \alpha B^u \delta h^k$
   (b) Compute λ^k with an exact line search along d^k
      i. Compute $\lambda_N^k = d^k \cdot \tilde{g}^k$ and $\lambda_D^k = d^k \cdot \delta g^k$
      ii. Compute $\lambda^k = \frac{\lambda_N^k}{\lambda_D^k}$
   (c) Update $u^{k+1} = u^k + \lambda^k d^k$ and $g^{k+1} = g^k + \lambda^k \delta g^k$
   (d) Apply preconditioner $\tilde{g}^{k+1} = \mathbb{M}^{-1} g^{k+1}$
   (e) Set $\beta_D^{k+1} = \beta_N^k$ and compute $\beta_N^{k+1} = g^{k+1} \cdot \tilde{g}^{k+1}$
   (f) Compute $\beta_{k+1} = \frac{\beta_N^{k+1}}{\beta_D^{k+1}}$
   (g) Update $d^{k+1} = -\tilde{g}^{k+1} + \beta_{k+1} d^k$
   (h) $k = k + 1$
6. Compute the solution h^k using u^k in i
   (a) $\leftarrow$ GPU to CPU communication of h^k;

The symbol δ is the increment of a variable inside the iterations, e.g. $\delta h^k = h^{k+1} - h^k$. The matrix $\mathbb{M}$ is an appropriate preconditioner, if no preconditioning is applied $\mathbb{M}$ is the identity matrix.

To solve a sparse triangular linear system we have inserted in CUDASolver one of the solvers `cusparseDcsrsv` implemented in Cusparse. This function is called first with argument L, then with the same argument and the option to use its transpose. Before the solving step of the linear system it is required to perform an analysis operation to define the pattern of L. This analysis is performed once at the upload step and it is reused at each Conjugate Gradient iteration. In order to suitably deal with the permutations of the matrix A applied by the AMD algorithm we implement two kernels, one for the direct and one for the inverse permutation; guided by the Nvidia Visual Profiler the unrolling loop technique, set to eight, is applied to improve performances both in the memory access and in the thread utilization. The direct permutation kernel is shown in Listing 1. The permutation matrix is stored in a vector whose i-th component contains the permuted position of the corresponding vector element.

Listing 1 Kernel CUDA to compute vector permutation

```
__global__ void permutation8(int* perm, double* b, double* pB,unsigned int dimension)
{
    unsigned int idx = blockIdx.x * blockDim.x * 8 + threadIdx.x;
    if(idx + 7 * blockDim.x < dimension){
        pB[perm[idx]] = b[idx];
        pB[perm[idx+1*blockDim.x]] = b[idx+1*blockDim.x];
        pB[perm[idx+2*blockDim.x]] = b[idx+2*blockDim.x];
        pB[perm[idx+3*blockDim.x]] = b[idx+3*blockDim.x];
        pB[perm[idx+4*blockDim.x]] = b[idx+4*blockDim.x];
        pB[perm[idx+5*blockDim.x]] = b[idx+5*blockDim.x];
        pB[perm[idx+6*blockDim.x]] = b[idx+6*blockDim.x];
        pB[perm[idx+7*blockDim.x]] = b[idx+7*blockDim.x];
        }
    else{
            if(idx < dimension )
              pB[perm[idx]] = b[idx];
            if(idx+ blockDim.x < dimension )
              pB[perm[idx+blockDim.x]] = b[idx+blockDim.x];
            if(idx+2*blockDim.x < dimension )
              pB[perm[idx+2*blockDim.x]] = b[idx+2*blockDim.x];
            if(idx+3*blockDim.x < dimension )
              pB[perm[idx+3*blockDim.x]] = b[idx+3*blockDim.x];
            if(idx+4*blockDim.x < dimension )
               pB[perm[idx+4*blockDim.x]] = b[idx+4*blockDim.x];
            if(idx+5*blockDim.x < dimension )
                pB[perm[idx+5*blockDim.x]] = b[idx+5*blockDim.x];
            if(idx+6*blockDim.x < dimension )
              pB[perm[idx+6*blockDim.x]] = b[idx+6*blockDim.x];
    }
}
```

Furthermore, we have considered the option of a direct computation of the inverse L^{-1} of the block diagonal matrix L, thus allowing the solution of a linear system via a matrix vector product. This approach results to be easily parallelizable on the GPU, even if the matrices are sparse. The computation of the inverse matrix is performed by a suitable Eigen function, similarly to the matrix factorization. The choice to store L^{-1} instead of directly A^{-1} is aimed at reducing the GPU memory requests in order to tackle larger DFNs with the limited memory of the device. The matrix–vector product between a sparse matrix and a dense vector is performed by the function `cusparseDcsrmv` of the CuSparseLibrary, whereas the dot product is performed by `cublasDdot` of the library CuBlas, and the scalar-vector product is implemented by the `cublasDscal` of the CuBlas Library. The update operation of vectors u^k, g^k and d^k is performed by the `cublasDaxpy` function of the CuBlas library.

4 Numerical results

In this section we first describe the geometrical properties of the DFNs considered in the numerical tests and then we provide a description of the performances reached employing the GPU. For all the tests we impose a Dirichlet boundary condition $H_D = 1000$ on the edges of the fractures on the left-hand side of the DFN and $H_D = 0$ on the right-hand side of the DFN. All the other boundaries of the fractures have a no-flux condition.

In the numerical experiments we mainly compare the performances obtained in the linear algebra phase, which carries the most predominant cost; namely we measure performances in solving the block diagonal linear system and matrix–vector products at each gradient iteration. The comparison is made between the GPU and a single core of the CPU. In the CPU tests we use the Eigen Library v3.3 to perform the same operations performed by the GPU. First, we test a classical couple of forward and backward operations resorting to the LL^T Cholesky factorization; moreover, we investigate the GPU performances computing the solution of the linear systems by the direct application of L^{-1}. This last test aims at showing that the matrix–vector product is an operation much more conforming to be parallelized than the forward and backward solver. In the following, we define the speedup parameter, *SpUp*, as the ratio between the elapsed time required by the CPU to perform a specific task and the elapsed time needed by the GPU for the same task.

4.1 DFNs description

In the following we consider four DFN configurations identified by the number of fractures as: DFN1425, DFN4937, DFN12000, DFN33334. In Figs. 1, 2, 3 and 4 we report a visualization of the DFNs, and in Table 1 for each DFN we report the number

Fig. 1 DFN1425

Fig. 2 DFN4937

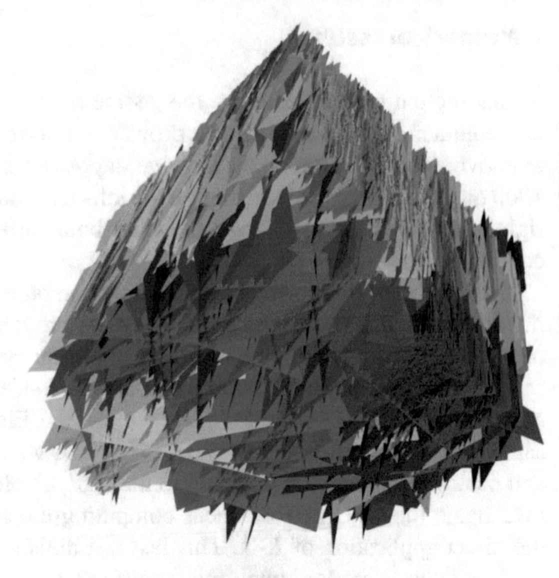

Fig. 3 DFN12000

of connected fractures, the total number of traces, the minimum, the maximum and the mean value of traces per fracture. Among the considered DFNs we have highly connected DFNs (DFN4937 and DFN12000) and moderately connected DFNs (DFN1425 and DFN33334). The last example (DFN33334) is taken with a small connectivity in order to have a relatively large DFN fitting the GPU memory constraints.

For each DFN we consider different meshes to investigate the behaviour of the GPU performances with respect to the number of DOFs for the discretization of both the hydraulic head and the control variable. Due to the GPU memory limits,

Fig. 4 DFN33334

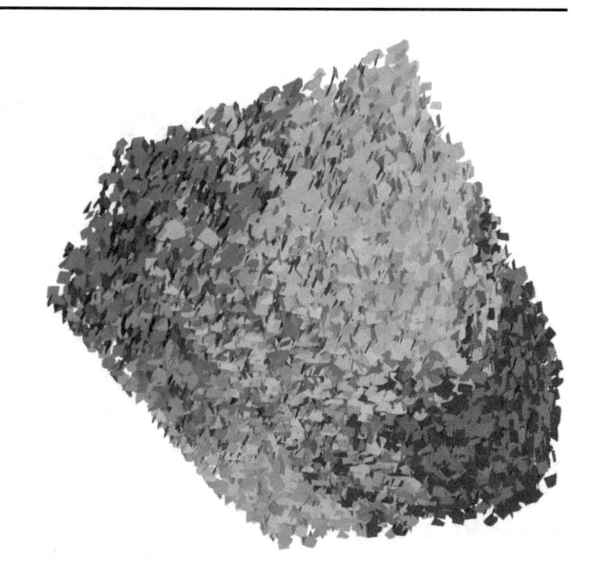

Table 1 Number of fractures and traces for each DFN

DFN	#Fractures	#Traces	Min #Tr	Max #Tr	$\frac{\#Traces}{\#Fractures}$
1425	1425	13,086	1	92	9.18
4937	4937	346,196	1	1040	70.12
12,000	12,000	991,638	18	1607	82.64
33,334	33,334	78,713	1	22	2.36

for some DFNs we can not perform all the tests. The mesh used for the discretization of the hydraulic head is obtained with *Triangle* (Shewchuk 1996, 2002), imposing an approximate number of elements on each fracture (#Tri). The mesh generator can introduce a slightly larger number of elements. The mesh for the discretization of the control variable on each trace of each fracture is induced by the intersection of the trace with the mesh on the fracture. We first intersect each trace with the edges of the mesh on the fracture; then, we define the elements for u as the segments obtained with this partitions (Berrone et al. 2015). Figures 5 and 6 show an example of two meshes applied to DFN1425; Fig. 5 represents a coarse mesh, and Fig. 6 shows a finer one. Thanks to the optimization approach, conformity to the traces is not required and the mesh on each fracture can be produced independently avoiding any local over-refinement usually induced by conformity constraints.

In all tests the discretization of the hydraulic head is performed using standard first order finite elements (P1) on each fracture. In order to investigate the impact of the number of DOFs for the control variable (uDofs) on the GPU parallel performances, we consider two different discretizations on the traces: piece-wise constant approximations (P0) and continuous piece-wise linear approximations (P1).

In Tables 2, 3, 4 and 5 we report the mesh parameter used on each fracture (#Tri), the polynomial order for the control variable (DegT), the number of DOFs for the

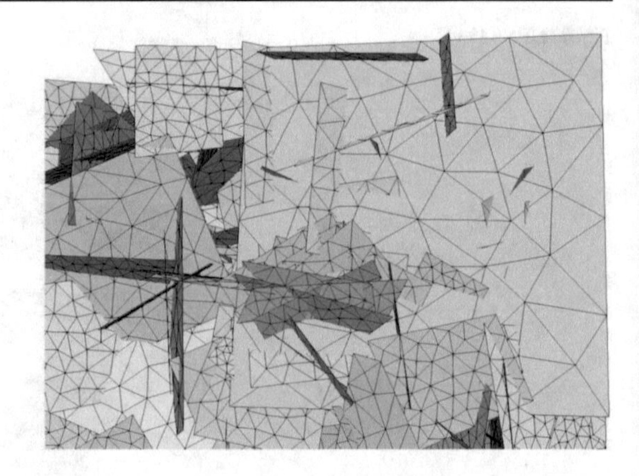

Fig. 5 DFN1425 mesh examples: #Tri = 50

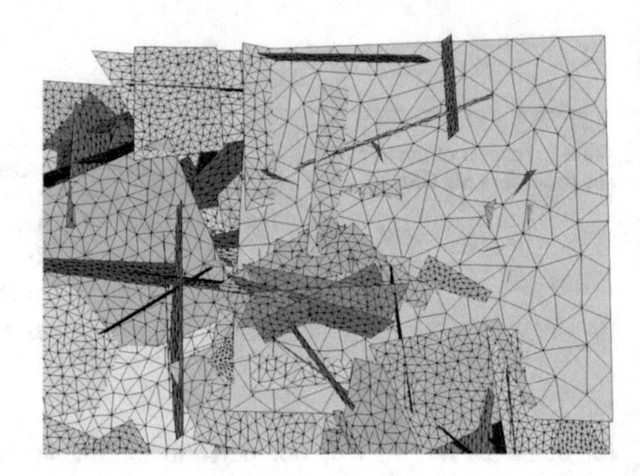

Fig. 6 DFN1425 mesh examples: #Tri = 200

Table 2 DFN1425: degrees of freedom on fractures and traces

#Tri	DegF	DegT	hDofs	uDofs	$\frac{\text{uDofs}}{\text{hDofs}}$
50	P1	P0	70,378	171,399	2.44
100	P1	P0	127,755	230,176	1.8
200	P1	P0	248,467	324,543	1.31
300	P1	P0	364,969	392,034	1.07
50	P1	P1	70,378	197,571	2.81
100	P1	P1	127,755	256,348	2.01
200	P1	P1	248,467	350,715	1.41
300	P1	P1	364,969	418,206	1.15

hydraulic head (hDofs), the number of DOFs for the control variable on the traces (uDofs) and the ratio between the last two values, (uDofs/hDofs). This ratio strongly affects the GPU performances, as shown in the sequel.

Table 3 DFN4937: degrees of freedom on fractures and traces

#Tri	DegF	DegT	hDofs	uDofs	$\frac{\text{uDofs}}{\text{hDofs}}$
50	P1	P0	244,251	4,898,113	20.05
100	P1	P0	445,013	6,622,551	14.88
200	P1	P0	862,838	9,351,907	10.84
300	P1	P0	1,271,911	11,346,019	8.92
50	P1	P1	244,251	5,590,505	22.89
100	P1	P1	445,013	7,314,943	16.44
200	P1	P1	862,838	10,044,299	11.64
300	P1	P1	1,271,911	12,038,411	9.46

Table 4 DFN12000: degrees of freedom on fractures and traces

#Tri	DegF	DegT	hDofs	uDofs	$\frac{\text{uDofs}}{\text{hDofs}}$
50	P1	P0	583,067	13,231,759	22.69
100	P1	P0	1,099,632	18,187,858	16.54
50	P1	P1	583,067	15,215,035	26.09
100	P1	P1	1,099,632	20,171,134	18.34

Table 5 DFN33334: degrees of freedom on fractures and traces

#Tri	DegF	DegT	hDofs	uDofs	$\frac{\text{uDofs}}{\text{hDofs}}$
50	P1	P0	1,665,169	929,225	0.56
100	P1	P0	2,995,257	1,234,638	0.41
200	P1	P0	5,876,163	1,742,790	0.30
300	P1	P0	8,617,381	2,106,485	0.24
50	P1	P1	1,665,169	1,086,651	0.65
100	P1	P1	2,995,257	1,392,064	0.46
200	P1	P1	5,876,163	1,900,216	0.32
300	P1	P1	8,617,381	2,263,911	0.26

4.2 Performance analysis using the standard Cholesky factorization

In this section we focus on the elapsed time needed to perform 1000 conjugate gradient iterations, comparing the time on a single CPU and the time required in the GPU. For the investigation here performed, we do not consider all the operations required by the setting up of the data structures describing the DFN geometry, the setting up of the matrices and i/o operations, because during the solution process most of the time is spent in the conjugate gradient operations. This is confirmed by the tests performed in Sect. 4.5, where we consider the full cost of the solution of the problem. We consider the non-preconditioned version of the Conjugate Gradient, whereas for the solution of the full problem we introduce a simple diagonal preconditioner. This will not change the results of the investigation.

Table 6 NVIDIA GTX TITAN X specification

Modal name	GTX TITAN X
RAM	12 GB GDDR5
Memory bandwidth	336.5 GB/s
Architecture	Maxwell
Streaming multiprocessor (SMX)	24
CUDA cores	128 × SMX 1000 MHz
CUDA capability	5.2

Table 7 DFN1425: speedup with Cholesky solver

#Tri	DegT	Cpu (s)	Gpu (s)	SpUp
50	P0	12.25	2.63	4.66
100	P0	20.27	4.32	4.70
200	P0	36.41	7.33	4.97
300	P0	52.53	10.66	4.93
50	P1	15.37	2.78	5.54
100	P1	23.66	4.56	5.19
200	P1	41.16	7.54	5.46
300	P1	57.69	10.34	5.58

Table 8 DFN4937: speedup with Cholesky solver

#Tri	DegT	Cpu (s)	Gpu (s)	SpUp
50	P0	251.38	19.36	12.99
100	P0	355.93	26.98	13.19
200	P0	538.22	41.92	12.84
300	P0	688.85	59.01	11.67
50	P1	337.93	22.89	14.76
100	P1	460.76	32.53	14.16
200	P1	671.94	48.19	13.94
300	P1	844.52	62.34	13.55

The computations are performed on a workstation with Intel(R) Xeon(R) CPU E5-2690 v2 @ 3.00 GHz, 256 GB RAM DDR3 and NVIDIA GTX TITAN X; more technical details of the hardware are reported in Table 6. The workstation is running a 64 bit Linux operating system Mandriva with gcc compiler version 5.4.0.

In Tables 7, 8, 9 and 10 we report the elapsed times required by the serial CPU and by the GPU. The meaning of #Tri and DegT labels is as in Tables 2, 3, 4 and 5. The following comments can be driven by the analysis of the previous Tables 7, 8, 9 and 10.

DFN1425, Table 7, has a low connectivity and the ratio (uDofs/hDofs) is approximately 2 for all the considered discretizations. The speedup approximately increases while increasing the mesh parameter (#Tri) and reaches the maximum value 5.58, with

Table 9 DFN12000: speedup with Cholesky solver

#Tri	DegT	Cpu (s)	Gpu (s)	SpUp
50	P0	735.33	48.46	15.17
100	P0	1120.59	72.13	15.53
50	P1	1201.51	61.87	19.42
100	P1	1586.14	83.23	19.06

Table 10 DFN33334: speedup with Cholesky solver

#Tri	DegT	Cpu (s)	Gpu (s)	SpUp
50	P0	171.92	25.84	6.65
100	P0	299.88	47.50	6.31
200	P0	590.89	99.53	5.94
300	P0	872.50	148.28	5.88
50	P1	188.75	26.74	7.06
100	P1	320.33	48.48	6.61
200	P1	618.81	100.58	6.15
300	P1	910.84	149.56	6.09

a mesh parameter 300 and P1 elements on the traces; the speedup is larger increasing the number of DOFs on the traces.

DFN4937, Table 8, displays a very large connectivity and the ratio (uDofs/hDofs) is strongly decreasing with respect to the mesh parameter, see Table 3. The speedup decreases from the maximum value, 14.76, with mesh parameter 50, to the minimum value, 11.67, with mesh parameter 300.

DFN12000, Table 9, is the most connected in our simulations. Due to memory constraints we provide solutions for the case up to #Tri=100. Also in this case the ratio (uDofs/hDofs) is decreasing refining the mesh on the fractures. For this test case the speedup reaches a value close to 20.

DFN33334, Table 10, is the largest DFN even if it is the less connected, with a ratio (uDofs/hDofs) very low. Also in this case increasing the order of the finite elements on the traces, the speedup increases.

We see that the speedup is larger for the tests with a larger number of DOFs on the traces and a larger connectivity; indeed increasing the uDofs number, the number of columns of the matrices $\mathcal{B}$, G^u, B^h, in the right hand side of Equations 5(a)i–5(a)iii increases. In these cases the number of operations required by the matrix–vector products involving the matrices $\mathcal{B}$, G^u, B^h are more relevant with respect to the operations performed in the forward/backward steps and the matrix–vector product is more suitable to be parallelized by the GPU than the forward/backward solver.

4.3 Analysis using inverse Cholesky factor

Forward–backward algorithms are usually less suitable to be parallelized, for this reason we propose another approach which computes the inverse of the L factor, denoted

Table 11 Non-zero elements of Cholesky factor and inverse Cholesky factor, #Tri = 100

DFN	nnz$L$	nnzL^{-1}	dim$A$
1425	943,179	2,591,354	127,755
4937	3,255,476	9,059,149	445,013
12,000	8,347,498	22,562,306	1,099,824
33,334	21,812,837	60,193,086	2,995,257

Table 12 Ratio of non-zero elements of Cholesky factor and inverse Cholesky factor

#Tri	nnzL^{-1}/nnz$L$			
	DFN1425	DFN4937	DFN12000	DFN33334
50	2.41	2.43	2.25	2.44
100	2.75	2.78	2.70	2.76
200	3.33	3.39	–	3.37
300	3.84	3.86	–	3.94

Table 13 DFN1425: speedup with the inverse matrix

#Tri	DegT	Gpu (s)	SpUp-R
50	P0	2.48	1.06
100	P0	2.44	1.77
200	P0	4.59	1.60
300	P0	7.56	1.42
50	P1	2.40	1.16
100	P1	2.65	1.72
200	P1	4.70	1.60
300	P1	7.75	1.33

by L^{-1}, and replaces the forward–backward substitution with two matrix vector multiplications. With this approach all the operations performed in each conjugate gradient iteration are matrix–vector products and we can take advantage of the parallel capabilities of the GPU.

For the case #Tri=100, in Table 11 we report the number of non-zero elements for the matrices L (nnz$L$), for the matrices L^{-1} (nnzL^{-1}) and the number of rows of the matrices A (dim$A$) to help the reader in assessing the data. Table 12 stores the ratio between nnzL^{-1} and nnz$L$, the ranges are between 2 and 4 in all the DFNs considered. This ratio has an important impact on the speedup, similarly to the ratio uDofs/hDofs seen in the previous section.

In Tables 13, 14, 15 and 16 we report the performances obtained; SpUp-R in the last column is the ratio between the times needed by the GPU for performing the computations using the triangular solver (Tables 7, 8, 9, 10) and using the inverse of the triangular factor (Gpu column in Tables 13, 14, 15, 16).

Almost all the speedups are better than those obtained using the triangular solver, and they are slightly decreasing when the number of elements for the discretization of h increases. The DFNs with larger SpUp-R using the factor L^{-1} are those with less

Table 14 DFN4937: speedup with the inverse matrix

#Tri	DegT	Gpu (s)	SpUp-R
50	P0	17.34	1.12
100	P0	23.39	1.15
200	P0	37.77	1.11
300	P0	59.41	0.99
50	P1	20.73	1.10
100	P1	28.88	1.13
200	P1	44.13	1.09
300	P1	59.83	1.04

Table 15 DFN12000: speedup with the inverse matrix

#Tri	DegT	Gpu (s)	SpUp-R
50	P0	45.26	1.07
100	P0	68.49	1.05
50	P1	61.18	1.01
100	P1	83.24	1.00

Table 16 DFN33334: speedup with the inverse matrix

#Tri	DegT	Gpu (s)	SpUp-R
50	P0	18.59	1.39
100	P0	33.06	1.44
200	P0	84.27	1.18
300	P0	169.88	0.88
50	P1	19.42	1.38
100	P1	34.01	1.43
200	P1	85.79	1.17
300	P1	155.14	0.96

DOFs on the traces, i.e. DFN1425 and DFN33334, because in these tests the solution of the linear systems have a larger impact and the new approach can provide a larger improvement of performances. The downgrade of the performances increasing the mesh parameter can be justified considering the faster increasing of non-zero values of the inverse of the matrices and the increasing cost of the matrix vector product with respect to the triangular solve, further increased by sparse storage of the matrices.

In Table 17 we report the number of non-zero elements for the matrices $\mathcal{B}$, G^h, B^h, G^u and B^u for all the DFNs considered, in the case of mesh parameter 100 and P0 elements on the traces. The number of non-zero elements for the inverse of the factors of the matrix A is reported in Table 11. Using these quantities we can provide an evaluation of the number of floating point operations per second (FLOPs) performed by the single core of the CPU and by the GPU. Table 18 reports these estimates. The total number of operations per iteration, #Operations, is computed using the points 5(a)i, 5(a)ii, 5(a)iii, 5(b)i, 5c and 5g of the conjugate gradient algorithm. Considering

Table 17 Non-zeros element of the conjugate gradient matrices, #Tri = 100

DFN	$\mathcal{B}$	G^h	B^h	G^u	B^u
1425	690,379	532,038	1,585,037	664,356	1,585,037
4937	19,867,417	2,720,503	45,664,763	19,175,261	45,664,763
12,000	54,044,833	6,952,939	125,329,252	52,580,298	125,329,252
33,334	3,703,898	5,552,626	8,484,939	3,546,488	8,484,939

Table 18 Floating point operations per second (FLOPs) with the inverse matrix, #Tri = 100

DFN	#Operations	GFLOPs (CPU)	GFLOPs (GPU)
1425	1.84×10^7	0.91	7.53
4937	23.78×10^7	0.67	10.17
12,000	63.79×10^7	0.57	9.31
33,334	29.79×10^7	0.99	9.01

the performances of the CPU, we can see that the FLOPs decreases increasing the number of traces per fracture of the DFN, whereas for the GPU the performances are approximately increased increasing the number of traces and the number of DOFs on the traces.

We remark that the obtained FLOPs for the GPU are conditioned by the use of the library Cusparse and conform to results known for matrix–vector products with sparse matrices in CSR format (Merrill and Garland 2016). The matrix vector product with matrices in CRS format results to be an algorithm very memory bounded (Merrill and Garland 2016). A large research activity is currently devoted to circumvent this limitations. In order to improve the computational performances we test a different storage format with an enlarged data proximity of the matrix elements.

4.4 Performance analysis using hybrid sparse storage format

In this section we investigate the effect on performances of Hybrid Format (HYB) storage (Nvidia 2008d) of sparse matrices applied to our formulation of the problem following ideas presented in Merrill and Garland (2016). The function `cusparseDcsr2hyb` provided in the CuSparse library converts matrices from the CSR format to the HYB format. Sparse matrices in hybrid format store a number of zero elements in order to make the matrix–vector product more efficient. The number of non-zero elements depends on an input parameter, and in our tests we use two different options in order to maximize the performances or balance performances and memory use: `CUSPARSE_HYB_PARTITION_MAX` and `CUSPARSE_HYB_PARTITION_AUTO`. The first one generates a matrix with a constant number of non-zero elements equal to the largest number of non-zero elements in the rows. This choice corresponds to the ELL format for sparse matrices, see Nvidia (2008d). With the second choice, the number of constant number of non-zero elements

Table 19 DFN1425: speedup applying hybrid format with inverse matrix

#Tri	DegT	CSR–HYB		HYB–HYB			
		Time_{M}	SpUp-R_M	Time_{M}	SpUp-R_M	Time_{A}	SpUp-R_A
50	P0	1.18	2.09	1.26	1.96	1.39	1.78
100	P0	1.86	1.31	2.02	1.21	2.25	1.08
200	P0	3.70	1.24	4.16	1.10	4.53	1.01
300	P0	6.29	1.19	7.17	1.05	7.49	1.00
50	P1	1.32	1.82	1.47	1.64	1.58	1.52
100	P1	2.07	1.28	2.25	1.18	2.51	1.05
200	P1	3.93	1.19	4.47	1.05	4.83	0.97
300	P1	6.55	1.18	7.58	1.02	7.80	0.99

Table 20 DFN4937: speedup applying hybrid format with inverse matrix

#Tri	DegT	CSR–HYB		HYB–HYB			
		Time_{M}	SpUp-R_M	Time_{M}	SpUp-R_M	Time_{A}	SpUp-R_A
50	P0	16.17	1.07	–	–	20.54	0.84
100	P0	22.31	1.05	–	–	28.78	0.81
200	P0	35.06	1.08	–	–	44.84	0.84
300	P0	52.27	1.14	–	–	61.62	0.96
50	P1	19.68	1.05	–	–	27.68	0.75
100	P1	27.84	1.04	–	–	37.49	0.77
200	P1	41.37	1.07	–	–	55.25	0.80
300	P1	55.79	1.07	–	–	73.64	0.81

per row is lower and automatically selected, and the possible exceeding non-zero elements are stored in COO format.

We present the effect of the HYB memorization approach only on two of the previous networks: DFN1425 and DFN4937, low and highly connected networks, respectively. Similar results can be obtained for the other networks with similar connectivity properties.

Tables 19 and 20 show the results of the speedups in the case of the use of inverse Cholesky Factor L^{-1} to solve equations 5(a)i, 5(a)ii. The two columns labelled CRS-HYB concern the simulations performed using the CRS format for the factor L^{-1} and the HYB format for the matrices that are used in the multiplications performed in the right-hand sides of the equations 5(a)i–5(a)iii; on the other hand, HYB–HYB are the simulations performed storing all the matrices with the hybrid format. Columns Time_{M} and Time_{A} report the time spent to perform the conjugate gradient iterations using hybrid format with `PARTITION_MAX` and `PARTITION_AUTO` partition type, respectively; SpUp-R_M and SpUp-R_A are the ratios between the speedup measured in these simulations and the speedups (SpUp) of the simulations in Tables 13 and 14. The simulations for the DFN4937 in the HYB–HYB `PARTITION_MAX` partition

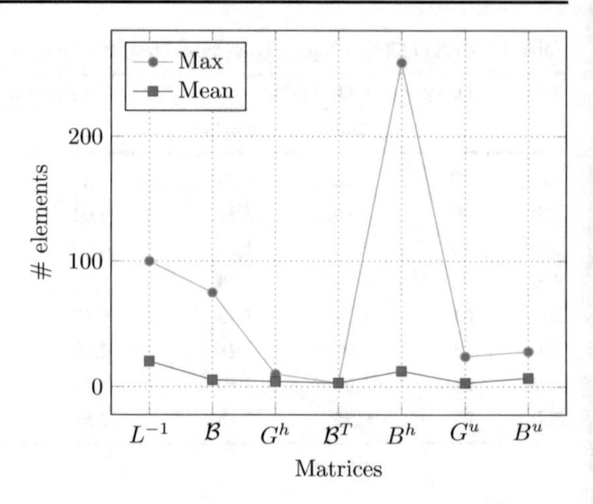

Fig. 7 DFN1425: non-zeros elements per rows

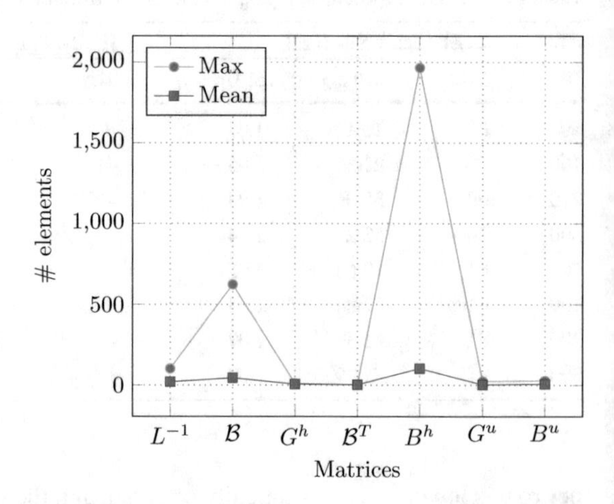

Fig. 8 DFN4937: non-zeros elements per rows

type case were not performed due to the GPU memory limit. Nevertheless, in the CSR–HYB format we use only `PARTITION_MAX` because, when it is viable, this parameter provides the best results.

In general we have no relevant improvement of the speedups. The reasons of these slight improvements is due to a non-suitable pattern of the matrices for the GPU parallelization due to the large variation of the number of non-zero elements among the rows. In Figs. 7 and 8 we report the maximum and the mean number of non-zero elements per row for each matrix used in the computations; it is clear that the maximum number of non-zeros per row, used to store the matrix in the hybrid format, is very far from the mean value of non-zeros, especially for $\mathcal{B}$ and B^h. This means that, when a matrix–vector product is performed with the matrix stored in HYB format, a large number of products by zeros is performed with a negative impact on the performances. These results show that, even changing the matrix storage format, few improvements can be achieved.

Table 21 Speedup on DFN1425 and DFN4937 upon fulfilment of stopping criterion

DFN	Cpu (s)	Gpu-L (s)	Gpu-L^{-1} (s)	SpUp	SpUp-R
1425	4408	940	541	4.69	1.74
4937	403,249	30,289	26,467	13.31	1.14

4.5 Scalabilty results for the full solution

As expected from the results presented in Tables 7, 8, 9 and 10, we see that the speedup for the full solution of the problem mainly depends on the conjugate gradient iterations. We discuss here the performances of the previous approaches with CSR memorization in the full solution of a problem: DFN input, matrices set-up, matrices factorization or inversion, host-device communications and Conjugate Gradient operations. The stopping criterion of the conjugate gradient algorithm is based on the relative residual, with a tolerance of 10^{-6}. The number of iterations needed to satisfy this criterion mainly depends on the total number of uDOFs. To confirm the validity of the investigations performed on the conjugate gradient operations we present the results on the two tests DFN1425 and DFN4937. The discretization used in both DFNs is P1 on the fracture with a mesh parameter of 100 triangles per fracture and P0 on the traces. Moreover, we apply a diagonal preconditioner, $\mathbb{M}^{-1} = \text{diag}\,(G^u)^{-1}$, defined in Berrone et al. (2016b), to reduce the total number of iterations. Table 21 reports for each DFN the time elapsed by the CPU and by the GPU using Cholesky factorization (GpuL) and the elapsed time required by the GPU when the inverse of the Cholesky Factor (GpuL^{-1}) is used. Finally, in the last two columns the corresponding speedups (SpUp = Cpu/GpuL, SpUp-R = GpuL^{-1}/GpuL) are reported. Figures 9 and 10 show the solutions on DFN1425 and on DFN4937, respectively. The solutions are reached after 2×10^5 and 9×10^5 iterations.

5 Conclusions

In this paper, we have shown a possible use of a single GPU to simulate the Darcy flow in a network of fractures and we have compared the performances of a GPU against a CPU using a single core. We have discussed two different approaches to suitably apply on a GPU an algorithm first presented in Berrone et al. (2013b) and two different sparse storage formats. For all the approaches the performances of the GPU in terms of FLOPs conform to the documented GPU performances for sparse matrix–vector multiplications, and the results on performances are not excessively influenced by the fact that they are obtained on a gaming device, being the performances limitations mainly due to the sparse matrix structure and device memory speed, rather than to the double precision FLOPs peaks of the device. This occurs because the sparse matrices originated by the method have a large number of rows with a number of non-zero elements much smaller than the GPU warp dimension and, in general, the number of non-zero elements is far from an integer multiple of the warp dimension, thus yielding

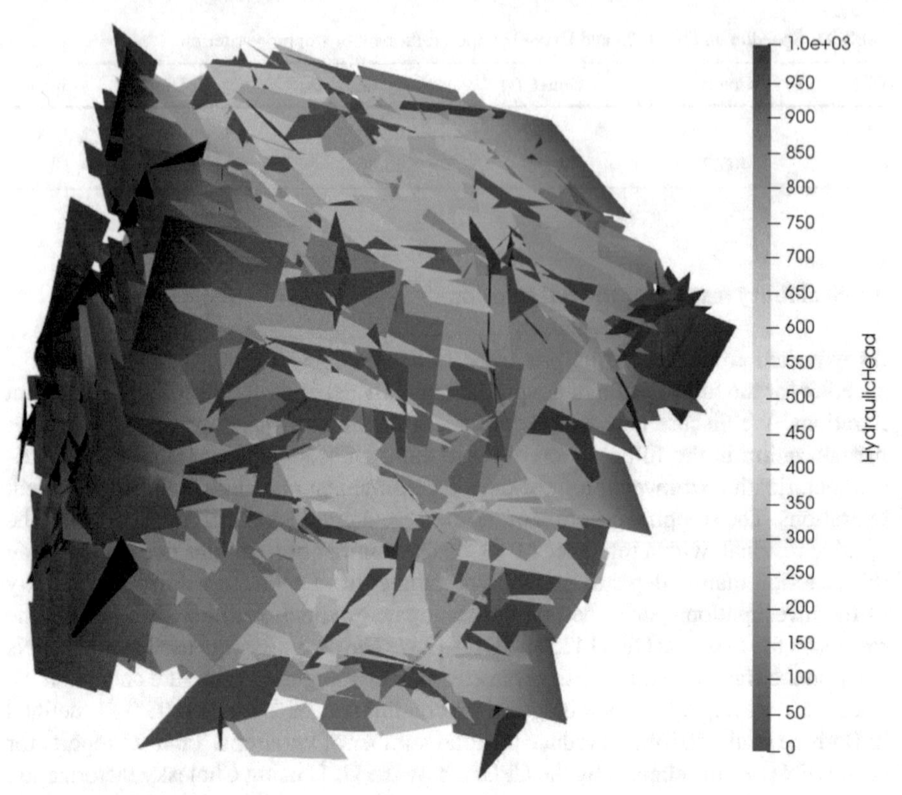

Fig. 9 DFN1425: solution

to a very difficult efficient hardware scheduling of the operations. Moreover, dealing with sparse matrices requires several device memory operations to set up the data to be processed by the warps in the matrix operations. The results of our performances analysis can be extended to many different contexts, either originated by the application of a similar optimization approach to other linear flow and mechanical models, or to different finite element formulations, yielding matrices with similar structures. Concerning the two tested linear algebra approaches, the storing of the inverse of the Cholesky factors usually yields to an improvement of the performances parallel to a larger use of memory resources. For very large DFNs the classical approach based on the local application of the Cholesky method for solving the linear systems is usually preferable yielding to good performances with a limited use of memory resources. Furthermore, among the sparse matrix storage approaches here tested, the CSR approach is the more reliable in term of performances and memory use. Sparse matrix format designed to save the memory use has the drawback to make computational operations memory bounded, reducing the computational power by a factor of ten in both GPU and CPU tests. The developing time required to modify a code for its application on a GPU is reasonably moderate, if compared to the improvement of performances which is, anyway, quite relevant. In the DFN simulations we observe that the speedup depends on the DFN connectivity and on the discretization of the hydraulic head and

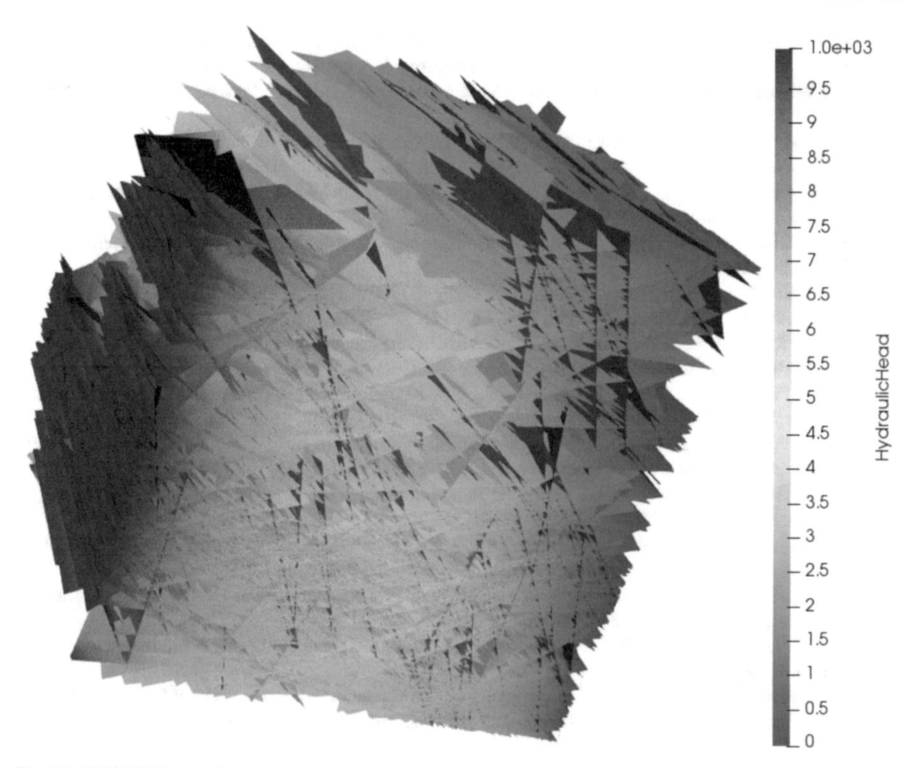

Fig. 10 DFN4937: solution

of the control variable in the fractures and on the traces, respectively. The limited GPU memory binds the dimension of the DFNs that can be considered, even though remarkably large. In order to overcome this limitation, the implementation of the algorithm on a multi-GPU system can be considered as well as its implementation together with a parallel MPI approach to exploit all the computational power of distributed memory HPC system. An efficient implementation of this approach requires a deep investigation of the communications structure and hardware loading in order to fully exploit the computational facilities and can not be considered as a straightforward extension of the present work.

References

Ahmed, R., Edwards, M.G., Lamine, S., Huisman, B.A., Pal, M.: CVD-MPFA full pressure support, coupled unstructured discrete fracture–matrix Darcy-flux approximations. J. Comput. Phys. **349**, 265–299 (2017)

Al-Hinai, O., Singh, G., Pencheva, G., Almani, T., Wheeler, M.F., et al.: Modeling multiphase flow with nonplanar fractures. In: SPE Reservoir Simulation Symposium. Society of Petroleum Engineers (2013)

Al-Hinai, O., Wheeler, M.F., Yotov, I.: A generalized mimetic finite difference method and two-point flux schemes over Voronoi diagrams. ESAIM Math. Model. Numer. Anal. **51**(2), 679–706 (2017)

Antonietti, P., Formaggia, L., Scotti, A., Verani, M., Verzott, N.: Mimetic finite difference approximation of flows in fractured porous media. ESAIM Math. Model. Numer. Anal. **50**(3), 809–832 (2016)

Beirão da Veiga, L., Brezzi, F., Cangiani, A., Manzini, G., Marini, L.D., Russo, A.: Basic principles of virtual element methods. Math. Models Methods Appl. Sci. **23**(01), 199–214 (2013)
Beirão da Veiga, L., Lipnikov, K., Manzini, G.: The Mimetic Finite Difference Method for Elliptic Problems, vol. 11. Springer, Berlin (2014)
Benedetto, M.F., Berrone, S., Pieraccini, S., Scialò, S.: The virtual element method for discrete fracture network simulations. Comput. Methods Appl. Mech. Eng. **280**, 135–156 (2014)
Benedetto, M.F., Berrone, S., Borio, A., Pieraccini, S., Scialò, S.: A hybrid mortar virtual element method for discrete fracture network simulations. J. Comput. Phys. **306**, 148–166 (2016)
Benedetto, M.F., Borio, A., Scialò, S.: Mixed virtual elements for discrete fracture network simulations. Finite Elem. Anal. Des. **134**, 55–67 (2017)
Berrone, S., Borio, A.: Orthogonal polynomials in badly shaped polygonal elements for the virtual element method. Finite Elem. Anal. Des. **129**, 14–31 (2017)
Berrone, S., Pieraccini, S., Scialò, S.: On simulations of discrete fracture network flows with an optimization-based extended finite element method. SIAM J. Sci. Comput. **35**(2), A908–A935 (2013a)
Berrone, S., Pieraccini, S., Scialò, S.: A PDE-constrained optimization formulation for discrete fracture network flows. SIAM J. Sci. Comput. **35**(2), B487–B510 (2013b)
Berrone, S., Pieraccini, S., Scialò, S.: An optimization approach for large scale simulations of discrete fracture network flows. J. Comput. Phys. **256**, 838–853 (2014)
Berrone, S., Pieraccini, S., Scialò, S., Vicini, F.: A parallel solver for large scale DFN flow simulations. SIAM J. Sci. Comput. **37**(3), C285–C306 (2015)
Berrone, S., Borio, A., Scialò, S.: A posteriori error estimate for a PDE-constrained optimization formulation for the flow in DFNs. SIAM J. Numer. Anal. **54**(1), 242–261 (2016a)
Berrone, S., Pieraccini, S., Scialò, S.: Towards effective flow simulations in realistic discrete fracture networks. J. Comput. Phys. **310**, 181–201 (2016b)
Berrone, S., Pieraccini, S., Scialò, S.: Flow simulations in porous media with immersed intersecting fractures. J. Comput. Phys. **345**, 768–791 (2017a)
Berrone, S., Pieraccini, S., Scialò, S.: Non-stationary transport phenomena in networks of fractures: effective simulations and stochastic analysis. Comput. Methods Appl. Mech. Eng. **315**, 1098–1112 (2017b)
Berrone, S., Canuto, C., Pieraccini, S., Scialò, S.: Uncertainty quantification in discrete fracture network models: stochastic geometry. Water Resour. Res. **54**, 1338–1352 (2018)
Brenner, K., Hennicker, J., Masson, R., Samier, P.: Gradient discretization of hybrid-dimensional Darcy flow in fractured porous media with discontinuous pressures at matrix–fracture interfaces. IMA J. Numer. Anal. **37**(3), 1551–1585 (2017)
Cacas, M.C., Ledoux, E., de Marsily, G., Tillie, B., Barbreau, A., Durand, E., Feuga, B., Peaudecerf, P.: Modeling fracture flow with a stochastic discrete fracture network: calibration and validation: 1. The flow model. Water Resour. Res. **26**, 479–489 (1990)
Chave, F., Di Pietro, D.A., Formaggia, L.: A hybrid high-order method for Darcy flows in fractured porous media. SIAM J. Sci. Comput. **40**(2), A1063–A1094 (2018)
de Dreuzy, J.R., Davy, P., Bour, O.: Hydraulic properties of two-dimensional random fracture networks following a power law length distribution: 2. Permeability of networks based on log-normal distribution of apertures. Water Resour. Res. **37**(8), 2079–2095 (2001)
de Dreuzy, J.R., Pichot, G., Poirriez, B., Erhel, J.: Synthetic benchmark for modeling flow in 3D fractured media. Comput. Geosci. **50**, 59–71 (2013)
Dershowitz, W.S., Einstein, H.H.: Characterizing rock joint geometry with joint system models. Rock Mech. Rock Eng. **1**, 21–51 (1988)
Dershowitz, W.S., Fidelibus, C.: Derivation of equivalent pipe networks analogues for three-dimensional discrete fracture networks by the boundary element method. Water Resour. Res. **35**, 2685–2691 (1999)
Fidelibus, C.: The 2D hydro-mechanically coupled response of a rock mass with fractures via a mixed BEM–FEM technique. Int. J. Numer. Anal. Methods Geomech. **31**(11), 1329–1348 (2007)
Fidelibus, C., Cammarata, G., Cravero, M.: Hydraulic characterization of fractured rocks. In: Abbie, M., Bedford, J.S. (eds.) Rock Mechanics: New Research. Nova Science Publishers Inc., New York (2009)
Flemisch, B., Berre, I., Boon, W., Fumagalli, A., Schwenck, N., Scotti, A., Stefansson, I., Tatomir, A.: Benchmarks for single-phase flow in fractured porous media. Adv. Water Resour. **111**, 239–258 (2018)
Franc, J., Jeannin, L., Debenest, G., Masson, R.: FV-MHMM method for reservoir modeling. Comput. Geosci. **21**(5), 895–908 (2017)
Fumagalli, A., Keilegavlen, E.: Dual virtual element method for discrete fractures networks. SIAM J. Sci. Comput. **40**(1), B228–B258 (2018)

Fumagalli, A., Keilegavlen, E., Scialò, S.: Conforming, non-conforming and non-matching discretization couplings in discrete fracture network simulations. https://doi.org/10.1016/j.jcp.2018.09.048 (2018)
Fumagalli, A., Scotti, A.: An efficient XFEM approximation of Darcy flows in arbitrarily fractured porous media. Oil Gas Sci. Technol. Rev. dIFP Energ. Nouv. **69**(4), 555–564 (2014)
Garipov, T.T., Karimi-Fard, M., Tchelepi, H.A.: Discrete fracture model for coupled flow and geomechanics. Comput. Geosci. **20**(1), 149–160 (2016)
Guennebaud, G., Jacob, B., et al.: Eigen v3 documentation. http://eigen.tuxfamily.org (2010). Accessed 10 Oct 2018
Hadgu, T., Karra, S., Kalinina, E., Makedonska, N., Hyman, J.D., Klise, K., Viswanathan, H.S., Wang, Y.: A comparative study of discrete fracture network and equivalent continuum models for simulating flow and transport in the far field of a hypothetical nuclear waste repository in crystalline host rock. J. Hydrol. **553**, 59–70 (2017)
Hyman, J.D., Gable, C.W., Painter, S.L., Makedonska, N.: Conforming Delaunay triangulation of stochastically generated three dimensional discrete fracture networks: a feature rejection algorithm for meshing strategy. SIAM J. Sci. Comput. **36**(4), A1871–A1894 (2014)
Jaffré, J., Roberts, J.E.: Modeling flow in porous media with fractures; discrete fracture models with matrix–fracture exchange. Numer. Anal. Appl. **5**(2), 162–167 (2012)
Lenti, V., Fidelibus, C.: A BEM solution of steady-state flow problems in discrete fracture networks with minimization of core storage. Comput. Geosci. **29**(9), 1183–1190 (2003)
Manzoor, S., Edwards, M.G., Dogru, A.H., Al-Shaalan, T.M.: Interior boundary-aligned unstructured grid generation and cell-centered versus vertex-centered CVD-MPFA performance. Comput. Geosci. **22**(1), 195–230 (2018)
Merrill, D., Garland, M.: Merge-based parallel sparse matrix–vector multiplication. In: Proceedings of the International Conference for High Performance Computing, Networking, Storage and Analysis, SC'16, pp. 58:1–58:12. IEEE Press, Piscataway (2016)
Mustapha, H., Mustapha, K.: A new approach to simulating flow in discrete fracture networks with an optimized mesh. SIAM J. Sci. Comput. **29**(4), 1439–1459 (2007)
Ngo, T.D., Fourno, A., Noetinger, B.: Modeling of transport processes through large-scale discrete fracture networks using conforming meshes and open-source software. J. Hydrol. **554**, 66–79 (2017)
Noetinger, B.: A quasi steady state method for solving transient Darcy flow in complex 3D fractured networks accounting for matrix to fracture flow. J. Comput. Phys. **283**, 205–223 (2015)
Noetinger, B., Jarrige, N.: A quasi steady state method for solving transient Darcy flow in complex 3D fractured networks. J. Comput. Phys. **231**(1), 23–38 (2012)
Nvidia, C: Cublas documentation. http://docs.nvidia.com/cuda/cublas (2008a). Accessed 10 Oct 2018
Nvidia, C.: Cuda toolkit. https://developer.nvidia.com/cuda-toolkit (2008b). Accessed 10 Oct 2018
Nvidia, C.: Cuda toolkit documentation. http://docs.nvidia.com/cuda-toolkit (2008c). Accessed 10 Oct 2018
Nvidia, C.: Cusparse documentation. http://docs.nvidia.com/cuda/cusparse (2008d). Accessed 10 Oct 2018
Parramore, E., Edwards, M.G., Pal, M., Lamine, S.: Multiscale finite-volume CVD-MPFA formulations on structured and unstructured grids. Multiscale Model. Simul. **14**(2), 559–594 (2016)
Pichot, G., Erhel, J., de Dreuzy, J.R.: A mixed hybrid mortar method for solving flow in discrete fracture networks. Appl. Anal. **89**(10), 1629–1643 (2010)
Pichot, G., Erhel, J., de Dreuzy, J.R.: A generalized mixed hybrid mortar method for solving flow in stochastic discrete fracture networks. SIAM J. Sci. Comput. **34**(1), B86–B105 (2012)
Pichot, G., Poirriez, B., Erhel, J., de Dreuzy, J.R.: A mortar BDD method for solving flow in stochastic discrete fracture networks. In: Erhel, J., Gander, M., Halpern, L., Pichot, G., Sassi, T., Widlund, O. (eds.) Domain Decomposition Methods in Science and Engineering XXI, pp. 99–112. Springer, Cham (2014)
Sentís, M.L., Gable, C.W.: Coupling LaGrit unstructured mesh generation and model setup with TOUGH2 flow and transport: a case study. Comput. Geosci. **108**, 42–49 (2017)
Shewchuk, J.R.: Triangle: engineering a 2D quality mesh generator and Delaunay triangulator. In: Lin, M.C., Manocha, D. (eds.) Applied Computational Geometry: Towards Geometric Engineering. Lecture Notes in Computer Science, vol. 1148, pp. 203–222. Springer, Berlin (1996). (from the first ACM workshop on applied computational geometry)
Shewchuk, J.R.: Delaunay refinement algorithms for triangular mesh generation. Comput. Geom. **22**(1), 21–74 (2002). (16th ACM symposium on computational geometry)

Vohralík, M., Maryška, J., Severýn, O.: Mixed and nonconforming finite element methods on a system of polygons. Appl. Numer. Math. **51**, 176–193 (2007)

Xing, F., Masson, R., Lopez, S.: Parallel vertex approximate gradient discretization of hybrid dimensional Darcy flow and transport in discrete fracture networks. Comput. Geosci. **21**(4), 595–617 (2017)

Publisher's Note Springer Nature remains neutral with regard to jurisdictional claims in published maps and institutional affiliations.

GEM - International Journal on Geomathematics
https://doi.org/10.1007/s13137-019-0119-5

ORIGINAL PAPER

Flow and transport in fractured poroelastic media

Ilona Ambartsumyan[1] · Eldar Khattatov[1] · Truong Nguyen[1] · Ivan Yotov[1]

Received: 13 April 2018 / Accepted: 4 December 2018

Abstract

We study flow and transport in fractured poroelastic media using Stokes flow in the fractures and the Biot model in the porous media. The Stokes–Biot model is coupled with an advection–diffusion equation for modeling transport of chemical species within the fluid. The continuity of flux on the fracture-matrix interfaces is imposed via a Lagrange multiplier. The coupled system is discretized by a finite element method using Stokes elements, mixed Darcy elements, conforming displacement elements, and discontinuous Galerkin for transport. The stability and convergence of the coupled scheme are analyzed. Computational results verifying the theory as well as simulations of flow and transport in fractured poroelastic media are presented.

Keywords Fluid-poroelastic structure interaction · Stokes-Biot model · Coupled flow and transport · Fractured poroelastic media

Mathematics Subject Classification 76S05 · 76D07 · 74F10 · 65M60 · 65M12

1 Introduction

Flow and transport in fractured poroelastic media occur in many applications, including enhanced oil and gas recovery, hydraulic fracturing, groundwater hydrology, and

---

Partially supported by DOE Grant DE-FG02-04ER25618 and NSF Grants DMS 1418947 and DMS 1818775.

✉ Ivan Yotov
yotov@math.pitt.edu

Ilona Ambartsumyan
ila6@pitt.edu

Eldar Khattatov
elk58@pitt.edu

Truong Nguyen
tqn4@pitt.edu

[1] Department of Mathematics, University of Pittsburgh, Pittsburgh, PA 15260, USA

subsurface waste repositories. These are challenging multiphysics processes involving interaction between a free fluid in the fractures with a fluid within the porous medium. The fluid flow may cause and be affected by solid deformation. For example, geomechanics effects are critical in hydraulic fracturing, as well as in modeling phenomena such as subsidence and compaction. Furthermore, the flow process may be coupled with transport phenomena, with the substance of interest propagating both through the fracture network and the porous matrix. Typical examples include tracking and cleaning up groundwater contaminants, leakage of subsurface radioactive waste, and proppant injection in hydraulic fracturing.

We use the Stokes equations to model the free fluid in the fractures and the Biot poroelasticity model (Biot 1941) for the fluid in the poroelastic region. The latter is based on a linear stress-strain constitutive relationship for the porous solid, and Darcy's law, which describes the average velocity of the fluid in the pores. The interaction across the fracture-matrix interfaces exhibits features of both Stokes–Darcy coupling (Discacciati et al. 2002; Girault and Rivière 2009; Layton et al. 2003; Rivière and Yotov 2005; Vassilev et al. 2014) and fluid–structure interaction (FSI) (Galdi and Rannacher 2010; Bazilevs et al. 2013; Bungartz and Schäfer 2006; Formaggia et al. 2010; Richter 2017). We refer to the Stokes–Biot coupling considered in this paper as fluid–poroelastic structure interaction (FPSI). There has been growing interest in such models in the literature. The well-posedness of the mathematical model was studied in Showalter (2005). Numerical studies include variational multiscale methods for the monolithic system and iterative partitioned scheme (Badia et al. 2009), a non-iterative operator-splitting method (Bukac et al. 2015), a partitioned method based on Nitsche's coupling (Bukac et al. 2015), and a Lagrange multiplier formulation for the continuity of flux (Ambartsumyan et al. 2018b).

To simplify the presentation we consider a fixed domain in time. As presented, the model is suitable for deformations that are small relative to the width of the fractures. This is valid for scales that are zoomed-in on the fractures or for meso-scale inclusions such as cavities. The model can be extended to account for the motion of the fluid domain by using the Arbitrary Lagrangian-Eulerian (ALE) approach, which has been done in Badia et al. (2009) and Bukac et al. (2015).

In this work we employ a monolithic scheme for the full-dimensional Stokes–Biot problem to model flow in fractured poroelastic media. We note that an alternative approach is based on a reduced-dimension fracture model, including the Reynolds lubrication equation (Ganis et al. 2014; Girault et al. 2015; Lee et al. 2016b; Mikelić et al. 2015) and an averaged Brinkman equation (Bukac et al. 2017). Works that do not account for elastic deformation of the media include averaged Darcy models (Martin et al. 2005; Frih et al. 2012; Morales and Showalter 2010; D'Angelo and Scotti 2012; Fumagalli and Scotti 2012; Boon et al. 2018; Flemisch et al. 2018), Forchheimer models (Frih et al. 2008), Brinkman models (Lesinigo et al. 2011), and an averaged Stokes model that results in a Brinkman model for the fracture flow (Morales and Showalter 2017).

For the discretization of the full-dimensional Stokes–Biot problem we consider the mixed formulation for Darcy flow in the Biot system, which provides a locally mass conservative flow approximation and an accurate Darcy velocity. This formulation results in the continuity of normal velocity condition being of essential type, which is enforced through a Lagrange multiplier (Ambartsumyan et al. 2018b). The discretization allows for the use of any stable Stokes spaces in the fracture region and any stable mixed Darcy spaces (Boffi et al. 2013). For the elasticity equation we employ a displacement formulation with continuous Lagrange elements.

The Stokes–Biot system is coupled with an advection diffusion equation for modeling transport of chemical species within the fluid. The transport equation is discretized by a discontinuous Galerkin (DG) method. DG methods (Arnold et al. 2001; Oden et al. 1998; Rivière et al. 1999; Cockburn and Shu 1998; Sun and Wheeler 2005b) exhibit local mass conservation, reduced numerical diffusion, variable degrees of approximation, and accurate approximations for problems with discontinuous coefficients. Due to their low numerical diffusion, DG methods are especially suited for advection–diffusion problems (Cockburn and Shu 1998; Cockburn and Dawson 2002; Sun and Wheeler 2005b; Dawson et al. 2004; Wheeler and Darlow 1980; Aizinger et al. 2000). Coupled Darcy flow and transport problems utilizing DG for transport have been studied in Sun and Wheeler (2005a), Sun et al. (2002), Dawson (1999) and Wheeler and Darlow (1980). Coupling of Stokes–Darcy flow with transport using a local discontinuous Galerkin scheme was developed in Vassilev and Yotov (2009). A coupled phase field-transport model for proppant-filled fractures is studied in Lee et al. (2016a). A flow-transport reduced fracture model using Darcy flow in the fracture and the matrix is developed in Fumagalli and Scotti (2013). To the best of our knowledge, the coupled Stokes–Biot-transport problem has not been studied in the literature. Here we follow the approach from Sun et al. (2002) for miscible displacement in porous media and employ the non-symmetric interior penalty Galerkin (NIPG) method for the transport problem. We note that the dispersion tensor in the transport equation is a nonlinear function of the velocity. The work in Sun et al. (2002) handles this difficulty by utilizing a cut-off operator. Here we avoid the need for the cut-off operator by establishing an L^∞-bound for the computed Stokes–Biot velocity. As a result, the velocity is directly incorporated into the transport scheme. We present a stability bound and an error estimate for the solution of the transport equation. The analysis in this paper is presented for saturated flow and linear transport. Extensions to unsaturated flow in poroelastic media and nonlinear transport can also be studied, using for example techniques developed in Both et al. (2018) and Radu et al. (2010).

The rest of the paper is organized as follows. The coupled Stokes–Darcy-transport problem and its variational formulation are presented in Sect. 2. The semi-discrete continuous-in-time approximation is developed in Sect. 3 and analyzed in Sect. 4. Computational experiments confirming the convergence of the method and illustrating its performance for a range of applications of flow in fractured poroelastic media are presented in Sect. 5.

2 Model problem

We consider a simulation domain $\Omega \subset \mathbb{R}^d$, $d = 2, 3$ which is a union of non-overlapping and possibly non-connected regions Ω_f and Ω_p, where Ω_f is a fracture region and Ω_p is a poroelasticity region, see Fig. 1. We denote by $\Gamma_{fp} = \partial\Omega_f \cap \partial\Omega_p$ the interface between Ω_f and Ω_p. We further denote by $(\mathbf{u}_\star, p_\star)$ the velocity-pressure

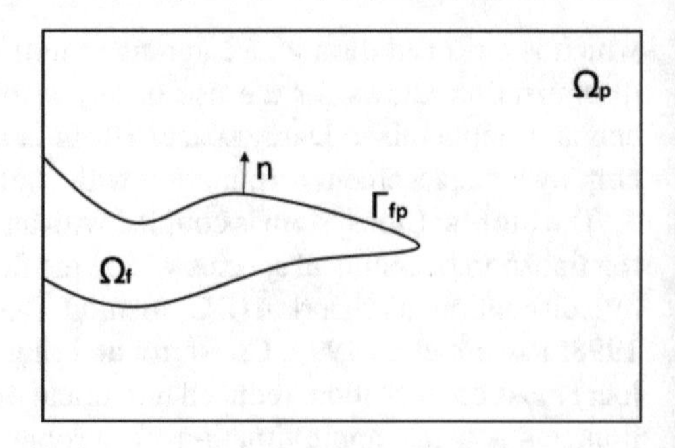

Fig. 1 Schematic representation of the physical domain

pairs in $\Omega_\star$, $\star = f, p$, by $\boldsymbol{\eta}_p$ the displacement in Ω_p and by $(\mathbf{f}_\star, q_\star)$ the body force and the external source or sink terms. The flow in the fracture region Ω_f is governed by the Stokes equations

$$-\nabla \cdot \boldsymbol{\sigma}_f(\mathbf{u}_f, p_f) = \mathbf{f}_f \qquad \text{in } \Omega_f \times (0, T], \tag{2.1}$$

$$\nabla \cdot \mathbf{u}_f = q_f \qquad \text{in } \Omega_f \times (0, T], \tag{2.2}$$

where the deformation and stress tensors, $\boldsymbol{\epsilon}(\mathbf{u})$ and $\boldsymbol{\sigma}_f(\mathbf{u}_f, p_f)$, are given by

$$\boldsymbol{\epsilon}(\mathbf{u}) = \frac{1}{2}\left(\nabla \mathbf{u}_f + \nabla \mathbf{u}_f^T\right), \quad \boldsymbol{\sigma}_f(\mathbf{u}_f, p_f) = -p_f \mathbf{I} + 2\nu \boldsymbol{\epsilon}(\mathbf{u}_f),$$

and ν denotes the fluid viscosity.

Let $\boldsymbol{\sigma}_e(\boldsymbol{\eta})$ and $\boldsymbol{\sigma}_p(\boldsymbol{\eta}, p)$ be the elasticity and poroelasticity stress tensors, respectively:

$$\boldsymbol{\sigma}_e(\boldsymbol{\eta}) = \lambda_p(\nabla \cdot \boldsymbol{\eta})\mathbf{I} + 2\mu_p \boldsymbol{\epsilon}(\boldsymbol{\eta}), \quad \boldsymbol{\sigma}_p(\boldsymbol{\eta}, p) = \boldsymbol{\sigma}_e(\boldsymbol{\eta}) - \alpha p_p \mathbf{I},$$

where α_p is the Biot–Willis constant and λ_p, μ_p are the Lamé coefficients.

The poroelasticity region Ω_p is governed by the quasi-static Biot system (Biot 1941)

$$-\nabla \cdot \boldsymbol{\sigma}_p(\boldsymbol{\eta}_p, p_p) = \mathbf{f}_p, \quad \nu K^{-1}\mathbf{u}_p + \nabla p_p = 0 \qquad \text{in } \Omega_p \times (0, T], \tag{2.3}$$

$$\frac{\partial}{\partial t}(s_0 p_p + \alpha \nabla \cdot \boldsymbol{\eta}_p) + \nabla \cdot \mathbf{u}_p = q_p \qquad \text{in } \Omega_p \times (0, T], \tag{2.4}$$

where s_0 is a storage coefficient and K is a symmetric and uniformly positive definite permeability tensor.

Following Ambartsumyan et al. (2018b), Badia et al. (2009), and Showalter (2005), on the fluid–poroelasticity interface Γ_{fp} we prescribe the following *interface conditions*: *mass conservation*, *balance of normal stress*, *conservation of momentum*, and the Beavers–Joseph–Saffman (BJS) condition modeling *slip with friction* (Beavers and Joseph 1967; Saffman 1971):

$$\mathbf{u}_f \cdot \mathbf{n}_f + \left(\frac{\partial \boldsymbol{\eta}_p}{\partial t} + \mathbf{u}_p\right) \cdot \mathbf{n}_p = 0 \qquad \text{on } \Gamma_{fp} \times (0, T], \tag{2.5}$$

$$-(\boldsymbol{\sigma}_f\mathbf{n}_f)\cdot\mathbf{n}_f = p_p, \qquad \boldsymbol{\sigma}_f\mathbf{n}_f + \boldsymbol{\sigma}_p\mathbf{n}_p = 0 \qquad \text{on } \Gamma_{fp}\times(0,T], \quad (2.6)$$

$$-(\boldsymbol{\sigma}_f\mathbf{n}_f)\cdot\boldsymbol{\tau}_{f,j} = \nu\alpha_{BJS}\sqrt{K_j^{-1}}\left(\mathbf{u}_f - \frac{\partial\boldsymbol{\eta}_p}{\partial t}\right)\cdot\boldsymbol{\tau}_{f,j} \quad \text{on } \Gamma_{fp}\times(0,T], \quad (2.7)$$

where $\mathbf{n}_f$ and $\mathbf{n}_p$ are the outward unit normal vectors to $\partial\Omega_f$ and $\partial\Omega_p$, respectively, $\boldsymbol{\tau}_{f,j}$, $1\le j\le d-1$, is an orthogonal system of unit tangent vectors on Γ_{fp}, $K_j = (K\boldsymbol{\tau}_{f,j})\cdot\boldsymbol{\tau}_{f,j}$ and $\alpha_{BJS} > 0$ is an experimentally determined friction coefficient.

The above system of equations is complemented by a set of boundary and initial conditions. Let $\Gamma_f = \partial\Omega_f\backslash\Gamma_{fp}$, $\Gamma_p = \partial\Omega_p\backslash\Gamma_{fp} = \Gamma_p^N\cup\Gamma_p^D$. For simplicity we assume homogeneous boundary conditions

$$\mathbf{u}_f = 0 \text{ on } \Gamma_f\times(0,T], \quad \mathbf{u}_p\cdot\mathbf{n}_p = 0 \text{ on } \Gamma_p^N\times(0,T], \quad p_p = 0 \text{ on } \Gamma_p^D\times(0,T],$$
$$\boldsymbol{\eta}_p = 0 \text{ on } \Gamma_p\times(0,T].$$

We further set the initial conditions

$$p_p(\mathbf{x},0) = p_{p,0}(\mathbf{x}), \quad \boldsymbol{\eta}_p(\mathbf{x},0) = \boldsymbol{\eta}_{p,0}(\mathbf{x}) \text{ in } \Omega_p.$$

Throughout the paper we will use the following standard notation. For a domain $G\subset\mathbb{R}^d$, the $L^2(G)$ inner product and norm for scalar and vector valued functions are denoted by $(\cdot,\cdot)_G$ and $\|\cdot\|_G$, respectively. The norms and seminorms of the Sobolev spaces $W^{k,p}(G)$, $k\in\mathbb{R}$, $p>0$ are denoted by $\|\cdot\|_{k,p,G}$ and $|\cdot|_{k,p,G}$, respectively. Conventionally, the norms and seminorms of Hilbert spaces $H^k(G)$ are denoted by $\|\cdot\|_{k,G}$ and $|\cdot|_{k,G}$, respectively. For a section of the domain or element boundary $S\subset\mathbb{R}^{d-1}$ we write $\langle\cdot,\cdot\rangle_S$ and $\|\cdot\|_S$ for the $L^2(S)$ inner product (or duality pairing) and norm, respectively. We will also use the space

$$H(\text{div};G) = \{\mathbf{v}\in L^2(G) : \nabla\cdot\mathbf{v}\in L^2(G)\}$$

equipped with the norm

$$\|\mathbf{v}\|_{\text{div},G} = \left(\|\mathbf{v}\|^2 + \|\nabla\cdot\mathbf{v}\|^2\right)^{1/2}.$$

For the weak formulation of the coupled Stokes–Biot equations we introduce the following function spaces:

$$\mathbf{V}_f = \{\mathbf{v}_f\in H^1(\Omega_f)^d : \mathbf{v}_f = 0 \text{ on } \Gamma_f\}, \qquad W_f = L^2(\Omega_f), \quad (2.8)$$
$$\mathbf{V}_p = \{\mathbf{v}_p\in H(\text{div};\Omega_p) : \mathbf{v}_p\cdot\mathbf{n}_p = 0 \text{ on } \Gamma_p^N\}, \qquad W_p = L^2(\Omega_p), \quad (2.9)$$
$$\mathbf{X}_p = \{\boldsymbol{\xi}_p\in H^1(\Omega_p)^d : \boldsymbol{\xi}_p = 0 \text{ on } \Gamma_p\}, \quad (2.10)$$

equipped with the norms

$$\|\mathbf{v}_f\|_{\mathbf{V}_f} = \|\mathbf{v}_f\|_{1,\Omega_f}, \qquad \|w_f\|_{W_f} = \|w_f\|_{\Omega_f},$$
$$\|\mathbf{v}_p\|_{\mathbf{V}_p} = \|\mathbf{v}_p\|_{\text{div},\Omega_p}, \qquad \|w_p\|_{W_p} = \|w_p\|_{\Omega_p},$$

$$\|\boldsymbol{\eta}_p\|_{\mathbf{X}_p} = \|\boldsymbol{\eta}_p\|_{1,\Omega_p}.$$

The weak formulation is obtained by multiplying the equations in each region by the corresponding test functions, integrating by parts the second order terms in space, and utilizing the interface and boundary conditions. The integration by parts in (2.1) and (2.3) leads to the bilinear forms, corresponding to the Stokes, Darcy and the elasticity operators:

$$\begin{aligned} a_f(\cdot,\cdot) &: \mathbf{V}_f \times \mathbf{V}_f \longrightarrow \mathbb{R}, \quad a_f(\mathbf{u}_f, \mathbf{v}_f) := (2\nu\boldsymbol{\epsilon}(\mathbf{u}_f), \boldsymbol{\epsilon}(\mathbf{v}_f))_{\Omega_f}, \\ a_p^d(\cdot,\cdot) &: \mathbf{V}_p \times \mathbf{V}_p \longrightarrow \mathbb{R}, \quad a_p^d(\mathbf{u}_p, \mathbf{v}_p) := (\nu K^{-1}\mathbf{u}_p, \mathbf{v}_p)_{\Omega_p}, \\ a_p^e(\cdot,\cdot) &: \mathbf{X}_p \times \mathbf{X}_p \longrightarrow \mathbb{R}, \quad a_p^e(\boldsymbol{\eta}_p, \boldsymbol{\xi}_p) := (2\mu_p\boldsymbol{\epsilon}(\boldsymbol{\eta}_p), \boldsymbol{\epsilon}(\boldsymbol{\xi}_p))_{\Omega_p} \\ &+ (\lambda_p \nabla\cdot\boldsymbol{\eta}_p, \nabla\cdot\boldsymbol{\xi}_p)_{\Omega_p}, \end{aligned}$$

the bilinear forms

$$b_\star(\cdot,\cdot) : \mathbf{V}_\star \times W_\star \longrightarrow \mathbb{R}, \quad b_\star(\mathbf{v}, w) := -(\nabla\cdot\mathbf{v}, w)_{\Omega_\star}, \quad \star = f, p,$$

and the interface term

$$I_{\Gamma_{fp}} = -\langle \boldsymbol{\sigma}_f \mathbf{n}_f, \mathbf{v}_f \rangle_{\Gamma_{fp}} - \langle \boldsymbol{\sigma}_p \mathbf{n}_p, \boldsymbol{\xi}_p \rangle_{\Gamma_{fp}} + \langle p_p, \mathbf{v}_p \cdot \mathbf{n}_p \rangle_{\Gamma_{fp}}.$$

To handle the interface term, we introduce a Lagrange multiplier λ with a meaning of Darcy pressure on the interface (Ambartsumyan et al. 2018b)

$$\lambda = -(\boldsymbol{\sigma}_f \mathbf{n}_f) \cdot \mathbf{n}_f = p_p \quad \text{on } \Gamma_{fp}.$$

Using (2.6)–(2.7), we obtain

$$I_{\Gamma_{fp}} = a_{BJS}(\mathbf{u}_f, \partial_t \boldsymbol{\eta}_p; \mathbf{v}_f, \boldsymbol{\xi}_p) + b_\Gamma(\mathbf{v}_f, \mathbf{v}_p, \boldsymbol{\xi}_p; \lambda),$$

where

$$\begin{aligned} a_{BJS}(\mathbf{u}_f, \boldsymbol{\eta}_p; \mathbf{v}_f, \boldsymbol{\xi}_p) &= \sum_{j=1}^{d-1} \langle \nu_I\, \alpha_{BJS} \sqrt{K^{-1}} (\mathbf{u}_f - \boldsymbol{\eta}_p) \cdot \boldsymbol{\tau}_{f,j}, (\mathbf{v}_f - \boldsymbol{\xi}_p) \cdot \boldsymbol{\tau}_{f,j} \rangle_{\Gamma_{fp}}, \\ b_\Gamma(\mathbf{v}_f, \mathbf{v}_p, \boldsymbol{\xi}_p; \mu) &= \langle \mathbf{v}_f \cdot \mathbf{n}_f + (\boldsymbol{\xi}_p + \mathbf{v}_p) \cdot \mathbf{n}_p, \mu \rangle_{\Gamma_{fp}}. \end{aligned}$$

We note that for the well-posedness of b_Γ term, we require $\lambda \in (\mathbf{V}_p \cdot \mathbf{n}_p|_{\Gamma_{fp}})'$. The normal trace theorem for $\mathbf{v}_p \in \mathbf{V}_p \subset H(\text{div}; \Omega_p)$ implies that $\mathbf{v}_p \cdot \mathbf{n}_p \in H^{-1/2}(\partial\Omega_p)$. With our choice of boundary conditions, one can verify that $\mathbf{v}_p \cdot \mathbf{n}_p \in H^{-1/2}(\Gamma_{fp})$, see Ambartsumyan et al. (2018b). Therefore, we take $\Lambda = H^{1/2}(\Gamma_{fp})$.

Stokes–Biot variational formulation: given $p_p(0) = p_{p,0} \in W_p$, $\boldsymbol{\eta}_p(0) = \boldsymbol{\eta}_{p,0} \in \mathbf{X}_p$, find, for $t \in (0, T]$, $\mathbf{u}_f(t) \in \mathbf{V}_f$, $p_f(t) \in W_f$, $\mathbf{u}_p(t) \in \mathbf{V}_p$, $p_p(t) \in W_p$,

$\eta_p(t) \in \mathbf{X}_p$, and $\lambda(t) \in \Lambda$ such that for all $\mathbf{v}_f \in \mathbf{V}_f$, $w_f \in W_f$, $\mathbf{v}_p \in \mathbf{V}_p$, $w_p \in W_p$, $\boldsymbol{\xi}_p \in \mathbf{X}_p$, and $\mu \in \Lambda$,

$$
\begin{aligned}
& a_f(\mathbf{u}_f, \mathbf{v}_f) + a_p^d(\mathbf{u}_p, \mathbf{v}_p) + a_p^e(\boldsymbol{\eta}_p, \boldsymbol{\xi}_p) + a_{BJS}(\mathbf{u}_f, \partial_t \boldsymbol{\eta}_p; \mathbf{v}_f, \boldsymbol{\xi}_p) \\
& \quad + b_f(\mathbf{v}_f, p_f) + b_p(\mathbf{v}_p, p_p) + \alpha_p b_p(\boldsymbol{\xi}_p, p_p) \\
& \quad + b_\Gamma(\mathbf{v}_f, \mathbf{v}_p, \boldsymbol{\xi}_p; \lambda) = (\mathbf{f}_f, \mathbf{v}_f)_{\Omega_f} + (\mathbf{f}_p, \boldsymbol{\xi}_p)_{\Omega_p}, \qquad (2.11) \\
& \left(s_0 \partial_t p_p, w_p\right)_{\Omega_p} - \alpha_p b_p\left(\partial_t \boldsymbol{\eta}_p, w_p\right) - b_p(\mathbf{u}_p, w_p) - b_f(\mathbf{u}_f, w_f) \\
& \quad = (q_f, w_f)_{\Omega_f} + (q_p, w_p)_{\Omega_p}, \qquad (2.12) \\
& b_\Gamma\left(\mathbf{u}_f, \mathbf{u}_p, \partial_t \boldsymbol{\eta}_p; \mu\right) = 0. \qquad (2.13)
\end{aligned}
$$

The well-posedness of the above problem has been established in Ambartsumyan et al. (2018a).

Theorem 2.1 *For each* $\mathbf{f}_f \in W^{1,1}(0, T; \mathbf{V}_f')$, $\mathbf{f}_p \in W^{1,1}(0, T; \mathbf{X}_p')$, $q_f \in W^{1,1}(0, T; W_f')$, $q_p \in W^{1,1}(0, T; W_p')$, *and* $p_p(0) = p_{p,0} \in W_p$, $\boldsymbol{\eta}_p(0) = \boldsymbol{\eta}_{p,0} \in \mathbf{X}_p$, *there exists a unique solution* $(\mathbf{u}_f(t), p_f(t), \mathbf{u}_p(t), p_p(t), \boldsymbol{\eta}_p(t), \lambda(t)) \in L^\infty(0, T; \mathbf{V}_f) \times L^\infty(0, T; W_f) \times L^\infty(0, T; \mathbf{V}_p) \times W^{1,\infty}(0, T; W_p) \times W^{1,\infty}(0, T; \mathbf{X}_p) \times L^\infty(0, T; \Lambda)$ *of* (2.11)–(2.13).

The Stokes–Biot problem is coupled with the transport equation in Ω:

$$
\phi c_t + \nabla \cdot (c\mathbf{u} - \mathbf{D}\nabla c) = q\tilde{c}, \quad \text{in } \Omega \times (0, T], \qquad (2.14)
$$

where $c(\mathbf{x}, t)$ is the concentration of some chemical component, $0 < \phi_* \leq \phi(\mathbf{x}) \leq \phi^*$ is the porosity of the medium in Ω_p (it is set to 1 in Ω_f), $\mathbf{u}$ is the velocity field over $\Omega = \Omega_f \cup \Omega_p$, defined as $\mathbf{u}|_{\Omega_f} = \mathbf{u}_f$, $\mathbf{u}|_{\Omega_p} = \mathbf{u}_p$, q is the source term given by $q\big|_{\Omega_f} = q_f$ and $q\big|_{\Omega_p} = q_p$, and

$$
\tilde{c} = \begin{cases} \text{injected concentration } c_w, & q > 0, \\ \text{resident concentration } c, & q < 0. \end{cases}
$$

The diffusion/dispersion tensor $\mathbf{D}$, which combines the effects of molecular diffusion and mechanical dispersion, is a nonlinear function of the velocity, given by (Peaceman 1977)

$$
\mathbf{D}(\mathbf{u}) = d_m \mathbf{I} + |\mathbf{u}|\{\alpha_l \mathbf{E}(\mathbf{u}) + \alpha_t(\mathbf{I} - \mathbf{E}(\mathbf{u}))\}, \qquad (2.15)
$$

where $d_m = \phi \tau D_m$, τ is the tortuosity coefficient, D_m is the molecular diffusivity, $\mathbf{E}(\mathbf{u})$ is the tensor that projects onto the $\mathbf{u}$ direction with $(\mathbf{E}(\mathbf{u}))_{ij} = \frac{u_i u_j}{|\mathbf{u}|^2}$, and α_l, α_t are the longitudinal and transverse dispersion, respectively.

The model is complemented by the initial condition

$$
c(\mathbf{x}, 0) = c_0(\mathbf{x}) \quad \text{in } \Omega, \qquad (2.16)
$$

and the boundary conditions

$$(c\mathbf{u} - \mathbf{D}\nabla c) \cdot \mathbf{n} = (c_{in}\mathbf{u}) \cdot \mathbf{n} \qquad \text{on } \Gamma_{in} \times (0, T], \tag{2.17}$$

$$(\mathbf{D}\nabla c) \cdot \mathbf{n} = 0 \qquad \text{on } \Gamma_{out} \times (0, T], \tag{2.18}$$

where $\Gamma_{in} := \{\mathbf{x} \in \partial\Omega : \mathbf{u} \cdot \mathbf{n} < 0\}$, $\Gamma_{out} := \{\mathbf{x} \in \partial\Omega : \mathbf{u} \cdot \mathbf{n} \geq 0\}$ and $\mathbf{n}$ is the unit outward normal vector to $\partial\Omega$.

Remark 2.1 We note that the coupling between the flow and transport problems is one way. In particular, the transport equation uses the Stokes–Biot velocity, but the flow problem does not depend on the concentration.

3 Semi-discrete continuous-in-time formulation

Let $\mathcal{T}_h^f$ and $\mathcal{T}_h^p$ be shape-regular and quasi-uniform affine element partitions of Ω_f and Ω_p (Ciarlet 2002), respectively, both consisting of elements with maximal element diameter h. The two partitions may be non-matching at the interface Γ_{fp}. We also consider a shape-regular and quasi-uniform affine element partition of Ω, denoted by $\mathcal{T}_h$. We note that $\mathcal{T}_h$ may be different from $\mathcal{T}_h^f$ and $\mathcal{T}_h^p$. We denote by E_h the set of all interior facets of $\mathcal{T}_h$ and on each facet we arbitrarily fix a unit normal vector $\mathbf{n}_e$. We further denote E_h^{out} and E_h^{in} the set of facets on Γ_{out} and Γ_{in}, for which $\mathbf{n}_e$ coincides with the outward unit normal vector.

For the discretization of the fluid velocity and pressure we choose finite element spaces $\mathbf{V}_{f,h} \subset \mathbf{V}_f$ and $W_{f,h} \subset W_f$, which are assumed to be inf-sup stable. Examples of such spaces include the MINI elements, the Taylor–Hood elements and the conforming Crouzeix–Raviart elements (Boffi et al. 2013). For the discretization of the porous medium problem we choose $\mathbf{V}_{p,h} \subset \mathbf{V}_p$ and $W_{p,h} \subset W_p$ to be any inf-sup stable mixed finite element spaces, such as the Raviart-Thomas or the Brezzi–Douglas–Marini spaces (Boffi et al. 2013). We employ a conforming Lagrange finite element space $\mathbf{X}_{p,h} \subset \mathbf{X}_p$ to approximate the structure displacement. For the discretization of the Lagrange multiplier variable we set

$$\Lambda_h = \mathbf{V}_{p,h} \cdot \mathbf{n}_p|_{\Gamma_{fp}},$$

which allows for optimal order approximation on non-matching grids (Layton et al. 2003). We note that this is a non-conforming choice, since $\Lambda_h \not\subset H^{1/2}(\Gamma_{fp})$. The space is equipped with the discrete $H^{1/2}$-norm analogue, $\|\mu_h\|_{\Lambda_h}^2 = \|\mu_h\|_{\Gamma_{fp}}^2 + |\mu_h|_{\Lambda_h}^2$ (Ambartsumyan et al. 2018b; Galvis and Sarkis 2007) with the semi-norm

$$|\mu_h|_{\Lambda_h}^2 = a_p^d(\mathbf{u}_{p,h}^*(\mu_h), \mathbf{u}_{p,h}^*(\mu_h)), \tag{3.1}$$

where $(\mathbf{u}_{p,h}^*(\mu_h), p_{p,h}^*(\mu_h)) \in \mathbf{V}_{p,h} \times W_{p,h}$ is the mixed finite element solution to the Darcy problem with Dirichlet data μ_h on Γ_{fp}:

$$a_p^d(\mathbf{u}_{p,h}^*(\mu_h), \mathbf{v}_{p,h}) + b_p(\mathbf{v}_{p,h}, p_h^*(\mu_h)) = -\langle \mathbf{v}_{p,h} \cdot \mathbf{n}_p, \mu_h \rangle_{\Gamma_{fp}}, \quad \forall \mathbf{v}_{p,h} \in \mathbf{V}_{p,h},$$

$$b_p(\mathbf{u}_{p,h}^*(\mu_h), w_{p,h}) = 0, \quad \forall\, w_{p,h} \in W_{p,h}.$$

We denote by $k_f \geq 1$ and $s_f \geq 1$ the degrees of polynomials in the spaces $\mathbf{V}_{f,h}$ and $W_{f,h}$ respectively. Let $k_p \geq 0$ and $s_p \geq 0$ be the degrees of polynomials in the spaces $\mathbf{V}_{p,h}$ and $W_{p,h}$ respectively. Finally, let $k_s \geq 1$ be the polynomial degree in $\mathbf{X}_{p,h}$.

Semi-discrete Stokes–Biot problem: given $p_{p,h}(0)$ and $\boldsymbol{\eta}_{p,h}(0)$, for $t \in (0, T]$, find $\mathbf{u}_{f,h}(t) \in \mathbf{V}_{f,h}$, $p_{f,h}(t) \in W_{f,h}$, $\mathbf{u}_{p,h}(t) \in \mathbf{V}_{p,h}$, $p_{p,h}(t) \in W_{p,h}$, $\boldsymbol{\eta}_{p,h}(t) \in \mathbf{X}_{p,h}$, and $\lambda_h(t) \in \Lambda_h$ such that for all $\mathbf{v}_{f,h} \in \mathbf{V}_{f,h}$, $w_{f,h} \in W_{f,h}$, $\mathbf{v}_{p,h} \in \mathbf{V}_{p,h}$, $w_{p,h} \in W_{p,h}$, $\boldsymbol{\xi}_{p,h} \in \mathbf{X}_{p,h}$, and $\mu_h \in \Lambda_h$,

$$\begin{aligned}
& a_f(\mathbf{u}_{f,h}, \mathbf{v}_{f,h}) + a_p^d(\mathbf{u}_{p,h}, \mathbf{v}_{p,h}) + a_p^e(\boldsymbol{\eta}_{p,h}, \boldsymbol{\xi}_{p,h}) + a_{BJS}(\mathbf{u}_{f,h}, \partial_t \boldsymbol{\eta}_{p,h}; \mathbf{v}_{f,h}, \boldsymbol{\xi}_{p,h}) \\
&\quad + b_f(\mathbf{v}_{f,h}, p_{f,h}) + b_p(\mathbf{v}_{p,h}, p_{p,h}) + \alpha b_p(\boldsymbol{\xi}_{p,h}, p_{p,h}) \\
&\quad + b_\Gamma(\mathbf{v}_{f,h}, \mathbf{v}_{p,h}, \boldsymbol{\xi}_{p,h}; \lambda_h) = (\mathbf{f}_f, \mathbf{v}_{f,h})_{\Omega_f} + (\mathbf{f}_p, \boldsymbol{\xi}_{p,h})_{\Omega_p}, \qquad (3.2) \\
& (s_0 \partial_t p_{p,h}, w_{p,h})_{\Omega_p} - \alpha b_p(\partial_t \boldsymbol{\eta}_{p,h}, w_{p,h}) - b_p(\mathbf{u}_{p,h}, w_{p,h}) - b_f(\mathbf{u}_{f,h}, w_{f,h}) \\
&\quad = (q_f, w_{f,h})_{\Omega_f} + (q_p, w_{p,h})_{\Omega_p}, \qquad (3.3) \\
& b_\Gamma(\mathbf{u}_{f,h}, \mathbf{u}_{p,h}, \partial_t \boldsymbol{\eta}_{p,h}; \mu_h) = 0. \qquad (3.4)
\end{aligned}$$

We take $p_{p,h}(0) = Q_{p,h} p_{p,0}$ and $\boldsymbol{\eta}_{p,h}(0) = I_{s,h} \boldsymbol{\eta}_{p,0}$, where the operators $Q_{p,h}$ and $I_{s,h}$ are defined in Sect. 4.

It was shown in Ambartsumyan et al. (2018b) that the above problem has a unique solution satisfying

$$\begin{aligned}
& \|\boldsymbol{\eta}_p - \boldsymbol{\eta}_{p,h}\|_{L^\infty(0,T;H^1(\Omega_p))} + \sqrt{s_0}\|p_p - p_{p,h}\|_{L^\infty(0,T;L^2(\Omega_p))} \\
&\quad + \|\mathbf{u}_f - \mathbf{u}_{f,h}\|_{L^2(0,T;H^1(\Omega_f))} + \|\mathbf{u}_p - \mathbf{u}_{p,h}\|_{L^2(0,T;L^2(\Omega_p))} \\
&\quad + \left|(\mathbf{u}_f - \partial_t \boldsymbol{\eta}_p) - (\mathbf{u}_{f,h} - \partial_t \boldsymbol{\eta}_{p,h})\right|_{L^2(0,T;a_{BJS})} \\
&\quad + \|p_f - p_{f,h}\|_{L^2(0,T;L^2(\Omega_f))} + \|p_p - p_{p,h}\|_{L^2(0,T;L^2(\Omega_p))} + \|\lambda - \lambda_h\|_{L^2(0,T;\Lambda_h)} \\
& \leq C\Bigg(h^{r_{k_f}} \|\mathbf{u}_f\|_{L^2(0,T;H^{r_{k_f}+1}(\Omega_f))} \\
&\quad + h^{r_{s_f}} \|p_f\|_{L^2(0,T;H^{r_{s_f}}(\Omega_f))} + h^{r_{k_p}} \|\mathbf{u}_p\|_{L^2(0,T;H^{r_{k_p}}(\Omega_p))} \\
&\quad + h^{\tilde{r}_{k_p}} \left(\|\lambda\|_{L^2(0,T;H^{\tilde{r}_{k_p}}(\Gamma_{fp}))} + \|\lambda\|_{L^\infty(0,T;H^{\tilde{r}_{k_p}}(\Gamma_{fp}))} + \|\partial_t \lambda\|_{L^2(0,T;H^{\tilde{r}_{k_p}}(\Gamma_{fp}))} \right) \\
&\quad + h^{r_{s_p}} \left(\|p_p\|_{L^\infty(0,T;H^{r_{s_p}}(\Omega_p))} + \|p_p\|_{L^2(0,T;H^{r_{s_p}}(\Omega_p))} + \|\partial_t p_p\|_{L^2(0,T;H^{r_{s_p}}(\Omega_p))} \right) \\
&\quad + h^{r_{k_s}} \Big(\|\boldsymbol{\eta}_p\|_{L^\infty(0,T;H^{r_{k_s}+1}(\Omega_p))} + \left\|\boldsymbol{\eta}_p\right\|_{L^2(0,T;H^{r_{k_s}+1}(\Omega_p))} \\
&\quad + \left\|\partial_t \boldsymbol{\eta}_p\right\|_{L^2(0,T;H^{r_{k_s}+1}(\Omega_p))} \Big) \Bigg), \\
& 0 \leq r_{k_f} \leq k_f, \; 0 \leq r_{s_f} \leq s_f + 1, \; 1 \leq \{r_{k_p}, \tilde{r}_{k_p}\} \leq k_p + 1, \\
& 0 \leq r_{s_p} \leq s_p + 1, \; 0 \leq r_{k_s} \leq k_s, \qquad (3.5)
\end{aligned}$$

where, for $\mathbf{v}_f \in \mathbf{V}_f, \boldsymbol{\xi}_p \in \mathbf{X}_p$,

$$|\mathbf{v}_f - \boldsymbol{\xi}_p|^2_{a_{BJS}} = a_{BJS}(\mathbf{v}_f, \boldsymbol{\xi}_p; \mathbf{v}_f, \boldsymbol{\xi}_p)$$
$$= \sum_{j=1}^{d-1} \mu \alpha_{BJS} \| K_j^{-1/4} (\mathbf{v}_f - \boldsymbol{\xi}_p) \cdot \boldsymbol{\tau}_{f,j} \|^2_{L^2(\Gamma_{fp})}.$$

We note that the result was derived under the assumption $|\Gamma_p^D| \neq 0$, but can be extended to the case of full Neumann boundary condition on Γ_p by restricting the mean value of the pressure over the entire domain to be zero.

Next, we derive the numerical method for the transport problem. Following Sun et al. (2002), we adopt the DG scheme known as the non-symmetric interior penalty Galerkin (NIPG) (Rivière et al. 1999).

For $s \geq 0$, we define the space

$$H^s(\mathcal{T}_h) = \{\phi \in L^2(\Omega) : \phi \in H^s(E), E \in \mathcal{T}_h\}.$$

The jump and average for $\phi \in H^s(\mathcal{T}_h)$, $s > 1/2$ are defined as follows. Let $E_i, E_j \in \mathcal{T}_h$ and $e = \partial E_i \cap \partial E_j \in E_h$, with $\mathbf{n}_e$ exterior to E_i. Let

$$[\phi] = (\phi|_{E_i})|_e - (\phi|_{E_j})|_e, \quad \{\phi\} = \frac{(\phi|_{E_i})|_e + (\phi|_{E_j})|_e}{2}.$$

For $\phi \in H^1(\mathcal{T}_h)$ we define the broken seminorm

$$|||\nabla \phi|||_\Omega = \left(\sum_{E \in \mathcal{T}_h} \|\nabla \phi\|^2_E \right)^{1/2}.$$

We consider the finite element space

$$\mathcal{D}_r(\mathcal{T}_h) = \{\phi \in L^2(\Omega) : \phi|_E \in \mathcal{P}_r(E), \ E \in \mathcal{T}_h\},$$

where $\mathcal{P}_r(E)$ denotes the space of polynomials of degree less than or equal to r on E.

Let the bilinear form $B_{\mathbf{u}_h}(c_h, \psi_h)$ and the linear functional $L_h(\psi_h)$ be defined as follows:

$$B_{\mathbf{u}_h}(c_h, \psi_h)$$
$$= \sum_{E \in \mathcal{T}_h} \int_E (\mathbf{D}(\mathbf{u}_h) \nabla c_h - c_h \mathbf{u}_h) \cdot \nabla \psi_h - \sum_{e \in E_h} \int_e \{\mathbf{D}(\mathbf{u}_h) \nabla c_h \cdot \mathbf{n}_e\} [\psi_h]$$
$$+ \sum_{e \in E_h} \int_e \{\mathbf{D}(\mathbf{u}_h) \nabla \psi_h \cdot \mathbf{n}_e\} [c_h] + \sum_{e \in E_h} \int_e c_h^* \mathbf{u}_h \cdot \mathbf{n}_e [\psi_h]$$

$$+ \sum_{e \in E_h^{out}} \int_e c_h \mathbf{u}_h \cdot \mathbf{n}_e \psi_h - \int_\Omega c_h q^- \psi_h + J_0^\sigma(c_h, \psi_h),$$

$$L_h(\psi_h) = \int_\Omega c_w q^+ \psi_h - \sum_{e \in E_h^{in}} \int_e c_{in} \mathbf{u}_h \cdot \mathbf{n}_e \psi_h. \tag{3.6}$$

Here $q^+ = \max(q, 0)$ is the injection part of the source term and $q^- = \min(q, 0)$ is the extraction part, $c_h^*|_e$ is the upwind value of concentration, defined as

$$c_h^*|_e = \begin{cases} c_h|_{E_1} & \text{if } \mathbf{u}_h \cdot \mathbf{n}_e > 0, \\ c_h|_{E_2} & \text{if } \mathbf{u}_h \cdot \mathbf{n}_e < 0, \end{cases} \tag{3.7}$$

and $J_0^\sigma(c_h, \psi_h)$ is the interior penalty term

$$J_0^\sigma(c_h, \psi_h) = \sum_{e \in E_h} \frac{\sigma_e}{h_e} \int_e [c_h][\psi_h], \tag{3.8}$$

where, σ is a discrete positive function that takes constant value σ_e on the edge and is bounded below by $\sigma_* > 0$ and above by σ^*, and h_e is the diameter of side of facet e.

Semi-discrete DG transport problem: find $c_h(t) \in \mathcal{D}_r(\mathcal{T}_h)$ such that $\forall \psi_h \in \mathcal{D}_r(\mathcal{T}_h)$,

$$(\phi \partial_t c_h, \psi_h) + B_{\mathbf{u}_h}(c_h, \psi_h) = L_h(\psi_h), \tag{3.9}$$

with initial condition $c_h(0)$ a suitable approximation of c_0.

It is easy to verify that, if the solution to (2.14) is sufficiently regular, it satisfies (3.9) with $B_{\mathbf{u}_h}$ replaced by $B_{\mathbf{u}}$.

4 Analysis of the semi-discrete problem

In this section we discuss the stability and error estimates for the transport problem (3.9). We note that a similar scheme has been used and analyzed in details in Sun et al. (2002). The main difference and improvement in this work is the fact that the numerically computed velocity field $\mathbf{u}_h$ is directly incorporated into the scheme for transport (3.9), while Sun et al. (2002) used a cut-off operator in order to ensure stability of their method. We avoid the need for a cut-off operator by establishing pointwise stability of the velocity solution in space and time, which is done in the next two lemmas. We first establish an error estimate for the fluid velocity in $L^\infty(0, T)$. The result requires control of $\mathbf{u}_{f,h}(0)$ and $\mathbf{u}_{p,h}(0)$. To simplify the analysis, we assume that the initial pressure $p_{p,0}$ and displacement $\boldsymbol{\eta}_{p,0}$ are constants.

Lemma 4.1 *Assume that $p_{p,0}$ and $\boldsymbol{\eta}_{p,0}$ are constants. If the solution of* (2.11)–(2.13) *is sufficiently regular, there exists a positive constant C independent of h such that*

$$
\begin{aligned}
&\|\mathbf{u}_f - \mathbf{u}_{f,h}\|_{L^\infty(0,T;H^1(\Omega_f))} + \|\mathbf{u}_p - \mathbf{u}_{p,h}\|_{L^\infty(0,T;L^2(\Omega_p))} \\
&\le C\Big[h^{r_{k_f}} \Big(\|\mathbf{u}_f\|_{L^2(0,T;H^{r_{k_f}+1}(\Omega_f))} + \|\mathbf{u}_f\|_{L^\infty(0,T;H^{r_{k_f}+1}(\Omega_f))} + \|\partial_t \mathbf{u}_f\|_{L^2(0,T;H^{r_{k_f}+1}(\Omega_f))} \Big) \\
&+ h^{r_{s_f}} \Big(\|p_f\|_{L^2(0,T;H^{r_{s_f}}(\Omega_f))} + \|p_f\|_{L^\infty(0,T;H^{r_{s_f}}(\Omega_f))} + \|\partial_t p_f\|_{L^2(0,T;H^{r_{s_f}}(\Omega_f))} \Big) \\
&+ h^{r_{k_p}} \Big(\|\mathbf{u}_p\|_{L^2(0,T;H^{r_{k_p}}(\Omega_p))} + \|\mathbf{u}_p\|_{L^\infty(0,T;H^{r_{k_p}}(\Omega_p))} + \|\partial_t \mathbf{u}_p\|_{L^2(0,T;H^{r_{k_p}}(\Omega_p))} \Big) \\
&+ h^{\tilde{r}_{k_p}} \Big(\|\lambda\|_{L^2(0,T;H^{\tilde{r}_{k_p}}(\Gamma_{fp}))} + \|\lambda\|_{L^\infty(0,T;H^{\tilde{r}_{k_p}}(\Gamma_{fp}))} + \|\partial_t \lambda\|_{L^2(0,T;H^{\tilde{r}_{k_p}}(\Gamma_{fp}))} \Big) \\
&+ h^{r_{s_p}} \Big(\|p_p\|_{L^\infty(0,T;H^{r_{s_p}}(\Omega_p))} + \|p_p\|_{L^2(0,T;H^{r_{s_p}}(\Omega_p))} + \|\partial_t p_p\|_{L^2(0,T;H^{r_{s_p}}(\Omega_p))} \Big) \\
&+ h^{r_{k_s}} \Big(\|\boldsymbol{\eta}_p\|_{L^\infty(0,T;H^{r_{k_s}+1}(\Omega_p))} + \|\boldsymbol{\eta}_p\|_{L^2(0,T;H^{r_{k_s}+1}(\Omega_p))} + \|\partial_t \boldsymbol{\eta}_p\|_{L^2(0,T;H^{r_{k_s}+1}(\Omega_p))} \\
&+ \|\partial_t \boldsymbol{\eta}_p\|_{L^\infty(0,T;H^{r_{k_s}+1}(\Omega_p))} + \|\partial_{tt} \boldsymbol{\eta}_p\|_{L^2(0,T;H^{r_{k_s}+1}(\Omega_p))} \Big) \Big]. \\
&0 \le r_{k_f} \le k_f,\ 0 \le r_{s_f} \le s_f + 1,\ 1 \le \{r_{k_p}, \tilde{r}_{k_p}\} \le k_p + 1, \\
&0 \le r_{s_p} \le s_p + 1,\ 0 \le r_{k_s} \le k_s. \qquad (4.1)
\end{aligned}
$$

Proof We introduce the errors for all variables and split them into approximation and discretization errors:

$$
\begin{aligned}
\mathbf{e}_f &:= \mathbf{u}_f - \mathbf{u}_{f,h} = (\mathbf{u}_f - I_{f,h}\mathbf{u}_f) + (I_{f,h}\mathbf{u}_f - \mathbf{u}_{f,h}) := \boldsymbol{\chi}_f + \boldsymbol{\phi}_{f,h}, \\
\mathbf{e}_p &:= \mathbf{u}_p - \mathbf{u}_{p,h} = (\mathbf{u}_p - I_{p,h}\mathbf{u}_p) + (I_{p,h}\mathbf{u}_p - \mathbf{u}_{p,h}) := \boldsymbol{\chi}_p + \boldsymbol{\phi}_{p,h}, \\
\mathbf{e}_s &:= \boldsymbol{\eta}_p - \boldsymbol{\eta}_{p,h} = (\boldsymbol{\eta}_p - I_{s,h}\boldsymbol{\eta}_p) + (I_{s,h}\boldsymbol{\eta}_p - \boldsymbol{\eta}_{p,h}) := \boldsymbol{\chi}_s + \boldsymbol{\phi}_{s,h}, \\
e_{fp} &:= p_f - p_{f,h} = (p_f - Q_{f,h}p_f) + (Q_{f,h}p_f - p_{f,h}) := \chi_{fp} + \phi_{fp,h}, \\
e_{pp} &:= p_p - p_{p,h} = (p_p - Q_{p,h}p_p) + (Q_{p,h}p_p - p_{p,h}) := \chi_{pp} + \phi_{pp,h}, \\
e_\lambda &:= \lambda - \lambda_h = (\lambda - Q_{\lambda,h}\lambda) + (Q_{\lambda,h}\lambda - \lambda_h) := \chi_\lambda + \phi_{\lambda,h}, \qquad (4.2)
\end{aligned}
$$

where the operator $I = \big(I_{f,h}, I_{p,h}, I_{s,h}\big)$ satisfies, see Ambartsumyan et al. (2018b) for details,

$$b_\Gamma\big(I_{f,h}\mathbf{v}_f, I_{p,h}\mathbf{v}_p, I_{s,h}\boldsymbol{\xi}_p; \mu_h\big) = 0, \qquad \forall \mu_h \in \Lambda_h, \qquad (4.3)$$

$$b_f(I_{f,h}\mathbf{v}_f - \mathbf{v}_f, w_{f,h}) = 0, \qquad \forall w_{f,h} \in W_{f,h}, \qquad (4.4)$$

$$b_p(I_{p,h}\mathbf{v}_p - \mathbf{v}_p, w_{p,h}) = 0, \qquad \forall w_{p,h} \in W_{p,h}, \qquad (4.5)$$

and $Q_{f,h}$, $Q_{p,h}$ and $Q_{\lambda,h}$ are the L^2–projection operators such that

$$(p_f - Q_{f,h}p_f, w_{f,h})_{\Omega_f} = 0, \qquad \forall\, w_{f,h} \in W_{f,h}, \qquad (4.6)$$

$$(p_p - Q_{p,h}p_p, w_{p,h})_{\Omega_p} = 0, \qquad \forall\, w_{p,h} \in W_{p,h}, \qquad (4.7)$$

$$\langle \lambda - Q_{\lambda,h}\lambda, \mu_h \rangle_{\Gamma_{fp}} = 0, \qquad \forall\, \mu_h \in \Lambda_h. \qquad (4.8)$$

The operators have the following approximation properties:

$$\|p_f - Q_{f,h}p_f\|_{L^2(\Omega_f)} \le Ch^{r_{s_f}}\|p_f\|_{H^{r_{s_f}}(\Omega_f)}, \qquad 0 \le r_{s_f} \le s_f + 1, \qquad (4.9)$$

$$\|p_p - Q_{p,h}p_p\|_{L^2(\Omega_p)} \le Ch^{r_{s_p}} \|p_p\|_{H^{r_{s_p}}(\Omega_p)}, \qquad 0 \le r_{s_p} \le s_p + 1, \tag{4.10}$$

$$\|\lambda - Q_{\lambda,h}\lambda\|_{L^2(\Gamma_{fp})} \le Ch^{\tilde{r}_{k_p}} \|\lambda\|_{H^{\tilde{r}_{k_p}}(\Gamma_{fp})}, \qquad 0 \le \tilde{r}_{k_p} \le k_p + 1, \tag{4.11}$$

$$\|\mathbf{v}_f - I_{f,h}\mathbf{v}_f\|_{H^1(\Omega_f)} \le Ch^{r_{k_f}} \|\mathbf{v}_f\|_{H^{r_{k_f}+1}(\Omega_f)}, \qquad 0 \le r_{k_f} \le k_f, \tag{4.12}$$

$$\|\boldsymbol{\xi}_p - I_h^s\boldsymbol{\xi}_p\|_{H^m(\Omega_p)} \le Ch^{r_{k_s}-m} \|\boldsymbol{\xi}_p\|_{H^{r_{k_s}}(\Omega_p)}, \quad m = 0, 1, \quad 1 \le r_{k_s} \le k_s + 1, \tag{4.13}$$

$$\begin{aligned} &\|\mathbf{v}_p - I_{p,h}\mathbf{v}_p\|_{L^2(\Omega_p)} \\ &\quad \le C\left(h^{r_{k_p}} \|\mathbf{v}_p\|_{H^{r_{k_p}}(\Omega_p)} + h^{r_{k_f}} \|\mathbf{v}_f\|_{H^{r_{k_f}+1}(\Omega_f)} + h^{r_{k_s}} \|\boldsymbol{\xi}_p\|_{H^{r_{k_s}+1}(\Omega_p)}\right), \\ &\quad 1 \le r_{k_p} \le k_p + 1, \; 0 \le r_{k_f} \le k_f, \; 0 \le r_{k_s} \le k_s. \end{aligned} \tag{4.14}$$

To obtain a velocity bound in $L^\infty(0, T)$, we differentiate (2.11) and (3.2) in time, and then subtract (3.2)–(3.3) from (2.11)–(2.12) to form the error equation

$$\begin{aligned} &a_f(\partial_t \mathbf{e}_f, \mathbf{v}_{f,h}) + a_p^d(\partial_t \mathbf{e}_p, \mathbf{v}_{p,h}) + a_p^e(\partial_t \mathbf{e}_s, \boldsymbol{\xi}_{p,h}) + a_{BJS}(\partial_t \mathbf{e}_f, \partial_{tt}\mathbf{e}_s; \mathbf{v}_{f,h}, \boldsymbol{\xi}_{p,h}) \\ &\quad + b_f(\mathbf{v}_{f,h}, \partial_t e_{fp}) + b_p(\mathbf{v}_{p,h}, \partial_t e_{pp}) + \alpha b_p(\boldsymbol{\xi}_{p,h}, \partial_t e_{pp}) \\ &\quad + b_\Gamma(\mathbf{v}_{f,h}, \mathbf{v}_{p,h}, \boldsymbol{\xi}_{p,h}; \partial_t e_\lambda) + \left(s_0\, \partial_t e_{pp}, w_{p,h}\right) \\ &\quad - \alpha b_p(\partial_t e_s, w_{p,h}) - b_p(\mathbf{e}_p, w_{p,h}) - b_f(\mathbf{e}_f, w_{f,h}) = 0. \end{aligned}$$

Setting $\mathbf{v}_{f,h} = \boldsymbol{\phi}_{f,h}, \mathbf{v}_{p,h} = \boldsymbol{\phi}_{p,h}, \boldsymbol{\xi}_{p,h} = \partial_t \boldsymbol{\phi}_{s,h}, w_{f,h} = \partial_t \phi_{fp,h}$, and $w_{p,h} = \partial_t \phi_{pp,h}$, we have

$$\begin{aligned} &a_f(\partial_t \boldsymbol{\chi}_f, \boldsymbol{\phi}_{f,h}) + a_f(\partial_t \boldsymbol{\phi}_{f,h}, \boldsymbol{\phi}_{f,h}) + a_p^d(\partial_t \boldsymbol{\chi}_p, \boldsymbol{\phi}_{p,h}) \\ &\quad + a_p^d(\partial_t \boldsymbol{\phi}_{p,h}, \boldsymbol{\phi}_{p,h}) + a_p^e\left(\partial_t \boldsymbol{\chi}_s, \partial_t \boldsymbol{\phi}_{s,h}\right) \\ &\quad + a_p^e\left(\partial_t \boldsymbol{\phi}_{s,h}, \partial_t \boldsymbol{\phi}_{s,h}\right) + a_{BJS}\left(\partial_t \boldsymbol{\chi}_f, \partial_{tt} \boldsymbol{\chi}_s; \boldsymbol{\phi}_{f,h}, \partial_t \boldsymbol{\phi}_{s,h}\right) \\ &\quad + a_{BJS}\left(\partial_t \boldsymbol{\phi}_{f,h}, \partial_{tt} \boldsymbol{\phi}_{s,h}; \boldsymbol{\phi}_{f,h}, \partial_t \boldsymbol{\phi}_{s,h}\right) \\ &\quad + b_f(\boldsymbol{\phi}_{f,h}, \partial_t \chi_{fp}) + b_f(\boldsymbol{\phi}_{f,h}, \partial_t \phi_{fp,h}) + b_p(\boldsymbol{\phi}_{p,h}, \partial_t \chi_{pp}) \\ &\quad + b_p(\boldsymbol{\phi}_{p,h}, \partial_t \phi_{pp,h}) + \alpha b_p\left(\partial_t \boldsymbol{\phi}_{s,h}, \partial_t \chi_{pp}\right) \\ &\quad + \alpha b_p\left(\partial_t \boldsymbol{\phi}_{s,h}, \partial_t \phi_{pp,h}\right) \\ &\quad + b_\Gamma\left(\boldsymbol{\phi}_{f,h}, \boldsymbol{\phi}_{p,h}, \partial_t \boldsymbol{\phi}_{s,h}; \partial_t \chi_\lambda\right) + b_\Gamma\left(\boldsymbol{\phi}_{f,h}, \boldsymbol{\phi}_{p,h}, \partial_t \boldsymbol{\phi}_{s,h}; \partial_t \phi_{\lambda,h}\right) \\ &\quad + \left(s_0\, \partial_t \chi_{pp}, \partial_t \phi_{pp,h}\right) + \left(s_0\, \partial_t \phi_{pp,h}, \partial_t \phi_{pp,h}\right) \\ &\quad - \alpha b_p\left(\partial_t \boldsymbol{\chi}_s, \partial_t \phi_{pp,h}\right) - \alpha b_p\left(\partial_t \boldsymbol{\phi}_{s,h}, \partial_t \phi_{pp,h}\right) \\ &\quad - b_p(\boldsymbol{\chi}_p, \partial_t \phi_{pp,h}) - b_p(\boldsymbol{\phi}_{p,h}, \partial_t \phi_{pp,h}) \\ &\quad - b_f(\boldsymbol{\chi}_f, \partial_t \phi_{fp,h}) - b_f(\boldsymbol{\phi}_{f,h}, \partial_t \phi_{fp,h}) = 0. \end{aligned} \tag{4.15}$$

The following terms simplify, due to the projection operators properties (4.7),(4.8), (4.4), and (4.5):

$$b_f(\boldsymbol{\chi}_f, \partial_t \phi_{fp,h}) = b_p(\boldsymbol{\chi}_p, \partial_t \phi_{pp,h}) = b_p(\boldsymbol{\phi}_{p,h}, \partial_t \chi_{pp}) = 0,$$

$$\left(s_0\,\partial_t \chi_{pp}, \partial_t \phi_{pp,h}\right) = \langle \boldsymbol{\phi}_{p,h} \cdot \mathbf{n}_p, \partial_t \chi_\lambda \rangle_{\Gamma_{fp}} = 0, \tag{4.16}$$

where we also used that $\Lambda_h = \mathbf{V}_{p,h} \cdot \mathbf{n}_p|_{\Gamma_{fp}}$ for the last equality. We also have

$$b_\Gamma\left(\boldsymbol{\phi}_{f,h}, \boldsymbol{\phi}_{p,h}, \partial_t \boldsymbol{\phi}_{s,h}; \partial_t \phi_{\lambda,h}\right) = 0, \quad b_\Gamma\left(\boldsymbol{\phi}_{f,h}, \boldsymbol{\phi}_{p,h}, \partial_t \boldsymbol{\phi}_{s,h}; \partial_t \chi_\lambda\right)$$
$$= \left\langle \boldsymbol{\phi}_{f,h} \cdot \mathbf{n}_f + \partial_t \boldsymbol{\phi}_{s,h} \cdot \mathbf{n}_p, \partial_t \chi_\lambda \right\rangle_{\Gamma_{fp}},$$

where we have used (4.3) and (3.4) for the first equality and the last equality in (4.16) for the second equality. Using these results, the error equation (4.15) becomes

$$\begin{aligned}
&\frac{1}{2}\partial_t \left(a_f(\boldsymbol{\phi}_{f,h}, \boldsymbol{\phi}_{f,h}) + a_p^d(\boldsymbol{\phi}_{p,h}, \boldsymbol{\phi}_{p,h}) + \left|\boldsymbol{\phi}_{f,h} - \partial_t \boldsymbol{\phi}_{s,h}\right|^2_{a_{BJS}}\right) \\
&\quad + a_p^e(\partial_t \boldsymbol{\phi}_{s,h}, \partial_t \boldsymbol{\phi}_{s,h}) + s_0 \|\partial_t \phi_{pp,h}\|^2_{L^2(\Omega_p)} \\
&= a_f(\partial_t \boldsymbol{\chi}_f, \boldsymbol{\phi}_{f,h}) + a_p^d(\partial_t \boldsymbol{\chi}_p, \boldsymbol{\phi}_{p,h}) + a_p^e\left(\partial_t \boldsymbol{\chi}_s, \partial_t \boldsymbol{\phi}_{s,h}\right) \\
&\quad + \sum_{j=1}^{d-1} \left\langle \nu \alpha_{BJS} \sqrt{K_j^{-1}} \partial_t (\boldsymbol{\chi}_f - \partial_t \boldsymbol{\chi}_s) \cdot \boldsymbol{\tau}_{f,j}, (\boldsymbol{\phi}_{f,h} - \partial_t \boldsymbol{\phi}_{s,h}) \cdot \boldsymbol{\tau}_{f,j} \right\rangle_{\Gamma_{fp}} \\
&\quad - b_f(\boldsymbol{\phi}_{f,h}, \partial_t \chi_{fp}) - \alpha b_p(\partial_t \boldsymbol{\phi}_{s,h}, \partial_t \chi_{pp}) + \alpha b_p(\partial_t \boldsymbol{\chi}_s, \partial_t \phi_{pp,h}) \\
&\quad - \langle \boldsymbol{\phi}_{f,h} \cdot \mathbf{n}_f + \partial_t \boldsymbol{\phi}_{s,h} \cdot \mathbf{n}_p, \partial_t \chi_\lambda \rangle_{\Gamma_{fp}} \\
&\le C\left(\|\boldsymbol{\phi}_{f,h}\|^2_{H^1(\Omega_f)} + \|\boldsymbol{\phi}_{p,h}\|^2_{L^2(\Omega_p)} + \left|\boldsymbol{\phi}_{f,h} - \partial_t \boldsymbol{\phi}_{s,h}\right|^2_{a_{BJS}}\right) + \epsilon \|\partial_t \boldsymbol{\phi}_{s,h}\|^2_{H^1(\Omega_p)} \\
&\quad + C\left(\|\partial_t \boldsymbol{\chi}_f\|^2_{H^1(\Omega_f)} + \|\partial_t \boldsymbol{\chi}_p\|^2_{L^2(\Omega_p)} + \|\partial_t \boldsymbol{\chi}_s\|^2_{H^1(\Omega_p)} + \|\partial_{tt} \boldsymbol{\chi}_s\|^2_{H^1(\Omega_p)}\right. \\
&\quad \left. + \alpha b_p(\partial_t \boldsymbol{\chi}_s, \partial_t \phi_{pp,h}) + \|\partial_t \chi_{fp}\|^2_{L^2(\Omega_f)} + \|\partial_t \chi_{pp}\|^2_{L^2(\Omega_p)} + \|\partial_t \chi_\lambda\|^2_{L^2(\Gamma_{fp})}\right),
\end{aligned} \tag{4.17}$$

where we used the Cauchy–Schwartz, Young's and trace inequalities. Using the coercivity of the bilinear forms $a_f(\cdot,\cdot)$, $a_p^d(\cdot,\cdot)$, and $a_p^e(\cdot,\cdot)$, choosing ϵ small enough, and integrating (4.17) in time from 0 to an arbitrary $t \in (0, T]$ gives

$$\begin{aligned}
&\|\boldsymbol{\phi}_{f,h}(t)\|^2_{H^1(\Omega_f)} + \|\boldsymbol{\phi}_{p,h}(t)\|^2_{L^2(\Omega_p)} + \left|\boldsymbol{\phi}_{f,h}(t) - \partial_t \boldsymbol{\phi}_{s,h}(t)\right|^2_{a_{BJS}} \\
&\quad + \int_0^t \left(\|\partial_t \boldsymbol{\phi}_{s,h}\|^2_{H^1(\Omega_p)} + s_0 \|\partial_t \phi_{pp,h}\|^2_{L^2(\Omega_p)}\right) ds \\
&\le \|\boldsymbol{\phi}_{f,h}(0)\|^2_{H^1(\Omega_f)} + \|\boldsymbol{\phi}_{p,h}(0)\|^2_{L^2(\Omega_p)} + \left|\boldsymbol{\phi}_{f,h}(0) - \partial_t \boldsymbol{\phi}_{s,h}(0)\right|^2_{a_{BJS}} \\
&\quad + C \int_0^t \left(\|\boldsymbol{\phi}_{f,h}\|^2_{H^1(\Omega_f)} + \|\boldsymbol{\phi}_{p,h}\|^2_{L^2(\Omega_p)} + \left|\boldsymbol{\phi}_{f,h} - \partial_t \boldsymbol{\phi}_{s,h}\right|^2_{a_{BJS}}\right. \\
&\quad + \|\partial_t \boldsymbol{\chi}_f\|^2_{H^1(\Omega_f)} + \|\partial_t \boldsymbol{\chi}_p\|^2_{L^2(\Omega_p)} + \|\partial_t \boldsymbol{\chi}_s\|^2_{H^1(\Omega_p)} + \|\partial_{tt} \boldsymbol{\chi}_s\|^2_{H^1(\Omega_p)} \\
&\quad \left. + \|\partial_t \chi_{fp}\|^2_{L^2(\Omega_f)} + \|\partial_t \chi_{pp}\|^2_{L^2(\Omega_p)} + \|\partial_t \chi_\lambda\|^2_{L^2(\Gamma_{fp})} + \alpha b_p(\partial_t \boldsymbol{\chi}_s, \partial_t \phi_{pp,h})\right) ds.
\end{aligned} \tag{4.18}$$

Using integration by parts for the last term, we get

$$\int_0^t \alpha b_p(\partial_t \boldsymbol{\chi}_s, \partial_t \phi_{pp,h})\, ds = \alpha b_p(\partial_t \boldsymbol{\chi}_s(t), \phi_{pp,h}(t)) - \alpha b_p(\partial_t \boldsymbol{\chi}_s(0), \phi_{pp,h}(0))$$
$$- \int_0^t \alpha b_p(\partial_{tt} \boldsymbol{\chi}_{s,h}, \phi_{pp,h})\, ds$$
$$\leq \epsilon \left(\|\phi_{pp,h}(t)\|^2_{L^2(\Omega_p)} + \int_0^t \|\phi_{pp,h}\|^2_{L^2(\Omega_p)} \right)$$
$$+ C \left(\|\partial_t \boldsymbol{\chi}_s(t)\|^2_{H^1(\Omega_p)} + \|\phi_{pp,h}(0)\|^2_{L^2(\Omega_p)} + \|\partial_t \boldsymbol{\chi}_s(0)\|^2_{H^1(\Omega_p)} \right.$$
$$\left. + \int_0^t \|\partial_{tt} \boldsymbol{\chi}_s\|^2_{H^1(\Omega_p)}\, ds \right). \tag{4.19}$$

Next, using an inf-sup condition for the Stokes–Darcy problem (Galvis and Sarkis 2007; Ambartsumyan et al. 2018b) and the error equation obtained by subtracting (3.2) from (2.11) and taking $\boldsymbol{\xi}_{p,h} = 0$, we obtain

$$\|(\phi_{fp,h}, \phi_{pp,h}, \phi_{\lambda,h})\|_{W_f \times W_p \times \Lambda_h}$$
$$\leq C \sup_{0 \neq \mathbf{v}_h \in \mathbf{V}_h} \frac{b_f(\mathbf{v}_{f,h}, \phi_{fp,h}) + b_p(\mathbf{v}_{p,h}, \phi_{pp,h}) + b_\Gamma(\mathbf{v}_{f,h}, \mathbf{v}_{p,h}, 0; \phi_{\lambda,h})}{\|\mathbf{v}_h\|_{\mathbf{V}}}$$
$$= C \sup_{0 \neq \mathbf{v}_h \in \mathbf{V}_h} \left(\frac{-a_f(\mathbf{e}_f, \mathbf{v}_{f,h}) - a_p^d(\mathbf{e}_p, \mathbf{v}_{p,h}) - a_{BJS}(\mathbf{e}_f, \partial_t \mathbf{e}_s; \mathbf{v}_{f,h}, 0)}{\|\mathbf{v}_h\|_{\mathbf{V}}} \right.$$
$$\left. + \frac{-b_f(\mathbf{v}_{f,h}, \chi_{fp}) - b_p(\mathbf{v}_{p,h}, \chi_{pp}) - b_\Gamma(\mathbf{v}_{f,h}, \mathbf{v}_{p,h}, 0; \chi_\lambda)}{\|\mathbf{v}_h\|_{\mathbf{V}}} \right).$$

We have $b_p(\mathbf{v}_{p,h}, \chi_{pp}) = 0$ and $\langle \mathbf{v}_{p,h} \cdot \mathbf{n}_p, \chi_\lambda \rangle_{\Gamma_{fp}} = 0$. Then, using the continuity of the bilinear forms and the trace inequality, we get

$$\epsilon(\|\phi_{fp,h}\|^2_{L^2(\Omega_f)} + \|\phi_{pp,h}\|^2_{L^2(\Omega_p)} + \|\phi_{\lambda,h}\|^2_{L^2(\Gamma_{fp})})$$
$$\leq C\epsilon \left(\|\boldsymbol{\phi}_{f,h}\|^2_{H^1(\Omega_f)} + \|\boldsymbol{\phi}_{p,h}\|^2_{L^2(\Omega_p)} + \|\boldsymbol{\phi}_{s,h}\|^2_{H^1(\Omega_p)} + \left|\boldsymbol{\phi}_{f,h} - \partial_t \boldsymbol{\phi}_{s,h}\right|^2_{a_{BJS}} \right.$$
$$+ \|\boldsymbol{\chi}_f\|^2_{H^1(\Omega_f)} + \|\boldsymbol{\chi}_p\|^2_{L^2(\Omega_p)} + \left\|\boldsymbol{\chi}_s\right\|^2_{H^1(\Omega_p)} + \left\|\partial_t \boldsymbol{\chi}_s\right\|^2_{H^1(\Omega_p)}$$
$$\left. + \|\chi_{fp}\|^2_{L^2(\Omega_f)} + \|\chi_{pp}\|^2_{L^2(\Omega_p)} + \|\chi_\lambda\|_{L^2(\Gamma_{fp})} \right). \tag{4.20}$$

Finally, to control the error at $t = 0$, we note that the assumed solution regularity on the right hand side of (4.1) implies that (2.11)–(2.13) and (3.2)–(3.4) hold at $t = 0$. We subtract (3.2)–(3.3) from (2.11)–(2.12) at $t = 0$, sum the two equations, and take $\mathbf{v}_{f,h} = \boldsymbol{\phi}_{f,h}$, $\mathbf{v}_{p,h} = \boldsymbol{\phi}_{p,h}$, $\boldsymbol{\xi}_{p,h} = \partial_t \boldsymbol{\phi}_{s,h}$, $w_{f,h} = \phi_{fp,h}$, and $w_{p,h} = \phi_{pp,h}$, to obtain

$$a_f(\boldsymbol{\phi}_{f,h}(0), \boldsymbol{\phi}_{f,h}(0)) + a_p^d(\boldsymbol{\phi}_{p,h}(0), \boldsymbol{\phi}_{p,h}(0)) + \left|\boldsymbol{\phi}_{f,h}(0) - \partial_t \boldsymbol{\phi}_{s,h}(0)\right|^2_{a_{BJS}}$$
$$= -a_p^e(\boldsymbol{\phi}_{s,h}(0), \partial_t \boldsymbol{\phi}_{s,h}(0)) - s_0(\partial_t \phi_{pp,h}(0), \phi_{pp,h}(0))_{\Omega_p}$$

$$+ a_f(\boldsymbol{\chi}_f(0), \boldsymbol{\phi}_{f,h}(0)) + a_p^d(\boldsymbol{\chi}_p(0), \boldsymbol{\phi}_{p,h}(0)) + a_p^e\left(\boldsymbol{\chi}_s(0), \partial_t \boldsymbol{\phi}_{s,h}(0)\right)$$
$$+ \sum_{j=1}^{d-1} \left\langle \mu \alpha_{BJS} \sqrt{K_j^{-1}} (\boldsymbol{\chi}_f(0) - \partial_t \boldsymbol{\chi}_s(0)) \cdot \boldsymbol{\tau}_{f,j}, (\boldsymbol{\phi}_{f,h}(0) - \partial_t \boldsymbol{\phi}_{s,h}(0)) \cdot \boldsymbol{\tau}_{f,j} \right\rangle_{\Gamma_{fp}}$$
$$- b_f(\boldsymbol{\phi}_{f,h}(0), \chi_{fp}(0)) + \alpha b_p(\partial_t \boldsymbol{\phi}_{s,h}(0), \chi_{pp}(0)) + \alpha b_p(\partial_t \boldsymbol{\chi}_s(0), \phi_{pp,h}(0))$$
$$+ \langle \boldsymbol{\phi}_{f,h}(0) \cdot \mathbf{n}_f + \partial_t \boldsymbol{\phi}_{s,h}(0) \cdot \mathbf{n}_p, \chi_\lambda(0) \rangle_{\Gamma_{fp}}.$$

Since $p_{p,h}(0) = Q_{p,h} p_{p,0}$ and $\boldsymbol{\eta}_{p,h}(0) = I_{s,h} \boldsymbol{\eta}_{p,0}$, we have that $\phi_{pp,h}(0) = 0$ and $\boldsymbol{\phi}_{s,h}(0) = 0$. Since $p_{p,0}$ and $\boldsymbol{\eta}_{p,0}$ are constants, we also have that $\boldsymbol{\chi}_s = 0$, $\chi_{pp} = 0$, and $\chi_\lambda = 0$. It is then easy to see that

$$\|\boldsymbol{\phi}_{f,h}(0)\|_{H^1(\Omega_f)}^2 + \|\boldsymbol{\phi}_{p,h}(0)\|_{L^2(\Omega_p)}^2 + \left|\boldsymbol{\phi}_{f,h}(0) - \partial_t \boldsymbol{\phi}_{s,h}(0)\right|_{a_{BJS}}^2$$
$$\leq C(\|\boldsymbol{\chi}_f\|_{H^1(\Omega_f)}^2 + \|\boldsymbol{\chi}_p\|_{L^2(\Omega_p)}^2 + \|\chi_{fp}\|_{L^2(\Omega_f)}^2). \tag{4.21}$$

The assertion of the lemma follows from combining (4.18)–(4.21) and using Gronwall's inequality, the triangle inequality, and the approximation properties (4.9)–(4.14). □

Lemma 4.2 *Under the assumptions of Lemma* 4.1, *for any choice of stable spaces when* $d = 2$, *and for* $k_f \geq 2$, $k_p \geq 1$, $s_p \geq 1$, *and* $k_s \geq 2$ *when* $d = 3$, *there exists a positive constant* $M = M(\mathbf{u}_f, p_f, \mathbf{u}_p, p_p, \boldsymbol{\eta}_p, \lambda)$, *such that, for* $t \in (0, T]$, *the solution* $\mathbf{u}_h$ *of* (3.2)–(3.4) *satisfies*

$$\|\mathbf{u}_h\|_{L^\infty(\Omega)} \leq M. \tag{4.22}$$

Proof We recall that by definition

$$\mathbf{u}_h = \begin{cases} \mathbf{u}_{f,h} & \text{in } \Omega_f, \\ \mathbf{u}_{p,h} & \text{in } \Omega_p. \end{cases}$$

Therefore, we prove (4.22) separately for $\mathbf{u}_{f,h}$ in the fluid domain and for $\mathbf{u}_{p,h}$ in the poroelastic domain. Let $S_{f,h}$ be the Scott-Zhang interpolant onto $\mathbf{V}_{f,h}$ (Scott and Zhang 1990), satisfying

$$\|S_{f,h}\mathbf{v}_f\|_{\infty,\Omega_f} \leq C(\|\mathbf{v}_f\|_{\infty,\Omega_f} + h\|\nabla \mathbf{v}_f\|_{\infty,\Omega_f}), \qquad \forall \mathbf{v}_f \in W^{1,\infty}(\Omega_f), \tag{4.23}$$

$$\|\mathbf{v}_f - S_{f,h}\mathbf{v}_f\|_{\Omega_f} \leq C h^{r_{k_f}} \|\mathbf{v}_f\|_{r_{k_f},\Omega_f}, \quad 1 \leq r_{k_f} \leq k_f + 1, \quad \forall \mathbf{v}_f \in H^{r_{k_f}}(\Omega_f). \tag{4.24}$$

By the triangle inequality,

$$\|\mathbf{u}_{f,h}\|_{L^\infty(\Omega_f)} \leq \|\mathbf{u}_{f,h} - S_{f,h}\mathbf{u}_f\|_{L^\infty(\Omega_f)} + \|S_{f,h}\mathbf{u}_f\|_{L^\infty(\Omega_f)}. \tag{4.25}$$

To obtain a bound on $\|\mathbf{u}_{f,h} - S_{f,h}\mathbf{u}_f\|_{L^\infty(\Omega_f)}$, we use a scaling argument via mapping to the reference element $\hat{E}$. Recall that, due to shape regularity, the determinant of the Jacobian matrix satisfies $\|J_E\|_{\infty,\hat{E}} \sim h^d$. Therefore, for any $E \in \mathcal{T}_h^f \cup \mathcal{T}_h^p$ and any polynomial $\mathbf{v}_h$ on E, using norm equivalence on $\hat{E}$, we have

$$\|\mathbf{v}_h\|_{\infty,E} \leq \|\hat{\mathbf{v}}_h\|_{\infty,\hat{E}} \leq \|\hat{\mathbf{v}}_h\|_{\hat{E}} \leq Ch^{-d/2}\|\mathbf{v}_h\|_E.$$

Using the above bound, we have

$$\begin{aligned}\|\mathbf{u}_{f,h} - S_{f,h}\mathbf{u}_f\|_{L^\infty(\Omega_f)} &\leq Ch^{-d/2}\|\mathbf{u}_{f,h} - S_{f,h}\mathbf{u}_f\|_{L^2(\Omega_f)} \\ &\leq Ch^{-d/2}(\|\mathbf{u}_f - \mathbf{u}_{f,h}\|_{L^2(\Omega_f)} + \|\mathbf{u}_f - S_{f,h}\mathbf{u}_f\|_{L^2(\Omega_f)}),\end{aligned}$$

which, combined with (4.25), implies

$$\begin{aligned}\|\mathbf{u}_{f,h}\|_{L^\infty(\Omega_f)} \leq\, & Ch^{-d/2}(\|\mathbf{u}_f - \mathbf{u}_{f,h}\|_{L^2(\Omega_f)} + \|\mathbf{u}_f - S_{f,h}\mathbf{u}_f\|_{L^2(\Omega_f)}) \\ &+\|S_{f,h}\mathbf{u}_f\|_{L^\infty(\Omega_f)}.\end{aligned} \tag{4.26}$$

Next, we consider the MFE interpolant $\Pi_{p,h}$ onto $\mathbf{V}_{p,h}$ that satisfies (Acosta et al. 2011)

$$\|\Pi_{p,h}\mathbf{v}_p\|_{\infty,\Omega_p} \leq C\left(\|\mathbf{v}_p\|_{\infty,\Omega_p} + h\|\nabla\mathbf{v}_p\|_{\infty,\Omega_p}\right), \qquad \forall \mathbf{v}_p \in W^{1,\infty}(\Omega_p), \tag{4.27}$$

$$\|\mathbf{v}_p - \Pi_{p,h}\mathbf{v}_p\|_{\Omega_p} \leq Ch^{r_{k_p}}\|\mathbf{v}_p\|_{r_{k_p},\Omega_p}, \quad 1 \leq r_{k_p} \leq k_p + 1, \quad \forall \mathbf{v}_p \in H^{r_{k_p}}(\Omega_p). \tag{4.28}$$

Similarly to (4.26) we obtain

$$\begin{aligned}\|\mathbf{u}_{p,h}\|_{L^\infty(\Omega_p)} \leq\, & Ch^{-d/2}(\|\mathbf{u}_p - \mathbf{u}_{p,h}\|_{L^2(\Omega_p)} + \|\mathbf{u}_p - \Pi_{p,h}\mathbf{u}_p\|_{L^2(\Omega_f)}) \\ &+\|\Pi_{p,h}\mathbf{u}_p\|_{L^\infty(\Omega_p)}.\end{aligned} \tag{4.29}$$

The proof is completed by combining (4.26), (4.29), (4.23)–(4.24), (4.27)–(4.28), and (4.1). □

Remark 4.1 We note that the above result assumes sufficient regularity of the solution, as indicated by (4.23)–(4.24), (4.27)–(4.28), and (4.1).

We will utilize the following positive definite property of the dispersion tensor, proved in Sun et al. (2002).

Lemma 4.3 *Assume that for* $\mathbf{D}(\mathbf{u})$ *defined in* (2.15), $d_m(\mathbf{x}) \geq d_{m,*} > 0$, $\alpha_l(\mathbf{x}) \geq 0$ *and* $\alpha_t(\mathbf{x}) \geq 0$ *uniformly in* Ω. *Then* $\mathbf{D}(\mathbf{u})$ *is uniformly positive definite and for all* $\mathbf{x} \in \Omega$,

$$\mathbf{D}(\mathbf{u})\nabla c \cdot \nabla c \geq d_{m,*}|\nabla c|^2. \tag{4.30}$$

We next prove a Gårding's inequality for the bilinear form $B_{\mathbf{u}_h}(\cdot,\cdot)$. To simplify the analysis we assume velocity boundary condition for the Darcy problem.

Lemma 4.4 *Under the assumptions of Lemma 4.2, and if $\Gamma_p^D = \emptyset$, then the bilinear form $B_{\mathbf{u}_h}(\cdot,\cdot)$ defined in (3.6) satisfies, $\forall t \in (0,T]$,*

$$B_{\mathbf{u}_h}(\psi_h,\psi_h) \geq C\left(|||\nabla\psi_h|||_\Omega^2 + J^\sigma(\psi_h,\psi_h) - \|\psi_h\|_\Omega^2\right), \quad \forall \psi_h \in \mathcal{D}_f(\mathcal{T}_h). \tag{4.31}$$

Proof For any $\psi_h \in \mathcal{D}_f(\mathcal{T}_h)$ we have

$$\begin{aligned} B_{\mathbf{u}_h}(\psi_h,\psi_h) &= \sum_{E\in\mathcal{T}_h}\int_E (\mathbf{D}(\mathbf{u}_h)\nabla\psi_h - \psi_h\mathbf{u}_h)\cdot\nabla\psi_h - \sum_{e\in E_h}\int_e \{\mathbf{D}(\mathbf{u}_h)\nabla\psi_h\cdot\mathbf{n}_e\}[\psi_h] \\ &+ \sum_{e\in E_h}\int_e \{\mathbf{D}(\mathbf{u}_h)\nabla\psi_h\cdot\mathbf{n}_e\}[\psi_h] + \sum_{e\in E_h}\int_e \psi_h^*\mathbf{u}_h\cdot\mathbf{n}_e[\psi_h] \\ &+ \int_{\Gamma_{out}} \psi_h\mathbf{u}_h\cdot\mathbf{n}\,\psi_h - \int_\Omega \psi_h q^-\psi_h + J_0^\sigma(\psi_h,\psi_h). \end{aligned} \tag{4.32}$$

Next we introduce the notation

$$J_1 := \sum_{E\in\mathcal{T}_h}\int_E (\mathbf{D}(\mathbf{u}_h)\nabla\psi_h - \psi_h\mathbf{u}_h)\cdot\nabla\psi_h, \quad J_2 := \sum_{e\in E_h}\int_e \psi_h^*[\psi_h]\mathbf{u}_h\cdot\mathbf{n}_e,$$

$$J_3 := \int_{\Gamma_{out}} \psi_h^2\mathbf{u}_h\cdot\mathbf{n} - \int_\Omega q^-\psi_h^2 + J_0^\sigma(\psi_h,\psi_h),$$

and rewrite (4.32) as

$$B_{\mathbf{u}_h}(\psi_h,\psi_h) = J_1 + J_2 + J_3. \tag{4.33}$$

Using (4.22) and (4.30), we bound J_1 as

$$\begin{aligned} J_1 &= \sum_{E\in\mathcal{T}_h}\int_E \mathbf{D}(\mathbf{u}_h)\nabla\psi_h\cdot\nabla\psi_h - \sum_{E\in\mathcal{T}_h}\int_E \psi_h\mathbf{u}_h\cdot\nabla\psi_h \geq d_{m,*}|||\nabla\psi_h|||_\Omega^2 \\ &- M\sum_{E\in\mathcal{T}_h}\|\psi_h\|_E\|\nabla\psi_h\|_E \\ &\geq d_{m,*}|||\nabla\psi_h|||_\Omega^2 - C\epsilon^{-1}\|\psi_h\|_\Omega^2 - \epsilon|||\nabla\psi_h|||_\Omega^2. \end{aligned} \tag{4.34}$$

For J_2 we have

$$\begin{aligned} J_2 &= \sum_{e\in E_h}\int_e \psi_h^*[\psi_h]\mathbf{u}_h\cdot\mathbf{n}_e \geq -M\left|\sum_{e\in E_h}\int_e \psi_h^*[\psi_h]\right| \geq -M\sum_{e\in E_h}\|\psi_h^*\|_e\|[\psi_h]\|_e \\ &\geq -\sum_{e\in E_h}\left(\frac{\epsilon\sigma_e}{h_e}\|[\psi_h]\|_e^2 + \frac{Ch_e}{\epsilon}\|\psi_h^*\|_e^2\right) \geq -\epsilon J^\sigma(\psi_h,\psi_h) - \frac{Ch}{\epsilon}\sum_{E\in\mathcal{T}_h} h^{-1}\|\psi_h\|_E^2 \\ &\geq -\epsilon J^\sigma(\psi_h,\psi_h) - C\epsilon^{-1}\|\psi_h\|_\Omega^2. \end{aligned} \tag{4.35}$$

We bound J_3, using that $\mathbf{u}_h \cdot \mathbf{n} = \mathbf{u} \cdot \mathbf{n} \geq 0$ on Γ_{out} and that $q^- \leq 0$,

$$J_3 = \int_{\Gamma_{out}} \psi_h^2 \mathbf{u} \cdot \mathbf{n} - \int_\Omega q^- \psi_h^2 + J_0^\sigma(\psi_h, \psi_h) \geq J_0^\sigma(\psi_h, \psi_h). \tag{4.36}$$

Combining (4.33)–(4.36) we obtain

$$B_{\mathbf{u}_h}(\psi_h, \psi_h) \geq (d_{m,*} - \epsilon)|||\nabla \psi_h|||_\Omega^2 - C\epsilon^{-1}\|\psi_h\|_\Omega^2 + (1-\epsilon)J^\sigma(\psi_h, \psi_h).$$

Choosing ϵ small enough completes the proof. □

We are now ready to prove a stability bound for the solution of (3.9).

Theorem 4.1 *Under the assumptions of Lemma 4.4, there exists a positive constant C independent of h such that the solution $c_h(t)$ of (3.9) satisfies, $\forall t \in (0, T]$,*

$$\|c_h(t)\|_\Omega^2 + \int_0^t |||\nabla c_h(s)|||_\Omega^2 \, ds \leq C \left(\|c_h(0)\|_\Omega^2 + \int_0^t \left(\|c_w(s) q^+(s)\|_\Omega^2 \right.\right.$$
$$\left.\left. + \|c_{in}(s)\|_{\Gamma_{in}}^2 \right) ds \right).$$

Proof With the choice $\psi_h = c_h$, (3.9) reads

$$\int_\Omega \phi c_h \partial_t c_h + B_{\mathbf{u}_h}(c_h, c_h) = L_h(c_h).$$

Using (4.31), the definition (3.6) of L_h, and that $\mathbf{u}_h \cdot \mathbf{n} = \mathbf{u} \cdot \mathbf{n}$, we obtain

$$\phi \frac{1}{2} \partial_t \|c_h\|_\Omega^2 + C \left(|||\nabla c_h|||_\Omega^2 + J^\sigma(c_h, c_h) \right)$$
$$\leq C\|c_h\|_\Omega^2 + \int_\Omega c_w q^+ c_h - \int_{\Gamma_{in}} c_{in} \mathbf{u} \cdot \mathbf{n} \, c_h.$$

Integrating in time from $s = 0$ to $s = t$ for $0 < t \leq T$ and using Cauchy–Schwarz and Young's inequalities, we obtain

$$\|c_h(t)\|_\Omega^2 + \int_0^t \left(|||\nabla c_h(s)|||_\Omega^2 + J^\sigma(c_h, c_h) \right) ds$$
$$\leq C \left(\|c_h(0)\|_\Omega^2 + \int_0^t \left(\|c_h(s)\|_\Omega^2 + \|c_w(s) q^+(s)\|_\Omega^2 \right.\right.$$
$$\left.\left. + \epsilon^{-1} \|c_{in}(s)\|_{\Gamma_{in}}^2 + \epsilon \|c_h(s)\|_{\Gamma_{in}}^2 \right) ds \right). \tag{4.37}$$

The last term above is bounded as

$$\|c_h\|_{\Gamma_{in}}^2 \leq C \left(\|c_h\|_\Omega^2 + |||\nabla c_h|||_\Omega^2 + J^\sigma(c_h, c_h) \right), \tag{4.38}$$

which can be shown following the argument presented in Brenner (2003). The theorem follows from (4.37)–(4.38), taking ϵ small enough, and using Gronwall's inequality. □

In the next theorem we state the error estimate for the transport problem (3.9). Derivation of the bound follows the steps in the proof of Theorem 4.1 in Sun et al. (2002), using the estimate (4.22), rather than a boundedness property of the "cut-off" operator. For the sake of space, we omit the proof and the reader is referred to Sun et al. (2002) for the details.

Theorem 4.2 *Under the assumptions of Lemma 4.4, and assuming further that the solution of* (2.14)–(2.18) *satisfies* $c \in L^\infty(0, T; W^{1,\infty}(\Omega)) \cap L^2(0, T; H^{r+1}(\Omega))$, *there exists a positive constant C independent of h such that,* $\forall t \in (0, T]$,

$$\|c(t) - c_h(t)\|_\Omega + \left(\int_0^t \|\!|\nabla(c(s) - c_h(s))|\!\|_\Omega^2 \, ds \right)^{1/2} \leq C h^{\min\{k_f, s_f+1, k_p+1, s_p+1, k_s, r\}}.$$

5 Numerical results

In this section, we present results from several computational experiments in two dimensions. The computations are performed using a fully discrete scheme with Backward Euler time discretization. The method is implemented using the finite element package FreeFem++ (Hecht 2012). We use a monolithic solver for the Stokes–Biot system at each time step. It is possible to design a non-overlapping domain decomposition algorithm, similar to the Stokes–Darcy problem, see e.g. Vassilev et al. (2014). One can also use various splitting schemes for the Biot system (Mikelić and Wheeler 2013; Kim et al. 2011b, a). We first present a numerical test that confirms the theoretical convergence rates for the Biot-Stokes-transport problem using an analytical solution, followed by five examples with simulations of fluid flow in a fractured reservoir with physically realistic parameters.

5.1 Convergence test

In this test we study the convergence of the spatial discretization using an analytical solution. The domain is a rectangle $\Omega = [0, 1] \times [-1, 1]$. We associate the upper half with the Stokes flow, while the lower half represents the flow in the poroelastic structure governed by the Biot system. The appropriate interface conditions are enforced along the interface $y = 0$. Following the example from Ambartsumyan et al. (2018b), the solution in the Stokes region is

$$\mathbf{u}_f = \pi \cos(\pi t) \begin{pmatrix} -3x + \cos(y) \\ y + 1 \end{pmatrix}, \quad p_f = e^t \sin(\pi x) \cos\left(\frac{\pi y}{2}\right) + 2\pi \cos(\pi t).$$

Table 1 Relative numerical errors and convergence rates

$\mathcal{P}_1^b - \mathcal{P}_1$, $\mathcal{RT}_0 - \mathcal{P}_0$, $\mathcal{P}_1$, $\mathcal{P}_0$ and $\mathcal{P}_1^{dc}$								
	$\|\mathbf{u}_f - \mathbf{u}_{f,h}\|_{l^2(H^1(\Omega_f))}$		$\|\mathbf{u}_p - \mathbf{u}_{p,h}\|_{l^2(L^2(\Omega_p))}$		$\|\|\|c - c_h\|\|\|_{l^2(H^1(\Omega))}$		$\|c - c_h\|_{l^\infty(L^2(\Omega))}$	
h	Error	Rate	Error	Rate	Error	Rate	Error	Rate
1/4	1.79E-02	–	2.10E−01	–	2.24E−01	–	2.52E−02	–
1/8	8.96E−03	1.0	1.05E−01	1.0	1.14E−01	1.0	6.17E−03	2.0
1/16	4.47E−03	1.0	5.23E−02	1.0	5.71E−02	1.0	1.56E−03	2.0
1/32	2.24E−03	1.0	2.61E−02	1.0	2.87E−02	1.0	3.96E−04	2.0
1/64	1.12E−03	1.0	1.31E−02	1.0	1.44E−02	1.0	1.00E−04	2.0

The Biot solution is chosen accordingly to satisfy the interface conditions (2.5)–(2.7):

$$\mathbf{u}_p = \pi e^t \begin{pmatrix} \cos(\pi x)\cos(\frac{\pi y}{2}) \\ \frac{1}{2}\sin(\pi x)\sin(\frac{\pi y}{2}) \end{pmatrix}, \quad p_p = e^t \sin(\pi x)\cos\left(\frac{\pi y}{2}\right),$$
$$\boldsymbol{\eta}_p = \sin(\pi t)\begin{pmatrix} -3x + \cos(y) \\ y + 1 \end{pmatrix}.$$

The right hand side functions $\mathbf{f}_f$, q_f, $\mathbf{f}_p$ and q_p are computed from (2.1)–(2.4) using the above solution. The model problem is then complemented with the appropriate Dirichlet boundary conditions and initial data. The total simulation time for this test case is $T = 10^{-3}$ and the time step is $\Delta t = 10^{-4}$. The time step is chosen sufficiently small, so that the time discretization error does not affect the convergence rates. The transport solution is set to

$$c = t\,(\cos(\pi x) + \cos(\pi y))\,/\pi,$$

with the diffusion tensor $\mathbf{D} = 10^{-3}\mathbf{I}$ and porosity $\phi = 1$.

For the spatial discretization we use the MINI elements $\mathcal{P}_1^b - \mathcal{P}_1$ for Stokes, the Raviart-Thomas $\mathcal{RT}_0 - \mathcal{P}_0$ for Darcy, continuous Lagrangian $\mathcal{P}_1$ elements for the displacement, and piecewise constant Lagrange multiplier $\mathcal{P}_0$. The transport equation is discretized using discontinuous piecewise linears, $\mathcal{P}_1^{dc}$. For simplicity the Stokes and Biot meshes are made matching along the interface and the transport mesh is the same as the flow mesh. Theorem 4.2 predicts first order convergence for all variables, which is confirmed by the results reported in Table 1. We also observe second order convergence for the concentration in the $l^\infty(L^2(\Omega))$ norm. Here the notation $l^2(\cdot)$ and $l^\infty(\cdot)$ refers to discrete-in-time norms.

5.2 Applications to coupled flow and transport through fractured poroelastic media

We present five examples with simulations of fluid flow in a fractured reservoir with physically realistic parameters. The examples are designed to illustrate the robustness

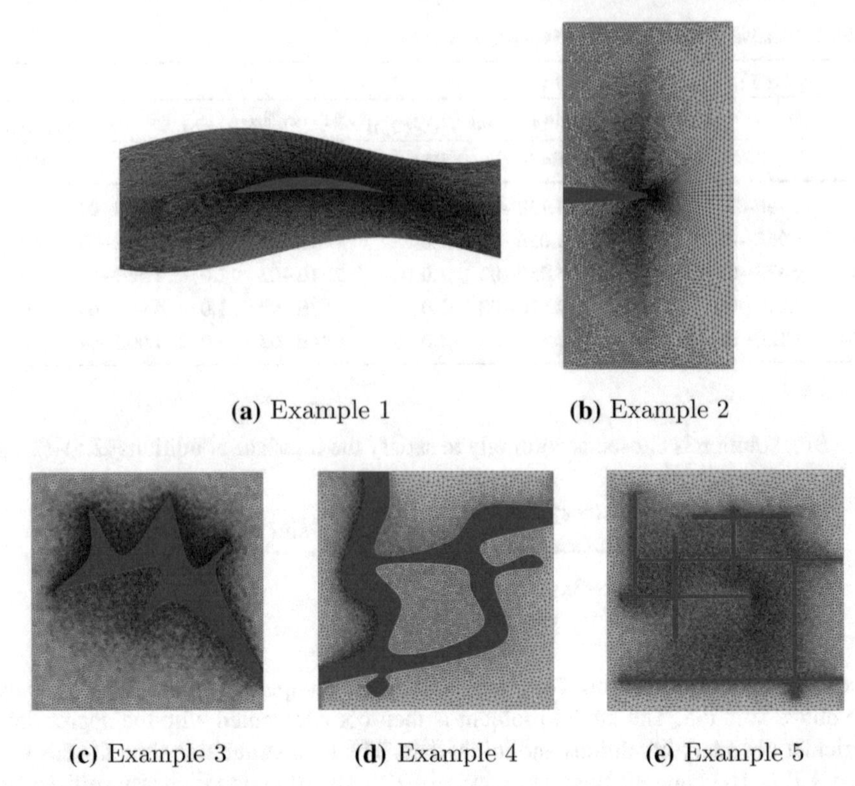

Fig. 2 Computational domains

of the method with respect to reservoir and fracture geometry, rock heterogeneities, and various flow and transport scenarios. The computational domains for the five examples are shown in Fig. 2. Examples 1 and 2 are motivated by hydraulic fracturing. Example 1 features irregularly shaped fracture and reservoir, with fluid injected in the center of the fracture. Example 2 has heterogeneous permeability, porosity, and Young's modulus, with fluid injected into the fracture via inflow boundary condition. Examples 3-5 model flow and transport through vuggy or naturally fractured poroelastic media. The flow is induced by a pressure drop between the left and right boundaries. The transport equation models the concentration of a tracer, which enters the domain with the fluid along the inflow boundary. The reservoir in Example 3 has a large irregularly shaped cavity. Examples 4 and 5 consider a network of channels and fractures, respectively. The latter is the computationally most challenging example, due to the small fracture thickness and sharp angles at the fracture intersections. For all examples in this section the physical units are meters for length, seconds for time, and kPa for pressure.

5.2.1 Example 1: fluid and tracer injection into a fracture

For this example, we first introduce the reference domain $\hat{\Omega}$ given by a square $[-1, 1] \times [-1, 1]$. A fracture, representing the reference fluid domain $\hat{\Omega}_f$, is then described by

Table 2 Poroelasticity and fluid parameters in Example 1

Parameter	Symbol	Units	Values
Young's modulus	E	(kPa)	10^7
Poisson's ratio	ν		0.2
Lamé coefficient	λ_p	(kPa)	$5/18 \times 10^7$
Lamé coefficient	μ_p	(kPa)	$5/12 \times 10^7$
Dynamic viscosity	μ	(kPa s)	10^{-6}
Permeability	K	(m^2)	$diag(200, 50) \times 10^{-12}$
Mass storativity	s_0	(kPa^{-1})	6.89×10^{-2}
Biot–Willis constant	α		1.0
Beavers–Joseph–Saffman coefficient	α_{BJS}		1.0

its top and bottom boundaries, as follows

$$\hat{y}^2 = 8^2(\hat{x} - 0.35)^2(\hat{x} + 0.35)^2, \quad \hat{x} \in [-0.35, 0.35].$$

The physical domain, see Fig. 2a, is obtained from the reference one $\hat{\Omega}$ via the mapping

$$\begin{bmatrix} x \\ y \end{bmatrix} = \begin{bmatrix} \hat{x} \\ 8\left(\cos\left(\frac{\pi\hat{x}+\hat{y}}{100}\right) + \frac{\hat{y}}{4}\right) \end{bmatrix}.$$

This example models the interaction between a fracture filled with fluid and a surrounding poroelastic reservoir. The physical parameters are given in the Table 2. They are taken from Girault et al. (2015) and are realistic for hydraulic fracturing.

The Lamé coefficients are determined from the Young's modulus E and the Poisson's ratio ν via the well-known relationships

$$\lambda_p = \frac{E\nu}{(1+\nu)(1-2\nu)}, \quad \mu_p = \frac{E}{2(1+\nu)}.$$

The boundary conditions are

$$p_p = 1000, \quad \boldsymbol{\eta}_p = 0 \text{ on } \Gamma_p.$$

The initial conditions are set accordingly to $\boldsymbol{\eta}_p(0) = 0$ and $p_p(0) = 1000$. The initial concentration is $c(0) = 0$. The total simulation time is $T = 100$ s with a time step of size $\Delta t = 1$s.

The flow is driven by the injection of the fluid into the fracture with the constant rate $5 \cdot 10^{-3}$ kg/s. The fluid is injected into a region obtained from mapping a disk of radius 0.017m at the center of the reference fracture $\hat{\Omega}_f$. A tracer is injected in this same region, continuously over the entire simulation period, i.e. $c_w = 1$ in the region specified above.

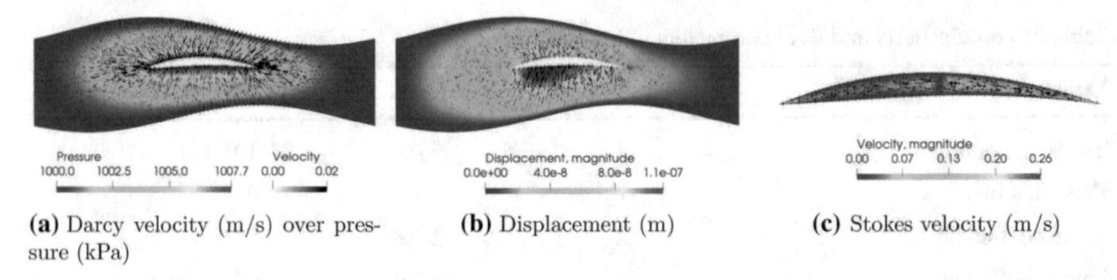

(a) Darcy velocity (m/s) over pressure (kPa) **(b)** Displacement (m) **(c)** Stokes velocity (m/s)

Fig. 3 Example 1, computed Stokes–Biot solution at $t = 100$ s

Recall, see (2.15), that the diffusion tensor is given as

$$\mathbf{D}(\mathbf{u}) = d_m\mathbf{I} + |\mathbf{u}|\{\alpha_l\mathbf{E} + \alpha_t(\mathbf{I} - \mathbf{E})\}.$$

For all examples in this section, in Ω_f we set $d_m = 10^{-6}$ m/s, $\alpha_l = \alpha_t = 0$, i.e., $\mathbf{D} = 10^{-6}\mathbf{I}$ m/s. In Ω_p we set $d_m = 10^{-4}$ m/s, $\alpha_l = \alpha_t = 10^{-4}$. The porosity is set to $\phi = 0.4$.

In all examples we use the Taylor–Hood $\mathcal{P}_2 - \mathcal{P}_1$ elements for the fluid velocity and pressure in the fracture region, the Raviart-Thomas $\mathcal{RT}_1 - \mathcal{P}_1^{dc}$ elements for the Darcy velocity and pressure, continuous Lagrangian $\mathcal{P}_1$ elements for the displacement, and the $\mathcal{P}_1^{dc}$ elements for the Lagrange multiplier. We use discontinuous piecewise linears $\mathcal{P}_1^{dc}$ for the concentration.

Figure 3 shows the computed velocity and pressure in the reservoir, the displacement, and the velocity in the fracture at the final time $T = 100$s. We observe channel-like flow in the fracture region, from the center to the tips. The leak-off into the reservoir is highest at the fracture tips, but there is also a noticeable leak-off along the fracture length. The structure displacement is small, but it is highest in the vicinity of the fracture and indicates a slight opening of the fracture, as expected.

Figure 4 shows the solution obtained for the concentration at various time moments. At early times, the tracer propagates in accordance with the Stokes velocity field, moving preferentially in horizontal directions towards the tips of the fracture. At later times, despite the small permeability, the tracer penetrates into the reservoir and it is further transported/diffused in it, following the Darcy velocity. We note that, due to the singular shape of the fracture tips, the Darcy velocity at the tips is slightly higher than the Stokes velocity inside the fracture near the tips. This acceleration effect leads to slightly lower concentration values at the fracture tips. This example demonstrates the ability of the method to handle irregularly shaped domains and fractures with a computationally challenging set of parameters.

5.2.2 Example 2: flow and transport through a fractured heterogeneous reservoir

As in the previous example, this example is motivated by hydraulic fracturing, while we illustrate the ability of the method to handle heterogeneous permeability, porosity, and Young's modulus. The domain Ω is given by the rectangle $[0, 1]m \times [-1, 1]m$. A fracture, which represents the fluid domain Ω_f is then positioned in the middle of

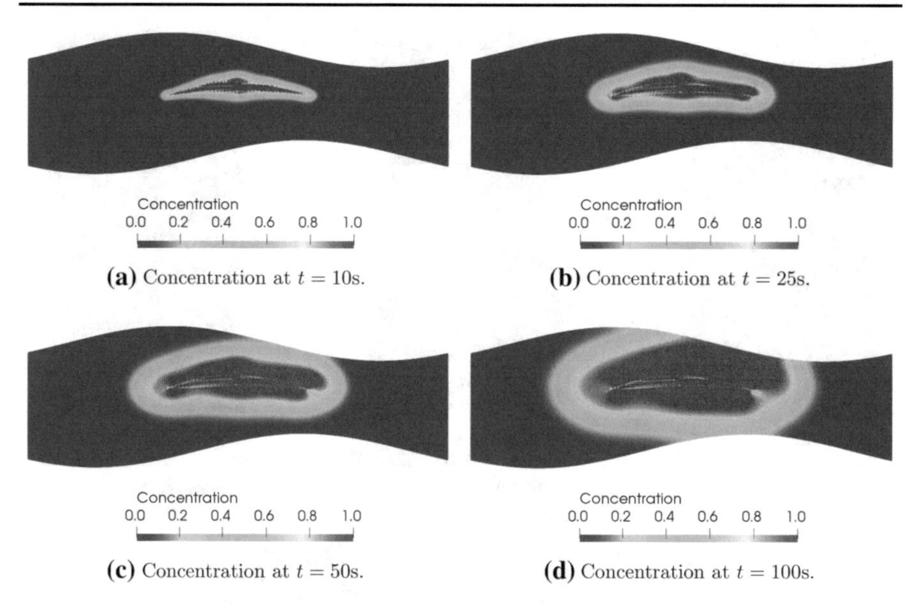

(a) Concentration at $t = 10$s.

(b) Concentration at $t = 25$s.

(c) Concentration at $t = 50$s.

(d) Concentration at $t = 100$s.

Fig. 4 Example 1, computed concentration solution

the rectangle, with the boundaries defined by

$$x^2 = 200(0.05 - y)(0.05 + y), \quad y \in [-0.05, 0.05],$$

see Fig. 2b. Fluid is injected into the opening of the fracture on the left boundary. The external boundary of Ω_p is split into $\Gamma_{p,\star}$, where $\star \in \{left, right, top, bottom\}$. The boundary conditions are

$$\begin{aligned}
&\mathbf{u}_f \cdot \mathbf{n}_f = 10, \quad \mathbf{u}_f \cdot \boldsymbol{\tau}_f = 0 && \text{on } \Gamma_{f,inflow}, \\
&\mathbf{u}_p \cdot \mathbf{n}_p = 0 && \text{on } \Gamma_{p,left}, \\
&p_p = 1000 && \text{on } \Gamma_{p,top} \cup \Gamma_{p,right} \cup \Gamma_{p,bottom}, \\
&\boldsymbol{\eta}_p \cdot \mathbf{n}_p = 0 && \text{on } \Gamma_{p,top} \cup \Gamma_{p,right} \cup \Gamma_{p,bottom}, \\
&(\boldsymbol{\sigma}_p \mathbf{n}_p) \cdot \boldsymbol{\tau}_p = 0 && \text{on } \Gamma_{p,top} \cup \Gamma_{p,right} \cup \Gamma_{p,bottom}, \\
&\boldsymbol{\sigma}_p \mathbf{n}_p = 0 && \text{on } \Gamma_{p,left}, \\
&(c\mathbf{u} - \mathbf{D}\nabla c) \cdot \mathbf{n} = (c_{in}\mathbf{u}) \cdot \mathbf{n}, \quad c_{in} = 1 && \text{on } \Gamma_{f,inflow}, \\
&(\mathbf{D}\nabla c) \cdot \mathbf{n} = 0 && \text{on } \partial\Omega \backslash \Gamma_{f,inflow}.
\end{aligned}$$

The initial conditions are $\boldsymbol{\eta}_p(0) = 0$ and $p_p(0) = 1000$. The same physical parameters as in Example 1 from Table 2 are used, except for the porosity ϕ, the permeability K, and the Young's modulus E. The permeability and porosity data is taken from a two-dimensional cross-section of the data provided by the Society of Petroleum Engineers (SPE) Comparative Solution Project 10.[1] The SPE data, which is given on

[1] http://www.spe.org/csp.

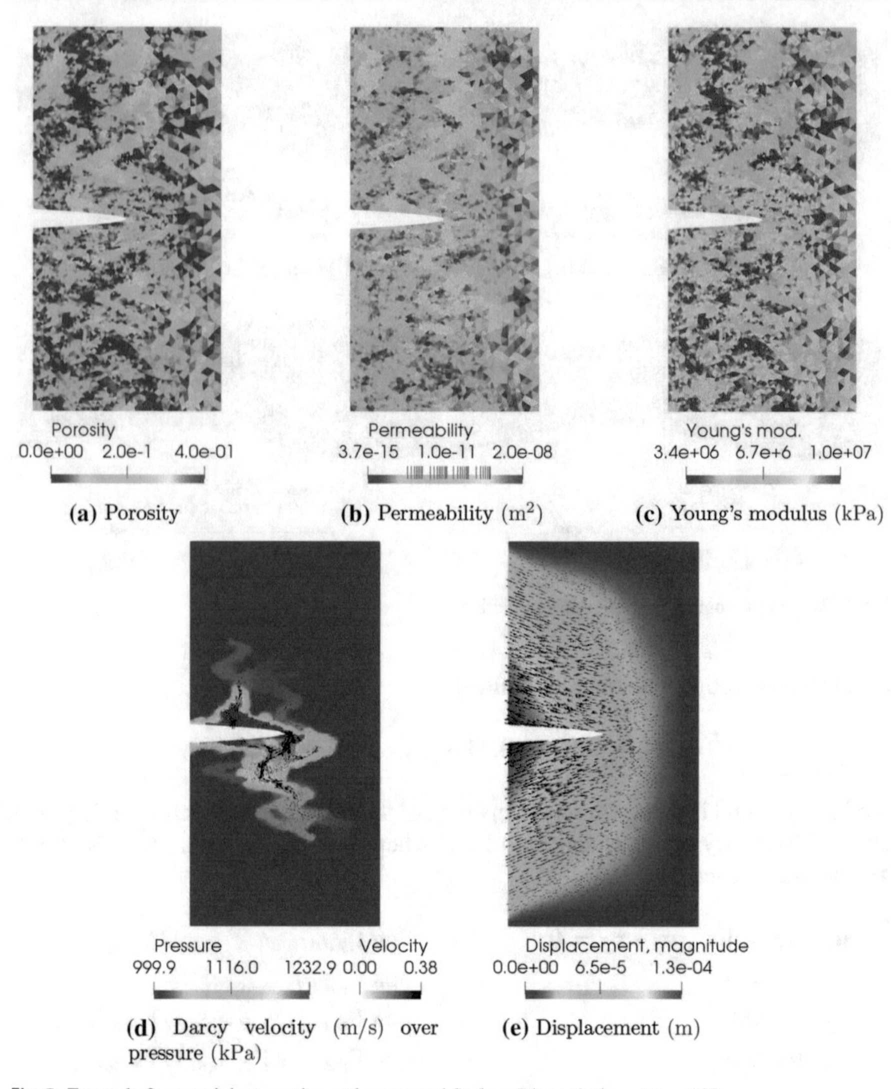

(a) Porosity **(b)** Permeability (m^2) **(c)** Young's modulus (kPa)

(d) Darcy velocity (m/s) over pressure (kPa) **(e)** Displacement (m)

Fig. 5 Example 2, material properties and computed Stokes–Biot solution at $t = 100\,\mathrm{s}$

a rectangular 60×220 grid is projected onto the triangular grid on the domain Ω, and visualized in Fig. 5a–c. We note that the permeability tensor is isotropic in this example. Given porosity ϕ, the Young's modulus is determined from the relationship (Kovacik 1999)

$$E = E_0 \left(1 - \frac{\phi}{\beta}\right)^{2.1},$$

where $E_0 = 10^7$ is the Young's modulus for the non-porous material and the constant $\beta = 0.5$ represents the porosity at which the effective Young's modulus becomes zero.

The computed Darcy velocity, pressure, and displacement at the final time are shown in Fig. 5d, e, respectively. We observe that most of the leak-off is through the fracture tip and the Darcy velocity is largest in a channel-like high permeability region near the tip of the fracture. The displacement field indicates that the fracture opens up as fluid is injected. We also note that the heterogeneities featuring higher permeability and porosity, and correspondingly less stiff material, result in overall larger displacement compared to the previous homogeneous example. Five snapshots of the concentration solution at various time steps are given in Fig. 6. At the early times the tracer propagates along the fracture following the Stokes velocity and penetrates into the high permeability reservoir regions near the middle top, middle bottom, and tip of the fracture. At later times, the tracer is diffused in the poroelastic region; however the overall profile of the concentration front roughly resembles the underlying permeability field.

5.2.3 Example 3: irregularly shaped fluid-filled cavity

The next two examples feature highly irregularly shaped fractures and grids that conform to the fracture geometries. They are motivated by modeling flow and transport through vuggy or naturally fractured reservoirs or aquifers. The domain in this example has a large fluid-filled cavity, see Fig. 2c. The flow is driven from left to right via a pressure drop of 1 kPa. We take the Darcy pressure boundary condition values to be 1 and 0, which can be considered as an offset from a reference pressure. We note that including the reference pressure in the boundary conditions for Darcy and Stokes, the latter being set through $(\boldsymbol{\sigma}_f \mathbf{n}_f) \cdot \mathbf{n}_f$, produces similar results. The boundary conditions are

$$
\begin{aligned}
&p_p = 1 && \text{on } \Gamma_{p,left},\\
&p_p = 0 && \text{on } \Gamma_{p,right},\\
&\mathbf{u}_p \cdot \mathbf{n}_p = 0 && \text{on } \Gamma_{p,top} \cup \Gamma_{p,bottom},\\
&\boldsymbol{\sigma}_p \mathbf{n}_p = 0 && \text{on } \Gamma_{p,left},\\
&\boldsymbol{\eta}_p = 0 && \text{on } \Gamma_{p,right},\\
&(\boldsymbol{\sigma}_p \mathbf{n}_p) \cdot \mathbf{n}_p = 0, \quad \boldsymbol{\eta}_p \cdot \boldsymbol{\tau}_p = 0 && \text{on } \Gamma_{p,top} \cup \Gamma_{p,bottom},\\
&(\boldsymbol{\sigma}_f \mathbf{n}_f) \cdot \mathbf{n}_f = 0, \quad \mathbf{u}_f \cdot \boldsymbol{\tau}_f = 0 && \text{on } \Gamma_{f,right},\\
&(c\mathbf{u} - \mathbf{D}\nabla c) \cdot \mathbf{n} = (c_{in}\mathbf{u}) \cdot \mathbf{n}, \quad c_{in} = 1 && \text{on } \Gamma_{p,left},\\
&(\mathbf{D}\nabla c) \cdot \mathbf{n} = 0 && \text{on } \partial\Omega \backslash \Gamma_{p,left}.
\end{aligned}
$$

The physical parameters for this test case are chosen as in the previous example, except for the permeability, which is $K = 10^{-8}\mathbf{I}$ m^2. The total simulation time is 10s, with time step size $\Delta t = 0.1$s.

The velocity fields in the poroelastic and fracture regions are shown in Fig. 7a, c, respectively, while the rock displacement is given in Fig. 7b. The Darcy velocity is largest in the region between the left inflow boundary and the cavity. This results in a larger displacement in this region. The Stokes velocity in the cavity is an order of magnitude larger than in the poroelastic region. A channel-like flow profile is clearly

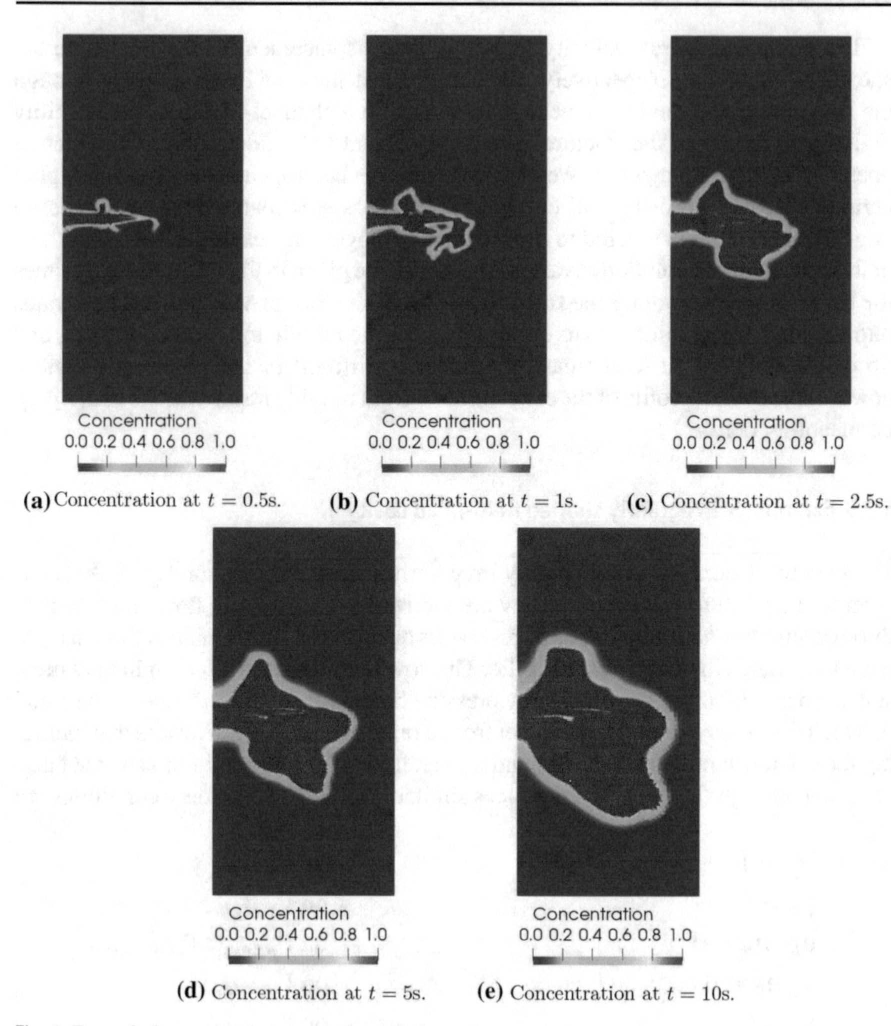

(a) Concentration at $t = 0.5$s. **(b)** Concentration at $t = 1$s. **(c)** Concentration at $t = 2.5$s. **(d)** Concentration at $t = 5$s. **(e)** Concentration at $t = 10$s.

Fig. 6 Example 2, computed concentration solution

visible within the cavity, with the largest velocity along a central path away from the cavity walls. Four snapshots of the concentration solution at different time moments are shown in Fig. 8. As expected, the tracer follows the flow, and tends to get into the free fluid region through the nearest fracture tip. After that, it is transported quickly toward the opening in the right boundary, following the Stokes velocity profile and with very little diffusion. In particular, the tracer follows a narrow central path within the cavity away from the walls. This behavior agrees qualitatively with the parameters in the transport equation.

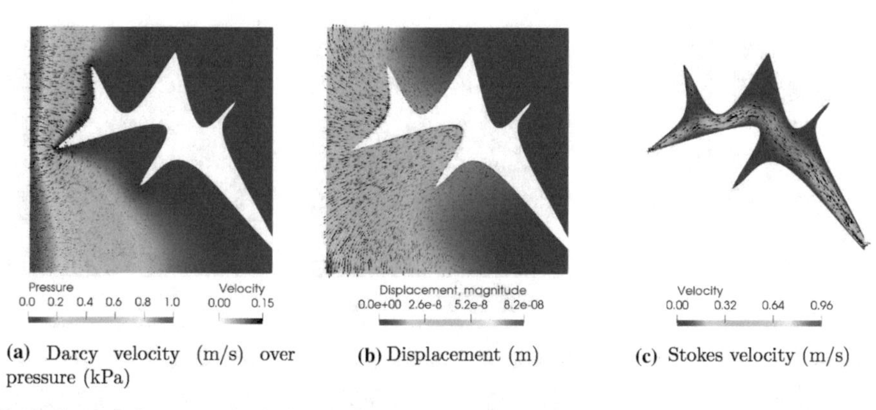

(a) Darcy velocity (m/s) over pressure (kPa) (b) Displacement (m) (c) Stokes velocity (m/s)

Fig. 7 Example 3, computed velocity, pressure, and displacement fields

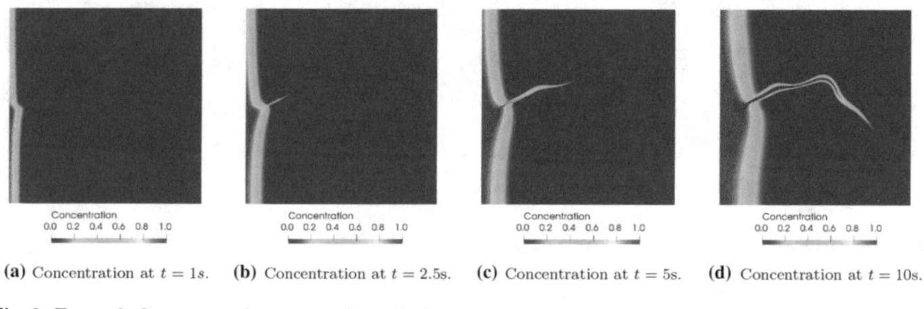

(a) Concentration at $t = 1s$. (b) Concentration at $t = 2.5s$. (c) Concentration at $t = 5s$. (d) Concentration at $t = 10s$.

Fig. 8 Example 3, computed concentration solution

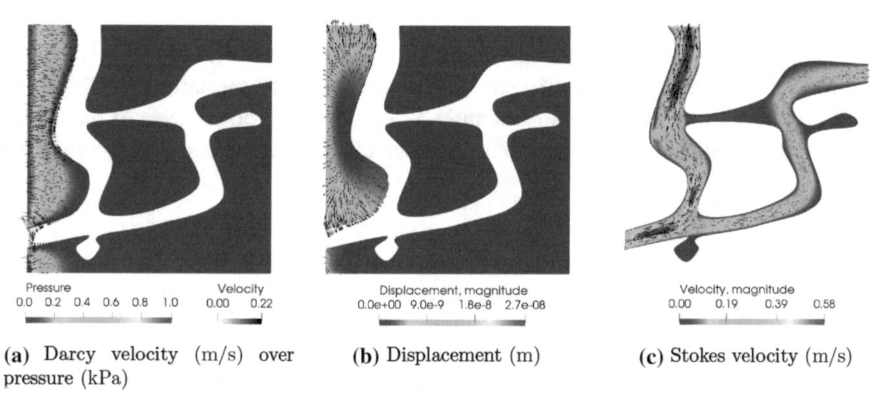

(a) Darcy velocity (m/s) over pressure (kPa) (b) Displacement (m) (c) Stokes velocity (m/s)

Fig. 9 Example 4, computed velocity, pressure, and displacement fields

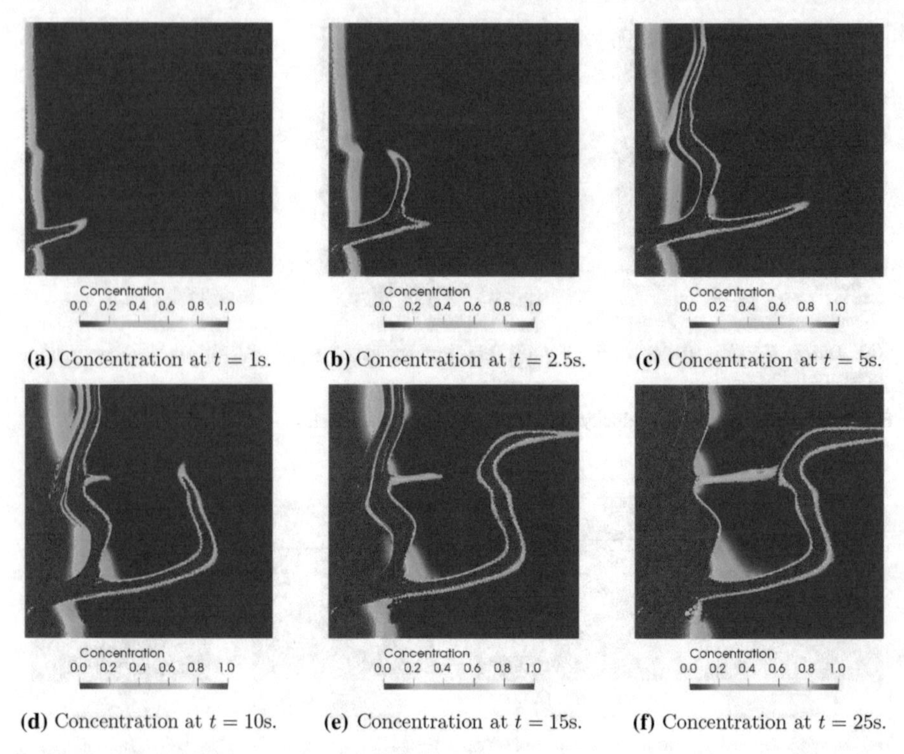

(a) Concentration at $t = 1$s. **(b)** Concentration at $t = 2.5$s. **(c)** Concentration at $t = 5$s.

(d) Concentration at $t = 10$s. **(e)** Concentration at $t = 15$s. **(f)** Concentration at $t = 25$s.

Fig. 10 Example 4, computed concentration solution

5.2.4 Example 4: flow through poroelastic media with channel network

The domain for this example is given in Fig. 2d. It features an irregularly shaped network of channels. The physical parameters and boundary and initial conditions for both flow and transport are as in Example 3 except for the boundaries of the Stokes region. Since the channel network has openings at both the left and right boundaries, we set

$$\begin{aligned} &\mathbf{u}_f \cdot \mathbf{n}_f = 0.2, \quad \mathbf{u}_f \cdot \tau_f = 0 && \text{on } \Gamma_{f,left}, \\ &(\sigma_f \mathbf{n}_f) \cdot \mathbf{n}_f = 0, \quad \mathbf{u}_f \cdot \tau_f = 0 && \text{on } \Gamma_{f,right} \cup \Gamma_{f,top}. \end{aligned}$$

We present the computed velocity fields in the poroelastic and fracture regions in Fig. 9a, c and the structure displacement in Fig. 9b. Four snapshots of the concentration solution at different times are shown in Fig. 10. The qualitative behavior of the flow and transport solution is similar to Example 3, with channel-like flow profile and higher velocity within the channel network. The tracer propagates much faster through the channel network, following the widest channel as a preferential path to reach the outflow boundary. Another interesting feature is that in some channels, the tracer enters both from the channel network and from the porous media. Since the diffusion in the channels is very low, this results in two coexisting tracer streams in close proximity to

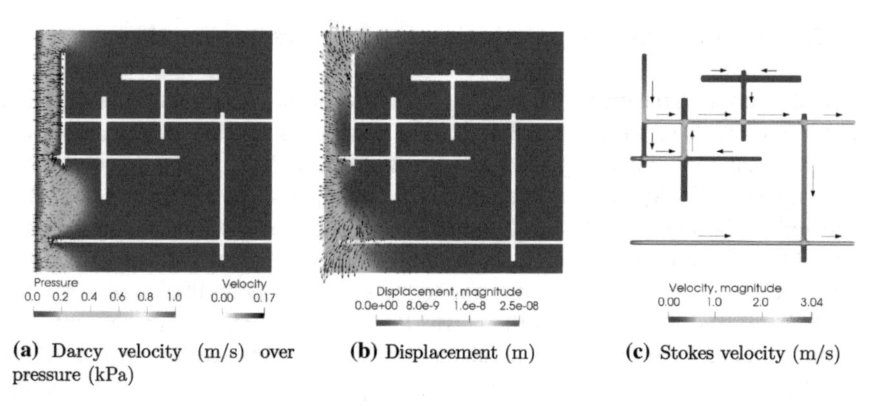

(a) Darcy velocity (m/s) over pressure (kPa) **(b)** Displacement (m) **(c)** Stokes velocity (m/s)

Fig. 11 Example 5, computed velocity, pressure, and displacement fields

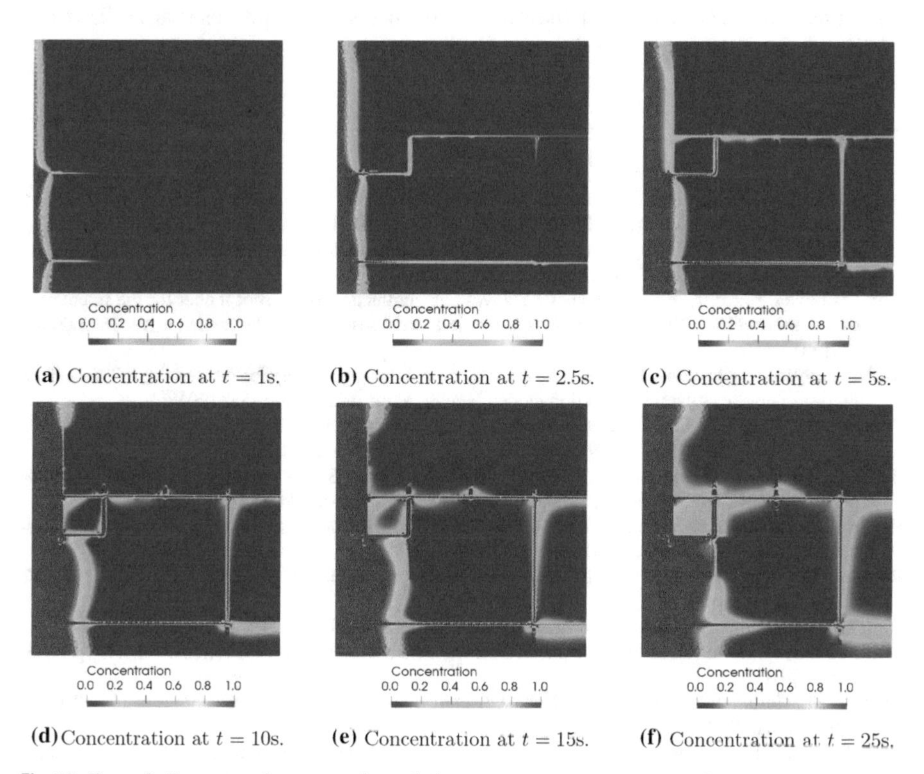

(a) Concentration at $t = 1$s. **(b)** Concentration at $t = 2.5$s. **(c)** Concentration at $t = 5$s.

(d) Concentration at $t = 10$s. **(e)** Concentration at $t = 15$s. **(f)** Concentration at $t = 25$s.

Fig. 12 Example 5, computed concentration solution

each other, but not mixing, being transported by the free fluid, see the upper outflow at time $t = 5$s.

5.2.5 Example 5: flow through poroelastic media with fracture network

The final example is the most challenging, since it involves a network of thin fractures that intersect at sharp angles, see Fig. 2e. The setting for this test case matches the one from Example 4, including physical parameters, initial and boundary conditions, The computed velocity fields in the poroelastic region and the fracture network are visualized in Fig. 11a, c, while the displacement of the porous media skeleton is shown in Fig. 11b. We note that the velocity in the fractures is higher than the velocity in the channels in the previous example, due to the smaller fracture thickness. Also, the velocity is highest in branches of the network where fluid enters from two different branches and that have connection to the outflow boundary. As seen in the plots of the concentration solution in Fig. 12, the tracer is transported quickly through the fractures toward the outflow boundary. Initially it follows well the fracture network geometry, despite the sharp angles between the branches, However, due to relatively small size of the fracture outflow boundaries, the concentration eventually builds up in the fracture region and the tracer starts to penetrate and diffuse into the porous matrix. This can be seen at the later times near the right boundary.

References

Acosta, G., Apel, T., Durán, R., Lombardi, A.: Error estimates for Raviart–Thomas interpolation of any order on anisotropic tetrahedra. Math. Comput. **80**(273), 141–163 (2011)

Aizinger, V., Dawson, C.N., Cockburn, B., Castillo, P.: The local discontinuous Galerkin method for contaminant transport. Adv. Water Resour. **24**, 73–87 (2000)

Ambartsumyan, I., Ervin, V.J., Nguyen, T., Yotov, I.: A nonlinear Stokes–Biot model for the interaction of a non-Newtonian fluid with poroelastic media I: well-posedness of the model. arXiv:1803.00947 (2018a)

Ambartsumyan, I., Khattatov, E., Yotov, I., Zunino, P.: A Lagrange multiplier method for a Stokes–Biot fluid-poroelastic structure interaction model. Numer. Math. **140**(2), 513–553 (2018b)

Arnold, D.N., Brezzi, F., Cockburn, B., Marini, L.D.: Unified analysis of discontinuous Galerkin methods for elliptic problems. SIAM J. Numer. Anal. **39**(5), 1749–1779 (2001/02)

Badia, S., Quaini, A., Quarteroni, A.: Coupling Biot and Navier–Stokes equations for modelling fluid-poroelastic media interaction. J. Comput. Phys. **228**(21), 7986–8014 (2009)

Bazilevs, Y., Takizawa, K., Tezduyar, T.E.: Computational Fluid–Structure Interaction: Methods and Applications. Wiley, New York (2013)

Beavers, G., Joseph, D.: Boundary conditions at a naturally permeable wall. J. Fluid Mech. **30**(1), 197–207 (1967)

Biot, M.: General theory of three-dimensional consolidation. J. Appl. Phys. **12**, 155–164 (1941)

Boffi, D., Brezzi, F., Fortin, M., et al.: Mixed Finite Element Methods and Applications, vol. 44. Springer, Berlin (2013)

Boon, W.M., Nordbotten, J.M., Yotov, I.: Robust discretization of flow in fractured porous media. SIAM J. Numer. Anal. **56**(4), 2203–2233 (2018)

Both, J., Kumar, K., Nordbotten, J.M., Radu, F.A.: Anderson accelerated fixed-stress splitting schemes for consolidation of unsaturated porous media. Comput. Math. Appl. https://doi.org/10.1016/j.camwa.2018.07.033 (2018)

Brenner, S.C.: Poincaré–Friedrichs inequalities for piecewise H^1 functions. SIAM J. Numer. Anal. **41**(1), 306–324 (2003)

Bukac, M., Yotov, I., Zakerzadeh, R., Zunino, P.: Partitioning strategies for the interaction of a fluid with a poroelastic material based on a Nitsche's coupling approach. Comput. Methods Appl. Mech. Eng. **292**, 138–170 (2015)

Bukac, M., Yotov, I., Zunino, P.: An operator splitting approach for the interaction between a fluid and a multilayered poroelastic structure. Numer. Methods Partial Differ. Equ. **31**(4), 1054–1100 (2015)

Bukac, M., Yotov, I., Zunino, P.: Dimensional model reduction for flow through fractures in poroelastic media. ESAIM Math. Model. Numer. Anal. **51**(4), 1429–1471 (2017)

Bungartz, H.-J., Schäfer, M.: Fluid–Structure Interaction: Modelling, Simulation, Optimisation, vol. 53. Springer, Berlin (2006)

Ciarlet, P.: The finite element method for elliptic problems. Class. Appl. Math. **40**, 1–511 (2002)

Cockburn, B., Dawson, C.: Approximation of the velocity by coupling discontinuous Galerkin and mixed finite element methods for flow problems. Comput. Geosci. **6**(3–4), 505–522 (2002). Locally conservative numerical methods for flow in porous media

Cockburn, B., Shu, C.-W.: The local discontinuous Galerkin method for time-dependent convection–diffusion systems. SIAM J. Numer. Anal. **35**(6), 2440–2463 (1998)

D'Angelo, C., Scotti, A.: A mixed finite element method for Darcy flow in fractured porous media with non-matching grids. ESAIM Math. Model. Numer. Anal. **46**(2), 465–489 (2012)

Dawson, C.: Conservative, shock-capturing transport methods with nonconservative velocity approximations. Comput. Geosci. **3**(3–4), 205–227 (1999)

Dawson, C., Sun, S., Wheeler, M.F.: Compatible algorithms for coupled flow and transport. Comput. Methods Appl. Mech. Eng. **193**(23–26), 2565–2580 (2004)

Discacciati, M., Miglio, E., Quarteroni, A.: Mathematical and numerical models for coupling surface and groundwater flows. Appl. Numer. Math. **43**(1–2), 57–74 (2002)

Flemisch, B., Berre, I., Boon, W., Fumagalli, A., Schwenck, N., Scotti, A., Stefansson, I., Tatomir, A.: Benchmarks for single-phase flow in fractured porous media. Adv. Water Res. **111**, 239–258 (2018)

Formaggia, L., Quarteroni, A., Veneziani, A.: Cardiovascular Mathematics: Modeling and Simulation of the Circulatory System, vol. 1. Springer, Berlin (2010)

Frih, N., Roberts, J.E., Saada, A.: Modeling fractures as interfaces: a model for Forchheimer fractures. Comput. Geosci. **12**(1), 91–104 (2008)

Frih, N., Martin, V., Roberts, J.E., Saada, A.: Modeling fractures as interfaces with nonmatching grids. Comput. Geosci. **16**(4), 1043–1060 (2012)

Fumagalli, A., Scotti, A.: A reduced model for flow and transport in fractured porous media with non-matching grids. In: Numerical Mathematics and Advanced Applications 2011, pp. 499–507. Springer, Heidelberg (2013)

Fumagalli, A., Scotti, A.: Numerical modelling of multiphase subsurface flow in the presence of fractures. Commun. Appl. Ind. Math. **3**(1), e–380, 23 (2012)

Galdi, G.P., Rannacher, R. (eds.): Fundamental Trends in Fluid–Structure Interaction. Contemporary Challenges in Mathematical Fluid Dynamics and Its Applications, vol. 1. World Scientific Publishing Co. Pte. Ltd., Hackensack, NJ (2010)

Galvis, J., Sarkis, M.: Non-matching mortar discretization analysis for the coupling Stokes–Darcy equations. Electron. Trans. Numer. Anal. **26**(20), 07 (2007)

Ganis, B., Mear, M.E., Sakhaee-Pour, A., Wheeler, M.F., Wick, T.: Modeling fluid injection in fractures with a reservoir simulator coupled to a boundary element method. Comput. Geosci. **18**(5), 613–624 (2014)

Girault, V., Rivière, B.: DG approximation of coupled Navier–Stokes and Darcy equations by Beaver–Joseph–Saffman interface condition. SIAM J. Numer. Anal. **47**(3), 2052–2089 (2009)

Girault, V., Wheeler, M.F., Ganis, B., Mear, M.E.: A lubrication fracture model in a poro-elastic medium. Math. Models Methods Appl. Sci. **25**(4), 587–645 (2015)

Hecht, F.: New development in FreeFem++. J. Numer. Math. **20**(3–4), 251–265 (2012)

Kim, J., Tchelepi, H.A., Juanes, R.: Stability and convergence of sequential methods for coupled flow and geomechanics: drained and undrained splits. Comput. Methods Appl. Mech. Eng. **200**(23–24), 2094–2116 (2011a)

Kim, J., Tchelepi, H.A., Juanes, R.: Stability and convergence of sequential methods for coupled flow and geomechanics: fixed-stress and fixed-strain splits. Comput. Methods Appl. Mech. Eng. **200**(13–16), 1591–1606 (2011b)

Kovacik, J.: Correlation between Young's modulus and porosity in porous materials. J. Mater. Sci. Lett. **18**(13), 1007–1010 (1999)

Layton, W.J., Schieweck, F., Yotov, I.: Coupling fluid flow with porous media flow. SIAM J. Numer. Anal. **40**(6), 2195–2218 (2003)

Lee, S., Mikelić, A., Wheeler, M.F., Wick, T.: Phase-field modeling of proppant-filled fractures in a poroelastic medium. Comput. Methods Appl. Mech. Eng. **312**, 509–541 (2016a)

Lee, S., Wheeler, M.F., Wick, T.: Pressure and fluid-driven fracture propagation in porous media using an adaptive finite element phase field model. Comput. Methods Appl. Mech. Eng. **305**, 111–132 (2016b)

Lesinigo, M., D'Angelo, C., Quarteroni, A.: A multiscale Darcy–Brinkman model for fluid flow in fractured porous media. Numer. Math. **117**(4), 717–752 (2011)

Martin, V., Jaffre, J., Roberts, J.E.: Modeling fractures and barriers as interfaces for flow in porous media. SIAM J. Sci. Comput. **26**(5), 1667–1691 (2005)

Mikelić, A., Wheeler, M.F.: Convergence of iterative coupling for coupled flow and geomechanics. Comput. Geosci. **17**(3), 455–461 (2013)

Mikelić, A., Wheeler, M.F., Wick, T.: Phase-field modeling of a fluid-driven fracture in a poroelastic medium. Comput. Geosci. **19**(6), 1171–1195 (2015)

Morales, F.A., Showalter, R.E.: The narrow fracture approximation by channeled flow. J. Math. Anal. Appl. **365**(1), 320–331 (2010)

Morales, F.A., Showalter, R.E.: A Darcy–Brinkman model of fractures in porous media. J. Math. Anal. Appl. **452**(2), 1332–1358 (2017)

Oden, J.T., Babuska, I., Baumann, C.E.: A discontinuous *hp* finite element method for diffusion problems. J. Comput. Phys. **146**(2), 491–519 (1998)

Peaceman, D.W.: Fundamentals of Numerical Reservoir Simulation. Elsevier, Amsterdam (1977)

Radu, F.A., Pop, I.S., Attinger, S.: Analysis of an Euler implicit-mixed finite element scheme for reactive solute transport in porous media. Numer. Methods Partial Differ. Equ. **26**(2), 320–344 (2010)

Richter, T.: Fluid–Structure Interactions: Models, Analysis and Finite Elements, vol. 118. Springer, Berlin (2017)

Rivière, B., Yotov, I.: Locally conservative coupling of Stokes and Darcy flows. SIAM J. Numer. Anal. **42**(5), 1959–1977 (2005)

Rivière, B., Wheeler, M.F., Girault, V.: Improved energy estimates for interior penalty, constrained and discontinuous Galerkin methods for elliptic problems. I. Comput. Geosci. **3**(3–4), 337–360 (1999)

Saffman, P.: On the boundary condition at the surface of a porous medium. Stud. Appl. Math. **50**(2), 93–101 (1971)

Scott, R., Zhang, S.: Finite element interpolation of nonsmooth functions satisfying boundary conditions. Math. Comput. **54**(190), 483–493 (1990)

Showalter, R.E.: Poroelastic filtration coupled to Stokes flow. In: Control Theory of Partial Differential Equations, vol. 242. Lecture Notes in Pure and Applied Mathematics, pp. 229–241. Chapman & Hall/CRC, Boca Raton, FL (2005)

Sun, S., Rivière, B., Wheeler, M.F.: A combined mixed finite element and discontinuous Galerkin method for miscible displacement problem in porous media. In: Recent Progress in Computational and Applied PDEs (Zhangjiajie, 2001), pp. 323–351. Kluwer/Plenum, New York (2002)

Sun, S., Wheeler, M.F.: Discontinuous Galerkin methods for coupled flow and reactive transport problems. Appl. Numer. Math. **52**(2–3), 273–298 (2005a)

Sun, S., Wheeler, M.F.: Symmetric and nonsymmetric discontinuous Galerkin methods for reactive transport in porous media. SIAM J. Numer. Anal. **43**(1), 195–219 (2005b)

Vassilev, D., Yotov, I.: Coupling Stokes–Darcy flow with transport. SIAM J. Sci. Comput. **31**(5), 3661–3684 (2009)

Vassilev, D., Wang, C., Yotov, I.: Domain decomposition for coupled Stokes and Darcy flows. Comput. Methods Appl. Mech. Eng. **268**, 264–283 (2014)

Wheeler, M.F., Darlow, B.L.: Interior penalty Galerkin procedures for miscible displacement problems in porous media. In: Computational Methods in Nonlinear Mechanics (Proceedings of Second International Conference, University of Texas, Austin, TX, 1979), pp. 485–506. North-Holland, Amsterdam-New York (1980)

Publisher's Note Springer Nature remains neutral with regard to jurisdictional claims in published maps and institutional affiliations.

GEM - International Journal on Geomathematics
https://doi.org/10.1007/s13137-019-0113-y

ORIGINAL PAPER

Phase-field modeling through iterative splitting of hydraulic fractures in a poroelastic medium

A. Mikelić[1] · M. F. Wheeler[2] · T. Wick[3]

Received: 6 December 2017 / Accepted: 9 October 2018

Abstract

We study the propagation of hydraulic fractures using the fixed stress splitting method. The phase field approach is applied and we study the *mechanics step* involving displacement and phase field unknowns, with a given pressure. We present a detailed derivation of an incremental formulation of the phase field model for a hydraulic fracture in a poroelastic medium. The mathematical model represents a linear elasticity system with fading elastic moduli as the crack grows that is coupled with an elliptic variational inequality for the phase field variable. The convex constraint of the variational inequality assures the irreversibility and entropy compatibility of the crack formation. We establish existence of a minimizer of an energy functional of an incremental problem and convergence of a finite dimensional approximation. Moreover, we prove that the fracture remains small in the third direction in comparison to the first two principal directions. Computational results of benchmark problems are provided that demonstrate the effectiveness of this approach in treating fracture propagation. Another novelty is the treatment of the mechanics equation with mixed boundary conditions of Dirichlet and Neumann types. We finally notice that the corresponding pressure step was studied by the authors in Mikelić et al. (SIAM Multiscale Model Simul 13(1):367–398, 2015a).

Keywords Hydraulic fracturing · Phase field formulation · Nonlinear elliptic system · Computer simulations · Poroelasticity

Mathematics Subject Classification 35Q74 · 35J87 · 49J45 · 65K15 · 74R10

A.M. would like to thank *Institute for Computational Engineering and Science (ICES)*, UT Austin for hospitality during his sabbatical stays. The research by M. F. Wheeler was partially supported by the U.S. Department of Energy, Office of Science, Office of Basic Energy Sciences through *DOE Energy Frontier Research Center: The Center for Frontiers of Subsurface Energy Security (CFSES)* under Contract No. DE-FG02-04ER25617, MOD. 005. The work of T. Wick was supported through an ICES Postdoc fellowship, the Humboldt foundation with a Feodor-Lynen fellowship and through the JT Oden faculty research program. Currently T. Wick is supported by the DFG-SPP 1748 program.

Extended author information available on the last page of the article

1 Introduction

The coupling of flow and geomechanics in porous media is a major research topic in energy and environmental modeling. Of specific interest is induced hydraulic fracturing or hydrofracturing commonly known as fracking. This technique is used to release petroleum and natural gas that includes shale gas, tight gas, and coal seam gas for extraction. Here fracking creates fractures from a wellbore drilled into reservoir rock formations. In 2012, more than one million fracturing jobs were performed on oil and gas wells in the United States and this number continues to grow. Clearly there are economic benefits of extracting vast amounts of formerly inaccessible hydrocarbons. In addition, there are environmental benefits of producing natural gas, much of which is produced in the United States from fracking. Opponents to fracking point to environmental impacts such as contamination of ground water, risks to air quality, migration of fracturing chemical and surface contamination from spills to name a few. For these reasons, hydraulic fracturing is being heavily scrutinized resulting in the need for accurate and robust mathematical and computational models for treating fluid field fractures surrounded by a poroelastic medium.

Even in the most basic formulation, hydraulic fracturing is complicated to model since it involves the coupling of (i) mechanical deformation; (ii) the flow of fluids within the fracture and in the reservoir; and (iii) fracture propagation. Generally, rock deformation is modeled using the theory of linear elasticity, i.e. they are modeled as an impermeable elastic medium. Using Green's function, an integral equation that determines a relationship between fracture width and the fluid pressure can be adopted. Fluid flow in the fracture is modeled using lubrication theory that relates fluid flow velocity, fracture width and the gradient of pressure.

Fluid flow in the reservoir is modeled as a Darcy flow and the respective fluids are coupled through a leakage term. The experiments show an analogy between hydraulic fracture propagation and crack propagation in fracture mechanics of solids. The criterion for fracture propagation is usually given by the conventional energy-release rate approach of linear elastic fracture mechanics (LEFM) theory; that is the fracture propagates if the stress intensity factor at the tip matches the rock toughness. Detailed discussions of the development of hydraulic fracturing models for use in petroleum engineering can be found in Adachi et al. (2007), Dean and Schmidt (2014), Hwang and Sharma (2013), McClure and Kang (2017) and in mechanical engineering and hydrology in de Borst et al. (2006), Gupta and Duarte (2014), Irzal et al. (2013), Schrefler et al. (2006) and in references cited therein.

In the literature, numerical models of fracture can be classified into two categories: discrete and continuum approaches. The discrete approach treats fractures as discontinuities. Its positive side is the simplicity in terms of modeling. One disadvantage is to consider topology changes in the implementation and mesh dependent fracture propagation is restricted to follow mesh lines. Some of these approaches, however, have difficulties with joining or branching fracture or with heterogeneous materials. For a detailed literature overview of various fracture propagation models, we refer the reader to Wick et al. (2016). In the following, we restrict our focus to a specific continuum approach, which has received a lot of attention.

In the last two decades, variational phase field models of brittle fracture gained popularity in fracture mechanics. In fracture mechanics, the fracture is a lower dimensional manifold. The phase-field variable is a smoothed indicator function that smoothly interpolates between the broken and unbroken regions. The change from the intact medium to a fracture takes place in a narrow mushy region. Francfort and Marigo developed in (1998) a variational formulation for quasi-static fracture evolution in a brittle material based on the minimization of the combined elastic energy in the bulk material and the fracture energy. It has been serving as a basis for a vast literature with numerical simulations and theoretical developments. The approach allows simulation of complex fracturing processes, such as branching and joining. Handling heterogeneous media does not pose major additional difficulties and updating the fracture shape is automatically contained in the model.

Because of the above mentioned analogy between fracture mechanics and hydraulic fracturing, it is appealing to simply borrow the approach of Francfort and Marigo. Nevertheless, it is important to recall that contrary to fracture mechanics, where Griffiths criterion has a deep physical meaning, using a phase field approach in hydraulic fracturing corresponds to a phenomenological overall behavior. The fractures are slender flow domains, but, nevertheless, their width is much bigger than the typical pore size of a porous medium. The mechanical interactions of the fracture interacting with the pore structure is not well understood and an open question. Furthermore, we do not consider the equations of fluid–structure, posed at the pore level, but their upscaled simplified form. Consequently, flow and deformation are described by Biot's equations (e.g., Tolstoy 1992), which are not fundamental physical equations. This is a very important modeling aspect because one wants to couple the upscaled poroelastic medium possibly with Stokes or Navier–Stokes flow in the fracture, which are first principle equations. Here, interface conditions have to be carefully derived. An alternative is a lubrication approximation, which does not contain enough information about (the prediction) of the tip velocity. Hence there are difficulties with including a correct description of the interaction between the hydraulic fracture and the surrounding poroelastic medium. More physical reasoning based on the energy could be necessary. For correct modeling of these processes, only experiments can indicate how far the analogy of fracture behavior with solid mechanics works.

In this paper, we present our phase-field fracture model for **pressurized fractures** in a **porous medium**. Our approach is based on the observation that phase field models require the energy functional in the case of elasticity (Francfort and Marigo 1998) and a free energy in the case of poroelasticity (Mikelić et al. 2015b). The Biot equations are obtained by upscaling and for a hydraulic fracture being a lower dimensional manifold we do not know how to formulate such an energy functional. But in the elastic case, where the pressure is given, we can modify the regularized phase field elastic energy of Francfort and Marigo, and study the corresponding phase field system. We note that now the fracture is a three dimensional slender body. Similarly, in the case of the full Biot system, one would modify Biot's free energy by inserting the phase field function.

Parts of this work are based on two preprints ICES-1315 (Mikelić et al. 2013) and ICES-1418 (Mikelić et al. 2014) that were published in the years 2013 and 2014 at the Institute of Computational Engineering and Sciences at the University of Texas at Austin.

The outline of this paper is as follows: First in Sect. 2, we provide general background information. In Sect. 3, we introduce an incremental formulation of a phase-field model for a pressurized crack. Here, the crack-pressure is incorporated with an interface law. In Sect. 4, we present a mathematical analysis of the incremental problem. In Sect. 5, a numerical formulation is briefly described. Finally in Sect. 6 we provide numerical experiments for classical benchmark cases, e.g. Sneddon's pressurized crack with constant fluid pressure (see Sect. 6.1 and Sneddon 1946). Here, we also focus on the behavior when working with mixed boundary conditions of Dirichlet and Neumann types.

2 Fundamental background information

In this section, we explain the idea of our approach and provide background information. We finish with a current literature overview of phase-field models used for hydraulic facturing.

2.1 The Biot system and fixed-stress iterative coupling

Major difficulties in simulating hydraulic fracturing in a deformable porous medium are treating crack propagation induced by high-pressure slick water injection and later the coupling to a multiphase reservoir simulator for production. A computational effective procedure in modeling coupled multiphase flow and geomechanics is to apply an iterative coupling algorithm as described in Mikelić and Wheeler (2012), and Mikelić et al. (2014).

Iterative coupling is a sequential procedure where either the flow or the mechanics is solved first followed by solving the other problem using the latest solution information. At each time step the procedure is iterated until the solution converges within an acceptable tolerance. There are four well-known iterative coupling procedures and we are interested primarily in one referred to as the **fixed stress split iterative method**.

In order to fix ideas we address the simplest model of real applied importance, namely, the quasi-static single phase Biot system. Let $\mathcal{C}$ denote any open set homeomorphic to an ellipsoid strictly contained in $(0, L)^3 \subset \mathbb{R}^3$ (a crack set). Its boundary is a closed surface $\partial\mathcal{C}$. In most applications $\mathcal{C}$ is a curved 3d domain, with two dimensions significantly smaller than the dominant one. Nevertheless, we consider $\mathcal{C}$ as a **3d domain** and use its particular geometry only when discussing the stress interface conditions. The boundary of $(0, L)^3$ is denoted by $\partial(0, L)^3 = \partial\Omega \backslash \partial\mathcal{C}$ divided into Dirichlet and Neumann parts, $\partial_D\Omega$ and $\partial_N(0, L)^3$ respectively. We assume that $\text{meas}(\partial_D\Omega) > 0$. Boundary conditions on $\partial(0, L)^3 = \partial_D\Omega \cup \partial_N(0, L)^3$ for the above situation involve displacements and tractions as well as pressure and flux.

Remark 1 We notice that in many references on fracture propagation, the crack $\mathcal{C}$ is considered as a lower dimensional manifold and the lubrication theory is applied to describe the fluid flow (see e.g. Adachi et al. 2007; Ganis et al. 2014; Girault et al.

Table 1 Unknowns and effective coefficients

Symbol	Quantity	Unit
$\mathbf{u}$	Displacement	m
p	Fluid pressure	Pa
σ^{por}	Total poroelasticity tensor	Pa
$e(\mathbf{u}) = (\nabla \mathbf{u} + \nabla^{\tau} \mathbf{u})/2$	Linearized strain tensor	Dimensionless
$\mathcal{K}$	Permeability	Darcy
α	Biot's coefficient	Dimensionless
ρ_b	Bulk density	kg/m^3
η	Fluid viscosity	kg/m s
M	Biot's modulus	Pa
$\mathcal{G}$	Gassman rank-4 tensor	Pa
K_{dr}	Drained bulk modulus	Pa

2015). We recall that the 3d flow in $\mathcal{C}$ can be reconstructed from a lower dimensional lubrication approximation (see Mikelić et al. 2015a), except at the tips where a law for their displacements has to be added separately.

The quasi-static Biot equations (see e.g. Tolstoy 1992) are an elliptic-parabolic system of PDEs, valid in the poroelastic domain $\Omega = (0, L)^3 \backslash \overline{\mathcal{C}}$, where for every $t \in (0, T)$ we have

$$\sigma^{por} - \sigma_0 = \mathcal{G}e(\mathbf{u}) - \alpha p I; \quad - \text{div } \{\sigma^{por}\} = \rho_b \mathbf{g}; \tag{1}$$

$$\partial_t \left(\frac{1}{M} p + \text{ div } (\alpha \mathbf{u}) \right) + \text{ div } \left\{ \frac{\mathcal{K}}{\eta} (\rho_f \mathbf{g} - \nabla p) \right\} = f, \tag{2}$$

where σ_0 is the reference state total stress and $\mathbf{g}$ is the gravity and f represents volume sources/sinks, respectively. By I, be denote the identity matrix. In the following, we set $\mathbf{g} = 0$ and $\sigma_0 = 0$. The important parameters and unknowns are given in Table 1.

The fixed stress split iterative method consists in imposing constant volumetric mean total stress σ_v. This means that the stress $\sigma_v = K_{dr} \text{ div } \mathbf{u} I - \alpha p I$ is kept constant at the half-time step.

The iterative process reads as follows:

$$\left(\frac{1}{M} + \frac{\alpha^2}{K_{dr}} \right) \partial_t p^{n+1} + \text{ div } \left\{ \frac{\mathcal{K}}{\eta} (\rho_f \mathbf{g} - \nabla p^{n+1}) \right\}$$

$$= -\frac{\alpha}{K_{dr}} \partial_t \sigma_v^n + f = f - \alpha \text{ div } \partial_t \mathbf{u}^n + \frac{\alpha^2}{K_{dr}} \partial_t p^n; \tag{3}$$

$$- \text{div } \{\mathcal{G}e(\mathbf{u}^{n+1})\} + \alpha \nabla p^{n+1} = 0. \tag{4}$$

Remark 2 We remark that the fixed stress approach is useful in employing existing reservoir simulators in that (3) can be extended to treat the mass balance equations arising in black oil or compositional flows and allows decoupling of multiphase flow and elasticity.

Remark 3 Interest in the system (3)–(4) is based on its robust numerical convergence. Under mild hypothesis on the data, the convergence of the iterations was studied in Mikelić and Wheeler (2012) and it was proven that the solution operator $\mathcal{S}$, mapping $\{\mathbf{u}^n, p^n\}$ to $\{\mathbf{u}^{n+1}, p^{n+1}\}$ is a contraction on appropriate functional spaces with the contraction constant $\gamma_{FS} = \frac{M\alpha^2}{K_{dr}+M\alpha^2} < 1$. The corresponding unique fixed point satisfies equations (1)–(2). Further important recent studies on the fixed-stress scheme have been undertaken in Both et al. (2017), Castelletto et al. (2015), Gaspar and Rodrigo (2017). For phase-field fracture, a very detailed computational analysis of the fixed-stress scheme was performed in Lee et al. (2017a).

Remark 4 We finally notice that the (discretized) Biot equations (without fractures) form a mixed system that is subject to a (discrete) inf-sup condition. Theoretical studies were undertaken in Murad and Loula (1992, 1994). Various finite element pairs have been investigated in Ferronato et al. (2010), Liu (2004), Philips and Wheeler (2003). More recent references can be found in Lee (2016), Rodrigo et al. (2016), Lee et al. (2017c), Hong and Kraus (2018). Important is the choice of the pressure space, which should be locally mass conservative, but can be still of lowest order. For these reasons, fluid-filled phase-field fractures using the entire Biot system have been formulated either with linear/linear elements for the displacements/pressure (see e.g., Mikelić et al. 2015a) or linear/enriched-linear elements (Lee et al. 2016a), where an enriched Galerkin formulation for the pressures ensures local mass conservation.

2.2 Focus on crack propagation in the fixed-stress elasticity step

Because of the complexity of this coupled nonlinear fluid-mechanics system, we follow the above splitting strategy and restrict our attention to a simplified model in which we assume that the pressure has been computed from the previous fixed-stress fluid iteration step. Our focus in this paper is therefore on **crack propagation** in the framework of the fixed-stress mechanics step (4) and we call this approach **a fluid filled crack with a given pressure**.

Remark 5 The extension to the full poroelastic system for crack propagation and therefore employing a phase-field formulation of the pressure step (3) is studied in Mikelić et al. (2015a) and called **fluid-filled crack propagation** in a poroelastic medium. In the fixed stress iterative splitting, the pressure is known in the mechanics step and can be included into the forcing terms. Then we arrive exactly in the situation studied in the current article.

In the following, we present an incremental formulation of the hydraulic fracture with a given pressure field surrounded by a poroelastic medium. The mathematical model involves the coupling of a linear elasticity system with an elliptic variational inequality for the phase field variable. With this approach, branching of fractures and heterogeneities in mechanical properties can be effectively treated as demonstrated numerically in Sect. 6.

Our formulation follows in Miehe et al. (2010) and is a thermodynamically consistent framework for phase-field models of quasi-static crack propagation in elastic solids, together with incremental variational principles. The work by Miehe et al. (2010) is further based on the variational approach to elastic fractures formulated by Francfort and Marigo (1998); see also Bourdin et al. (2008). Our contribution represents an extension to a phase-field pressurized fracture model in a poroelastic medium that we describe in the following in more detail.

Following Griffith's criterion, we suppose that the crack propagation occurs when the elastic energy restitution rate reaches its critical value G_c. In the classical setting the crack $\mathcal{C}$ is a lower dimensional manifold and for a traction force τ applied at the part of the boundary $\partial_N \Omega$, then we associate to the crack $\mathcal{C}$ the following total energy

$$E(\mathbf{u}, \mathcal{C}) = \int_{\Omega} \frac{1}{2} \mathcal{G} e(\mathbf{u}) : e(\mathbf{u}) \, dx - \int_{\partial_N \Omega} \tau \mathbf{u} \, dS - \int_{\Omega} \alpha p_B \mathrm{div}\, \mathbf{u} \, dx + G_c \mathcal{H}^2(\mathcal{C}), \quad (5)$$

where p_B is the poroelastic medium pressure calculated in the previous iterative coupling step and $\alpha \in (0, 1)$ is the Biot coefficient. In (5), the first three terms stem from (4) and the last term, $G_c \mathcal{H}^2(\mathcal{C})$ is the surface energy related to the fracture.

This energy functional is then minimized with respect to the kinematically admissible displacements $\mathbf{u}$ and any crack set satisfying a crack growth condition. The computational modeling of this minimization problem treats complex crack topologies and requires approximation of the crack location and of its length. This was overcome by regularizing the sharp crack surface topology in the solid by diffusive crack zones described by a scalar auxiliary variable. This variable is a phase-field that interpolates between the unbroken and the broken states of the material, which is introduced through a time-dependent φ (the crack phase field), defined on $(0, L)^3 \times (0, T)$. The functional from (5) is regularized using the phase field unknown and the new crack functional (the last term in (5) divided by G_c) reads

$$\Gamma_\varepsilon(\varphi) = \int_{(0,L)^3} \left(\frac{1}{2\varepsilon}(1 - \varphi)^2 + \frac{\varepsilon}{2} |\nabla \varphi|^2 \right) dx = \int_{(0,L)^3} \gamma(\varphi, \nabla \varphi) \, dx, \quad (6)$$

where γ is the crack surface density per unit volume. This regularization of $\mathcal{H}^2(\mathcal{C})$, in the sense of the Γ-limit when $\varepsilon \to 0$, was used in Bourdin et al. (2000).

The model proposed in this paper is a simple extension of the crack functional (6) that, after the time discretization, can be analyzed both as a minimization problem and as a variational PDE formulation. For simplicity the presentation of the time discretized (or the incremental problem) here is based on energy minimization, whereas our treatment of the corresponding variational formulation can be found in Mikelić et al. (2013) and the current paper. For the full quasi-static problem we refer to Mikelić et al. (2015c).

2.3 Energy minimization versus the variational PDE formulation

In the numerical analysis of fracture propagation in solid mechanics, solving the minimization problem (Francfort and Marigo 1998; Bourdin et al. 2008) by considering the variational formulation is for instance treated in Burke et al. (2010). Most other

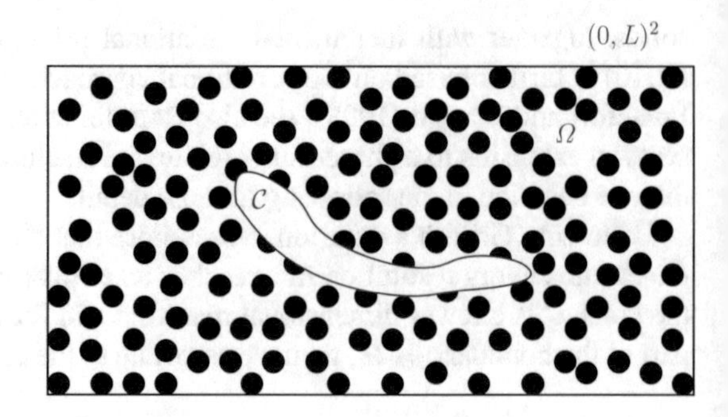

Fig. 1 Illustration of our approach for a 2d situation: a crack $\mathcal{C} \subset \mathbb{R}$ embedded in a porous medium $(0, L)^2$. Here, the dimensions of the crack are assumed to be much larger than the pore scale size (black dots) of the porous medium (color figure online)

works also start from the energy level. We base our computational framework on the variational PDE formulation since more additional realistic physical interfacial effects (see Fig. 1 and Adachi et al. 2007) and associated dissipative terms and nonlinear physical models can be employed. Moreover, the Biot system does not correspond to an energy minimization formulation in $\mathbf{u}$ and p, but has a free energy linked to a Lyapunov functional. For these reasons, the variational PDE formulations allow for more general settings, with the drawback that only stationary points are computed and not only the minimizers.

2.4 Literature on hydraulic phase-field fracture modeling

In our knowledge, applying the phase field approach to the simulation of propagation of pressurized fractures in an elastic medium was initiated by the SPE conference paper (Bourdin et al. 2012). The pressurized fracture was described through a boundary term $\int_{\mathcal{C}} p[\mathbf{u} \cdot \mathbf{n}]$, with the crack $\mathcal{C}$ being a surface and $[\mathbf{u} \cdot \mathbf{n}]$ the displacement jump across the crack (see also Remark 7 in Sect. 3.2). The phase field handling of such terms goes back to the work presented in Chambolle (2004).

In the years Mikelić et al. (2013) and (2014), the first model for pressurized fractures in porous media (including Biot's coefficient α) was proposed and rigorously investigated. Here the displacement phase-field system was modeled in a monolithic framework. Later, a decoupled model was investigated in Mikelić et al. (2015c) for which a corresponding robust numerical augmented Lagrangian approach was developed in Wheeler et al. (2014). We also notice the development of a sharp interface model for pressurized fractures using variational techniques in Almi et al. (2014). The efficient and robust numerical solution of pressurized phase-field models based on quasi-monolithic approaches was presented in Heister et al. (2015), Wick et al. (2015). Based on the first models (Mikelić et al. 2013, 2014) (and also the current paper), fully monolithic solution techniques for pressurized fractures have been developed in Wick (2017a, b). For a different treatment of the decoupled model (Mikelić et al. 2015c), using the discontinuous Galerkin (DG) formulation for the displacements, we refer to Engwer and Schumacher (2017). Various adaptive mesh refinement schemes for pressurized phase-field fracture, with focus on the crack path or other quantities of interest, have been proposed in Heister et al. (2015), Lee et al. (2016b), Wick (2016b).

The pressurized phase-field method has been then further extended to fluid-filled fractures in which a Darcy type equation is used for modeling fracture flow (Mikelić et al. 2015a) and similar studies have appeared simultaneously (Miehe and Mauthe 2016; Miehe et al. 2015; Markert and Heider 2015). A rigorous mathematical analysis including detailed numerical studies of a fully-coupled fluid geomechanics phase-field model in porous media was first presented in Mikelić et al. (2015b). Here, the important phenomenon of negative pressures at fracture tips was observed. This feature is known to appear in such configurations, but was not yet quantified using a phase-field method. In the year 2016, we note further contributions to fluid-filled phase-field fractures from Heider and Markert (2017), Lee et al. (2016b). To reduce the computational cost, we notice that parallel computation frameworks have been implemented in most groups, e.g., Bourdin et al. (2012), Heister et al. (2015), Miehe et al. (2015), Lee et al. (2016b). Coupling to other codes and reservoir simulators has been first accomplished in Wick et al. (2016). However, further research is necessary because the current modeling, the coupling algorithm, and the treatment of the multi-scale nature of the problem must be further improved.

Recent results concentrated on the extension to proppant flow (Lee et al. 2016a), two-phase flow inside the fracture (Lee et al. 2018), single phase-flow for nonlinear poroelastic media (van Duijn et al. 2018), fractures in partially saturated porous media (Cajuhi et al. 2017), fracture initialization with probability maps of fracture networks (Lee et al. 2017b), consequences on further multiphysics coupling of the pressurized fractures interface law (Wick 2016a), more accurate crack width computations and computational analysis of fixed-stress splitting (Lee et al. 2017a), a phase-field formulation (in elasticity) with a lower-dimensional lubrication formulation (Santillan et al. 2017), and a multirate analysis in which different time steps for different regimes are used (Almani et al. 2017).

3 An incremental phase field formulation

We introduce the time-dependent crack phase field φ, defined on $(0, L)^3 \times (0, T)$. The regularized crack functional is given by (6). Our further considerations are based on the fact that the evolution of cracks is fully dissipative in nature. First, the crack phase field φ is intuitively a regularization of $1 - \mathbb{1}_C$ and we impose its negative evolution

$$\partial_t \varphi \leq 0. \tag{7}$$

3.1 A global constitutive dissipation functional

Next we follow Miehe et al. (2010) and Bourdin et al. (2000) and replace the energy (5) by a global constitutive dissipation functional for a rate independent fracture process. That is

$$E_\varepsilon(\mathbf{u},\varphi) = \int_{(0,L)^3} \frac{1}{2}\Big((1-k)\varphi^2 + k\Big)\mathcal{G}e(\mathbf{u}) : e(\mathbf{u})\,dx - \int_{\partial_N\Omega} \tau\mathbf{u}\,dS$$
$$-\int_{(0,L)^3} \alpha\varphi^2 p_B \operatorname{div}\mathbf{u}\,dx + G_c \int_{(0,L)^3}\left(\frac{1}{2\varepsilon}(1-\varphi)^2 + \frac{\varepsilon}{2}|\nabla\varphi|^2\right)dx. \quad (8)$$

We remark that $\partial_N\Omega$ contains both the outer boundary and the fracture boundary. Moreover, k is a positive regularization parameter for the elastic energy, with $k \ll \varepsilon$, e.g., Braides (1998). We notice that $k > 0$ is necessary for quasi-static phase-field fracture models in order to avoid a singular discrete system. For dynamic fracture, $k = 0$ may be chosen, see e.g., Borden et al. (2012). Due to the presence of the acceleration term, in the limit $\varphi \to 0$ non-zero respective matrix entries are assured, removing the degeneracy.

We note that the pressure cross term reads

$$\int_{(0,L)^3} \alpha\varphi^2 p_B \operatorname{div}\mathbf{u}\,dx,$$

instead of

$$\int_{(0,L)^3} \alpha\varphi p_B \operatorname{div}\mathbf{u}\,dx.$$

This is linked to the behavior for negative values of the phase field variable. Moreover, if $\varphi \le 0$, there should be no contribution. Therefore instead of

$$\int_{(0,L)^3} \alpha\varphi^2 p_B \operatorname{div}\mathbf{u}\,dx,$$

we use

$$\int_{(0,L)^3} \alpha\varphi_+^2 p_B \operatorname{div}\mathbf{u}\,dx.$$

Using φ_+^2 yields a higher regularity and avoids difficulties in the differentiation since we need first and second order derivatives for Newton's method. We notice that for $0 \le \varphi \le 1$, using φ_+^2, instead of φ_+ in the pressure cross term should not affect the phase field approximation. If $\varphi = \mathbb{1}_C$, we do not see any difference.

We explain this choice in more detail in the following. In the incremental formulation, the entropy condition $\partial_t\varphi \le 0$ leads to a condition similar to the obstacle problem, which guarantees that $\varphi \le 1$. On the contrary, the presence of the pressure gradient can lead to negative values of the phase field variable. Later in Theorem 2, we show for the incremental, continuous in space problem that $\varphi \ge 0$. For the formulation which is discretized in space, the approximation for φ is not necessarily nonnegative. It becomes nonnegative only when passing to the space continuous problem. For this reason, working with φ_+ is a safeguard choice, which in the end does not modify the original problem and is numerically stable. There are formulations with good estimates for the time derivatives in which φ is nonnegative only in the space–time continuous formulation, which has been proven in Mikelić et al. (2015c).

In the following, we consider a quasi-static formulation where velocity changes are small. First, we derive an incremental form, i.e., we replace the time derivative in inequality (7) with a discretized version; more precisely

$$\partial_t \varphi \to \partial_{\Delta t} \varphi = (\varphi - \varphi_p)/(\Delta t),$$

where $\Delta t > 0$ is the time step and φ_p is the phase field from the previous time step. After time discretization, our quasistatic constrained minimization problem becomes a stationary problem, called the incremental problem.

3.2 Interface coupling of a pressurized crack with a porous medium

The crack is filled with a fluid and, consequently, it is pressurized. However, the energy E_ε given by (8) is incomplete though and we need to include the crack-pressure. To this end, we work with an internal interface between the fracture and the porous medium and derive appropriate interface conditions. A general description of a crack embedded in a porous medium is illustrated in Fig. 1. Here, we consider a setting in which the complex interface crack/pore structure is simplified. We notice that such a complex structure would require the solution of a variational problem since the formulation as energy minimization might not be well defined. Furthermore, we assume that the crack is a 3d thin domain with a width much less than its length, then lubrication theory can be applied. Hence, the leading order of the stress in $\mathcal{C}$ is $-p_f I$.

At the crack boundary, we assume the continuity of the pressures and the continuity of contact forces:

$$p_f = p_B, \qquad \sigma \mathbf{n} = (\mathcal{G}e(\mathbf{u}) - \alpha p_B I)\mathbf{n} = -p_f \mathbf{n}, \tag{9}$$

where p_f denotes the fracture fluid pressure and $\mathbf{n}$ the normal vector. We recall that $\partial \Omega$ consists of $\partial \mathcal{C}$, $\partial_N \Omega \backslash \partial \mathcal{C} = \partial_N (0, L)^3$ and $\partial_D \Omega = \partial_D (0, L)^3$. The Neumann and interface boundary parts can be written as $\partial_N \Omega = \partial_N (0, L)^3 \cup \partial \mathcal{C}$. On $\partial_D (0, L)^3$ we set the Dirichlet condition $\mathbf{u} = 0$ and on $\partial_N (0, L)^3$, we have $\sigma \mathbf{n} = \tau$.

Before introducing the phase field variable, we eliminate the traction crack surface integrals and obtain

$$\begin{aligned} &\int_\Omega \alpha p_B \text{div } \mathbf{w} \, dx + \int_{\partial \mathcal{C}} \sigma \mathbf{n} \mathbf{w} \, dS \\ &= \int_\Omega \alpha p_B \text{div } \mathbf{w} \, dx - \int_{\partial \mathcal{C}} p_f w_n \, dS \\ &= \int_\Omega \alpha p_B \text{div } \mathbf{w} \, dx - \int_\Omega \text{div } (p_B \mathbf{w}) \, dx + \int_{\partial_N (0,L)^3} p_B w_n \, dS \\ &= \int_\Omega (\alpha - 1) p_B \text{div } \mathbf{w} \, dx - \int_\Omega \nabla p_B \mathbf{w} \, dx + \int_{\partial_N (0,L)^3} p_B w_n \, dS, \end{aligned} \tag{10}$$

where $\mathbf{w} \cdot \mathbf{n}$ denotes the normal component of the vector function $\mathbf{w}$, where $\mathbf{n}$ is oriented towards interior of $\mathcal{C}$.

Remark 6 To date, in most studies dealing with pressurized fractures, the outer domain boundary conditions are of homogeneous Dirichlet type. Here, the test function $\mathbf{w}$ cancels $\int_{\partial_N(0,L)^3} p_B w_n \, dS$. However, this last integral is important when Neumann boundary conditions are prescribed, where the test function $\mathbf{w}$ does not vanish. We present such a case in Sect. 6.2.

In the above calculations, surface integrals are now treated with Gauss' divergence theorem:

$$
\begin{aligned}
&-\int_{\partial_N(0,L)^3} \tau \mathbf{w} \, dS + \int_{\partial \mathcal{C}} p w_n \, dS - \int_{\Omega} \alpha p \operatorname{div} \mathbf{w} \, dx \\
&= -\int_{\Omega} (\alpha - 1) p \operatorname{div} \mathbf{w} \, dx + \int_{\Omega} \nabla p \mathbf{w} \, dx - \int_{\Omega} \operatorname{div} (\mathcal{T} \mathbf{w}) \, dx \\
&= -\int_{\Omega} (\alpha - 1) p \operatorname{div} \mathbf{w} \, dx + \int_{\Omega} (\nabla p - \operatorname{div} \mathcal{T}) \mathbf{w} \, dx - \int_{\Omega} \mathcal{T} : e(\mathbf{w}) \, dx, \quad (11)
\end{aligned}
$$

where $\mathcal{T}$ is a smooth symmetric 3×3 matrix with compact support in a neighborhood of $\partial(0, L)^3$, such that $\mathcal{T}\mathbf{n} = \tau + p\mathbf{n}$ on $\partial_N(0, L)^3$. The tensor $\mathcal{T}$ is introduced in order to handle the phase field only in volume terms. Assuming that the crack $\mathcal{C}$ does not interact with $\partial_N \Omega$, it can be eliminated by using Green's formula. Hence the solution does not depend on the choice of $\mathcal{T}$. We set

$$
\mathcal{F} = -(\alpha - 1) p I - \mathcal{T}, \qquad \mathbf{f} = \nabla p - \operatorname{div} \mathcal{T}. \quad (12)
$$

In the case of $\partial_N \Omega = \emptyset$, we have $\mathcal{T} \equiv 0$. Then, the terms in (12) for $\mathcal{F}$ and $\mathbf{f}$ reduce to $\mathcal{F} = -(\alpha - 1) p I$ and $\mathbf{f} = \nabla p$.

After the above transformation and after taking $\sigma_0 = 0$ and neglecting the gravity term $\rho_b g$, the weak formulation of problem (1) reads as follows

$$
\begin{aligned}
0 &= \int_{\Omega} \sigma^{por} : e(\mathbf{w}) \, dx - \int_{\partial_N \Omega} \tau \cdot \mathbf{w} \, dS \\
&= \int_{\Omega} \mathcal{G} e(\mathbf{u}) : e(\mathbf{w}) \, dx - \int_{\Omega} (\alpha - 1) p \operatorname{div} \mathbf{w} \, dx + \int_{\Omega} \nabla p \cdot \mathbf{w} \, dx \\
&\quad - \int_{\partial_N \Omega \setminus \partial \mathcal{C}} (\tau + p\mathbf{n}) \cdot \mathbf{w} \, dS \\
&= \int_{\Omega} \mathcal{G} e(\mathbf{u}) : e(\mathbf{w}) \, dx + \int_{\Omega} (\mathcal{F} : e(\mathbf{w}) + \mathbf{f} \cdot \mathbf{w}) \, dx \quad (13)
\end{aligned}
$$

for all $\mathbf{w} \in \{\mathbf{w} \in H^1(\Omega)^3 | \ \mathbf{w} = 0 \text{ on } \partial_D \Omega\}$. To the variational equation (13) corresponds the following variant of the energy functional (8):

$$\tilde{E}_\varepsilon(\mathbf{u},\varphi) = \int_{(0,L)^3} \frac{1}{2}\Big((1-k)\varphi^2 + k\Big)\mathcal{G}e(\mathbf{u}) : e(\mathbf{u})\,dx$$
$$+ \int_{(0,L)^3} \varphi^2(\mathcal{F} : e(\mathbf{u}) + \mathbf{f}\cdot\mathbf{u})\,dx$$
$$+ G_c \int_{(0,L)^3} \left(\frac{1}{2\varepsilon}(1-\varphi)^2 + \frac{\varepsilon}{2}|\nabla\varphi|^2\right)\,dx. \tag{14}$$

Remark 7 We note that introduction of the phase field approximation of the pressured fracture in this section was introduced, differs from Bourdin et al. (2012). In fact, the presence of the term $\int_\Omega p\mathbf{u}\cdot\nabla\varphi\,dx$ must be treated carefully numerically and we have derived therefore a different phase field energy functional. The $\Gamma-$limit of our formulation was calculated for a particular one dimensional setting in Engwer and Schumacher (2017), which leads to the same formulation for the lower dimensional fracture as in Bourdin et al. (2012).

Remark 8 We emphasize that the previous choice $p_B = p_f$ on the fracture boundary is one possible modeling choice. It may be justified to assume that $p_f \gg p_B$ such that a discontinuous pressure could be more appropriate. Such a modeling is left for future studies.

Remark 9 In the fixed stress splitting $\mathcal{F}$ and f depend on the pressure. For details we refer the reader to Mikelić et al. (2015b).

3.3 The final energy functional

In the case of elastic cracks it can be shown that the phase field unknown satisfies $0 \le \varphi \le 1$. In order to establish this property for the spatially continuous incremental problem, we first modify (14) for negative values of φ. As previously discussed, we now use φ_+ instead of φ in terms where negative φ could lead to incorrect physics in the bulk energy, traction and pressure forces. With this modification, the final energy functional reads

$$\mathcal{E}_\varepsilon(\mathbf{u},\varphi) = \int_{(0,L)^3} \frac{1}{2}\Big((1-k)\varphi_+^2 + k\Big)\mathcal{G}e(\mathbf{u}) : e(\mathbf{u})\,dx$$
$$+ \int_{(0,L)^3} \varphi_+^2(\mathcal{F} : e(\mathbf{u}) + \mathbf{f}\cdot\mathbf{u})\,dx$$
$$+ G_c \int_{(0,L)^3} \left(\frac{1}{2\varepsilon}(1-\varphi)^2 + \frac{\varepsilon}{2}|\nabla\varphi|^2\right)\,dx. \tag{15}$$

As functional space of admissible displacements, we choose

$$V_U = \{\mathbf{z} \in H^1((0,L)^3)^3 \mid \mathbf{z} = 0 \text{ on } \partial_D\Omega\}.$$

The entropy condition (7) is imposed in its discretized form and we introduce a convex set K:

$$K = \{\psi \in H^1((0,L)^3) \mid \psi \le \varphi_p \le 1 \text{ a.e. on } (0,L)^3\}, \tag{16}$$

where $\varphi_p(x)$ is the value of the phase field from the previous time step. The incremental minimization problem now reads:

Definition 1 Find $\mathbf{u} \in V_U$ and a nonnegative $\varphi \in K$ such that

$$\mathcal{E}_\varepsilon(\mathbf{u}, \varphi) = \min_{\{\mathbf{v}, \psi\} \in V_U \times K} \mathcal{E}_\varepsilon(\mathbf{v}, \psi). \tag{17}$$

Note that the value of the phase field unknown φ from the previous time step enters only the convex set K, as the obstacle φ_p. The goal of Sect. 4 is to establish a solution to the minimization problem (17).

3.4 The Euler–Lagrange equations in strong form

From the energy functional, we obtain by differentiation and application of the fundamental lemma of calculus of variations the strong formulation: Find $u : (0, L)^3 \to \mathbb{R}^3$ and $\varphi : (0, L)^3 \to \mathbb{R}$ such that

$$-\operatorname{div}\left(\left((1-k)\varphi_+^2 + k\right)\mathcal{G}e(\mathbf{u})\right) + \varphi_+^2 \mathbf{f} - \operatorname{div}(\varphi_+^2 \mathcal{F}) = 0 \quad \text{in } (0, L)^3, \tag{18}$$

$$\mathbf{u} = 0 \quad \text{on } \partial_D (0, L)^3, \tag{19}$$

$$\left((1-k)\varphi_+^2 + k\right)\mathcal{G}e(\mathbf{u})\mathbf{n} = -\varphi_+^2 \mathcal{F}\mathbf{n} \quad \text{on } \partial_N (0, L)^3, \tag{20}$$

and

$$\partial_{\Delta t}\varphi \le 0 \quad \text{on } (0, L)^3 \quad \text{and} \quad \frac{\partial \varphi}{\partial \mathbf{n}} = 0 \quad \text{on} \quad \partial(0, L)^3, \tag{21}$$

$$\begin{aligned} &-G_c \varepsilon \Delta \varphi - \frac{G_c}{\varepsilon}(1-\varphi) + (1-k)\mathcal{G}e(\mathbf{u}) : e(\mathbf{u})\varphi_+ \\ &+2\varphi_+(\mathcal{F} : e(\mathbf{u}) + \mathbf{f} \cdot \mathbf{u}) \le 0 \quad \text{in } (0, L)^3, \end{aligned} \tag{22}$$

$$\begin{aligned} &\Big\{-G_c \varepsilon \Delta \varphi - \frac{G_c}{\varepsilon}(1-\varphi) + (1-k)\mathcal{G}e(\mathbf{u}) : e(\mathbf{u})\varphi_+ \\ &+2\varphi_+(\mathcal{F} : e(\mathbf{u}) + \mathbf{f} \cdot \mathbf{u})\Big\}\, \partial_{\Delta t}\varphi = 0 \ \text{in} \ (0, L)^3, \end{aligned} \tag{23}$$

where (23) is the strong form of Rice' condition (which is a well-known complementarity condition). This two-field formulation can be compared with the Model I formulation given in Miehe et al. (2010) (see p. 1289). The main difference is that the system (18)–(23) is a variational inequality; and in Miehe et al. (2010) a penalization term is used for solving the inequality.

4 Well-posedness of the model

4.1 Existence of a minimizer to the energy functional $\mathcal{E}_\varepsilon$

In this section, we seek for a solution to the minimization problem (17). The strategy is to consider the integrand of (15), using the notation $\mathbf{z} := (\mathbf{v}, \varphi)$, and ξ stands for the components of the gradient of the displacements and the gradient of the phase-field function. With z_4, we access the fourth component of $\mathbf{z}$, namely the phase-field function. Lastly, z_{4+} denotes the positive part of the phase-field unknown. Then,

$$g(x, \mathbf{z}, \xi) = \frac{1}{2}\Big((1-k)(\inf\{z_{4+}, 1\})^2 + k\Big) \sum_{i,j,k,r=1}^{3} \mathcal{G}_{ijkr}\xi_{kr}\xi_{ij} + G_c(\frac{1}{2\varepsilon}(1-z_4)^2 + \frac{\varepsilon}{2}|\nabla z_4|^2) + (\inf\{z_{4+}, 1\})^2 \left(\sum_{i,j=1}^{3} F_{ij}\xi_{ij} + \sum_{i=1}^{3} f_i z_i \right), \tag{24}$$

defined on $(0, L)^3 \times \mathbb{R}^4 \times \mathbb{R}^{12} \to \mathbb{R} \cup \{+\infty\}$. It is convex in ξ and we will prove that it is

(i) a Caratheodory function (i.e. a continuous function on $\mathbb{R}^4 \times \mathbb{R}^{12}$ for every x from $(0, L)^3$ and a measurable function on $(0, L)^3$ for every $\{\mathbf{z}, \xi\}$ from $\mathbb{R}^4 \times \mathbb{R}^{12}$);
(ii) the energy functional (15) is coercive.

Then Corollary 3.24, p. 97, from Dacorogna's monograph (2008) yields the lower semi-continuity of the energy functional. Proving existence of at least one point of minimum is then a classical task.

We start with a result which follows directly from the basic theory:

Lemma 1 *Let* $\mathbf{f}$ *and* $\mathbf{F} \in L^2$; *and* G_c, b *be nonnegative constants. Let* ε *be a positive small parameter. Then the integrand* $g(\cdot, \cdot, \cdot)$ *given by* (24) *is a Caratheodory function.*

Proposition 1 *Under the assumptions of* Lemma 1, *the functional*

$$\Phi(\mathbf{v}, \varphi) = \int_{(0,L)^3} g(x, \{\mathbf{v}, \varphi\}, \{e(\mathbf{v}), \nabla\varphi\})\, dx \tag{25}$$

is coercive over $V_U \times H^1((0, L)^3) \cap K$, *i.e.*

$$\lim \Phi(\mathbf{v}, \varphi) \to \infty, \qquad \text{when } ||\mathbf{v}||_{V_U} + ||\varphi||_{H^1} \to \infty. \tag{26}$$

Proof Let us introduce the abbreviation $\tilde{\varphi} = \inf\{\varphi_+, 1\}$. Let c be a generic constant. We estimate all terms one by one:

$$\left| \int_{(0,L)^3} (\tilde{\varphi})^2 (\mathbf{f}\mathbf{v} + \mathcal{F} : e(\mathbf{v})\, dx \right| \le ||\mathbf{v}||_{L^2}\, ||\mathbf{f}||_{L^2} + ||\tilde{\varphi} e(\mathbf{v})||_{L^2}\, ||\mathcal{F}||_{L^2}. \tag{27}$$

The elastic energy terms yield

$$\int_{(0,L)^3} \left((1-k)(\tilde{\varphi})^2 + k\right) \mathcal{G}e(\mathbf{v}) : e(\mathbf{v})\, dx \geq ck||e(\mathbf{v})||_{L^2}^2 + c(1-k)||\tilde{\varphi}e(\mathbf{v})||_{L^2}^2. \quad (28)$$

We recall that, by Korn's inequality,

$$||\mathbf{v}||_{H^1((0,L)^3)} \leq C_K ||e(\mathbf{v})||_{L^2((0,L)^3)}, \qquad \forall \mathbf{v} \in V_U. \quad (29)$$

Therefore, putting together (27) and (28), and using (29), yields

$$\Phi(\mathbf{v}, \varphi) \geq G_c \int_{(0,L)^3} \left(\frac{(1-\varphi)^2}{2\varepsilon} + \varepsilon |\nabla \varphi|^2 \right) dx + \frac{ck}{4} ||e(\mathbf{v})||_{L^2}^2$$
$$+ \frac{c(1-k)}{4} ||\tilde{\varphi}e(\mathbf{v})||_{L^2}^2 - \frac{||\mathcal{F}||_{L^2}^2}{c(1-k)} - \frac{C_K^2 ||\mathbf{f}||_{L^2}^2}{ck}. \quad (30)$$

The coerciveness property (26) follows from (30). □

Our goal is to prove the following theorem:

Theorem 1 (Existence of a minimizer to the incremental phase field problem) *Let* $\varepsilon, k > 0$ *and* $\mathcal{F}, \mathbf{f} \in L^2$, $\varphi_p \in H^1$, $0 \leq \varphi_p \leq 1$ *a.e. on* $(0, L)^3$. *Then the minimization problem* (17) *has at least one solution* $\{\mathbf{u}, \varphi\} \in V_U \times K$ *and* $\varphi \geq 0$ *a.e. on* $(0, L)^3$.

Proof Let $\{\mathbf{u}^k, \varphi^k\}_{k\in\mathbb{N}} \in V_U \times K$ be a minimizing sequence for the minimization problem (17) for Φ; that is a sequence of elements of $V_U \times K$ such that $\Phi(\mathbf{u}^k, \varphi^k) \to \inf_{V_U \times K} \Phi(\mathbf{v}, \varphi)$. By proposition (1) and the inequality (30) $\inf_{V_U \times K} \Phi(\mathbf{v}, \varphi) \neq -\infty$. The sequence $\{\mathbf{u}^k, \varphi^k\}_{k\in\mathbb{N}}$ is uniformly bounded in $V_U \times K$ and $\{\varphi_+^k\}_{k\in\mathbb{N}}$ is uniformly bounded in $L^\infty((0, L)^3)$. Therefore there exists $\{\mathbf{u}, \varphi\}$ and a subsequence, denoted by the same superscript, such that for $k \to \infty$

$$\begin{aligned} \{\mathbf{u}^k, \varphi^k\} \to \{\mathbf{u}, \varphi\} \quad & \text{weakly in } V_U \times H^1((0, L)^3), \\ & \text{strongly in } L^q((0, L)^3)^4, \ q < 6, \\ & \text{and a.e. on } (0, L)^3. \end{aligned} \quad (31)$$

Next, inequality (30) yields

$$g(x, \mathbf{v}, \xi) \geq \langle a(x), \xi \rangle + B, \qquad \text{for } \{\mathbf{v}, \xi\} \in \mathbb{R}^4 \times \mathbb{R}^{12} \text{ and a.e. } x \in (0, L)^3,$$

with $a \in L^2((0, L)^3)$ and $B \in \mathbb{R}$. Consequently, we are in a situation to apply Corollary 3.24, p. 97, from Dacorogna (2008). This result yields the weak lower semicontinuity of the functional Φ and hence

$$\Phi(\mathbf{u}, \varphi) \leq \liminf \Phi(\mathbf{u}^k, \varphi^k) = \inf_{V_U \times K} \Phi(\mathbf{v}, \varphi). \quad (32)$$

Since

$$\Phi(\mathbf{v}, \psi) = \mathcal{E}_\varepsilon(\mathbf{v}, \psi) \quad \text{on } V_u \times K,$$

we have proven that $\{\mathbf{u}, \varphi\} \in V_U \times H^1((0, L)^3) \cap K$ is a solution for the minimization problem.

It remains to prove that φ is nonnegative. We evaluate the functional Φ at the point $\{\mathbf{u}, \varphi_+\}$. Obviously $\varphi_+ \in K$. A direct calculation yields

$$\Phi(\mathbf{u}, \varphi_+) = \Phi(\mathbf{u}, \varphi) - \frac{G_c}{2\varepsilon} \int_{(0,L)^3} \varphi_-(\varphi_- - 2)\, dx - \frac{\varepsilon G_C}{2} \int_{(0,L)^3} |\nabla \varphi_-|^2\, dx. \quad (33)$$

Therefore $\{\mathbf{u}, \varphi\}$ can be a point of minimum only if $\varphi_- = 0$ and we conclude that $\varphi \geq 0$ a.e. on $(0, L)^3$. □

Corollary 1 (Euler–Lagrange weak PDE formulation) *Let the hypotheses of* Theorem 1 *be satisfied. Then the Euler–Lagrange equations corresponding to the minimization problem* (17)

$$\int_{(0,L)^3} \Big((1-k)\varphi^2 + k\Big) \mathcal{G}e(\mathbf{u}) : e(\mathbf{w})\, dx + \int_{(0,L)^3} \varphi^2(\mathcal{F} : e(\mathbf{w}) + \mathbf{f} \cdot \mathbf{w})\, dx = 0, \quad (34)$$

for all $\mathbf{w} \in V_U$, *and*

$$\int_{(0,L)^3} (1-k)\varphi\psi\mathcal{G}e(\mathbf{u}) : e(\mathbf{u})\, dx + G_c \int_{(0,L)^3} \left(-\frac{1}{\varepsilon}(1-\varphi)\psi + \varepsilon\nabla\varphi \cdot \nabla\psi\right) dx$$
$$+2\int_{(0,L)^3} \varphi(\mathbf{f} \cdot \mathbf{u} + \mathcal{F} : e(\mathbf{u}))\psi\, dx \leq 0, \quad (35)$$

for all $\psi \in H^1((0, L)^3)$, $\psi \geq 0$ *a.e. on* $(0, L)^3$, *and*

$$\int_{(0,L)^3} (1-k)\varphi(\varphi_p - \varphi)\mathcal{G}e(\mathbf{u}) : e(\mathbf{u})\, dx + G_c \int_{(0,L)^3} \Big(-\frac{1}{\varepsilon}(1-\varphi)(\varphi_p - \varphi)$$
$$+\varepsilon\nabla\varphi \cdot \nabla(\varphi_p - \varphi)\Big)\, dx + 2\int_{(0,L)^3} \varphi(\mathbf{f} \cdot \mathbf{u} + \mathcal{F} : e(\mathbf{u}))(\varphi_p - \varphi)\, dx = 0, \quad (36)$$

admit a solution $\{\mathbf{u}, \varphi\} \in V_U \times H^1((0, L)^3) \cap K$, *such that* $\varphi \geq 0$ *a.e. on* $(0, L)^3$. *We observe that equation* (36) *is the Rice condition* (*see e.g.* Francfort 2011).

In the next result, we show that our crack cannot become a 'fat' (balloon-like) crack, but remains tiny in the third direction:

Corollary 2 *Let the hypotheses of Theorem 1 be satisfied. Let in addition the previous phase-field values* φ_p *satisfy*

$$\int_{(0,L)^3} (1-\varphi_p)^2\,dx = ||1-\varphi_p||^2_{L^2((0,L)^3)} = C\varepsilon$$

and

$$\|\sqrt{\varepsilon}\nabla\varphi_p\|_{L^2((0,L)^3)} \leq C.$$

Then the current phase-field variable φ *satisfy the same estimates:*

$$\int_{(0,L)^3} (1-\varphi)^2\,dx = ||1-\varphi||^2_{L^2((0,L)^3)} = C\varepsilon$$

and

$$\int_{\{\varphi\leq q\}} dx = meas\{\varphi \leq q\} \leq \frac{C\varepsilon}{(1-q)^2} \quad \forall q \in [0,1).$$

Proof We evaluate

$$\begin{aligned}\Phi(0,\varphi_p) &= G_c \int_{(0,L)^3} \left(\frac{1}{2\varepsilon}(1-\varphi_p)^2 + \varepsilon|\nabla\varphi_p|^2\right) \\ &= \underbrace{G_c\varepsilon \int_{(0,L)^3} |\nabla\varphi_p|^2}_{\leq c} + \frac{G_c}{2\varepsilon}\int_{((0,L)^3)} (1-\varphi_p)^2\,dx \\ &\leq C.\end{aligned}$$

Since $\Phi(u,\varphi) \leq \Phi(0,\varphi_p)$ we use (30) and obtain the claimed estimates. □

Remark 10 This theoretical property in Corollary 2 has also been confirmed in our numerical simulations in Sect. 6 in which the crack stays tiny in the second (2d) or third direction (3d), but grows into the other (2d) or two other (3d) directions.

4.2 A finite dimensional approximation

The finite dimensional approximation serves for two purposes. First, we continue our well-posedness study. Secondly, by specifying the discrete basis function through finite element functions with small support, we obtain a numerical procedure for a computer implementation. Let $\{\psi_r\}_{r\in\mathbb{N}}$ be a basis for $H^1((0,L)^3)$ and $\{\mathbf{w}^r\}_{r\in\mathbb{N}}$ be a basis for V_U. We start by defining a finite dimensional approximation to the minimization problem (17).

Definition 2 **(of a penalized approximation)**
Let us suppose the assumptions of Theorem 1 and a penalization parameter $\delta \in \mathbb{R}$ and in particular, let $\delta := M \in \mathbb{N}$ in this section. Let $\tilde{\varphi} = \inf\{1, \varphi_+\}$. The pair

$\{\mathbf{u}^M, \varphi^M\}$, $\mathbf{u}^M = \sum_{r=1}^M a_r \mathbf{w}^r$ and $\varphi^M = \sum_{r=1}^M b_r \psi_r$, is a finite dimensional approximative solution for problem (17) if it is a minimizer to the problem

$$\inf_{V_U^M \times W^M} \left\{ \Phi(\mathbf{v}, \varphi) + \int_{(0,L)^3} \frac{\delta}{2} (\varphi - \varphi_p^M)_+^2 \, dx \right\}, \tag{37}$$

where $V_U^M = \text{span}\,\{\mathbf{w}^r\}_{r=1,\ldots,M}$, $W^M = \text{span}\,\{\psi_r\}_{r=1,\ldots,M}$ and φ_p^M is a projection of φ_p on W^M.

Formulation 1 (Discrete weak formulation) *Each solution for the problem* (37) *satisfies the discrete variational formulation*

$$\int_{(0,L)^3} \left((1-k)(\tilde{\varphi}^M)^2 + k \right) \mathcal{G} e(\mathbf{u}^M) : e(\mathbf{w}^r) \, dx$$
$$+ \int_{(0,L)^3} (\tilde{\varphi}^M)^2 (\mathcal{F} : e(\mathbf{w}^r) + \mathbf{f} \mathbf{w}^r) \, dx = 0, \quad \forall r = 1, \ldots, M, \tag{38}$$

$$G_c \int_{(0,L)^3} \left(-\frac{1}{\varepsilon} (1 - \varphi^M) \psi_r + \varepsilon \nabla \varphi^M \cdot \nabla \psi_r \right) dx + \int_{(0,L)^3} \delta (\varphi^M - \varphi_p^M)_+ \psi_r \, dx$$
$$+ 2 \int_{(0,L)^3} (\tilde{\varphi}^M)(\mathcal{F} : e(\mathbf{u}^M) + \mathbf{f} \cdot \mathbf{u}^M) \psi_r \, dx$$
$$+ \int_{(0,L)^3} (1-k) \tilde{\varphi}^M \psi_r \mathcal{G} e(\mathbf{u}^M) : e(\mathbf{u}^M) \, dx = 0, \quad \forall\, r = 1, \ldots, M. \tag{39}$$

Proposition 2 *We assume the hypotheses of* Theorem 1. *Then there exists a penalized finite dimensional approximation for problem* (37) *that satisfies the a priori estimate*

$$G_c \int_{(0,L)^3} \frac{(1 - \varphi^M)^2}{\varepsilon} \, dx + \int_{(0,L)^3} M (\varphi^M - \varphi_p^M)_+^2 \, dx$$
$$+ k ||e(\mathbf{u}^M)||_{L^2}^2 + ||\tilde{\varphi}^M e(\mathbf{u}^M)||_{L^2}^2 \leq c, \tag{40}$$

where c is independent of M.

Proof This is a consequence of (30) in Proposition (1) and the continuity of the integrand. □

Theorem 2 *Assume the hypotheses of* Theorem 1. *Then there exists a subsequence of* $\{\mathbf{u}^M, \varphi^M\} \in V_U^M \times W^M$, *denoted by the same symbol, and* $\{\mathbf{u}, \varphi\} \in V_U \times H^1((0, L)^3) \cap K$, $\varphi \geq 0$ *a.e., being a minimizer to the problem* (17) *and such that*

$$\{\mathbf{u}^M, \varphi^M\} \to \{\mathbf{u}, \varphi\} \quad in \quad V_U \times H^1((0, L)^3). \tag{41}$$

Proof By Proposition 2 there is a solution $\{\mathbf{u}^M, \varphi^M\}$ for problem (37), satisfying the a priori estimate (40). Therefore there exists $\{\mathbf{u}, \varphi\}$ and a subsequence, denoted by the same superscript, such that

$$\begin{aligned}
&\{\mathbf{u}^M, \varphi^M\} \to \{\mathbf{u}, \varphi\} \quad \text{weakly in } V_U \times H^1((0, L)^3), \\
&\quad \text{strongly in } L^q((0, L)^3)^4, \; q < 6, \\
&\quad \text{and a.e. on } \quad (0, L)^3, \quad \text{as } M \to \infty. \qquad (42)
\end{aligned}$$

Obviously $(\varphi^M - \varphi_p^M)_+ \to 0$, as $M \to \infty$, and $\varphi \in K$.

Next, let

$$\varphi \in K^N = \{z \in W^N : z(x) \leq \varphi_p^M(x) \text{ a.e. on } (0, L)^3\}, N \leq M.$$

Then we have

$$\Phi(\mathbf{u}^M, \varphi^M) + \int_{(0,L)^3} \frac{M}{2} (\varphi^M - \varphi_p^M)_+^2 \, dx \leq \Phi(\mathbf{v}, \varphi),$$

for all $\{\mathbf{v}, \varphi\} \in V_U^N \times W^N \cap K$. The limit $M \to \infty$ yields

$$\Phi(\mathbf{u}, \varphi) \leq \Phi(\mathbf{v}, \varphi), \quad \forall \{\mathbf{v}, \varphi\} \in V_U^N \times W^N \cap K.$$

After passing to the limit $N \to \infty$, we conclude that $\{\mathbf{u}, \varphi\} \in V_U \times H^1((0, L)^3) \cap K$ is a solution to problem (17). As before, it still can be shown that φ is nonnegative.

It remains to establish strong convergence of the gradients. Passing to the limit in equation (38) is straightforward and we conclude that $\{\mathbf{u}, \varphi\}$ satisfies equation (34). Next we choose $\mathbf{w} = \mathbf{u}^M$ as test function in (38) and pass to the limit $M \to \infty$. Thus,

$$\begin{aligned}
&\int_{(0,L)^3} \Big((1 - k)(\varphi_+)^2 + k \Big) \mathcal{G}e(\mathbf{u}) : e(\mathbf{u}) \, dx \\
&\quad + \int_{(0,L)^3} (\varphi_+)^2 (\mathcal{F} : e(\mathbf{u}) + \mathbf{f} \cdot \mathbf{u}) \, dx = 0. \qquad (43)
\end{aligned}$$

Therefore we have the convergence of the weighted elastic energies

$$\begin{aligned}
&\lim_{M \to \infty} \int_{(0,L)^3} \Big((1 - k)(\tilde{\varphi}^M)^2 + k \Big) \mathcal{G}e(\mathbf{u}^M) : e(\mathbf{u}^M) \, dx \\
&= \int_{(0,L)^3} \Big((1 - k)(\varphi_+)^2 + k \Big) \mathcal{G}e(\mathbf{u}) : e(\mathbf{u}) \, dx. \qquad (44)
\end{aligned}$$

Using Fatou's lemma we have

$$\begin{aligned}
&\int_{(0,L)^3} \liminf_{M \to \infty} \Big((1 - k)(\tilde{\varphi}^M)^2 + k \Big) \mathcal{G}e(\mathbf{u}^M) : e(\mathbf{u}^M) \, dx \\
&\leq \liminf_{M \to \infty} \int_{(0,L)^3} \Big((1 - k)(\tilde{\varphi}^M)^2 + k \Big) \mathcal{G}e(\mathbf{u}^M) : e(\mathbf{u}^M) \, dx \\
&= \int_{(0,L)^3} \Big((1 - k)(\varphi_+)^2 + k \Big) \mathcal{G}e(\mathbf{u}) : e(\mathbf{u}) \, dx. \qquad (45)
\end{aligned}$$

Consequently

$$\mathbf{u}^M \to \mathbf{u} \quad \text{strongly in} \quad V_U, \quad \text{as } M \to \infty. \tag{46}$$

For every $\psi \in L^\infty((0,L)^3) \cap H^1((0,L)^3)$, (46) implies

$$\lim_{M\to\infty} \left| \int_{(0,L)^3} \tilde{\varphi}^M \psi \mathcal{G} e(\mathbf{u}^M - \mathbf{u}) : e(\mathbf{u}^M - \mathbf{u})\, dx \right| \to 0, \quad \text{as } M \to \infty,$$

and

$$\begin{aligned}
&\int_{(0,L)^3} \tilde{\varphi}^M \psi \mathcal{G} e(\mathbf{u}^M) : e(\mathbf{u}^M)\, dx = \int_{(0,L)^3} \tilde{\varphi}^M \psi \mathcal{G} e(\mathbf{u}^M - \mathbf{u}) : e(\mathbf{u}^M - \mathbf{u})\, dx \\
&\quad +2 \int_{(0,L)^3} \tilde{\varphi}^M \psi \mathcal{G} e(\mathbf{u}^M) : e(\mathbf{u})\, dx - \int_{(0,L)^3} \tilde{\varphi}^M \psi \mathcal{G} e(\mathbf{u}) : e(\mathbf{u})\, dx \\
&\quad \to \int_{(0,L)^3} \varphi_+ \psi \mathcal{G} e(\mathbf{u}) : e(\mathbf{u})\, dx, \quad \text{as } M \to \infty.
\end{aligned} \tag{47}$$

Next we use Minty's lemma and write equation (39) in the equivalent form

$$\begin{aligned}
&\int_{(0,L)^3} (1-k) \inf\{\varphi_+^M, 1\} (\psi - \varphi^M) \mathcal{G} e(\mathbf{u}^M) : e(\mathbf{u}^M)\, dx \\
&\quad + G_c \int_{(0,L)^3} \left(\frac{(\psi - 1)}{\varepsilon} (\psi - \varphi^M) \right. \\
&\quad \left. + \varepsilon \nabla \psi \cdot \nabla (\psi - \varphi^M) \right) dx \\
&\quad +2 \int_{(0,L)^3} (\inf\{\varphi_+^M, 1\}) \, (\mathbf{f} \cdot \mathbf{u}^M + \mathcal{F} : e(\mathbf{u}^M))(\psi - \varphi^M) dx \\
&\quad + \int_{(0,L)^3} M(\psi - \varphi_p^M)_+ (\psi - \varphi^M)\, dx \geq 0, \quad \forall\, \psi \in W^M.
\end{aligned} \tag{48}$$

After taking $\psi = \varphi_p^M$, we use the convergence (47) pass to the limit $M \to \infty$ (see e. g. Kinderlehrer and Stampacchia 2000), and obtain

$$\begin{aligned}
&\lim_{M\to\infty} G_c \int_{(0,L)^3} \varepsilon |\nabla \varphi^M|^2\, dx = - \int_{(0,L)^3} (1-k) \varphi (\varphi - \varphi_p) \mathcal{G} e(\mathbf{u}) : e(\mathbf{u})\, dx \\
&\quad + \int_{(0,L)^3} \left(\frac{G_c}{\varepsilon} (1 - \varphi_p)(\varphi - \varphi_p) - \frac{G_c}{\varepsilon} (\varphi - \varphi_p)^2 + \varepsilon \nabla \varphi \cdot \nabla \varphi_p \right) dx \\
&\quad -2 \int_{(0,L)^3} \varphi_+ \, (\mathbf{f} \cdot \mathbf{u} + \mathcal{F} : e(\mathbf{u}))(\varphi - \varphi_p)\, dx = G_c \int_{(0,L)^3} \varepsilon |\nabla \varphi|^2\, dx.
\end{aligned} \tag{49}$$

This finishes the proof. □

5 Numerical approximation

We now formulate finite element approximations to (35)–(34), which are analogous to equations (38)–(39). For spatial discretization, we apply a standard Galerkin finite element method on quadrilaterals (2d) and hexahedra (3d), respectively. Specifically, we approximate displacements by continuous bilinears (2d) or trilinears (3d) and refer to the finite element space as V_h. Also, we take φ to be bilinears (2d) and trilinears (3d), and denote this space as W_h; see e.g. Ciarlet (1987). Here h represents the standard approximation parameter. We deal with:

Formulation 2 (Weak form) *Find* $\{u^h, \varphi^h\} \in V_h \times W_h$ *such that*

$$\int_{(0,L)^3} \Big((1-k)(\tilde{\varphi}^h)^2 + k\Big) \mathcal{G}e(\mathbf{u}^h) : e(\mathbf{w})\, dx + \int_{(0,L)^3} (\tilde{\varphi}^h)^2 (\mathcal{F} : e(\mathbf{w}) + \mathbf{f} \cdot \mathbf{w})\, dx = 0 \quad \forall \mathbf{w} \in V_h, \tag{50}$$

$$G_c \int_{(0,L)^3} \left(-\frac{1}{\varepsilon}(1-\varphi^h)\psi + \varepsilon \nabla\varphi^h \cdot \nabla\psi\right) dx + \int_{(0,L)^3} \delta(\partial_{\Delta t}\varphi^h)_+ \psi\, dx$$
$$+2\int_{(0,L)^3} (\tilde{\varphi}^h)(\mathcal{F} : e(\mathbf{u}^h) + \mathbf{f} \cdot \mathbf{u}^h)\psi\, dx$$
$$+\int_{(0,L)^3} (1-k)\tilde{\varphi}^h \psi \mathcal{G}e(\mathbf{u}^h) : e(\mathbf{u}^h)\, dx = 0 \quad \forall\, \psi \in W_h. \tag{51}$$

The incremental formulation (50)–(51) corresponds to the (pseudo-) time stepping scheme based on a difference quotient approximation with backward differences for the time derivatives. In the quasi-static model the time derivative $\delta[\partial_t\varphi]_+$ is present and is discretized as follows

$$\delta[\partial_t\varphi]_+ \rightarrow \delta[\partial_{\Delta t}\varphi]_+ = \delta\frac{[\varphi - \varphi^{n-1}]_+}{\Delta t},$$

with the time step size Δt, where $n-1$ is used to indicate the preceding time step. We then obtain for the weak form:

$$\delta(\varphi_+ - \varphi_+^{n-1}, \psi)_{L^2} + \Delta t(B, \psi)_{L^2} = 0, \quad \forall\psi \in W_h. \tag{52}$$

Here, $(\cdot,\cdot)$ denotes the discrete scalar product in L^2 and A and B denote the operators of all remaining terms for the present time step in the weak formulation, where the equation (52) is related to equations (50) and (51). Finally, the spatially discretized semi-linear form can be written in the following way:

Finite Element Formulation 1 *Find* $\mathbf{U}^h := \{\mathbf{u}^h, \varphi^h\} \in V_h \times W_h$ *such that:*

$$A(\mathbf{U}^h)(\mathbf{\Psi}) = \delta([\varphi^h - \varphi^{h,n-1}]_+, \psi)_{L^2} + \Delta t A_S(\mathbf{U}^h)(\mathbf{\Psi}) = 0,$$

with

$$
\begin{aligned}
A_S(\mathbf{U}^h)(\boldsymbol{\Psi}) &= \Big(((1-k)(\inf\{\varphi_+^h,1\})^2 + k)\mathcal{G}e(\mathbf{u}^h), e(\mathbf{w})\Big)_{L^2} - \langle \tau, \mathbf{w}\rangle_{\partial_N \Omega} \\
&- ((\inf\{\varphi_+^h,1\})^2(\alpha-1)p_B, \nabla\cdot\mathbf{w})_{L^2} + (\nabla p_B(\inf\{\varphi_+^h,1\})^2, \mathbf{w})_{L^2} \\
&+ \Big((1-k)\mathcal{G}e(\mathbf{u}^h) : e(\mathbf{u}^h)(\inf\{\varphi_+^h,1\}), \psi\Big)_{L^2} - \frac{G_c}{\epsilon}(1-\varphi^h,\psi)_{L^2} \\
&+ G_c\epsilon(\nabla\varphi^h, \nabla\psi)_{L^2} - 2\Big((\inf\{\varphi_+^h,1\})\Big((\alpha-1)p_B\nabla\cdot\mathbf{u}^h - \nabla p_B\cdot\mathbf{u}^h\Big),\psi\Big)_{L^2}
\end{aligned}
$$

for all $\boldsymbol{\Psi} = \{\mathbf{w},\psi\} \in V_h \times W_h$*, where* $A_S(\cdot)(\cdot)$ *is the sum of equations* (50) *and* (51) *and equality* (11) *is applied in the relation between* τ *and* $\mathcal{T}$.

5.1 Linearization and Newton's method

The nonlinear problem is solved with Newton's method. For the iteration steps $m = 0, 1, 2, \ldots$, it holds:

$$
A'(\mathbf{U}^{h,m})(\boldsymbol{\Delta}\mathbf{U}^h, \boldsymbol{\Psi}) = -A(\mathbf{U}^{h,m})(\boldsymbol{\Psi}), \quad \mathbf{U}^{h,m+1} = \mathbf{U}^{h,m} + \lambda\boldsymbol{\Delta}\mathbf{U}^h, \qquad (53)
$$

with $\boldsymbol{\Delta}\mathbf{U}^h = \{\boldsymbol{\Delta}\mathbf{U}^h, \Delta\varphi^h\}$, and a line search parameter $\lambda \in (0,1]$. Here, we need the (approximated) Jacobian of Finite Element Formulation 1 (defined without using the subscript h):

$$
A'(\mathbf{U})(\boldsymbol{\Delta}\mathbf{U}, \boldsymbol{\Psi}) = \delta(\Delta[\varphi - \varphi^{n-1}]_+, \psi)_{L^2} + \Delta t A_S'(\mathbf{U})(\boldsymbol{\Delta}\mathbf{U}, \boldsymbol{\Psi}),
$$

with

$$
\begin{aligned}
A_S'(\mathbf{U})(\boldsymbol{\Delta}\mathbf{U}, \boldsymbol{\Psi}) =& \Big(2(1-k)\inf\{\varphi_+,1\}H(1-\varphi)\Delta\varphi\mathcal{G}e(\mathbf{u}) + ((1-k)(\inf\{\varphi_+,1\})^2 \\
&+k)\mathcal{G}e(\Delta\mathbf{u}), e(\mathbf{w})\Big)_{L^2} - (2(\inf\{\varphi_+,1\})H(1-\varphi)\Delta\varphi(\alpha-1)p_B, \nabla\cdot\mathbf{w})_{L^2} \\
&+(2(\inf\{\varphi_+,1\})H(1-\varphi)\Delta\varphi\nabla p_B, \mathbf{w})_{L^2} + (2(1-k)\mathcal{G}e(\mathbf{u}) : e(\Delta\mathbf{u})\inf\{\varphi_+,1\} \\
&+(1-k)\mathcal{G}e(\mathbf{u}) : e(\mathbf{u})H(1-\varphi)\Delta\varphi, \psi)_{L^2} + \frac{G_c}{\epsilon}(\Delta\varphi,\psi)_{L^2} + G_c\epsilon(\nabla\Delta\varphi, \nabla\psi)_{L^2} \\
&-(\alpha-1)2(p_B(H(1-\varphi)\Delta\varphi\nabla\cdot\mathbf{u} + (\inf\{\varphi_+,1\})\nabla\cdot\Delta\mathbf{u}), \psi)_{L^2} \\
&+2\,(\nabla p_B\cdot(H(1-\varphi)\Delta\varphi\mathbf{u} + (\inf\{\varphi_+,1\})\Delta\mathbf{u}), \psi)_{L^2},
\end{aligned}
$$

for all $\boldsymbol{\Psi} = \{\mathbf{w},\psi\} \in V_h \times W_h$. Here, $H(\cdot)$ is Heaviside's function.

Remark 11 The realization of (53) is based on a modified Newton method with dynamic Jacobian modification developed in Wick (2017b), where the terms related to the nonconvex parts (i.e., in the displacement equation) are scaled accordingly. Other monolithic solvers worthy to mention are Gerasimov and Lorenzis (2016) and Wick (2017a) in which line-search assisted or error-oriented Newton methods were developed, respectively. Alternatively, a robust and efficient technique is to replace φ^h in the elasticity equation by a time-lagged extrapolated φ^h, which has been demonstrated computationally to provide a robust and stable numerical scheme Heister et al. (2015) (2d) and Lee et al. (2016b) (3d).

6 Numerical tests

We perform four numerical tests. The first test assumes a constant pressure $p_B = 10^{-3}$ that acts in the pressure [Sneddon's 2d benchmark (Sneddon and Lowengrub 1969)]. The second example considers again Sneddon's 2d benchmark, but with Neumann conditions on the bottom and top boundaries. In the third example, we study two interacting propagating fractures subject to a nonconstant pressure. In the fourth test, we address Sneddon's 3d benchmark in which a penny-shaped fracture is subject to a constant pressure, again $p_B = 10^{-3}$. The programming code is based on the finite element software `deal.II` (see Arndt et al. 2017; Bangerth et al. 2007) and the underlying monolithic numerical treatment is described in detail in Wick (2017a, b).

6.1 Constant pressure in a crack (Sneddons's 2d benchmark)

The first example is motivated by Bourdin et al. (2012), Wheeler et al. (2014) and is based on Sneddon's theoretical calculations (1969, 1946). Specifically, we consider a 2d problem where a (constant) pressure p_B is used to drive the deformation and crack propagation. We assume a dimensionless form of the equations.

The configuration is displayed in Fig. 2. We prescribe the initial crack implicitly (see e.g. Borden et al. 2012 and specifically for this setting Wheeler et al. 2014). Therefore, we deal with the following geometric data: $\Omega = (0, 4)^2$ and a (prescribed) initial crack with length $2l_0 = 0.4$ on $\Omega_C = (1.8, 2.2) \times (2 - h, 2 - h) \subset \Omega$ where h is the local mesh size. Thus, we deal with a $2d$ crack with a length much larger than its width. As boundary conditions we set the displacements zero on $\partial\Omega$. The test is

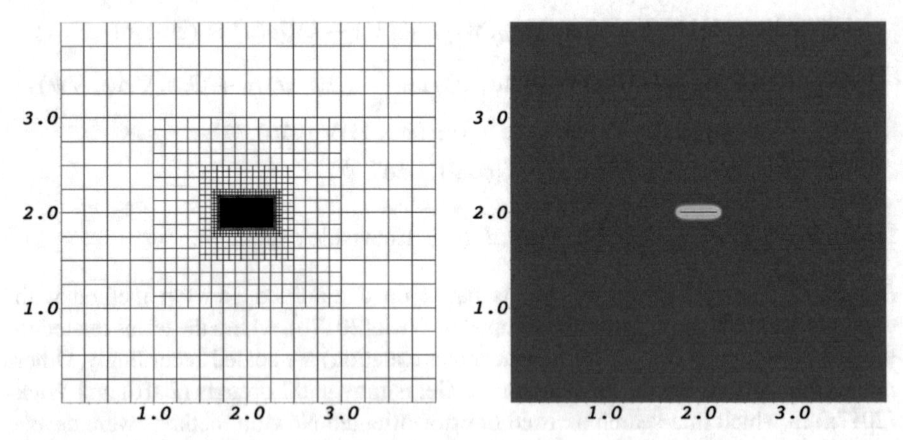

Fig. 2 Example 1: Configuration (left) and crack pattern (right)

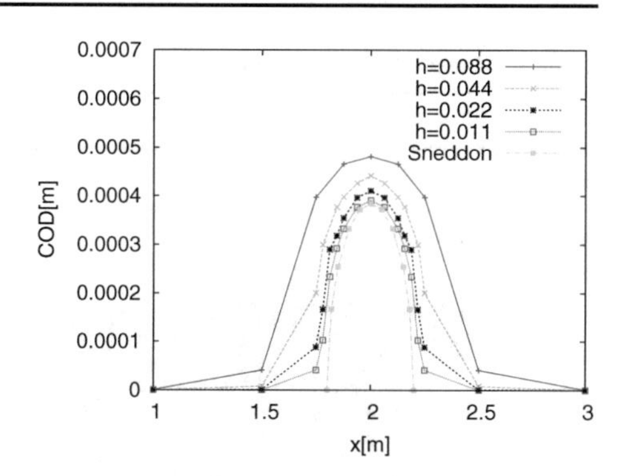

Fig. 3 Example 1: COD for different h. Sneddon's turquoise line with squares corresponds to his analytical solution. It is well observed that the crack tips must be resolved correctly as they are not well approximated on coarse meshes (color figure online)

stationary, but we perform two (pseudo) time steps in order to account for the crack irreversibility condition.

Applying the theory of Γ-convergence based on a related finite element analysis in Bourdin (1999), we choose $h \ll k \ll \epsilon$, i.e., $k = 0.25\sqrt{h}$ and $\epsilon = 0.5\sqrt{h}$. Furthermore, it is well-known that δ must depend on h, i.e., here, we choose $\delta = 100 \times h^{-2}$. The Biot coefficient and critical energy release rate are chosen as $\alpha = 0$ and $G_c = 1.0$, respectively. The mechanical parameters are Young's modulus and Poisson's ratio are set to be $E = 1.0$ and $\nu_s = 0.2$. The applied fracture pressure is $p_B = 10^{-3}$.

The goal is to measure the crack opening displacement (COD) and the volume of the crack under spatial mesh refinement. To this end, we observe u along Ω_C. Specifically, the width is determined as the jump of the normal displacements $COD := w := w(x, y) = [\mathbf{u} \cdot \mathbf{n}]$. This expression can be written in integral form as follows:

$$COD := w := w(x, y) = \int_{-\infty}^{\infty} \mathbf{u} \cdot \nabla\varphi \, dy. \tag{54}$$

We note that the integration is perpendicular to the crack direction. Here, the crack is aligned with the x-axis and therefore integration into the normal direction coincides with the y-direction.

The COD formula (54) is obvious since the phase-field variable φ can be related to a level-set function. This level-set can be used to compute the (unit) normal vector, e.g., Nguyen et al. (2016), Lee et al. (2017a). Here, the normal vector is in the y-direction and therefore, the above formula is obtained corresponding to $[\mathbf{u} \cdot \mathbf{n}]$ for $\varepsilon = 0$.

Second, following (Dean and Schmidt 2014, p. 710), the volume of the fracture is $V = \pi w l_0$. The analytical expression for the width (to which we compare) Dean and Schmidt (2014) is $w = 4\frac{(1-\nu_s^2)l_0 p}{E}$. Then, the analytical expression for the volume becomes

$$V = 2\pi \frac{(1 - \nu_s^2) l_0^2 p}{E}. \tag{55}$$

In contrast to Bourdin et al. (2012), we use the numerical approximation of the phase-field function instead of a synthetic choice of the crack indicator function.

Table 2 Example 1: Fracture volume

h	8.8×10^{-2}	4.4×10^{-2}	2.2×10^{-2}	1.1×10^{-2}	Exact
V	3.02×10^{-4}	2.77×10^{-4}	2.57×10^{-4}	2.49×10^{-4}	2.41×10^{-4}

The exact formula is given in (55)

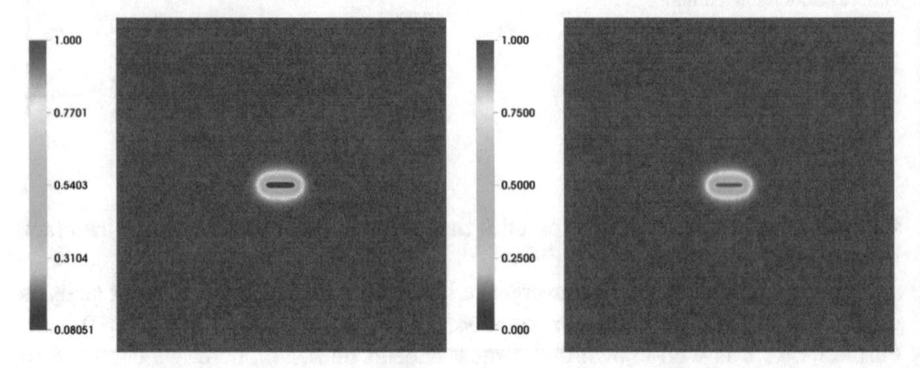

Fig. 4 Example 2: phase-field function for Case 1 and Case 2 at left and the Cases 3 and 4 at right

The crack pattern and the corresponding mesh are displayed in Fig. 2. Our findings for different spatial mesh parameters h are summarized in Fig. 3. Specifically, we observe overall convergence to Sneddon's analytical solution (Sneddon and Lowengrub 1969) as well as much better approximation of the crack tips under local mesh refinement. The obtained crack volumes are displayed in Table 2 in which the exact value is computed by Formula (55).

6.2 Sneddon's 2d-benchmark with mixed boundary conditions

In this second test, the domain and parameters are the same as in the previous example. The only (major) change concerns the boundary conditions. The top Γ_{top} and bottom Γ_{bottom} boundaries form now a Neumann boundary $\partial_N(0, L)^2 = \Gamma_{top} \cup \Gamma_{bottom}$. Here, we prescribe Neumann conditions τ of homogeneous and nonhomogeneous type. Then, we compare these results to the previous setting. In total we design the following tests:

- Case 1: $\partial_N(0, L)^2 = \emptyset$ and $u = 0$ on $\partial_D(0, L)^2$;
- Case 2: $\tau = (0, 0)^T$ on $\partial_N(0, L)^2$ and $u = 0$ on $\partial_D(0, L)^2$;
- Case 3: $\tau = (0, 0.001)^T$ on Γ_{top}, $\tau = (0, -0.001)^T$ on Γ_{bottom} and $u = 0$ on $\partial_D(0, L)^2$;
- Case 4: $\tau = (0, 0.1)^T$ on Γ_{top}, $\tau = (0, -0.1)^T$ on Γ_{bottom} and $u = 0$ on $\partial_D(0, L)^2$.

These computations are performed on a 7 times uniformly refined mesh with $h = 0.044$. The phase-field function is displayed in Fig. 4.

The maximal crack width openings w_{max} at $x = 2$ computed with the help of (54) in the middle of the fracture are:

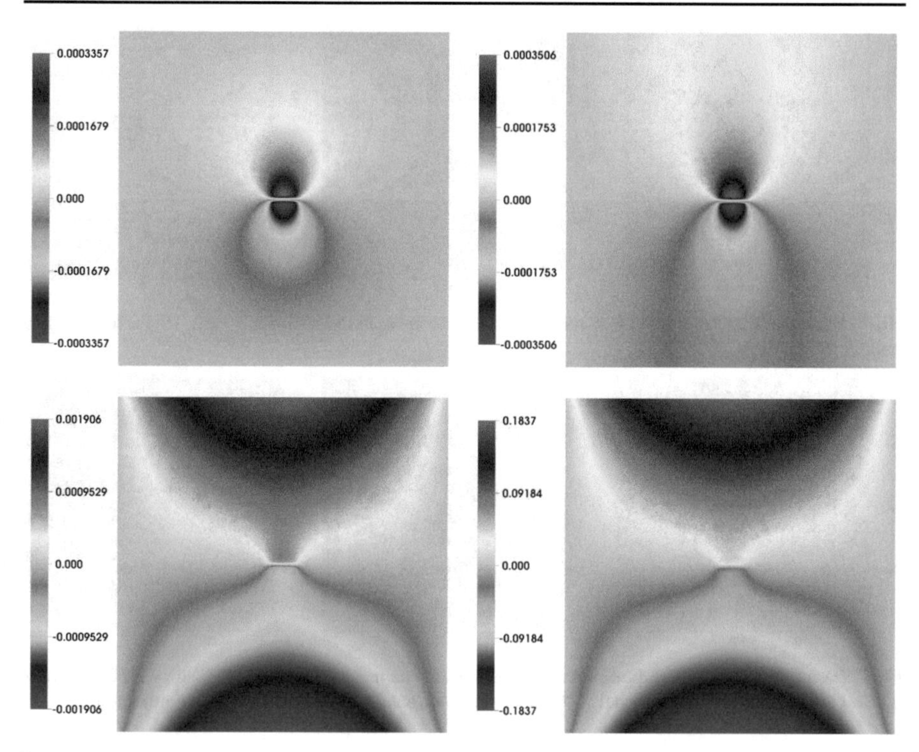

Fig. 5 Example 2: the y-displacements for all four cases

- Case 1: $w_{max}(x = 2;\ 0 \le y \le 4) = 5.25244 \times 10^{-4}$;
- Case 2: $w_{max}(x = 2;\ 0 \le y \le 4) = 5.52572 \times 10^{-4}$;
- Case 3: $w_{max}(x = 2;\ 0 \le y \le 4) = 1.31092 \times 10^{-3}$;
- Case 4: $w_{max}(x = 2;\ 0 \le y \le 4) = 7.67588 \times 10^{-2}$.

These findings are plausible: in Case 2 zero traction forces are applied on the top and the bottom boundaries and the fracture pressure keeps the fracture open. In addition, the maximal crack opening displacement is very similar (as expected) to Case 1. We further remark that the pressure boundary term $\int_{\partial_N(0,L)^2} p_B w_n \, dS$ in (10) is important when working with Neumann conditions. For instance, when $\int_{\partial_N(0,L)^2} p_B w_n \, dS$ is not used in Case 2, we obtain a negative width, which is of course nonphysical for this setting. The influence is significant for all cases when $\|\tau\| < \|p\mathbf{n}\|$ (Case 2) and $\|\tau\| \approx \|p\mathbf{n}\|$ (Case 3). In the Cases 3 and 4, the domain is pulled, since now the traction forces are strictly positive/negative, respectively. Consequently the fracture opens more than in the first two cases. Specifically, when the traction force is increased by a magnitude of order 2, the fracture width is also higher about a magnitude of order 2. The y-displacement fields (here directly to the crack opening displacements since the fracture is aligned with the x-axis) are displayed in Fig. 5.

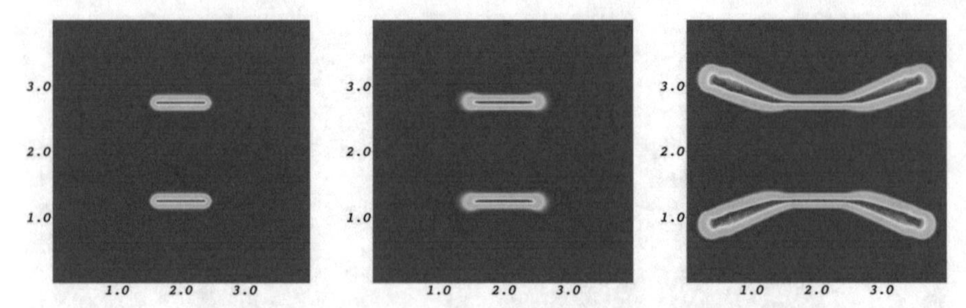

Fig. 6 Example 3: crack evolution in red in a homogeneous material at times $T = 0, 15, 30$ (color figure online)

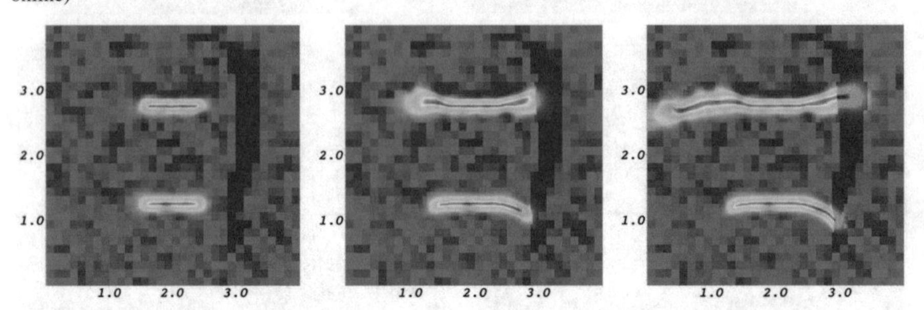

Fig. 7 Example 3: crack evolution in red in a heterogeneous material at times $T = 30, 40, 50$. The light blue regions denote smooth material $E \approx 1$ and dark blue stands for $E \approx 11.0$ (color figure online)

6.3 Two-crack interaction subject to non-constant pressure

In this third example, we extend the previous setting to study the interaction of two different fractures that are subject to a linearly increasing pressure p_B. In the first part, a homogeneous material is considered and in the second part a heterogenous material field. The pressure function is given by $p_B(t) = 0.1 + t \cdot 0.1$, where t denotes the total time, and Young's modulus is set to be $E = 1$ in the first part and it varies between 1.1 and 11.0 in the second part. Poisson ratio is 0.2. The penalization parameter is chosen as $\delta = 10h^{-2}$. The remaining parameters are chosen as in the previous example. Our results in the Figs. 6 and 7 show two propagating, interacting fractures. Specifically, they curve away due to stress-shadowing effects (see e.g., Castonguay et al. 2013). The extension to nonconstant pressure evolution using Darcy's law and application of the fixed-stress splitting is studied in Wick et al. (2016) and Mikelić et al. (2015a).

6.4 Sneddons's 3d benchmark with a constant pressure in a penny-shaped crack

The last example is again based on Sneddon's theoretical calculations (Sneddon and Lowengrub 1969, Section 3.3, pp. 138–139). Specifically, we consider a 3d problem where a (constant) pressure $p_B = 10^{-3}$ is used to open a penny-shaped fracture.

The configuration is displayed in Fig. 8 and $\Omega = (0, 10)^3$. We prescribe the initial crack implicitly by setting the intial value of the phase-field variable to zero in the $y = 5$-plane with origin $(5, 5, 5)$. The radius of the fracture is $\rho = 1$. As boundary conditions we set the displacements zero on $\partial\Omega$. We perform five (pseudo) time steps.

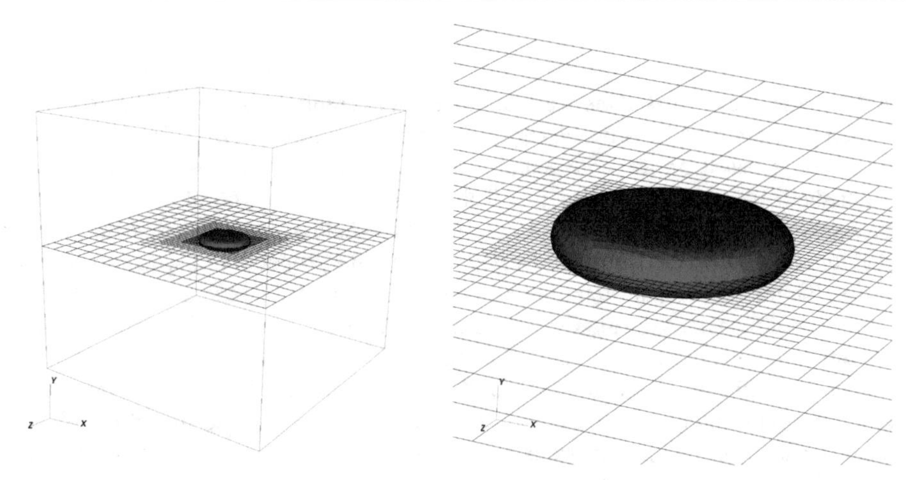

Fig. 8 Example 4: A penny-shaped fracture and locally refined mesh (left) and zoom-in at right. Specifically, the fracture remains thin in the third direction as shown theoretically in Corollary 2

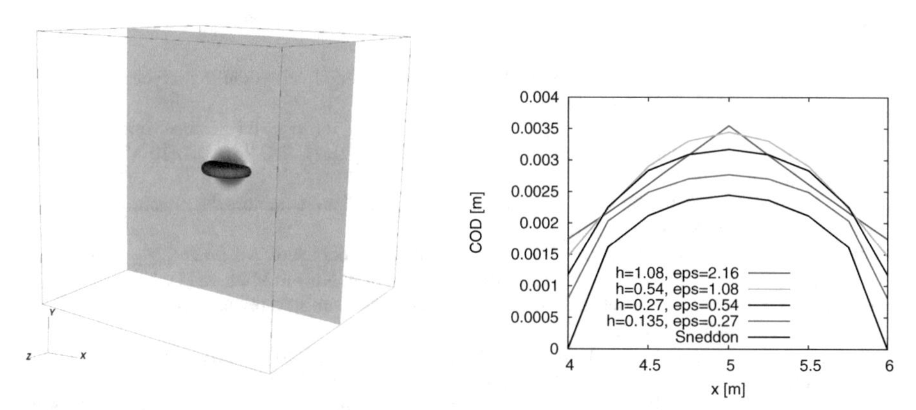

Fig. 9 Example 4: Crack opening displacements. Graphical illustration of the y displacements (left) and evaluation of the crack opening displacements for the four different h values. The reference curve of Sneddon has been computed with the formula given in Sneddon and Lowengrub (1969) on p. 139

We choose $k = 10^{-12}$ and $\varepsilon = 2h$ and $h_{min} = 1.08, 0.54, 0.27, 0.135$. The Biot coefficient and critical energy release rate are chosen as $\alpha = 0$ and $G_c = 1.0$, respectively. The mechanical parameters are Young's modulus and Poisson's ratio are set to be $E = 1.0$ and $\nu_s = 0.2$. The applied fracture pressure is $p_B = 10^{-3}$.

The locally refined mesh on the finest level and the penny-shaped fracture are shown in Fig. 8. Specifically, we observe that the fracture remains thin in the third direction as shown theoretically in Corollary 2. The crack opening displacement (here the displacements in y direction) and the corresponding plots for the four different h values are shown in Fig. 9. We notice that ε depends on h. For this reason, we cannot expect 'convergence' in the classical sense. Such results however have been shown in our other papers in which ε was fixed and only h was varied Lee et al. (2016b), Wheeler et al. (2014).

7 Conclusion

In this paper, we discussed the mechanics step of hydraulic phase-field fractures with a given pressure field for propagating cracks in a poroelastic medium. The phase-field algorithm is based on an incremental formulation and existence of a minimizer is established. We rigorously show that if the initial crack size was of order ε (with a reasonable control of the gradient of its initial phase-field description), then the phase-field function at the next (future) time step has the same property. Consequently, in our model, the initially slender fractures remain indeed thin in the second (in 2d) or third (3d) space dimensions during the incremental evolution. Numerical benchmarks are demonstrating the correctness of the theory. Specifically, a numerical test with mixed boundary conditions (Dirichlet and Neumann) was designed in which our modeling of the pressure interface conditions was further confirmed. The modeling of this paper forms the basis for extensions to crack growth in heterogeneous porous media, fluid-filled, and proppant-filled crack evolutions.

References

Adachi, J., Siebrits, E., Peirce, A., Desroches, J.: Computer simulation of hydraulic fractures. Int. J. Rock Mech. Min. Sci. **44**, 739–757 (2007)

Almani, T., Lee, S., Wheeler, M., Wick, T.: Multirate coupling forflow and geomechanics applied to hydraulic fracturing using anadaptive phase-field technique (2017). SPE RSC 182610-MS, Feb. 2017, Montgomery, Texas, USA

Almi, S., Maso, G.D., Toader, R.: Quasi-static crack growth in hydraulic fracture. Nonlinear Anal. Theory Methods Appl. **109**, 301–318 (2014)

Arndt, D., Bangerth, W., Davydov, D., Heister, T., Heltai, L., Kronbichler, M., Maier, M., Pelteret, J.P., Turcksin, B., Wells, D.: The `deal.II` library, version 8.5. J. Numer. Math. **25**(3), 137–146 (2017)

Bangerth, W., Hartmann, R., Kanschat, G.: deal.II—a general purpose object oriented finite element library. ACM Trans. Math. Softw. **33**(4), 24/1–24/27 (2007)

Borden, M.J., Verhoosel, C.V., Scott, M.A., Hughes, T.J.R., Landis, C.M.: A phase-field description of dynamic brittle fracture. Comput. Meth. Appl. Mech. Eng. **217**, 77–95 (2012)

de Borst, R., Rethoré, J., Abellan, M.: A numerical approach for arbitrary cracks in a fluid-saturated porous medium. Arch. Appl. Mech. 595–606 (2006)

Both, J., Borregales, M., Nordbotten, J., Kumar, K., Radu, F.: Robust fixed stress splitting for biots equations in heterogeneous media. Appl. Math. Lett. **68**, 101–108 (2017)

Bourdin, B.: Image segmentation with a finite element method. Math. Model. Numer. Anal. **33**(2), 229–244 (1999)

Bourdin, B., Chukwudozie, C., Yoshioka, K.: A variational approach to the numerical simulation of hydraulic fracturing. In: SPE Journal, Conference Paper 159154-MS (2012)

Bourdin, B., Francfort, G., Marigo, J.J.: Numerical experiments in revisited brittle fracture. J. Mech. Phys. Solids **48**(4), 797–826 (2000)

Bourdin, B., Francfort, G., Marigo, J.J.: The variational approach to fracture. J. Elast. **91**(1–3), 1–148 (2008)

Braides, A.: Approximation of Free-Discontinuity Problems. Springer, Berlin (1998)

Burke, S., Ortner, C., Süli, E.: An adaptive finite element approximation of a variational model of brittle fracture. SIAM J. Numer. Anal. **48**(3), 980–1012 (2010)

Cajuhi, T., Sanavia, L., De Lorenzis, L.: Phase-field modeling of fracture in variably saturated porous media. Comput. Mech. **61**(3), 299–318 (2018)

Castelletto, N., White, J.A., Tchelepi, H.A.: Accuracy and convergence properties of the fixedstress iterative solution of twoway coupled poromechanics. Int. J. Numer. Anal. Methods Geomech. **39**(14), 1593–1618 (2015)

Castonguay, S., Mear, M., Dean, R., Schmidt, J.: Predictions of the growth of multiple interacting hydraulic fractures in three dimensions. SPE-166259-MS pp. 1–12 (2013)

Chambolle, A.: An approximation result for special functions with bounded variations. J. Math. Pures Appl. **83**, 929–954 (2004)
Ciarlet, P.G.: The Finite Element Method for Elliptic Problems, 2 edn. North-Holland, Amsterdam (1987)
Dacorogna, B.: Direct Methods in the Calculus of Variations. Springer Verlag, Berlin (2008)
Dean, R., Schmidt, J.: Hydraulic-fracture predictions with a fully coupled reservoir simulator. SPE J. **14**(4), 707–714 (2014)
Engwer, C., Schumacher, L.: A phase field approach to pressurized fractures using discontinuous Galerkin methods. Math. Comput. Simul. **137**, 266–285 (2017)
Ferronato, M., Castelletto, N., Gambolati, G.: A fully coupled 3-d mixed finite element model of Biot consolidation. J. Comput. Phys. **229**(12), 4813–4830 (2010)
Francfort, G.: Un résumé de la théorie variationnelle de la rupture (2011). Séminaire Laurent Schwartz – EDP et applications, Institut des hautes études scientifiques, 2011–2012, Exposé no. XXII, 1-11. http://slsedp.cedram.org/slsedp-bin/fitem?id=SLSEDP_2011-2012
Francfort, G., Marigo, J.J.: Revisiting brittle fracture as an energy minimization problem. J. Mech. Phys. Solids **46**(8), 1319–1342 (1998)
Ganis, B., Girault, V., Mear, M., Singh, G., Wheeler, M.F.: Modeling fractures in a poro-elastic medium. Oil Gas Sci. Technol. **4**, 515–528 (2014)
Gaspar, F.J., Rodrigo, C.: On the fixed-stress split scheme as smoother in multigrid methods for coupling flow and geomechanics. Comput. Methods Appl. Mech. Eng. **326**, 526–540 (2017)
Gerasimov, T., Lorenzis, L.D.: A line search assisted monolithic approach for phase-field computing of brittle fracture. Comput. Methods Appl. Mech. Eng. **312**, 276–303 (2016)
Girault, V., Wheeler, M.F., Ganis, B., Mear, M.E.: A lubrication fracture model in a poro-elastic medium. Math. Models Methods Appl. Sci. **25**(04), 587–645 (2015)
Gupta, P., Duarte, C.: Simulation of non-planar three-dimensional hydraulic fracture propagation. Int. J. Numer. Anal. Meth. Geomech. **38**, 1397–1430 (2014)
Heider, Y., Markert, B.: A phase-field modeling approach of hydraulic fracture in saturated porous media. Mech. Res. Commun. **80**, 38–46 (2017)
Heister, T., Wheeler, M.F., Wick, T.: A primal-dual active set method and predictor–corrector mesh adaptivity for computing fracture propagation using a phase-field approach. Comput. Meth. Appl. Mech. Eng. **290**, 466–495 (2015)
Hong, Q., Kraus, J.: Parameter-robust stability of classical three-field formulation of Biot's consolidation model. Electron. Trans. Numer. Anal. **48**, 202–226 (2018)
Hwang, J., Sharma, M.: A 3-dimensional fracture propagation model for long-term water injection. In: 47th US Rock Mechanics/Geomechanics Symposium (2013)
Irzal, F., Remmers, J.J., Huyghe, J.M., de Borst, R.: A large deformation formulation for fluid flow in a progressively fracturing porous material. Comput. Methods Appl. Mech. Eng. **256**, 29–37 (2013)
Kinderlehrer, D., Stampacchia, G.: An introduction to variational inequalities and their applications. In: Classics in Applied Mathematics. Society for Industrial and Applied Mathematics (2000)
Lee, J.J.: Robust error analysis of coupled mixed methods for Biot's consolidation model. J. Sci. Comput. **69**(2), 610–632 (2016)
Lee, S., Mikelić, A., Wheeler, M.F., Wick, T.: Phase-field modeling of proppant-filled fractures in a poroelastic medium. Comput. Methods Appl. Mech. Eng. **312**, 509–541 (2016a)
Lee, S., Wheeler, M.F., Wick, T.: Pressure and fluid-driven fracture propagation in porous media using an adaptive finite element phase field model. Comput. Methods Appl. Mech. Eng. **305**, 111–132 (2016b)
Lee, S., Wheeler, M.F., Wick, T.: Iterative coupling of flow, geomechanics and adaptive phase-field fracture including level-set crack width approaches. J. Comput. Appl. Math. **314**, 40–60 (2017a)
Lee, S., Wheeler, M.F., Wick, T., Srinivasan, S.: Initialization of phase-field fracture propagation in porous media using probability maps of fracture networks. Mech. Res. Commun. **80**, 16–23 (2017b)
Lee, J.J., Mardal, K.A., Winther, R.: Parameter-robust discretization and preconditioning of Biot's consolidation model. SIAM J. Sci. Comput. **39**(1), A1–A24 (2017c)
Lee, S., Mikelić, A., Wheeler, M.F., Wick, T.: Phase-field modeling of two phase fluid filled fractures in a poroelastic medium (2018). SIAM Multiscale Model Simul. **16**(4), 1542–1580 (2018)
Liu, R.: Discontinuous galerkin finite element solution for poromechanics. Ph.D. thesis, The University of Texas at Austin (2004)
Markert, B., Heider, Y.: Recent Trends in Computational Engineering—CE2014: Optimization, Uncertainty, Parallel Algorithms, Coupled and Complex Problems, chap. Coupled Multi-Field Continuum Methods for Porous Media Fracture, pp. 167–180. Springer, Cham (2015)

McClure, M.W., Kang, C.A.: A three-dimensional reservoir, wellbore, and hydraulic fracturing simulator that is compositional and thermal, tracks proppant and water solute transport, includes non-darcy and non-newtonian flow, and handles fracture. SPE-182593-MS (2017)

Miehe, C., Mauthe, S.: Phase field modeling of fracture in multi-physics problems. Part III. Crack driving forces in hydro-poro-elasticity and hydraulic fracturing of fluid-saturated porous media. Comput. Methods Appl. Mech. Eng. **304**, 619–655 (2016)

Miehe, C., Mauthe, S., Teichtmeister, S.: Minimization principles for the coupled problem of Darcy-Biot-type fluid transport in porous media linked to phase field modeling of fracture. J. Mech. Phys. Solids **82**, 186–217 (2015)

Miehe, C., Welschinger, F., Hofacker, M.: Thermodynamically consistent phase-field models of fracture: variational principles and multi-field fe implementations. Int. J. Numer. Methods Eng. **83**, 1273–1311 (2010)

Mikelić, A., Wang, B., Wheeler, M.F.: Numerical convergence study of iterative coupling for coupled flow and geomechanics. Comput. Geosci. **18**(3), 325–341 (2014)

Mikelić, A., Wheeler, M., Wick, T.: A phase-field approach to the fluid filled fracture surrounded by a poroelastic medium. ICES Report 13-15 (2013)

Mikelić, A., Wheeler, M., Wick, T.: Phase-field modeling of pressurized fractures in a poroelastic medium. ICES Report 14-18 (2014)

Mikelić, A., Wheeler, M.F.: Convergence of iterative coupling for coupled flow and geomechanics. Comput. Geosci. **17**(3), 455–462 (2012)

Mikelić, A., Wheeler, M.F., Wick, T.: A phase-field method for propagating fluid-filled fractures coupled to a surrounding porous medium. SIAM Multiscale Model. Simul. **13**(1), 367–398 (2015a)

Mikelić, A., Wheeler, M.F., Wick, T.: Phase-field modeling of a fluid-driven fracture in a poroelastic medium. Comput. Geosci. **19**(6), 1171–1195 (2015b)

Mikelić, A., Wheeler, M.F., Wick, T.: A quasi-static phase-field approach to pressurized fractures. Nonlinearity **28**(5), 1371–1399 (2015c)

Murad, M.A., Loula, A.F.: Improved accuracy in finite element analysis of Biot's consolidation problem. Comput. Methods Appl. Mech. Eng. **95**(3), 359–382 (1992)

Murad, M.A., Loula, A.F.D.: On stability and convergence of finite element approximations of Biot's consolidation problem. Int. J. Numer. Methods Eng. **37**(4), 645–667 (1994)

Nguyen, T., Yvonnet, J., Zhu, Q.Z., Bornert, M., Chateau, C.: A phase-field method for computational modeling of interfacial damage interacting with crack propagation in realistic microstructures obtained by microtomography. Comput. Methods Appl. Mech. Eng. **312**, 567–595 (2016)

Philips, P., Wheeler, M.: A coupling of mixed and galerkin finite element methods for poro-elasticity. Comput. Geosci. **12**(4), 417–435 (2003)

Rodrigo, C., Gaspar, F., Hu, X., Zikatanov, L.: Stability and monotonicity for some discretizations of the Biots consolidation model. Comput. Methods Appl. Mech. Eng. **298**, 183–204 (2016)

Santillan, D., Juanes, R., Cueto-Felgueroso, L.: Phase field model of fluid-driven fracture in elastic media: immersed-fracture formulation and validation with analytical solutions. J. Geophys. Res. Solid Earth **122**, 2565–2589 (2017)

Schrefler, B.A., Secchi, S., Simoni, L.: On adaptive refinement techniques in multi-field problems including cohesive fracture. Comput. Meth. Appl. Mech. Eng. **195**, 444–461 (2006)

Sneddon, I.N.: The distribution of stress in the neighbourhood of a crack in an elastic solid. Proc. R. Soc. Lond. A **187**, 229–260 (1946)

Sneddon, I.N., Lowengrub, M.: Crack Problems in the Classical Theory of Elasticity. SIAM Series in Applied Mathematics. Wiley, Philadelphia (1969)

Tolstoy, I.: Acoustic, Elasticity, and Thermodynamics of Porous Media. Twenty-One Papers by M.A. Biot. Acoustical Society of America, New York (1992)

van Duijn, C.J., Mikelić, A., Wick, T.: A monolithic phase-field model of a fluid-driven fracture in a nonlinear poroelastic medium. Math. Mech. Solids (2018). https://doi.org/10.1177/1081286518801050

Wheeler, M., Wick, T., Wollner, W.: An augmented-Lagangrian method for the phase-field approach for pressurized fractures. Comput. Meth. Appl. Mech. Eng. **271**, 69–85 (2014)

Wick, T.: Coupling fluid–structure interaction with phase-field fracture. J. Comput. Phys. **327**, 67–96 (2016a)

Wick, T.: Goal functional evaluations for phase-field fracture using PU-based DWR mesh adaptivity. Comput. Mech. **57**(6), 1017–1035 (2016b)

Wick, T.: An error-oriented Newton/inexact augmented Lagrangian approach for fully monolithic phase-field fracture propagation. SIAM J. Sci. Comput. **39**(4), B589–B617 (2017)

Wick, T.: Modified Newton methods for solving fully monolithic phase-field quasi-static brittle fracture propagation. Comput. Methods Appl. Mech. Eng. **325**, 577–611 (2017)
Wick, T., Lee, S., Wheeler, M.: 3D phase-field for pressurizedfracture propagation in heterogeneous media. In: ECCOMAS and IACMCoupled Problems Proceedings, May 2015 at San Servolo, Venice, Italy (2015)
Wick, T., Singh, G., Wheeler, M.: Fluid-filled fracture propagation using a phase-field approach and coupling to a reservoir simulator. SPE J. **21**(03), 981–999 (2016)

Affiliations

A. Mikelić[1] · M. F. Wheeler[2] · T. Wick[3]

✉ T. Wick
thomas.wick@ifam.uni-hannover.de

A. Mikelić
andro.mikelic@univ-lyon1.fr

M. F. Wheeler
mfw@ices.utexas.edu

1 Univ Lyon, Université Claude Bernard Lyon 1, CNRS UMR 5208, Institut Camille Jordan, 43 blvd. du 11 novembre 1918, 69622 Villeurbanne Cedex, France

2 Center for Subsurface Modeling, The Institute for Computational Engineering and Sciences, The University of Texas at Austin, Austin, TX 78712, USA

3 Institut für Angewandte Mathematik, Leibniz Universität Hannover, Welfengarten 1, 30167 Hannover, Germany

GEM - International Journal on Geomathematics
https://doi.org/10.1007/s13137-019-0127-5

ORIGINAL PAPER

Numerical aspects of hydro-mechanical coupling of fluid-filled fractures using hybrid-dimensional element formulations and non-conformal meshes

Patrick Schmidt[1] · Holger Steeb[1]

Received: 31 March 2018 / Accepted: 22 January 2019

Abstract

In the field of porous and fractured media, subsurface flow provides insight into the characteristics of fluid storage and properties connected to underground matter and heat transport. Subsurface flow is precisely described by many diffusion based models in the literature. However, diffusion-based models lack to reproduce important hydro-mechanical coupling phenomena like inverse water-level fluctuations (Noordbergum effect). In theory, contemporary modeling approaches, such as direct numerical simulations (DNS) of surface-coupled fluid-solid (fracture) interactions or coarse-grained continuum approaches like Biot's theory, are capable of capturing such phenomena. Nevertheless, during modeling processes of fractures with high aspect ratios, DNS methods with the explicit discretization of the fluid domain fail, and coarse-grained continuum approaches require a non-linear formulation for the fracture deformation since large deformation can be reached easily within fractures. Hence a hybrid-dimensional approach uses a parabolic velocity profile to avoid an explicit discretization of the fluid domain within the fracture. For fracture flow, the primary variable is the pressure field only, and the fracture domain is reduced by one dimension. The interaction between the fracture and the surrounding matrix domain, respectively, is realized by modified balance equations. The coupled system is numerically stiff when fluids are described with a low compressibility modulus. Two algorithms are proposed within this work, namely the weak coupling scheme, which uses an implicit staggered-iterative algorithm to find the residual state and the strong coupling scheme which directly couples both domains by implementing interface elements. In the course of this work, a consistent implementation scheme for the coupling of hybrid-dimensional elements with a surrounding bulk matrix is proposed and validated and tested throughout different numerical experiments.

Keywords Fracture flow · Hydromechanical coupling · Deformation-induced flow · Pressure diffusion

✉ Patrick Schmidt
pschmidt@mechbau.uni-stuttgart.de

Extended author information available on the last page of the article

Mathematics Subject Classification 86-08

1 Introduction

Underground fluid flow in fractured and porous media exhibits a variety of hydro-mechanical phenomena, which have been intriguing the research community for several decades (Oliver and Chen 2011; Ehlers and Bluhm 2013). The long-established research interest in subsurface flow, such as studies of reservoir storage capacities, is regarding both theoretical/numerical and experimental investigations. Various hydro-mechanical coupling phenomena, like inverse pumping or Noordbergum effect (Rodrigues 1983; Kim and Parizek 1997), are observed in fracture networks, and apparently, affect the reservoir behavior during fluid pressure or deformation fluctuations. For fractures with a high aspect-ratio (length l vs. aperture δ, i.e. $l/\delta > 10^4$) filled by viscous fluids with low compressibility these effects become evident.
A considerable number of theoretical and numerical studies of subsurface flow in fractured media can be found in the literature. The first solution strategy to mention is the so-called diffusion-based approach. This approach uses the poroelastic framework with slight modifications to the diffusion formulation within the fracture and assumes that the mechanical response of the surrounding rock (bulk) material is decoupled from the governing equation of fluid flow. For specific combinations of material parameters or a constant mean stress (Renner and Steeb 2015), this decoupling assumption is valid and has successfully been applied in several investigations, to study pressure diffusion effects along fractures (Ortiz et al. 2011, 2013). Nevertheless, phenomena due to a strong hydro-mechanical coupling, such as the earlier mentioned Noordbergum effect, cannot be captured by even extended diffusion formulations.
Further strategies to describe hydro-mechanical phenomena are based on partly-analytical solutions. In a first approach, Sneddon and Elliot (1946) introduced a plane strain crack solution to find deformation states for various fluid pressure conditions. Based on Sneddon's formulation further models evolved such as the PK-model by Perkins and Kern (1961), the PKN model by Nordgren (1972) which takes fluid loss effects into account or the penny-shaped KGD model independently developed by Geertsma and De Klerk (1969) and Zheltov (1955). All models provide further insights on phenomena related to hydro-mechanical coupling. However, the pressure state described in all models is not associated with the deformation response of the surrounding bulk matrix, and furthermore, only relatively simple geometries can be analyzed. Partly-analytical formulations serve for a better understanding of the physics but are not suitable for fully coupled formulations and/or complex fracture geometries.
Most promising with regards to hydro-mechanical investigations of complex fracture networks are fully coupled numerical strategies. In direct numerical simulation (DNS), the fracture domain is explicitly discretized, e.g., with finite elements, and the inherent fluid is modeled by the Navier–Stokes equations, coupled to the deformation of the surrounding rock. Thus hydro-mechanical phenomena can be reproduced using different coupling algorithms, such as staggered or fully coupled schemes. However, in the case of creeping flow conditions, i.e., Poiseuille-type flows a specific discretization over the fracture height is required to reproduce the parabolic velocity field. Having

in mind fractures with high aspect ratios, the total number of elements increases drastically, and DNS are finally applicable to fracture ratios of $l/\delta < 10^4$ only due to technical limitations.
For fractures with higher aspect ratios than $l/\delta > 10^4$ and low Reynolds numbers, viscous shear stresses in the bulk fluid can be neglected and creeping flow conditions can be assumed. Coarse-grained continuum approaches like Biot's quasi-static poroelastic theory (Biot 1941), take these assumptions into account and couple fluid pressure to solid deformation in a smeared continuum-based formulation at the local material point. The set of governing equations are based on volume-averaged quantities (Coussy 2004) and treats materials homogeneous if inhomogeneities occur at the length scale of the averaged pore size. The classical poroelastic formulation relates the specific storage capacity as an inherent material property to each material point which is not valid anymore for structural properties with finite extensions like fractures. By exchanging the solid with the fracture surface deformation, this limitation can be circumvented. Still, an explicit discretization of the fracture domain is required. Since poroelastic formulations are designed explicitly for hydro-mechanical problems, the number of elements needed decreases compared to DNS strategies. Nevertheless, the explicit discretization of the fracture domain constraints the method to aspect ratios of $l/w < 10^5$. Concerning numerical efficiency, subsurface flow for conduits with aspect ratios of $l/\delta > 10^5$ demands a different numerical treatment, other than an explicit discretization of the fracture domain. The efficiency of the procedure can significantly be increased by using hybrid-dimensional formulations (Vinci et al. 2014, 2015; Kim et al. 2011a, b; Girault et al. 2016, 2015). Hybrid-dimensional approaches allow for a lower-dimensional discretization of fracture flow which reduces the discretization space of the fracture domain's dimension by assuming a priori parabolic velocity profiles of Poiseuille-type. Besides a storage analysis of fracture networks, comparable strategies have been used in the field of hydro-mechanical coupling (Taleghani 2009; Settgast et al. 2017; Segura and Carol 2008a, b; Hanowski and Sander 2016). Hybrid-dimensional approaches have been used in the past often for purely hydraulic problems related to fracture flow. Authors focused on non-conformal techniques (Tunc et al. 2012), the formulation of discontinuous pressure at fracture interfaces (Martin et al. 2005; Brenner et al. 2017) and the improvement of performance and stability (Sandve et al. 2012).
Another important limiting factor for numerical treatments of hydro-mechanical coupling is the stiffness of the overall coupled system. Contributions of Adachi et al. (2007) and Yew and Weng (2014) study the hydro-mechanical coupling by investigating high aspect-ratio fractures using a staggered coupling scheme. Nevertheless, it has been stated that the introduced scheme should preferably be used for fluids with low or moderate compressibility parameters otherwise the scheme requires high numbers of iterations or non-practical small time steps. In unfractured porous media different staggered schemes have been introduced to solve such stiff systems in a stable fashion. Kim et al. (2011a, b) proposed the drained/undrained (modifications throughout mechanical calculations) and the fixed-strain/fixed-stress (modifications throughout flow calculations) in their work. The fixed-stress scheme achieves numerical stability by varying the strain rate throughout the flow problem to fix the stress state, whereas stability is reached using the drained scheme by enforcing a constant fluid mass per

element in each iteration step (Castelletto et al. 2015). The fixed-stress scheme has been extended by Girault et al. (2016, 2015) for coupling flow in deformable fractures embedded in a biphasic porous medium.
Deformation dependent permeability changes within fractures influence the pressure prediction, particularly when high aspect ratios are investigated. In the literature mentioned above, a constant aperture is assumed for the calculations of the fracture permeability (Girault et al. 2015, 2016). Fixed fracture permeabilities might lead to incorrect predictions of the pressure state. This work will take permeability changes due to local deformations of the fracture into account and show the importance based on a three-dimensional boundary value problem. Furthermore, staggered algorithms are used in the literature to solve the coupled set of governing equations; hence iterations are needed even for linear problem formulations. Mathematically it has been shown, that a fully coupled, monolithically solved scheme is unconditionally stable (Girault et al. 2016), but has not been investigated concerning efficiency yet. Since no iterations are needed the performance of a monolithic scheme is potentially higher than a staggered scheme. Nevertheless, by using non-conformal discretizations for porous and fracture domain, the performance of a staggered algorithm can be increased. This work will investigate the performance improvement by reducing the total number of degrees of freedom (DoF) taking the loss of accuracy of the solution into account. Based on the number of iterations needed in combination with the reduced number of DoF the efficiency of the staggered will be compared to a fully coupled monolithic scheme.

2 Model formulation

This section introduces the continuum based mathematical continuum-based model for underground flow in fractured porous media. The bounded subsurface domain will be denoted as $\mathcal{B}$ and includes the poroelastic subdomain $\mathcal{B}^{Pe}$ with its outer boundary surface Γ^{Pe} and the embedded deformable fracture subdomain $\mathcal{B}^{Fr}$. The domain of the fracture is defined through its exterior outer normal vectors $\mathbf{n}^{\pm}$, cf. Fig. 1.

2.1 Partial differential equations in the poroelastic subdomain $\mathcal{B}^{Pe}$

The subdomain $\mathcal{B}^{Pe}$ is described as a linear biphasic poroelastic medium, see Biot (1941) or e.g. Wang (2000). The leading set of partial differential equations (PDEs)

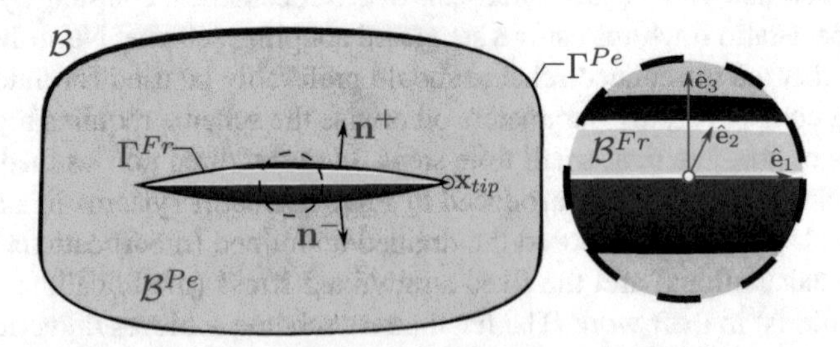

Fig. 1 Subsurface domain $\mathcal{B}$ including poroelastic subdomain $\mathcal{B}^{Pe}$, the fracture subdomain $\mathcal{B}^{Fr}$ and boundaries

of poroelasticity consists of the balance of momentum of the mixture and the balance of momentum of the fluid combined with the linearized form of the balance of mass. The system of equations is closed by constitutive assumptions for the extra stresses of the solid phase denoted as $\boldsymbol{\sigma}_E^{\mathfrak{s}}$, the momentum interaction between the viscous pore fluid and the solid matrix $\hat{\mathbf{p}}^{\mathfrak{s}}$, and an equation of state for the compressible pore fluid.

2.1.1 Balance of momentum of the mixture

The local form of the balance of linear momentum for a biphasic mixture with the solid constituent $\varphi^{\mathfrak{s}}$ and the pore fluid $\varphi^{\mathfrak{f}}$ reads

$$-\operatorname{div}(\boldsymbol{\sigma}) = \rho\,\mathbf{b} \tag{1}$$

in its (quasi-static) local form if inertia forces are neglected. The total stresses are denoted as $\boldsymbol{\sigma}$ and the body forces as $\rho\,\mathbf{b}$. The density of the mixture reads $\rho = \rho^{\mathfrak{s}} + \rho^{\mathfrak{f}}$. The partial densities are introduced as $\rho^{\mathfrak{s}} = \mathrm{d}m^{\mathfrak{s}}/\mathrm{d}v$ and $\rho^{\mathfrak{f}} = \mathrm{d}m^{\mathfrak{f}}/\mathrm{d}v$, respectively. Here, the mass elements of the constituents are denoted as $\mathrm{d}m^{\mathfrak{s}}$ and $\mathrm{d}m^{\mathfrak{f}}$. The volume element of the mixture is denoted as $\mathrm{d}v$. Effective densities are given by $\rho^{\mathfrak{s}R} = \mathrm{d}m^{\mathfrak{s}}/\mathrm{d}v^{\mathfrak{s}}$ and $\rho^{\mathfrak{f}R} = \mathrm{d}m^{\mathfrak{f}}/\mathrm{d}v^{\mathfrak{f}}$. Partial and effective densities are related through the porosity $\phi = \mathrm{d}v^{\mathfrak{f}}/\mathrm{d}v$, i.e. $\rho^{\mathfrak{s}} = (1-\phi)\,\rho^{\mathfrak{s}R}$ and $\rho^{\mathfrak{f}} = \phi\,\rho^{\mathfrak{f}R}$. In poroelasticity, the total stresses split into the effective stress and the pore pressure. Following (Biot 1941), we assume compressible constituents (i.e. a compressible skeleton with dry bulk modulus K, compressible grains which are composing the skeleton with bulk modulus $K^{\mathfrak{s}}$, and a compressible pore fluid with bulk modulus $K^{\mathfrak{f}}$). Thus, the effective stress principle composes the total stresses additively into the effective stresses and the pore pressure p

$$\boldsymbol{\sigma} = \boldsymbol{\sigma}_E^{\mathfrak{s}} - \alpha\,p\,\mathbf{I}. \tag{2}$$

The Biot parameter is related to the bulk moduli by $\alpha = 1 - K/K^{\mathfrak{s}}$. Thus, in case of $K^{\mathfrak{s}} \gg K$, the effective stress principle reduces to Terzaghi's effective stress principle with $\alpha = 1$. The effective stresses of the solid constituents are given by

$$\boldsymbol{\sigma}_E^{\mathfrak{s}} = 3\,K\,\mathrm{vol}(\boldsymbol{\varepsilon}_{\mathfrak{s}}) + 2\,G\,\mathrm{dev}(\boldsymbol{\varepsilon}_{\mathfrak{s}}) + (1-\alpha)\,p\,\mathbf{I}. \tag{3}$$

Here, we split the solid strain tensor $\boldsymbol{\varepsilon}_{\mathfrak{s}} = 1/2\,(\operatorname{grad}\mathbf{u}_{\mathfrak{s}} + \operatorname{grad}^T\mathbf{u}_{\mathfrak{s}})$ into its volumetric and deviatoric part. Further, the dry shear modulus of the skeleton is given by G.

2.1.2 Balance of mass and balance of momentum of the fluid

Two constitutive assumptions are inherent in the second PDE of poroelasticity. A linear equation of state for the barotropic pore fluid is introduced

$$p = K^{\mathfrak{f}}\left[\frac{\rho^{\mathfrak{f}R}}{\rho_0^{\mathfrak{f}R}} - 1\right]. \tag{4}$$

The effective density in the initial configuration at time $t = t_0$ is introduced as $\rho^{\mathfrak{f}R}(t_0) =: \rho_0^{\mathfrak{f}R}$.
The viscous momentum interactions between the solid and fluid constituent $\hat{\mathbf{p}}^{\mathfrak{f}} = -\hat{\mathbf{p}}^{\mathfrak{s}}$, or drag forces, are modeled as

$$\hat{\mathbf{p}}^{\mathfrak{f}} = p \,\mathrm{grad}\phi - \frac{\phi_0^2 \gamma_0^{\mathfrak{f}R}}{k^{\mathfrak{f}}} \mathbf{w}_{\mathfrak{f}}. \tag{5}$$

The first (equilibrium) term on the right hand side of Eq. 5 is a nonlinear equilibrium term which is vanishing in linear poroelasticity. The second non-equilibrium term describes the viscous momentum exchange and contains the initial porosity $\phi(t_0) =: \phi_0$, the effective weight of the fluid $\gamma_0^{\mathfrak{f}R}$ at $t = t_0$ and the Darcy permeability $k^{\mathfrak{f}}$ sometimes also denoted as hydraulic conductivity. Thus, the non-equilibrium momentum exchange is proportional to the seepage velocity $\mathbf{w}_{\mathfrak{f}} = \dot{\mathbf{u}}_{\mathfrak{f}} - \dot{\mathbf{u}}_{\mathfrak{s}}$. The solid and fluid displacements are given by $\mathbf{u}_{\mathfrak{s}}$ and $\mathbf{u}_{\mathfrak{f}}$. The hydraulic conductivity $k^{\mathfrak{f}}$ [m/s] can be related to the intrinsic permeability $k^{\mathfrak{s}}$ [m^2] by $k^{\mathfrak{f}} = \rho^{\mathfrak{f}R} g \, k^{\mathfrak{s}} / \eta^{\mathfrak{f}R}$. Here, $\eta^{\mathfrak{f}R}$ [Pa s] is the effective dynamic viscosity of the pore fluid and g is the gravitational constant. The constitutive equations (4) and (5) can be inserted into the balance of mass of the fluid and the balance of momentum of the fluid. The derivation of the resulting equations is straightforward but needs a proper linearization. We would therefore refer to the literature (e.g. the review Renner and Steeb 2015) and present the resulting pressure diffusion equation with the coupling term on the right hand side

$$\frac{\dot{p}}{M} - \frac{k^{\mathfrak{f}}}{\gamma_0^{\mathfrak{f}R}} \,\mathrm{div}\,\mathrm{grad}\, p = -\alpha \,\mathrm{div}\, \dot{\mathbf{u}}_s \tag{6}$$

Here, we have introduced the (local) storativity or inverse storage capacity $1/M = \phi_0/K^{\mathfrak{f}} + (\alpha - \phi_0)/K^{\mathfrak{s}}$, cf. Wang (2000). Furthermore, it should be noted that in linear poroelasticity, the "dot" derivative is identical to the partial time derivative.

2.1.3 Resulting set of PDEs

The resulting set of PDEs for the porelastic domain is then summarized $\forall \mathbf{x} \in \mathcal{B}^{Pe}$ as Eqs. (1) and (6)

$$\begin{aligned} -\,\mathrm{div}(\boldsymbol{\sigma}_E^{\mathfrak{s}} - \alpha\, p\, \mathbf{I}) &= \rho\, \mathbf{b}, \\ \dot{p} - \frac{k^{\mathfrak{f}} M}{\gamma_0^{\mathfrak{f}R}} \,\mathrm{div}\,\mathrm{grad}\, p &= -\,\alpha\, M \,\mathrm{div}\, \dot{\mathbf{u}}_s, \end{aligned} \tag{7}$$

with Dirichlet boundary conditions for the solid displacement $\bar{\mathbf{u}}_{\mathfrak{s}}$ and the pore pressure $\bar{p}$. Neumann boundary conditions are described for the fluxes of the total stresses $\bar{\mathbf{t}}$ and the pore fluid $\bar{w}_{\mathfrak{f}}$

$$\mathbf{u}_{\mathfrak{s}} = \bar{\mathbf{u}}_{\mathfrak{s}} \quad \text{on} \quad \Gamma_D^{\mathfrak{s}} \quad \text{and} \quad p = \bar{p} \quad \text{on} \quad \Gamma_D^{\mathfrak{f}}, \tag{8}$$

$$\boldsymbol{\sigma} \cdot \mathbf{n} = \bar{\mathbf{t}} \quad \text{on} \quad \Gamma_N^{\mathfrak{s}} \quad \text{and} \quad \mathbf{w}_{\mathfrak{f}} \cdot \mathbf{n} = \bar{w}_{\mathfrak{f}} \quad \text{on} \quad \Gamma_N^{\mathfrak{f}}. \tag{9}$$

The set of PDEs is formulated in a weak format as the basis for subsequent finite element investigations. Primary variables are then solid displacements and pore pressure. Note that we multiply the pressure diffusion equation with the storativity M. This guarantees consistency in dimensions with the governing equations of the fracture flow model (introduced in the following section) and holds for a monolithic assembly covering both domains.

2.2 Equations in $\mathcal{B}^{Fr}$: hybrid-dimensional fracture flow model

The hybrid-dimensional approach recently introduced by Vinci et al. (2014) has specifically been designed to investigate high aspect ratio fractures. In order to avoid an explicit discretization of the fracture domain $\mathcal{B}^{Fr}$, equilibrium conditions with respect to a deformable fracture are imposed on the fluid phase to derive the governing equations of fluid flow. The model for a biphasic porous medium introduced in Sect. 2.1 provides the local aperture from which the local fracture permeability is calculated; hence fluid flow and solid deformation are coupled by means of the fracture aperture δ. Later in this work numerical coupling schemes for solving the governing equations are introduced.

2.3 Balance of momentum of the fluid: fracture flow

The hybrid-dimensional approach is based on the hydraulic description for flow of compressible and viscous fluids ($Re \ll 1$) with compressibility $\beta^{\mathfrak{f}} = 1/K^{\mathfrak{f}}$, where $K^{\mathfrak{f}}$ denotes the fluid's bulk modulus and $\eta^{\mathfrak{f}R}$ is the fluid's effective dynamic viscosity. Fluid flow is investigated for high aspect, hydraulic transmissive fracture domains possessing low contact areas; hence the pressure-driven Poiseuille-type flow between two plates is a valid assumption. Due to its geometrical nature the flow process is predominant within the plane defined by the normal vector $\mathbf{n}^+$ of the fracture surface. For a planar fracture geometry with basis vector $\hat{\mathbf{e}}_1$ and $\hat{\mathbf{e}}_2$ and normal $\hat{\mathbf{e}}_3 = \mathbf{n}^+$ (Fig. 1), the parabolic velocity profile results in a relative fluid velocity

$$\hat{\mathbf{w}}_{\mathfrak{f}} = -\frac{\delta^2(\mathbf{x}, t)}{12\,\eta^{\mathfrak{f}R}} \operatorname{grad} \hat{p} = -\frac{k^{\mathfrak{s}}_{Fr}(\mathbf{x}, t)}{\eta^{\mathfrak{f}R}} \operatorname{grad} \hat{p}, \tag{10}$$

where the pressure of the fluid within the fracture is $\hat{p}$ in order to avoid confusion with the pore pressure p that has been introduced in course of the poroelastic formulation. Dependent on fracture surface roughness or in case of mechanically-closed fractures the proportionality factor $1/12$ might be adapted (Renshaw 1995). Since the relative velocity $\hat{\mathbf{w}}_{\mathfrak{f}}$ is space-resolved for coordinate $\mathbf{x}$, the local and time-dependent effective fracture permeability $k^{\mathfrak{s}}_{Fr}(\mathbf{x}, t)$ for high aspect ratio fractures is introduced. Finally, a hydraulic formulation for a pressure driven flow that allows for varying local permeabilities has been introduced.

2.4 Balance of mass of the fluid: fracture flow

The derivation of the balance of mass requires consideration of the fluid compressibility along with a varying aperture $\delta(\mathbf{x}, t)$ of a single fracture. Evaluation of the mass balance yield relations between fluid velocity, fluid density changes, injected fluid volume and fracture volume. Fulfilment of condition

$$(\rho^{\mathfrak{f}R}\,\delta)^{\dot{}} + \operatorname{div}\left(\hat{\mathbf{w}}_{\mathfrak{f}}\,\rho^{\mathfrak{f}R}\,\delta\right) = 0 \tag{11}$$

guarantees conservation of mass in a fluid-filled fracture. For investigations where leak-off is of interest the system needs to be modified by simply extending the right hand side of Eq. (11) by a source term that will explained in Sect. 2.5.

2.5 Governing equations

The governing equation are derived by evaluating the balance of mass Eq.(11) and the balance of momentum Eq. (10) with respect to the fluid within the deformable fracture. Assuming proportionality between the fluid pressure and the effective fluid density given by Eq. (4) finally leads to the governing scalar PDE

$$\begin{array}{r|ll}
\text{(I)} & \dot{\hat{p}} & \text{Transient} \\
\text{(II)} & -\frac{\delta^2}{12\,\eta^{\mathfrak{f}R}}\,\widehat{\operatorname{grad}}\,\hat{p}\cdot\widehat{\operatorname{grad}}\,\hat{p} & \text{Quadratic} \\
\text{(III)} & -\frac{\delta}{12\,\eta^{\mathfrak{f}R}\,\beta^{\mathfrak{f}}}\,\widehat{\operatorname{grad}}\,\delta\cdot\widehat{\operatorname{grad}}\,\hat{p} & \text{Convection} \\
\text{(IV)} & -\frac{1}{12\,\eta^{\mathfrak{f}R}\beta^{\mathfrak{f}}}\,\widehat{\operatorname{div}}\left(\delta^2\,\widehat{\operatorname{grad}}\,\hat{p}\right) & \text{Diffusion} \\
\text{(V)} & +\frac{1}{\delta\,\beta^{\mathfrak{f}}}\,\dot{\delta} & \text{Coupling} \\
\text{(VI)} & = \frac{w_{\mathfrak{f}}^N}{\delta\,\beta^{\mathfrak{f}}} & \text{Leak-Off}
\end{array} \tag{12}$$

where divergence $\widehat{\operatorname{div}}$ and gradient $\widehat{\operatorname{grad}}$ are evaluated with respect to the lower dimensional fracture domain $\mathcal{B}^{Fr}$. The non-linear storage Eq. (12) consists of a transient term I), a quadratic term II), a convection term III), a diffusion term IV) and a hydro-mechanical coupling term which takes into account the transient evolution of the fracture's aperture V). In the present case we are focusing on high aspect ratio fractures. Thus, based on the results of a dimensional analysis of Eq. (12) Vinci (2014), the quadratic and the convection term can be neglected due to their minor contribution to the solution.

An initial pressure of $\hat{p}(\hat{\mathbf{x}}, t_0) = 0$ relative to the confining pressure and an undrained (non-flux) condition at the fracture tip $q(\hat{\mathbf{x}}_{tip}, t) = 0$ (where $\hat{\mathbf{x}}_{tip}$ denotes the fracture tip position) define the Dirichlet and Neumann boundary conditions. In case of applications related to borehole investigations, the related pressure or flux can additionally be prescribed at the borehole intersection.

For a surrounding poroelastic matrix material, an exchange of mass occurs between the fracture and the surrounding matrix which is taken into account by a non-zero right-hand-side in Eq. (11). The sink or source term is considered by

$$q_{lk} = \frac{w_{\mathfrak{f}}^{N}}{\delta\,\beta^{\mathfrak{f}}}, \tag{13}$$

where $w_{\mathfrak{f}}^{N}$ is the leak-off induced seepage velocity determined by Darcy's law of the poroelastic matrix. In the course of this work, the leak-off treatment differs for the coupling strategies and is individually derived for each scheme in Sect. 3.

3 Numerical schemes for hybrid-dimensional formulations

Consistent implementation of the governing equations describing fluid flow, deformation and the possible exchange of fluid from the fracture to a poroelastic matrix material, require an implicitly coupled system of both domains, the fracture, and the matrix material. Suitable algorithms to solve the discrete system for pressure $\hat{p}$ and fracture width δ will be discussed in this section.

3.1 Weak form of the governing equations

In the course of this work, a classical Bubnov Galerkin finite element formulation for poroelasticity will be applied to the surrounding matrix material with domain $\mathcal{B}^{Pe}$. Analysis of the coupled response inside the fracture, i.e., the fluid pressure $\hat{p}$-fracture aperture δ interplay is analyzed. The numerical formulation for the biphasic porous matrix is based on the weak form of the set of governing equations (7) and reads

$$\begin{aligned}
&\int_{\mathcal{B}^{Pe}} \left(\boldsymbol{\sigma}_E^{\mathfrak{s}} - \alpha\, p\,\mathbf{I}\right) : \operatorname{grad} w_{\mathbf{u}}\, \mathrm{d}v = \int_{\Gamma_N^{\mathfrak{s}}} \bar{\mathbf{t}} \cdot w_{\mathbf{u}} \mathrm{d}a, \\
&\int_{\mathcal{B}^{Pe}} \left[\frac{\dot{p}}{M} w_p + \frac{k^{\mathfrak{f}}}{\gamma^{\mathfrak{f}R}} \operatorname{grad} p \cdot \operatorname{grad} w_p + \alpha \operatorname{div} \dot{\mathbf{u}}_{\mathfrak{s}}\, w_p\right] \mathrm{d}v = \int_{\Gamma_N^{\mathfrak{f}}} \bar{\mathrm{w}}_f\, w_p\, \mathrm{d}a.
\end{aligned} \tag{14}$$

Note that the test functions of the balance of momentum of the mixture $w_{\mathbf{u}}$ and the balance of mass of the fluid w_p are introduced. To investigate pressure fluctuations related to aperture changes, the poroelastic equation (14) of the surrounding matrix (domain $\mathcal{B}^{Pe}$) medium are coupled to the modified hybrid-dimensional formulation (domain $\mathcal{B}^{Fr}$) in its weak form

$$\int_{\mathcal{B}^{Fr}} \left[\dot{p}\, w_{\hat{p}} + \frac{\delta^2}{12\,\eta^{\mathfrak{f}R}\beta^{\mathfrak{f}}} \operatorname{grad} \hat{p} \cdot \operatorname{grad} w_{\hat{p}} + \frac{1}{\delta\,\beta^{\mathfrak{f}}}\frac{\partial \delta}{\partial t}\, w_{\hat{p}}\right] \mathrm{d}v = 0. \tag{15}$$

Note that advection and quadratic terms are neglected based on the dimensional analysis evaluated in Vinci (2014). The test functions for the hybrid-dimensional problem are given by w_p.

In Sect. 2.2, the fracture width δ has already been introduced as the hydro-mechanical coupling parameter. By assuming predominant deformation normal to the fracture surface the local fracture aperture

$$\delta = \mathbf{u}^{+} \cdot \mathbf{n}^{+} - \mathbf{u}^{-} \cdot \mathbf{n}^{-} \tag{16}$$

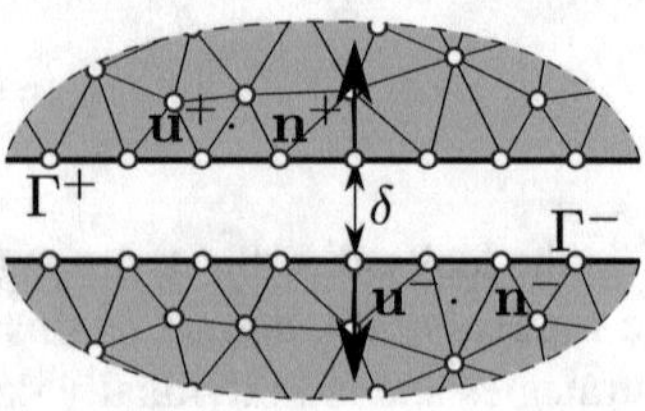

can be rewritten with respect to the fracture normal surface displacements $\mathbf{u}^{+}$ and $\mathbf{u}^{-}$ which are primary variables of the solid formulation [Eq. (14a)].

3.2 Characteristic of the coupled system

To obtain a deeper understanding of the coupled system, Adachi et al. (2007) studied the characteristic of hydro-mechanical coupled systems by assuming a 1D penny-shaped Kristianovich–Geertsma–de Klerk (KGD) fracture width description coupled to the Reynold's equation modified for the fluid flow in deformable fractures. Based on linear elastic solid mechanics coupled to a diffusive process, Adachi's simplified system is equivalent to a highly permeable poroelastic drained medium, since the deformation of the surrounding matrix is not damped by the pore pressure evolution; hence it represents the stiffest case of the coupled system due to possibly large fracture deformations. By analyzing the eigenvalues of the coupled matrices (Peirce and Siebrits 2005) it has been shown that the stability is guaranteed if the Courant–Friedrichs–Lewy (CFL) condition

$$\Delta t < \frac{\Delta x^3}{E' D} \tag{17}$$

is fulfilled. Equation (17) does not only clarify that explicit time-stepping using the explicit Euler method is not suitable due to inadequate small time steps, but also the extreme stiffness of the coupled system that requires an implicit coupling procedure. In the following an implicit time discretization in form of a backward Euler method will be used throughout this work. In Eq. (17) Δx denotes the element length, E' the elastic plane-strain modulus and D the conductivity.

3.3 Weak coupling: staggered coupling scheme

Non-conformal meshes and computations on two different domains ($\mathcal{B}^{Pe}$ and $\mathcal{B}^{Fr}$) are the motivation for the first proposed strategy, namely a staggered coupling scheme. Besides the reduction of DoFs by the lower-dimensional elements for pressure diffusion in the fracture, computational time can potentially be reduced by using an individual discretization for both domains. In the numerical study section an investigation of the error evolution for different numerical resolution combinations of the fracture (pressure diffusion) and poroelastic matrix domains will provide further insights of increasing computational efficiency based on non-conformal meshes

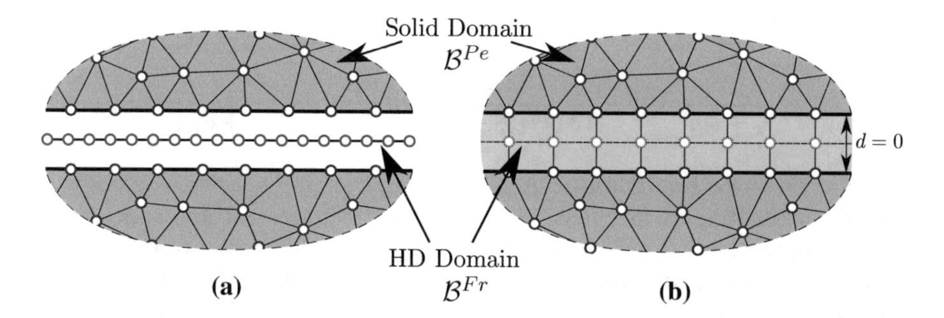

Fig. 2 Comparison of a weak coupling scheme using non-conformal meshes (**a**) and a strong coupling scheme with implemented interface elements (here interface elements, and auxiliary nodes are explicitly shown for presentation purposes only; in final discretization fracture surfaces align with d = 0) (**b**)

(reduction of DoF). Schematically the idea of non-conformal meshing using hybrid-dimensional elements is shown in Fig. 2a.

3.3.1 Coupling algorithm

In the field of non-fractured porous media different sequential implicit strategies for solving biphasic mixtures have been proposed. Depending on the sequence both problems are solved Kim et al. (2011a, b) differentiate between drained/undrained (mechanical followed by flow problem) and fixed-stress/fixed-strain splits (flow followed by the mechanical problem). The drained and undrained split differ concerning the consideration of the applied pressure within the mechanical problem. The drained split treats the pressure constant throughout the mechanical calculations, unlike the undrained split where the pressure can vary with the volumetric deformation enforcing a constant fluid mass in each element. In contrast, the fixed-strain and fixed-stress split differ concerning the consideration of the applied strain used in the flow calculations. For the fixed-strain split, the applied strain rate stays constant in contradistinction to the fixed-stress split where the strain rate can vary depending on the calculated fluid pressure enforcing a constant stress rate throughout the flow calculations. Girault et al. (2016) modified the fixed-stress method to achieve convergence and stability for fractured porous media applications.
In the course of this work, the staggered scheme implicitly couples the fracture with the porous domain allowing non-conformal numerical discretizations of the porous and lower dimensional fracture domain. Nevertheless, the porous domain does not explicitly discretize the fracture volume, and volumetric changes of the fracture throughout porous matrix calculations are not accessible. Hence the fixed-stress and fixed-strain algorithms suit the described discretization strategy best since no volumetric change information are required. An algorithm comparable to the fixed-strain scheme and the modified fracture formulation of the fixed-stress scheme is implemented and tested with regards to stability and convergence. In case of the fixed-stress algorithm the simplified governing equations (quadratic and convection terms are neglected; high aspect

ratio assumption) Eq. (12) is extended to achieve constant stress rates throughout the flow problem

$$\dot{\hat{p}}^{new} - \frac{\gamma_c}{\gamma_c + \delta\,\beta^{\mathfrak{f}}}\dot{\hat{p}}^{old} - \frac{\delta^3\,\beta^{\mathfrak{f}}}{12\,\eta^{\mathfrak{f}R}(\gamma_c + \delta\,\beta^{\mathfrak{f}})}\,\hat{\text{div}}\,\hat{\text{grad}}\,\hat{p} - \frac{1}{\gamma_c + \delta\,\beta^{\mathfrak{f}}}\dot{\delta} = q_{lk}. \quad (18)$$

Note that a term has been consistently added to the governing equation to fix the stress rate without limiting the functions, where $\gamma_c = M\beta^{\mathfrak{f}}\alpha^2/(\lambda + M\beta^f\lambda\phi)$ is an estimate to achieve convergence, $\hat{p}^{new} = \hat{p}^t_{k+1}$ is the solution of the current and $\hat{p}^{old} = \hat{p}^t_k$ the solution of the former iteration (the interested reader is referred to Girault et al. 2016). Since the time derivative is evaluated with respect to the fixed pressure $\hat{p}^{t-1}$ of the former time step the introduced term vanishes once the problem converges ($\hat{p}^t_{k+1} \approx \hat{p}^t_k$). The fixed-stress formulation can be classified as a preconditioned Richardson iteration (Castelletto et al. 2015).

$$\delta^t_{k+1} = \overline{\alpha}\delta^t_{k+1/2} + (1-\overline{\alpha})\,\delta^t_k \quad 0 < \overline{\alpha} < 1 \quad (19)$$

updates the fracture width to δ^t_{i+1}, where δ^n_i is the fracture width of the prior iteration step, $\delta^n_{k+1/2}$ an intermediate value and $\bar{\alpha}$ the Picard update constant. In case of the fixed-stress strategy the update constant is equal to one $\bar{\alpha} = 1$. Convergence of the procedure is checked by the error measure

$$\epsilon = \frac{\sum_{n=1}^{n_{total}} \left|\delta^t_{k+1,n} - \delta^t_{k,n}\right|}{\sum_{n=1}^{n_{total}} \left|\delta^t_{k+1,n}\right|}, \quad (20)$$

which is calculated by iterating over all nodes n, where n_{total} denotes the total number of nodes.

The Algorithm 1 schematically shows the implemented coupling scheme and is valid for both strategies. The numerical scheme contains two major loops; one time-stepping loop due to the transient (diffusive) nature of the problem and a second one for the residuum iterations of every time-step. For a better understanding the main steps of the algorithm are summarized below.

Residuum Iterations

1. The iteration procedure starts with an update of the pressure to $\hat{p}^t_{k+1}$ based on the fracture width of the prior iteration step δ^t_{k-1}. During the first iteration the initially calculated value δ^t_0 is used. Further, to guarantee consistency, the discrete form of fracture change velocity $\dot{\delta} = \frac{\delta^t_{k-1} - \delta^{t-1}}{\Delta t}$ is related to the fixed fracture width δ^{t-1} of the previous time-step.
2. With pressure $\hat{p}^t_{k+1}$ all necessary boundary conditions are provided to determine deformation $\mathbf{u}^t_{k+1}$.
3. In the following step the intermediate fracture width $\delta^t_{k+1/2}$ is obtained from deformation $\mathbf{u}^t_{k+1}$.

Algorithm 1: Staggered Coupling Procedure

```
while t <= t_end do
    while k < k_max do
        # Update pressure state dependent on fracture width changes
        p̂^t_{k+1} = eq. (15) using δ^t_k
        # Update deformation for new pressure state
        u^t_{k+1} = eq. (14) using p^t_{k+1}
        # Intermediate fracture width for
        δ^t_{k+1/2} = eq. (16) using u^t_{k+1}
        # Use Picard update to calculate new fracture width
        δ^t_{k+1} = eq. (19) using δ^t_{k+1/2} and δ^t_k
        # Calculate Error
        ϵ = eq. (20)
        if ϵ < ϵ_max then
            # Update solutions for time-step
            p̂^{t+1} = p̂^t_{k+1}
            δ^{t+1} = δ^t_{k+1}
            # Break iteration loop
            break
        else
            # Set pressure values to values of prior time-step
            p̂^t_{k+1} = p̂^{t-1}
            # Continue
            continue
        end
    end
end
```

4. The new fracture width δ^t_{k+1} (using the Picard update) can be considered as the result of each iteration step.
5. Error ϵ is calculated using fracture width δ^t_{k+1}. If the error is less than a predefined maximum error convergence is reached and the pressure $\hat{p}^t$ and fracture width δ^t of the active time-step t are updated by the values $\hat{p}^t_c$ and δ^t_c, where index c denotes the convergence iteration step. If the error criterion is not fulfilled the pressure $\hat{p}^t_{k+1}$ is set back to the fixed pressure $\hat{p}^{t-1}$ of the previous time-step to avoid transient artifacts and guarantee consistency, before the process is repeated within the next iteration.

In case of a low fluid compressibilities $\beta^{\mathfrak{f}} < 10^{-6}$ GPa^{-1} it is expected that the fixed-strain scheme requires small values for the Picard update constant (approx. $\bar{\alpha} < 5.0e^{-3}$) to reach convergence. Small values for $\bar{\alpha}$ result in an inadequate high number of iteration steps and reasonable values for $\bar{\alpha}$ in inadequate small time steps (Shen et al. 2014), respectively. Nevertheless, for moderate up to high compressibilities $\beta^{\mathfrak{f}} > 10^{-6}$ GPa^{-1}, the scheme converges in a low number of iterations. In contrast the fixed-stress approach is supposed to convergece even for a low compressibility in an adequate number of iterations. For both schemes it is intuitive to implement non-conformal mesh coupling in order to gain an advantage with respect to computation times.

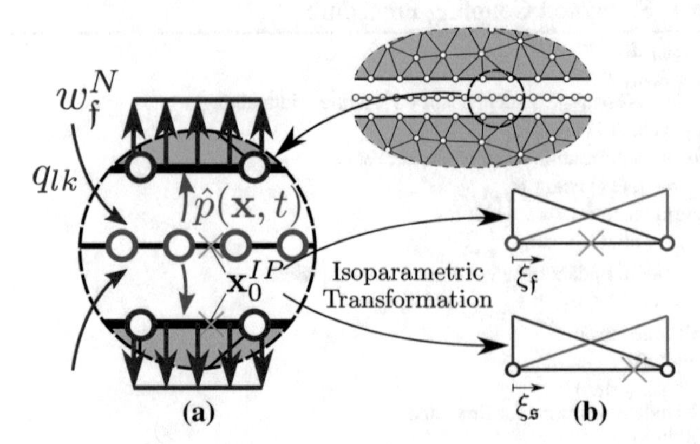

Fig. 3 Leak-off interaction of the poroelastic and fracture flow domain (**a**) by using a consistent interpolation scheme (**b**)

3.3.2 Leak-off: weak coupling

Leak-off in the weak coupling scheme is related to the normal seepage velocity determined by the poroelastic problem and is captured by an extension of the fracture flow formulation by source term q_{lk}. In order to determine q_{lk} in Eq. (2.5) the seepage velocity $w_{\mathfrak{f}}^N$ normal to the fracture surface Γ^{Fr} must be calculated based on Darcy's law

$$w_{\mathfrak{f}}^N(\mathbf{x}) = -\frac{k_{Fr}^{\mathfrak{s}}}{\eta^{\mathfrak{f}R}} \operatorname{grad} p \cdot \mathbf{n}^{\pm} \tag{21}$$

where $\mathbf{n}^{\pm}$ denotes the top and bottom normal vector. Once source term q_{lk} is calculated it is applied in the form of a Neumann boundary condition to the fracture flow problem. The pore pressure p on the fracture surfaces is obtained by the pressure state $\hat{p}$ of the fracture flow problem, and no discontinuities of the pressure state between both fracture surfaces are taken into account. Nevertheless, the leak-off into the porous medium can differ on both surfaces due to possibly varying permeabilities of the surrounding matrix and also allows fluid flow across the fracture. Throughout the equilibrium iterations of the staggered algorithm, the leak-off term is updated for every iteration step. The interaction procedure throughout the calculations is schematically shown in Fig. 3a.

3.3.3 Consistent interpolation

Information of all primary variables $\mathbf{u}$, p and $\hat{p}$ are required for element integration on integration point level of both problems. Since the lower dimensional elements are geometrically aligning with the fracture surfaces of the poroelastic problem the global coordinates of the integration points $\mathbf{x}^{IP}$ can be transformed to the isoparametric element representation of both domains. In Fig. 3b the transformation is arbitrarily shown for an integration point $\mathbf{x}_0^{IP}$ and linear element formulations in both domains.

Note, that the element order can independently be chosen for the fracture flow and poroelastic domain discretization.

3.4 Strong coupling: interface elements

Especially for fluids with a low compressibility $\beta^{\mathrm{f}} < 10^{-6}\ \mathrm{GPa}^{-1}$, the system of equations becomes extremely stiff. Besides the already proposed weakly coupled fixed-stress strategy an implicitly coupled formulation will be proposed to model such stiff systems. In the present approach, the strong coupling is realized by zero-thickness interface elements based on the work of Segura and Carol (2008a, b) which will be applied to the proposed hybrid dimensional formulation.

3.5 Strong coupling: interface element formulation

This section will introduce zero-thickness interface elements to build up a global assembly matrix concerning only two primary variables, namely pressure p and deformation $\mathbf{u}$, for both domains. The resulting global system is then monolithically solved. For linear cases, no iterations are needed, and convergence is always guaranteed (Girault et al. 2016); for non-linear cases, the Newton–Raphson scheme is implemented. In general interface elements possess no thickness, meaning that facing nodes are aligning. In the case of fracture flow the volume of the fracture is not captured geometrically, but through integration by using the fracture aperture to determine the fracture volume. Different interface elements have been proposed in the literature such as single node—interface elements (Woodbury and Zhang 2001) that capture longitudinal fracture flow only, doubled node—interface elements (Segura and Carol 2008b) with two aligning nodes where the element-integration takes place on mid-plane auxiliary nodes allowing transversal and longitudinal flow and triple node—interface elements (Guiducci et al. 2003) where a third aligning node is introduced in order to separate the transversal from the longitudinal flow. This work exclusively focuses on the double node—interface element since no additional DoF are introduced and flow in both directions is captured.
In contrast to the weak coupling scheme where only longitudinal flow was described in the lower dimensional elements and transversal flow captured by leak-off only, the proposed interface elements directly model longitudinal and transversal flow concerning the pressure state on the fracture surfaces of the porous matrix and allows a discontinuous pressure jump regarding both fracture surfaces. Modeling of fracture pressure $\hat{p}$ in direct relation to pressure p obtained on the fracture surface of the porous matrix requires an averaging across the fracture as proposed in Martin et al. (2005). Martin et al. treat the fracture as an interface separating two connected domains and is employed to derive the proposed interface elements. By averaging the conservation equation, a formulation governing the flow into, across and along the fracture is obtained. Both equations are then closed by boundary conditions for the pressure along the fracture surfaces.

3.5.1 Averaging of balance of mass of the fluid: fracture flow

An extension of the mentioned work of Martin et al. to capture deformable fractures is derived in the following. Balance of mass, given by Eq. (11), is rewritten by inserting the pressure density relationship for a barotropic fluid of Eq. (4)

$$\frac{\delta}{K^{\mathfrak{f}}}\dot{\hat{p}} + \dot{\delta} + \hat{\text{div}}\left(\hat{\mathbf{w}}_{\mathfrak{f}}\,\delta\right) = 0. \tag{22}$$

In order to capture a pressure jump along the fracture surfaces the seepage velocity $\hat{\mathbf{w}}_{\mathfrak{f}}$ is decomposed into a longitudinal $\hat{\mathbf{w}}_{\mathfrak{f}}^{l}$ and a transversal $\hat{\mathbf{w}}_{\mathfrak{f}}^{t}$ component

$$\hat{\mathbf{w}}_{\mathfrak{f}} = \hat{\mathbf{w}}_{\mathfrak{f}}^{l} + \hat{\mathbf{w}}_{\mathfrak{f}}^{t}. \tag{23}$$

Note that the transversal component is defined concerning the seepage velocity normal to the fracture interface $\hat{\mathbf{w}}_{\mathfrak{f}}^{t} = \hat{\mathbf{w}}_{\mathfrak{f}} \cdot \mathbf{n}$ with $\mathbf{n} = \mathbf{n}^{+} = -\mathbf{n}^{-}$. Since two flow directions are taken into account divergence $\hat{\text{div}}_{l}$ and gradient $\hat{\text{grad}}_{l}$ in longitudinal and divergence $\hat{\text{div}}_{t}$ and gradient $\hat{\text{grad}}_{t}$ in transversal direction are introduced. By inserting these in Eq. (22) the decomposed balance of mass of the fluid within the fracture

$$\frac{\delta}{K^{\mathfrak{f}}}\dot{\hat{p}} + \dot{\delta} + \hat{\text{div}}_{l}\left(\hat{\mathbf{w}}_{\mathfrak{f}}\,\delta\right) + \delta\hat{\text{div}}_{t}\left(\hat{\mathbf{w}}_{\mathfrak{f}}\right) = 0 \tag{24}$$

is derived. In order to formulate the fracture flow with respect to a line segment an averaging in transversal direction is necessary. Note that only pressure $\hat{p}$ and seepage velocity $\hat{\mathbf{w}}_{\mathfrak{f}}$ vary in tangential direction. The averaging results in

$$\frac{\delta}{K^{\mathfrak{f}}}\dot{\hat{P}} + \dot{\delta} + \hat{\text{div}}_{l}\left(\hat{\mathbf{W}}_{\mathfrak{f}}\delta\right) = -\hat{\mathbf{w}}_{\mathfrak{f}} \cdot \mathbf{n}|_{\Gamma_{+}^{Fr}} + \hat{\mathbf{w}}_{\mathfrak{f}} \cdot \mathbf{n}|_{\Gamma_{-}^{Fr}} \tag{25}$$

where $\hat{P} = \int_{-\delta/2}^{\delta/2} \hat{p}\,\mathbf{dn}$ defines the integral value of the fracture fluid pressure and $\hat{\mathbf{W}}_{\mathfrak{f}} = \int_{-\delta/2}^{\delta/2} \hat{\mathbf{w}}_{\mathfrak{f}}^{t}\,\mathbf{dn}$ the integral value of the longitudinal seepage velocity. Continuity between the fluxes of the porous medium and the fracture domain on the fracture surfaces $\Gamma_{\pm}^{Fr}$ is introduced by

$$w_{\mathfrak{f}}^{\pm} \cdot \mathbf{n} = \hat{\mathbf{w}}_{\mathfrak{f}} \cdot \mathbf{n}. \tag{26}$$

Inserting the continuity condition formulated in Eq. (26) in the averaged mass balance of Eq. (25)

$$\frac{\delta}{K^{\mathfrak{f}}}\dot{\hat{P}} + \dot{\delta} + \hat{\text{div}}_{l}\left(\hat{\mathbf{W}}_{\mathfrak{f}}\delta\right) = \mathbf{w}_{\mathfrak{f}}^{+} \cdot \mathbf{n}^{+} + \mathbf{w}_{\mathfrak{f}}^{-} \cdot \mathbf{n}^{-} \tag{27}$$

combines the surrounding matrix and the fracture domain, where the source term on the right hand side governs the leak-off into the surrounding matrix and across the fracture.

3.5.2 Averaging of balance of momentum of the fluid: fracture flow

In longitudinal direction Eq. (10) defines the relative fluid velocity. By integration over the aperture δ the averaged expression for the seepage velocity reads

$$\hat{\mathbf{W}}_{\mathfrak{f}} = -\frac{k^{\mathfrak{s}}_{Fr,l}(\mathbf{x},t)}{\eta^{\mathfrak{f}R}}\,\hat{\operatorname{grad}}_l\,\hat{P}. \tag{28}$$

The averaged mass balance given by Eq. (27) combined with the averaged seepage velocity defined by Eq. (28) provides a formulation for the flow within the fracture including leak-off. Pressure differences on both fracture surfaces are modelled by introducing another governing equation for the transversal flow. In transversal direction a Darcy type flow is introduced by

$$\hat{\mathbf{w}}^t_{\mathfrak{f}} = -\frac{k^{\mathfrak{s}}_{Fr,t}}{\eta^{\mathfrak{f}R}}\,\hat{\operatorname{grad}}_t\,\hat{p} \tag{29}$$

where $k^{\mathfrak{s}}_{Fr,t}$ defines the permeability in transversal direction. Integration over the aperture results in

$$\int_{-\delta/2}^{\delta/2} \hat{\mathbf{w}}^t_{\mathfrak{f}}\cdot\mathbf{n}\,\mathrm{d}\,\mathbf{n} = -\frac{k^{\mathfrak{s}}_{Fr,t}}{\eta^{\mathfrak{f}R}}\left(-\hat{p}|_{\Gamma^{Fr}_{+}} + \hat{p}|_{\Gamma^{Fr}_{-}}\right). \tag{30}$$

Integration of the left hand side $\int_{-\delta/2}^{\delta/2} \hat{\mathbf{w}}^t_{\mathfrak{f}}\cdot\mathbf{n}\,\mathrm{d}\,\mathbf{n}$ is not carried out explicitly, but approximated by the trapezoidal rule

$$\int_{-\delta/2}^{\delta/2} \hat{\mathbf{w}}^t_{\mathfrak{f}}\cdot\mathbf{n}\,\mathrm{dn} \approx \frac{\delta}{2}\left(\hat{\mathbf{w}}_{\mathfrak{f}}\cdot\mathbf{n}|_{\Gamma^{Fr}_{+}} + \hat{\mathbf{w}}_{\mathfrak{f}}\cdot\mathbf{n}|_{\Gamma^{Fr}_{-}}\right) = \frac{\delta}{2}\left(-\mathbf{w}^{+}_{\mathfrak{f}}\cdot\mathbf{n}^{+} + \mathbf{w}^{-}_{\mathfrak{f}}\cdot\mathbf{n}^{-}\right). \tag{31}$$

where continuity of fluid flow on the fracture surfaces is assumed. Finally, the approximation given in Eq. (31) is inserted in Eq. (30) to define the relation of pressure and flow in transversal direction

$$\frac{1}{2}\left(-\mathbf{w}^{+}_{\mathfrak{f}}\cdot\mathbf{n}^{+} + \mathbf{w}^{-}_{\mathfrak{f}}\cdot\mathbf{n}^{-}\right) = -\frac{k^{\mathfrak{s}}_{Fr,t}}{\eta^{\mathfrak{f}R}}\frac{-p^{+}+p^{-}}{\delta}. \tag{32}$$

Note that Eq. (32) made use of the continuity of the pressure on the fracture surface $p^{\pm} = \hat{p}|_{\Gamma_{\pm}}$. The formulation depends on parameters that can be expressed by the primary variables of the porous domain.

3.6 Interface element formulation

To close the system of equations and to derive the interface element formulation, a boundary condition for the averaged fracture domain pressure $\hat{P}$ is introduced. In contrast to the weak coupling scheme where no discretization of the surrounding matrix governs the fracture domain, the constitutive modelling of solid behaviour, such as closing and opening of rough fracture surfaces (Segura and Carol 2008a) can be realized within the framework of interface elements if needed. Extension of the existing set of governing equations by a formulation of the fracture surface interaction allows i.e. modelling of hydraulically open and mechanically closed fractures and motivates a discontinuous pressure assumption for the transversal fracture flow which is also valid for investigations of open fractures by choosing a suitable value for $k^{\mathfrak{s}}_{Fr,t}$. Throughout the averaging process of the fracture pressure $\hat{P}$ different domain decompositions are possible (Martin et al. 2005) leading to a general set of balance equations at both fracture surfaces

$$
\begin{aligned}
-\xi\, \mathbf{w}_{\mathfrak{f}}^{+} \cdot \mathbf{n}^{+} + \frac{2\, k^{\mathfrak{s}}_{Fr,t}}{\delta\, \eta^{\mathfrak{f}R}}\, p^{+} &= -(1-\xi)\, \mathbf{w}_{\mathfrak{f}}^{-} \cdot \mathbf{n}^{-} + \frac{2\, k^{\mathfrak{s}}_{Fr,t}}{\delta\, \eta^{\mathfrak{f}R}}\, \hat{P}, \\
-\xi\, \mathbf{w}_{\mathfrak{f}}^{-} \cdot \mathbf{n}^{-} + \frac{2\, k^{\mathfrak{s}}_{Fr,t}}{\delta\, \eta^{\mathfrak{f}R}}\, p^{-} &= -(1-\xi)\, \mathbf{w}_{\mathfrak{f}}^{+} \cdot \mathbf{n}^{+} + \frac{2\, k^{\mathfrak{s}}_{Fr,t}}{\delta\, \eta^{\mathfrak{f}R}}\, \hat{P},
\end{aligned}
\tag{33}
$$

where weighting parameter ξ is a positive real number within the range $\xi \in [1/2, 1]$. In this work ξ is chosen to be exactly the stability limit $\xi = 1/2$ (Segura and Carol 2004) which despite a high solution quality for most cases might lead to numerical instabilities for some numerical setups where ξ needs to be adjusted (Martin et al. 2005). Nevertheless, no such numerical instabilities occurred throughout the performed numerical studies. Hence the averaged fracture domain pressure $\hat{P}$ and the pressure difference along δ read

$$
\begin{aligned}
\hat{P} &= \frac{p^{+} + p^{-}}{2}, \\
\mathbf{w}_{\mathfrak{f}}^{-} \cdot \mathbf{n}^{-} - \mathbf{w}_{\mathfrak{f}}^{+} \cdot \mathbf{n}^{+} &= 2\, k^{\mathfrak{s}}_{Fr,t} \frac{p^{-} - p^{+}}{\delta},
\end{aligned}
\tag{34}
$$

where the transversal discontinuous pressure assumption is introduced by $\hat{P}^{t} = (p^{-} - p^{+})/\delta$. By summation/subtraction of Eqs. (25) and (34) a formulation governing longitudinal and transversal flow depending only on the primary variables of the porous matrix is found for the upper and lower fracture surface

$$
\begin{aligned}
\frac{1}{2}\left[\frac{1}{K^{\mathfrak{f}}}\dot{\hat{P}} + \dot{\delta} + \hat{\operatorname{div}}_{l}\left(\hat{\mathbf{W}}_{\mathfrak{f}}\delta\right)\right] - k^{\mathfrak{s}}_{Fr,t}\, \hat{P}^{t} &= \mathbf{w}_{\mathfrak{f}}^{+} \cdot \mathbf{n}^{+} \quad \text{on } \Gamma_{+}^{Fr}, \\
\frac{1}{2}\left[\frac{1}{K^{\mathfrak{f}}}\dot{\hat{P}} + \dot{\delta} + \hat{\operatorname{div}}_{l}\left(\hat{\mathbf{W}}_{\mathfrak{f}}\delta\right)\right] + k^{\mathfrak{s}}_{Fr,t}\, \hat{P}^{t} &= \mathbf{w}_{\mathfrak{f}}^{-} \cdot \mathbf{n}^{-} \quad \text{on } \Gamma_{-}^{Fr}.
\end{aligned}
\tag{35}
$$

The weak form of the governing set of equations given by (35) are simplified and evaluated for the case of high aspect ratio fractures similar to the derivation of the hybrid dimensional formulation used for the weak coupling scheme

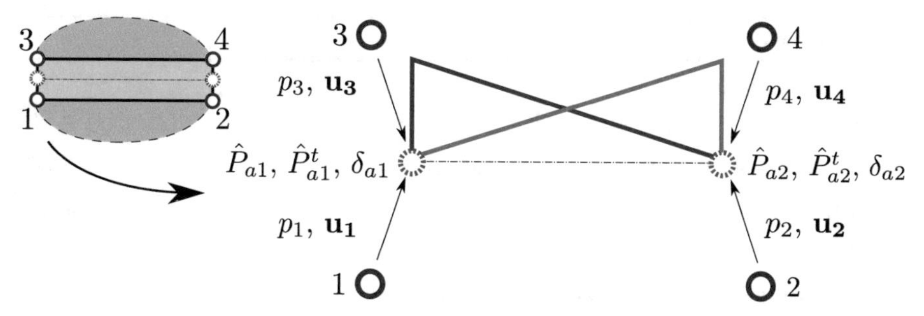

Fig. 4 Integration of linear auxiliary element, including the contribution of interface element nodes to averaged pressure $\hat{P}_{ai}$, pressure difference $\hat{P}^t_{ai}$ and aperture δ_{ai} on auxiliary element nodes

$$
\begin{aligned}
&\int_{\Gamma^{Fr}_{+}} \left[\frac{1}{2} \left(\dot{p}\, w_a + \frac{\delta^2}{12\, \eta^{\mathfrak{f}R} \beta^{\mathfrak{f}}} \hat{\mathrm{grad}}_l\, \hat{P} \cdot \hat{\mathrm{grad}}_l\, w_a + \frac{1}{\delta\, \beta^{\mathfrak{f}}} \frac{\partial \delta}{\partial t}\, w_a \right) - \frac{1}{\delta\, \beta^{\mathfrak{f}}} k^{\mathfrak{s}}_{Fr,t}\, \hat{P}^t\, w_a \right] \\
&\quad \mathrm{d}v = \frac{1}{\delta\, \beta^{\mathfrak{f}}} \mathbf{w}_{\mathfrak{f}}^{+} \cdot \mathbf{n}^{+}, \\
&\int_{\Gamma^{Fr}_{-}} \left[\frac{1}{2} \left(\dot{p}\, w_a + \frac{\delta^2}{12\, \eta^{\mathfrak{f}R} \beta^{\mathfrak{f}}} \hat{\mathrm{grad}}_l\, \hat{P} \cdot \hat{\mathrm{grad}}_l\, w_a + \frac{1}{\delta\, \beta^{\mathfrak{f}}} \frac{\partial \delta}{\partial t}\, w_a \right) + \frac{1}{\delta\, \beta^{\mathfrak{f}}} k^{\mathfrak{s}}_{Fr,t}\, \hat{P}^t\, w_a \right] \\
&\quad \mathrm{d}v = \frac{1}{\delta\, \beta^{\mathfrak{f}}} \mathbf{w}_{\mathfrak{f}}^{-} \cdot \mathbf{n}^{-}.
\end{aligned} \tag{36}
$$

where w_a denotes the lower dimensional test functions used for the auxiliary element.

3.7 Leak-off: strong coupling

Leak-off in the strong coupling scheme is described by the right-hand side of Eqs. (36). Nevertheless, the boundary term is naturally covered by the pore pressure evolution within the surrounding poroelastic matrix since the fluid pressure $\hat{p}$ within the fracture is calculated based on the pore pressure p of the fracture boundary. Leak-off from the fracture into the porous matrix can vary concerning to the upper and lower fracture surface $\Gamma^{Fr}_{\pm}$.

3.8 Interface element integration

Integration of auxiliary elements requires the averaged pressure $\hat{P}$, pressure difference $\hat{P}^t$ and aperture δ on auxiliary node level. The values are calculated from the pressure p and deformation state $\mathbf{u}$ of aligning nodes of the interface element which is exemplarily shown for a linear element formulation in Fig. 4. Note that the upper surface (node 3 and 4) is governed by the first equation and the lower surface (node 1 and 2) by the second equation of Eq. (36). The governing equations and boundary conditions are summarized in the following.

Governing Equations	
Surrounding matrix—$\mathcal{B}^{Pe}$	
Solid equilibrium: $-\operatorname{div}(\tilde{\boldsymbol{\sigma}}_E^{\mathfrak{s}} - \alpha\, p\, \mathbf{I}) = \rho \mathbf{b}$	in $\mathcal{B}^{Pe}$
Fluid equilibrium: $\dot{p} - \dfrac{M k^{\mathfrak{f}}}{\gamma^{\mathfrak{f}R}} \operatorname{div} \operatorname{grad} p + M\alpha \operatorname{div} \dot{\mathbf{u}}_{\mathfrak{s}} = M \bar{\mathrm{w}}_f$	in $\mathcal{B}^{Pe}$
with: $\boldsymbol{\sigma}_E^{\mathfrak{s}} = 3\, K \operatorname{vol}(\boldsymbol{\varepsilon}_{\mathfrak{s}}) + 2\, G \operatorname{dev}(\boldsymbol{\varepsilon}_{\mathfrak{s}}) + (1-\alpha)\, p\, \mathbf{I}$	in $\mathcal{B}^{Pe}$
$\varepsilon_{\mathfrak{s}} = \dfrac{1}{2}\left(\operatorname{grad} \mathbf{u} + \operatorname{grad}^{\mathrm{T}} \mathbf{u}\right)$	
Fluid filled fracture—$\mathcal{B}^{Fr}$ (high aspect ratio simplification)	
Fracture flow: $\dot{p} - \dfrac{\delta^2}{12\, \eta^{\mathfrak{f}R}} \hat{\operatorname{div}}\, \hat{\operatorname{grad}} \hat{p} + \dfrac{1}{\delta\, \beta^{\mathfrak{f}}} \dot{\delta} = q_{lk}$	in $\mathcal{B}^{Fr}$
Initial, boundary and coupling conditions	
Fracture surface force equilibrium: $\mathbf{t}^{\pm} = -\hat{p} \cdot \mathbf{n}^{\pm}$	on $\Gamma_{\pm}^{Fr}$
Fracture tip no flux condition: $\hat{q} = 0$	at $\Gamma_{tip}^{\mathfrak{f}}$
Fracture surface flux equilibrium: $q_{lk} = \dfrac{w_{\mathfrak{f}}^N}{\delta\, \beta^{\mathfrak{f}}}$	on $\Gamma_{\pm}^{Fr}$
Seepage velocity porous matrix: $w_{\mathfrak{f}}^N(\mathbf{x}) = -\dfrac{k^{\mathfrak{s}}}{\eta^{\mathfrak{f}R}} \operatorname{grad} p \cdot \mathbf{n}^{\pm}$	on $\Gamma_{\pm}^{Fr}$

Notation			
Poro-elastic domain $\mathcal{B}^{pe}$			
φ	Constituent	ρ^{φ} [kg/m^3]	Partial density
$\varphi^{\mathfrak{s}}$	Solid constituent	$\rho^{\varphi R}$ [kg/m^3]	Effective density
$\varphi^{\mathfrak{f}}$	Fluid constituent	$k^{\mathfrak{f}}$ [m/s]	Darcy permeability
n^{φ} [–]	Volume fraction	$k^{\mathfrak{s}}$ [m^2]	Intrinsic permeability
α [–]	Biot–Willis parameter	M [Pa]	Biot's coupling parameter
ϕ [–]	Porosity	$\boldsymbol{\sigma}_E^{\mathfrak{s}}$ [Pa]	Cauchy extra stress
œ [Pa]	Mixture stress tensor	$K^{\mathfrak{s}}$ [Pa]	Grain bulk modulus
p [Pa]	Pore pressure	K^f [Pa]	Fluid bulk modulus
$\gamma^{\mathfrak{f}R}$ [kg]	Effective fluid weight	$\mathbf{w}_{\mathfrak{f}}$ [m/s]	Relative fluid flow
K [Pa]	Dry frame bulk modulus	G [Pa]	Shear modulus
$\mathbf{u}_{\mathfrak{s}}$ [m]	Deformation		
Fracture domain $\mathcal{B}^f$			
δ [m]	Fracture aperture	$\eta^{\mathfrak{f}R}$ [Pa s]	Effective fluid viscosity
$\beta^{\mathfrak{f}}$ [1/Pa]	Fluid compressibility	$\hat{p}$ [Pa]	Fracture fluid pressure

4 Numerical studies

The derived hybrid model for fractured porous media is applied to several numerical studies in order to validate the implemented schemes and to investigate hydro-mechanical phenomena. The weak and strong coupling approaches are validated based on solutions obtained by Biot's poroelastic formulation using an explicit discretization (equidimensional model) of the fracture geometry. Since the weak coupling scheme allows for non-conformal meshes of the fracture and the surrounding porous domain, the error evolution for different discretization combinations is studied once the method is verified. Local permeability changes due to fracture aperture fluctuations influence the pressure state; hence a third study simplifies the problem assuming a rigid fracture

Table 1 Collection of required modelling parameters in the poroelastic domain $\mathcal{B}^{Pe}$ and fracture domain $\mathcal{B}^{f}$

Quantity	Value	Unit	Quantity	Value	Unit
Poroelastic domain $\mathcal{B}^{Pe}$					
Dry frame bulk modulus K	8.0×10^{9}	[Pa]	Grain bulk modulus $K^{\mathfrak{s}}$	3.0×10^{10}	[Pa]
Shear modulus μ	1.54×10^{10}	[Pa]	Porosity ϕ	1.0×10^{-2}	[–]
Intrinsic permeability $k^{\mathfrak{s}}$	*Varies*	[m^2]	Fluid compressibility β^{f}	*Varies*	[1/Pa]
Fracture domain $\mathcal{B}^{Fr}$					
Effective fluid viscosity η^{fR}	1.0×10^{-3}	[Pa s]	Initial fracture aperture δ_0	*Varies*	[m]
Fluid compressibility β^{f}	*Varies*	[1/Pa]			

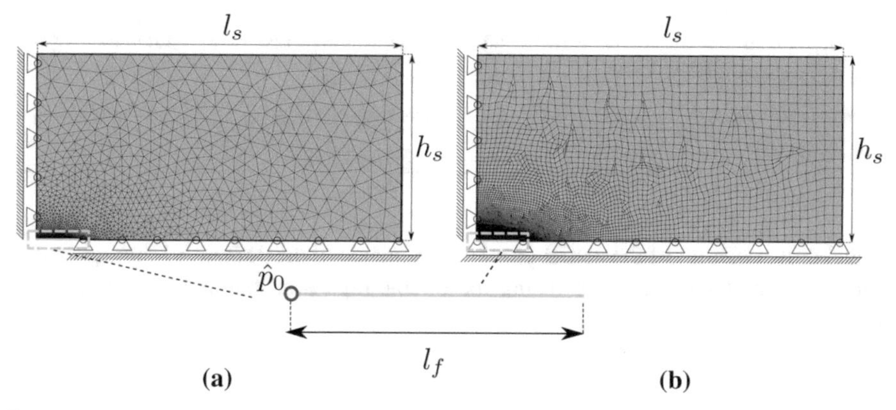

Fig. 5 Discretization and boundary conditions used for the weak coupling (**a**) and strong coupling scheme (**b**) throughout the validation

to investigate the pressure deviations due to a globally applied constant fracture change. For small fracture deformations (constant permeability) the hydro-mechanically triggered inverse pressure response is investigated for fractures embedded in different permeable porous matrices in two dimensions. Finally, the influence of local permeability changes is investigated by simulations of a three-dimensional fracture network. The decisive material parameters used throughout the simulations are given in Table 1. Note that varying material and numerical parameters along with values for boundary conditions are introduced for each study individually. The two-dimensional boundary value problems use a radial symmetric formulation of the governing equations.

4.1 Validation and efficiency comparison of weak and strong coupling schemes

The analysis of hydro-mechanical phenomena requires a validation of the weak and strong coupling scheme implementations. Reduction of the boundary value problem's geometrical complexity is achieved by using a single fracture shown in Fig. 5. The applied material and numerical parameters are given in Table 2 and chosen to reproduce the porous material behavior of sandstone. Symmetry for both directions is used to

Table 2 Collection of parameters used for validation of the weak and strong coupling schemes

Quantity	Value	Unit	Quantity	Value	Unit
Poroelastic domain $\mathcal{B}^{Pe}$					
Intrinsic permeability $k^{\mathfrak{s}}$	1.1×10^{-19}	$[m^2]$	Min fluid comp. β^f_{min}	4.5×10^{-10}	[1/Pa]
Max fluid comp. β^f_{max}	4.5×10^{-4}	[1/Pa]	Sample length l_{Pe}	1.0×10^3	[m]
Sample height h_{Pe}	5.0×10^2	[m]			
Fracture domain $\mathcal{B}^{Fr}$					
Min fluid comp. β^f_{min}	4.5×10^{-10}	[1/Pa]	Max fluid comp. β^f_{max}	4.5×10^{-4}	[1/Pa]
Fracture aperture δ_0	5.0×10^{-3}	[m]	Fracture length l^{Fr}	1.0×10^2	[m]
Pumping pressure p_0	2.0×10^4	[Pa]			
Numerical parameter					
Time step size Δt	1.0×10^{-2}	[s]	Fracture discret. in $\mathcal{B}^{Fr}$ Δx^{Fr}_{Fr}	1.0×10^{-1}	[m]
Fracture discret. in $\mathcal{B}^{Pe}$ Δx^{Fr}_{Pe}	1.0×10^{-1}	[m]	Evaluation time t_0	1.0×10^1	[s]
Evaluation position x_0	1.0×10^1	[m]	Error tolerance ϵ_{max}	1.0×10^{-6}	[-]
Number of DoF	1.4×10^5	[-]			

model only a quarter of the actual problem. Influence of the domain's boundary to the fracture zone is avoided by choosing reasonable high dimensions for its height and length.

4.1.1 Validition

Throughout the validation calculations the fracture permeability is constant (see 4.3); hence it is valid to use an equidimensional Biot model to govern the behavior of the surrounding matrix and the flow within the fracture. The reference Biot simulation requires an explicit discretization of the fracture geometry in which the intrinsic permeability of fracture elements is set equal to the permeability $k^{\mathfrak{s}}_{Fr}(\delta_0)$ calculated for the hybrid model. The numerically converged (deviation between results 1.0×10^{-4} by almost doubling the number of DoFs along the fracture; element size 0.0055 m) poroelastic solution serves as a reference for the hybrid-dimensional implementations. Assessing convergence of the weak coupling scheme iterations is based on the error measure ϵ given in Eq. (20) with the condition $\epsilon \leq \epsilon_{max} = 1.0 \times 10^{-6}$. The results are validated at $t_0 = 10.0$ s (1000 time steps) at position $x_0 = 10.0$ m using conformal meshes with element sizes of $\Delta x^{Fr}_{Fr} = \Delta x^{Fr}_{Pe} = 0.01$ m along the fracture for both domains. The varying fluid compressibility $\beta^{\mathfrak{f}}$ is investigated within the limits of 4.5×10^{-5} Pa^{-1} to 4.5×10^{-10} Pa^{-1} (compressibility of water at 25°C). The relative error

$$\epsilon_{rel} = \frac{\hat{p}(x_1, t_1) - \hat{p}_{cb}(x_1, t_1)}{\hat{p}_{cb}(x_1, t_1)}, \tag{37}$$

shown in Fig. 6a is continuously below 0.0035 for both schemes within the investigated limits. Note that in Eq. (37) $\hat{p}_{cb}$ is the pressure of the converged poroelastic solution and

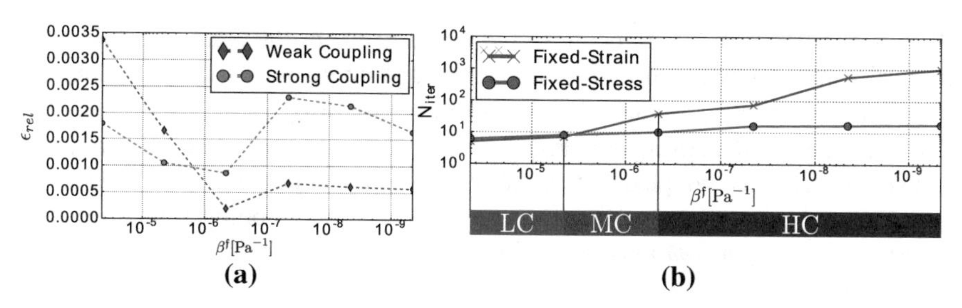

Fig. 6 Plot of the relative error ϵ_{rel} for the strong and weak coupling schemes (**a**) and recommended compressibility range of application for both methods based on the necessary number of iterations N_{iter} of the fixed-strain and fixed-stress scheme for **L**ow **C**ompressibilities, **M**oderate **C**ompressibilities and **H**igh **C**ompressibilities of the fluid (**b**)

$\hat{p}$ the pressure solution obtained by either weak or strong coupling. The errors calculated for the fixed-strain and fixed-stress split are identical since the same convergence criterion is used. Such small errors do confirm not only the correct implementation of the introduced schemes but also their suitability for such strong hydro-mechanically coupled problems. The implementation of the proposed weak and strong coupling procedures reproduce the results obtained by Biot's poroelastic formulation like expected earlier in this work and can equivalently be applied for fracture flow analysis in the field of hydromechanics.

4.1.2 Efficiency comparison

Since the weak coupling scheme relies on a staggered coupling of both domains, the number of iterations to reach a converged state is varying with the system's stiffness. Regarding the hybrid-dimensional formulation, the stiffness of the system varies with the fluid compressibility $\beta^{\mathfrak{f}}$. Hence ranges need to be defined in which the non-conformal mesh strategy has computational advantages over the strong formulation using interface elements and vice versa. The regions are defined based on the number of iterations (Fig. 6b) necessary for the weak-coupling scheme to converge. The number of iterations varies throughout the simulation since the convergence behavior is influenced by the pressure gradient between two time steps. In the current boundary value problem, an initial constant pressure $\hat{p}_0$ is applied. The first time step will require the most iterations to reach equilibrium due to the highest pressure gradient; hence it is adducted throughout the evaluation process as a numerical effort indicator. The number of iterations required for the fixed-strain algorithm to reach convergence correspond to the compressibility of the fluid. For fluid compressibilities between $\beta^{\mathfrak{f}} = 4.5 \times 10^{-4}$ Pa^{-1} and $\beta^{\mathfrak{f}} = 4.5 \times 10^{-5}$ Pa^{-1} the number of iterations is below 7. The low number of iterations in combination with the potentially reduced number of global DoFs guarantees high performance of the fixed-strain scheme for low fluid compressibilities. Within the range of moderate compressibilities $\beta^{\mathfrak{f}} = 4.5 \times 10^{-5}$ Pa^{-1} and $\beta^{\mathfrak{f}} = 4.5 \times 10^{-6}$ Pa^{-1} the number of iterations increases to 38. For any fluid compressibility lower than $\beta^{\mathfrak{f}} = 4.5 \times 10^{-6}$ Pa^{-1} the fixed-strain scheme is not suitable due to an unreasonably high number of iterations (1018 iterations for

$\beta^{\mathfrak{f}} = 4.5 \times 10^{-10}$ Pa^{-1}). In contrast, the convergence of the fixed-stress scheme is reached within a low number of iterations throughout the whole range of fluid compressibilities with a maximum of 18 iterations in case of $\beta^{\mathfrak{f}}_{min}$.
For stiff systems, the strong coupling scheme and fixed-stress algorithm are recommended to solve the coupled system of governing equations. Still, to compete with a directly solved strong coupling formulation concerning computational efficiency the fixed-stress algorithm must take advantage of the non-conformal mesh option to compensate the efficiency loss due to additional iterations.

4.2 Non-conformal mesh study: sensitivity analysis of varying discretization for fluid and porous domain

In Sect. 4.1 the potential computational performance increase of the weak coupling scheme based on the reduction of the global number of DoF by non-conformal meshing of the fluid $\mathcal{B}^{Fr}$ and porous domain $\mathcal{B}^{Pe}$ was mentioned. Nevertheless, the quality of the obtained solution for the primary variables should not decrease. This numerical study investigates the solution quality dependence on different discretization combinations. The aim is to define combinations that allow a reduction of the system's number of DoF while maintaining a high solution quality. Within the study two parameters vary to identify the error evolution, namely the fracture discretization for the hybrid, one-dimensional model Δx^{Fr}_{Fr} and the mesh density for the higher dimensional porous matrix Δx^{Fr}_{Pe} along the fracture. For investigations of the error evolution of different discretization combinations, the already introduced boundary value problem shown in Fig. 5a is adducted with a moderate fluid compressibility of $\beta^{\mathfrak{f}} = 4.5 \times 10^{-6}$ Pa^{-1}. The material parameters used for the non-conformal mesh study are given in Table 3. The mesh-dependent error is calculated on the basis of the aforementioned numerically converged poroelastic solution shown in Eq. (37). The mesh density varies between $N^{min}_E = 10$ and $N^{max}_E = 1800$ elements along the fracture. Calculated errors and discretization of the used meshes are displayed in Fig. 7. Note that the solutions obtained with the fixed-strain and fixed-stress approach are identical since the same convergence criterion is applied $\epsilon \leq \epsilon_{max} = 1.0 \times 10^{-6}$ and results of this study are valid for both schemes. Combinations involving coarse meshes ($N_E \leq 100$) produce large errors if both domains are coarse $\epsilon^{S^4F^4}_{rel} = 1.08$. An improvement of the error throughout the solution domain is investigated for a fine discretization of the fluid domain S^3F^1 more evidently than for a fine discretization of the porous domain S^1F^3. In case of first approximations of the exact solution combination S^3F^2 and S^3F^1 provide an opportunity to give a reasonable prediction with a low number of DoF. For a fine discretization of the domains ($N_E \geq 1000$) all possible combinations provide solutions with a high quality $\epsilon_{err} < 0.006$.
The weak coupling scheme has an advantage over the strong coupling whenever a first prediction of a boundary value problem is of interest. It is possible to reduce the number of DoFs in the surrounding matrix domain $\mathcal{B}^{Pe}$ by a factor of 10 compared to the discretization of the fracture domain $\mathcal{B}^{Fr}$ along the fracture surface which directly leads to a reduction of computational effort. Nevertheless, for precise solutions, the discretization of both domains must be adapted to the boundary value problem. Adap-

Table 3 Collection of parameters used throughout numerical studies on non-conformal meshes

Quantity	Value	Unit	Quantity	Value	Unit
Poroelastic domain $\mathcal{B}^{Pe}$					
Intrinsic permeability $k^{\mathfrak{s}}$	1.1×10^{-19}	[m^2]	Fluid comp. β^f	4.5×10^{-6}	[1/Pa]
Sample length $l^{\mathfrak{s}}$	1.0×10^3	[m]	Sample height $l^{\mathfrak{s}}$	5.0×10^2	[m]
Fracture domain $\mathcal{B}^{Fr}$					
Fluid comp. β^f_{min}	4.5×10^{-6}	[1/Pa]	Fracture aperture δ_0	5×10^{-3}	[m]
Fracture length l^{Fr}	1.0×10^2	[m]	Pumping pressure p_0	2.0×10^4	[Pa]
Numerical parameter					
Time step size Δt	1.0×10^{-2}	[s]	Fracture discret. in $\mathcal{B}^{Fr}$ $\Delta x^{Fr}_{Fr,\min}$	5.5×10^{-2}	[m]
Fracture discret. in $\mathcal{B}^{Pe}$ $\Delta x^{Fr}_{Fr,\max}$	1.0×10^1	[m]	Fracture discret. in $\mathcal{B}^{Fr}$ $\Delta x^{Fr}_{Pe,\min}$	5.5×10^{-2}	[m]
Fracture discret. in $\mathcal{B}^{Pe}$ $\Delta x^{Fr}_{Pe,\max}$	1.0×10^1	[m]	error tolerance ϵ_{max}	1.0×10^{-6}	[–]

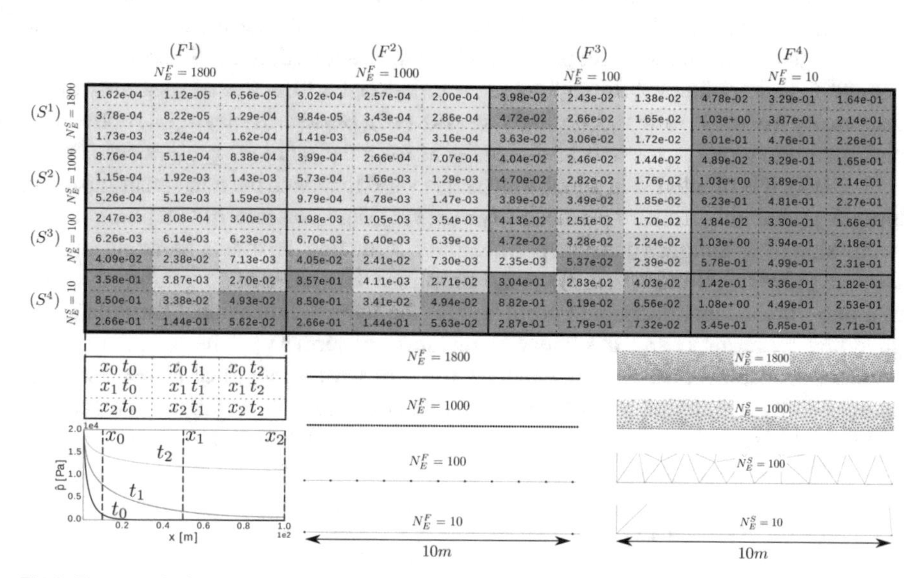

Fig. 7 Error matrix for discretization combinations with legend and visualized mesh densities for an interval of 10 m. The error is evaluated for for three positions x_0, x_1 and x_2 at three time-steps t_0, t_1 and t_2

tion of just one domain does not lead to satisfactory results. The weak and strong coupling schemes provide solutions of the same quality once similar discretizations of the domain are used throughout simulations. Similar discretizations lead to a comparable number of DoF, and effectively the non-conformal meshing strategy provides no advantage to the weak coupling. Hence the strong coupling scheme is chosen for the following numerical studies based on the efficiency of the method.

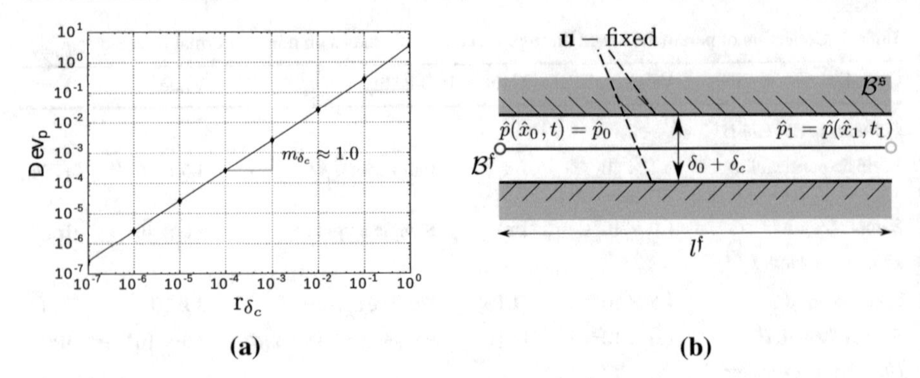

Fig. 8 Investigation of fracture change influence on diffusion process in a radial fracture; **a** double logarithmic plot of pressure derivation Der_p with respect to fracture aperture change ratio r_{δ_c}, **b** model parameters and sketch of simulation domain with boundary conditions and investigation position p_1

4.3 Influence of aperture change on fracture permeability

In case of hydro-mechanical investigations, the evolution of fracture aperture in time and space is directly related to the deformation state of the surrounding poroelastic matrix material and the effective permeability of the fracture domain. Since the aperture change δ_c varies with time throughout the simulation, the effective permeability of the fracture $k^{\mathfrak{s}}_{Fr}(\mathbf{u}(\mathbf{x}, t))$ varies accordingly, see Eq. (10). Besides the negligible convective and quadratic terms of the hybrid-dimensional formulation in the present case (Vinci et al. 2015), it is of interest in which ranges it is valid to reduce the non-linear diffusion term IV) (changing fracture permeability $\delta(\mathbf{u}(\mathbf{x}, t))$ and pressure $\hat{p}$) to a purely linear expression by assuming a constant permeability. Reducing the model premising a rigid fracture ($\partial\delta/\partial t = 0$ and $\delta(\mathbf{u}(\mathbf{x}, t)) = \delta_0$) simplifies the system of governing equations to a one-dimensional decoupled problem and the hybrid-dimensional formulation is governed by terms I) and the linear formulation of term IV). The boundary value problem is displayed in Fig. 8. A single fracture with a length of $l^f = 100$ m is stimulated by a pumping pressure $\hat{p} = 20$ kPa at position $\hat{x}_0 = 0.0$ m. The fracture aperture change δ_c is applied globally as $\delta = \delta_0 + \delta_c$ and deviation

$$\mathrm{Dev}_p = \frac{\hat{p}(a, t_1, \delta_0 + \delta_c) - \hat{p}(a, t_1, \delta_0)}{\hat{p}(a, t_1, \delta_0)}, \tag{38}$$

is determined concerning the solution obtained with $\delta = \delta_0$. The applied numerical and material parameters are given in Table 4. Fracture aperture changes in the limits of δ_c^{min} and δ_c^{max} are conducted to investigate their influence on the resulting pressure deviation. Equation (38) is evaluated after 1000 timesteps at $t_0 = 0.01$ s and position $\hat{x}_1$ concerning the fracture change ratio $r_{\delta_c} = \delta_c/\delta$. The results are plotted with a double logarithmic scale shown in Fig. 8a. As highlighted in the graph, the pressure deviation increases linearily (slope $m_{\delta_c} = 1$) with the fracture change ratio. Nevertheless, for ratios $r_{\delta_c} \leq 0.01$ the deviation is smaller than 2 % when compared to the reference solution received with aperture $\delta = \delta_0$.

Table 4 Collection of parameters used throughout numerical aperture change studies for a rigid fracture

Quantity	Value	Unit	Quantity	Value	Unit
Fracture domain $\mathcal{B}^{Fr}$					
Fluid comp. β^f_{min}	4.5×10^{-10}	[1/Pa]	Fracture aperture δ_0	5.0×10^{-3}	[m]
Fracture length l^{Fr}	1.0×10^2	[m]	Pressure boundary posit. $\hat{x}_0$	0.0	[m]
Pumping pressure p_0	2.0×10^4	[Pa]	Min. aperture change δ_{min}	5.0×10^{-10}	[m]
Max. aperture change δ_{max}	5.0×10^{-3}	[m]			
Numerical parameter					
Time step size Δt	1.0×10^{-5}	[s]	Fracture discret. in $\mathcal{B}^{Fr}$ Δx^{Fr}_{Fr}	1.0×10^{-2}	[m]
Evaluation time t_0	1.0×10^{-2}	[s]	Evaluation position $\hat{x}_1$	1.0×10^2	[m]
Number of DoF	1.0×10^4	[–]			

To strengthen the reduced prediction model a second boundary value problem is conducted using the set up introduced for validation in Sect. 4.1 with identical parameters given in Table 2 and a fluid compressibility of $\beta^f = 4.5 \times 10^{-10}\,\mathrm{Pa}^{-1}$. In contrast to the simplified rigid fracture assumption, the complexity is increased by assuming a deformable fracture and requires the hybrid dimensional formulation defined by terms (I), (IV) and (V), coupled to the poroelastic domain. The reference case is modeled with the linear formulation of term (IV) assuming a constant permeabilty $k^{\mathfrak{s}}_{Fr}(\delta_0)$ throughout the simulation. This solution is then compared after 1000 time steps at $t_2 = 10.0\,\mathrm{s}$ to the results calculated with the non-linear form of term (IV) by assuming a aperture dependent permeability $k^{\mathfrak{s}}_{Fr}(\mathbf{u}(\mathbf{x}, t))$. At time t_2 the maximum local aperture change within the fracture is $\delta^{max}_c = 1.81 \times 10^{-5}\,\mathrm{m}$ which is equivalent to a ratio of 0.004. Based on the introduced prediction model a deviation of approximately $\mathrm{Dev}_p \approx 0.0092$ is expected. Like stated above, the conservative assumptions result in an overestimation of the aperture change influence which is confirmed by the deviation $\mathrm{Dev}_p = 0.0039$ obtained with the hydro-mechanical model. Still, the conservative prediction model gives deviations in the same order of magnitude and is a sufficient estimator to determine whether fracture aperture changes are relevant for calculations regarding the change of fracture permeability. For small deformations ($r_{\delta_c} < 0.01$), it is reasonable to assume constant permeabilities throughout the following two-dimensional hydro-mechanical simulations. In case of large fracture deformations ($r_{\delta_c} > 0.01$) permeability changes need to be considered to reproduce the pressure state correctly.

4.4 Pressure induced hydro-mechanical response: fracture stimulation

Hydro-mechanical coupling of fluid-filled fractures in deformable poroelastic media includes pressure induced phenomena like the Noordbergum effect (Kim and Parizek 1997). Throughout fracture stimulations, the mechanical response in the form of

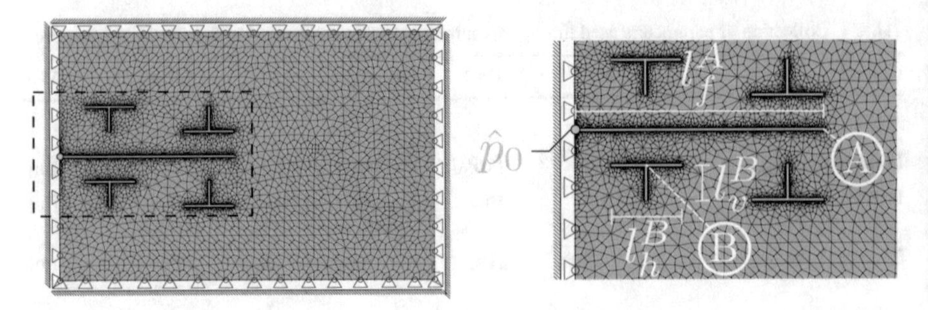

Fig. 9 Discretization and boundary conditions used for the hydro-mechanical response studies

Table 5 Collection of parameters used throughout hydro-mechanical response studies

Quantity	Value	Unit	Quantity	Value	Unit
Poroelastic domain $\mathcal{B}^{Pe}$					
Intrinsic permeability $k_1^{\mathfrak{s}}$	1.1×10^{-19}	[m^2]	Intrinsic permeability $k_2^{\mathfrak{s}}$	1.1×10^{-5}	[m^2]
Fluid comp. β^f	4.5×10^{-10}	[1/Pa]	Sample length l^{pe}	1.5×10^3	[m]
Sample height h^{pe}	1.0×10^3	[m]			
Fracture domain $\mathcal{B}^{Fr}$					
Fluid comp. β^f_{min}	4.5×10^{-10}	[1/Pa]	Fracture aperture δ_0	5.0×10^{-4}	[m]
Fracture A length $l^{Fr}_{A,l}$	7.0×10^2	[m]	Fracture B length $l^{Fr}_{B,l}$	2.0×10^2	[m]
Fracture B height $l^{Fr}_{B,h}$	1.0×10^2	[m]	Fracture discret. $\Delta x^{\mathfrak{Fr}}$	1.0×10^{-2}	[m]
Numerical parameter					
Time step size Δt	1.0×10^{-2}	[s]	Number of DoF	2.3×10^5	[–]

non-local deformations is faster than the pressure diffusion within the fracture. The volumetric change of the fracture influences the pressure state, e.g., in a borehole during well testing where inverse pressure responses can be measured. Calculating accurately the inverse pressure response poses high demands on numerical simulation techniques and will therefore be investigated in further detail to demonstrate the efficiency and accuracy of the proposed algorithms. The validated strongly coupled hydro-dimensional formulation is applied to a fracture stimulation problem with the boundary conditions shown in Fig. 9a. Corresponding material parameters are given in Table 5.

Two different phenomena are investigated. Besides the fluid pressure evolution within the main fracture A (inverse pressure response), shown in Fig. 9b, the pressure response within T-shape fracture B induced by the deformation of the surrounding medium, displayed in Fig. 10, are analyzed. Note that the pressure $\hat{p}$ in the graphs correspond to a pressure difference concerning an initial geostatic pressure. The fluid within the surrounding porous rock matrix $\mathcal{B}^{Pe}$ and fracture $\mathcal{B}^{Fr}$ possesses a compressibility of $\beta^{\mathfrak{f}} = 4.5 \times 10^{-10}$ Pa^{-1}. A high aspect fracture with an aperture of $\delta_0 = 5.0 \times 10^{-4}$ m is modeled. Examination of leak-off and its influence on the pressure response within the fluid-filled fracture is studied by assuming two different

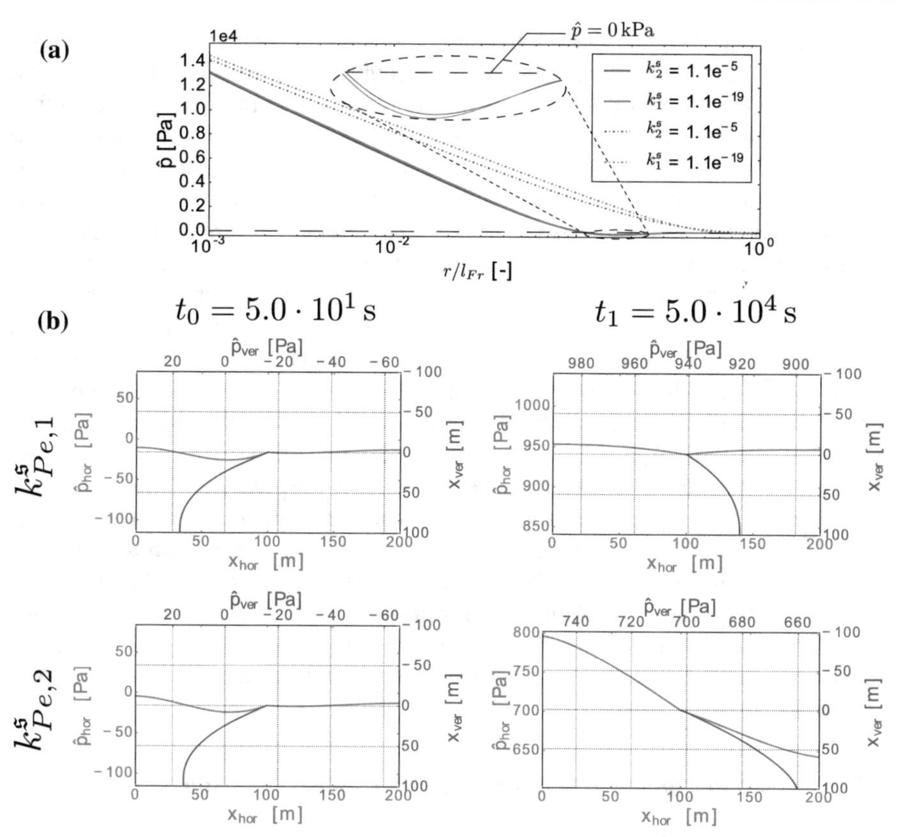

Fig. 10 Semi-logarithmic pressure distribution along fracture A for time steps $t_0 = 5.0 \times 10^1$ s and $t_1 = 5.0 \times 10^4$ s are displayed along with pressure distributions in horizontal (green) and vertical (red) direction within T-shape fracture B for two differing permeabilities of the surrounding matrix. The pressure solutions in fracture A obtained at time t_0 are plotted with plane lines and solutions obtained at t_1 with dashed lines (color figure online)

extreme cases of permeabilities $k_1^{\mathfrak{s}} = 1.1 \times 10^{-19}\,\mathrm{m}^2$ and $k_2^{\mathfrak{s}} = 1.1 \times 10^{-5}\,\mathrm{m}^2$ for the surrounding poroelastic matrix.

Focusing on the pressure distribution along fracture A the inverse pressure is evident at time step t_0. For both permeabilities of the surrounding poroelastic matrix, the pressure decreases when compared to the initial geostatic pressure due to non-local deformation of the fracture. For permeability $k_2^{\mathfrak{s}}$ the inverse response is less strong since the leak-off into the surrounding high-permeable medium leads to a smearing effect on the pressure distribution within the fluid filled fracture and is noticeable for an advanced time step t_1 where both pressure distributions differ evidently. The pressure state within a fracture surrounded by a medium with a low permeability $k_1^{\mathfrak{s}}$ is ahead of the fluid pressure within a fracture surrounded by more permeable (permeability $k_2^{\mathfrak{s}}$) porous matrix. Leak-off harms the diffusion process within the fracture which results in a lower pressure state.

Pressure changes in the T-shape fracture B are not directly induced by the pressure Dirichlet boundary conditions, but mainly through the deformation triggered by the pressure change within fracture A. Most interesting is that the Noordbergum effect causes a local closing of main fracture A at t_0 that translates in an opening of the horizontal part of T-Shape fracture B. The volume change of the horizontal fracture induces a negative pressure response. Due to the geometrical nature of a T-shape fracture, once the horizontal part opens up, the vertical fracture part closes, and a pressure rise along its direction is notable. At time t_1 the advanced fluid pressure along main fracture A leads to an opening orthogonal to its direction that causes a closing of the horizontal part of T-shape fracture B. Due to the negative volume change a pressure rise results. The vertical part is influenced by the pressure rise and opens up slightly. The positive volume change harms the pressure diffusion process along the vertical direction. Hence the pressure level is lower than its counterpart in horizontal direction. Again for different permeabilities the pressure levels differ for t_1. The leak-off into the surrounding medium reduces the pressure within fracture B. Reproduction of the pressure-induced hydro-mechanical phenomena within a fluid-filled fracture through passive (deformation) and active (pumping) stimulation using the hybrid-dimensional formulation has been numerically investigated by the introduced boundary value problem. The formulation is capable to model pumping experiments and to reproduce the key phenomena concerning hydro-mechanical coupling induced by pressure fluctuations for radial symmetric three-dimensional geometries.

4.5 Deformation dependent permeability: three-dimensional fracture stimulation

Large relative aperture changes in high aspect ratio fractures occur even for small absolute deformations. It is recommended to take aperture dependent permeabilities into account throughout pressure diffusion simulations once $r_{\delta_c} > 0.01$ (see 4.3). The aperture change is directly related to the pressure state and the geometrical stiffness of the fracture. Especially in three dimensions, the stiffness of fractures influences their opening and closing during pressure stimulations resulting in preferential flow paths due to inhomogeneous permeabilities. This numerical study investigates the deviation of pressure solutions of a constant compared to an aperture dependent permeability for three connected fractures in three dimensions. Parameters of the simulation are given in Table 6, and the geometry along with the discretization of the boundary value problem is shown in Fig. 11. The fractures are placed in the same plane, where fracture 1 is connected to fracture 2 and 3 via cross-sections Γ_1^{Con} and Γ_2^{Con}.

The displacement DoF on the poroelastic domain's boundaries are fixed in the direction normal to its surface and pressure $\hat{p}_0$ is induced in the center of fracture 1 on a length of 1.5 m. It is supposed that the apparent difference between cross-section areas $A_{\Gamma_1^{Con}}$ and $A_{\Gamma_2^{Con}}$ initiates a dominant flow from fracture 1 into fracture 3 rather than into fracture 2. Additionally, since cross-section Γ_2^{Con} weakens the surrounding geometric stiffness stronger than Γ_1^{Con} higher aperture changes and hence higher permeabilities in the direction of fracture 3 are expected.

Table 6 Collection of parameters used throughout deformation dependent permeability studies

Quantity	Value	Unit	Quantity	Value	Unit
Poroelastic domain $\mathcal{B}^{Pe}$					
Intrinsic permeability $k^{\mathfrak{s}}_{Pe}$	1.1×10^{-19}	[m^2]	Fluid comp. β^f	4.5×10^{-10}	[1/Pa]
Poroel. domain length in $l_{\mathbf{e}_1}$	4.0×10^2	[m]	Poroel. domain length in $l_{\mathbf{e}_2}$	2.0×10^2	[m]
Poroel. domain length in $l_{\mathbf{e}_3}$	4.0×10^2	[m]	Cross-section area $A_{\Gamma_1^{con}}$	4.2×10^{-4}	[m^2]
Connection cross-section $A_{\Gamma_2^{con}}$	2.5×10^{-5}	[m^2]			
Fracture domain $\mathcal{B}^{Fr}$					
Fluid comp. β^f_{min}	4.5×10^{-10}	[1/Pa]	Fracture aperture δ_0	5.0×10^{-4}	[m]
Fracture domain length $l^{Fr}_{\hat{\mathbf{e}}_1}$	2.0×10^2	[m]	Fracture domain length $l^{Fr}_{\hat{\mathbf{e}}_1}$	1.6×10^2	[m]
Fracture center height $c^{Fr}_{\mathbf{e}_1}$	2.0×10^2	[m]	Fracture center distance $c^{Fr}_{\mathbf{e}_3}$	2.0×10^2	[m]
Numerical parameter					
Time step size Δt	5.0×10^3	[s]	Error tolerance ϵ^{NL}_{max}	1.0×10^{-6}	[–]
DoF	4.3×10^5	[–]			

Note that the cross-sections $A_{\Gamma_1^{con}}$ and $A_{\Gamma_2^{con}}$ were determined based on the initial fracture aperture δ_0. ϵ^{NR} defines the error between iterations of the Newton–Raphson scheme and $\epsilon^{NR} < \epsilon^{NR}_{max}$ convergence criterion throughout the non-linear simulation

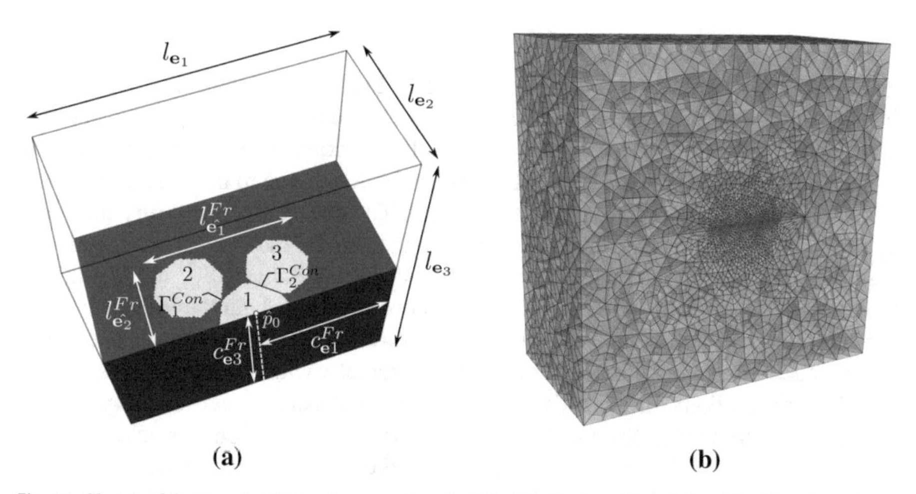

Fig. 11 Sketch of fracture (white) and porous domain (blue) is displayed in **a**. Discretization of the three-dimensional geometry is shown in **b** (color figure online)

The pressure states at times $t_0 = 7.5 \times 10^5$ s, $t_1 = 4.0 \times 10^6$ s and $t_2 = 8.25 \times 10^6$ s for pressure values $\hat{p}$ obtained by a constant and an aperture dependent permeability along with the permeability distribution are given in Fig. 12. For t_1 the pressure

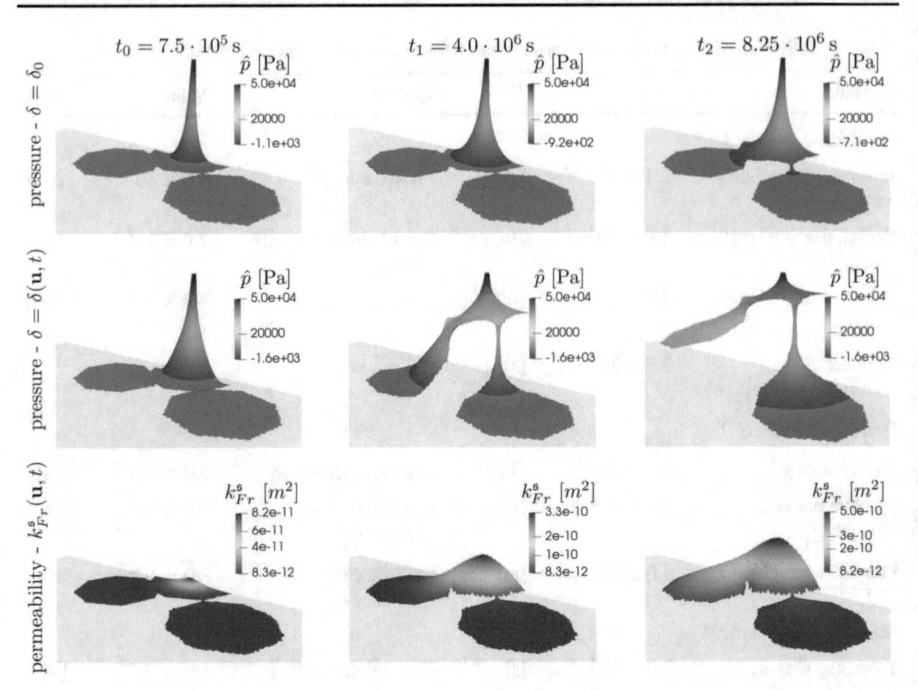

Fig. 12 Plots of the pressure $\hat{p}$ obtained with a constant and a deformation dependent permeability along with a distribution of the displacement dependent permeability are shown for three different times. Pressure $\hat{p}$ is the pressure difference concerning an initial geostatic pressure. Note that the grey isosurface in the pressure plot represents the pressure difference $\hat{p} = 0$ Pa and in the permeability plots the initial permeability obtained with δ_0

evolution takes mainly place within fracture 1. For both pressure distributions the inverse pressure response can be examined; for the deformation dependent permeability stronger than for the constant permeability solution. Even at an early stage, an apparent difference between both solutions can be found due to much higher permeabilities close to the pressure stimulation point. The constant permeability measures $k^{\mathfrak{s}}_c = 8.3 \times 10^{-12}\,\text{m}^2$ calculated based on δ_0 and is by the factor of approx. 10 lower than the deformation dependent permeability $k^{\mathfrak{s}}_{Fr,up} = 8.2 \times 10^{-11}\,\text{m}^2$ around the point of pressure initiation. Nevertheless, at t_0 no statement about the influence of the permeability distribution on the flow direction towards one of the connected fractures can be made since the pressure solution evolved radially within fracture 1. For time t_2 the pressure solutions differ vastly. The deformation dependent permeability already shows a clear trend towards a higher permeability in the direction of fracture 3. This might influence the flow path decisively. Nevertheless, since the pressure evolution for the constant permeability calculations is still limited to the region of fracture 1, no statement about the influence of the aperture dependent permeability on the effective flow is possible. The maximum permeability value is approx. 40 times higher than the constant permeability. For time t_3 the pressure diffusion advanced far into fracture 3 and barely into fracture 2 for the non-linear case. The solutions obtained with a constant permeability shows a slightly faster pressure distribution into fracture 3, but the difference is not evident. The difference of the pressure state of fracture 2 and fracture 3 is much higher for the calculations using a deformation-dependent permeability than for calculations with constant permeability and shows its influence on the pressure evolution. The absolute pressure difference between both solutions is explicit since the maximum permeability value is 60 times higher for the aperture dependent compared to the constant permeability value. A maximum number of 6 iterations was needed within the Newton–Raphson scheme to converge for the non-linear case in the first two time-steps. Throughout the rest of the simulation, convergence was reached within 3–5 iterations.

This study has shown the influence of aperture dependent permeabilities for the case of substantial aperture changes. Strongly differing solutions in the diffusion time and for the diffusion direction motivate the non-linear formulation of the permeability throughout simulations containing large aperture changes.

5 Technical details

Implementation work of the strong coupling scheme includes the extension of the Dune-PDELab package to lower dimensional element integrations within the DUNE framework (Bastian et al. 2010). Slight modifications to the assembly procedure of the global system and introduction of new local operators on element level guarantee the maintenance of the high computational efficiency of the C++ (template metaprogramming) based package. Non-conformal mesh calculations have been implemented into the FEniCS environment (Alnæs et al. 2015). Both weak coupling schemes, namely the fixed-strain and fixed-stress strategy have been embedded in a flexible Python-based framework that allows interaction of discretizations of different dimensions.

Fracture discretization is challenging especially when interface elements are introduced. Hence the GMSH API (Geuzaine and Remacle 2009) has been consulted to generate an environment that allows mesh creation of fracture networks including the generation of interface elements.

6 Conclusions

This work focused on the numerical formulation and validation of methods suitable to solve hydro-mechanical phenomena of compressible fluids using the proposed hybrid-dimensional formulation. An implicit weak coupling scheme, naturally able to perform non-conformal mesh calculations and a strong coupling scheme in the form of interface elements were introduced. Validation of both methods based on Biot's formulation showed the ability to reproduce hydro-mechanical phenomena of interest like the Noordbergum effect. Further advantages and disadvantages of both methods were studied. The fixed-strain scheme has been proven to be effective for high fluid compressibilities only. An error evolution for different fine discretization of fluid and porous domain has been investigated, and references for the preferred coupling scheme based on the relative error were given. The strong coupling and fixed-stress scheme were able to solve stiff systems with low fluid compressibility. Implementation of the strong coupling scheme is intuitive since an existent FEM framework requires small modifications for the integration of interface elements only. Since a displacement–pressure-formulation of the surrounding poroelastic material has been conducted, the same global assembly matrix is used for the hybrid-dimensional interface elements. The weak coupling and especially implementation of non-conformal mesh coupling requires more implementation work.

Pressure induced boundary value problems have been performed for different material parameters of the surrounding porous matrix in two and three dimensions. Both stimulation types showed the ability of the strong coupling scheme to solve research questions in the field of fractured porous media even for complex fracture geometries such as well testing and non-linear three-dimensional simulations. For large changes of the aperture, the importance of a deformation-dependent permeability throughout simulations have been shown. Both proposed methods in combination with the hybrid-dimensional formulation are capable of modeling high aspect ratio fractures and reproduce hydro-mechanical effects including leak-off where modeling approaches, that neglect the influence of fracture deformation would fail.

Acknowledgements The authors gratefully acknowledge the funding provided by the German Federal Ministry of Education and Research (BMBF) for the GeomInt project, Grant Number 03A0004E, within the BMBF Geoscientific Research Program "Geo:N Geosciences for Sustainability".

References

Adachi, J., Siebrits, E., Peirce, A., Desroches, J.: Computer simulation of hydraulic fractures. Int. J. Rock Mech. Min. Sci. **44**(5), 739–757 (2007)

Alnæs, M.S., Blechta, J., Hake, J., Johansson, A., Kehlet, B., Logg, A., Richardson, C., Ring, J., Rognes, M.E., Wells, G.N.: The FEniCS project version 1.5. Arch. Numer. Softw. **3**(100), 9–23 (2015)

Bastian, P., Heimann, F., Marnach, S.: Generic implementation of finite element methods in the distributed and unified numerics environment (DUNE). Kybernetika **46**(2), 294–315 (2010)

Biot, M.A.: General theory of three-dimensional consolidation. J. Appl. Phys. **12**(2), 155–164 (1941)

Brenner, K., Hennicker, J., Masson, R., Samier, P.: Gradient discretization of hybrid-dimensional darcy flow in fractured porous media with discontinuous pressures at matrix—fracture interfaces. IMA J. Numer. Anal. **37**(3), 1551–1585 (2017). https://doi.org/10.1093/imanum/drw044

Castelletto, N., White, J.A., Tchelepi, H.A.: Accuracy and convergence properties of the fixed-stress iterative solution of two-way coupled poromechanics. Int. J. Numer. Anal. Methods Geomech. **39**(14), 1593–1618 (2015). https://doi.org/10.1002/nag.2400

Coussy, O.: Poromechanics. Wiley, New York (2004)

Ehlers, W., Bluhm, J.: Porous Media: Theory, Experiments and Numerical Applications. Springer, Berlin (2013)

Geertsma, J., De Klerk, F.: A rapid method of predicting width and extent of hydraulically induced fractures. J. Pet. Technol. **21**(12), 1–571 (1969)

Geuzaine, C., Remacle, J.F.: Gmsh: a 3-D finite element mesh generator with built-in pre- and post-processing facilities. Int. J. Numer. Methods Eng. **79**(11), 1309–1331 (2009)

Girault, V., Wheeler, M.F., Ganis, B., Mear, M.E.: A lubrication fracture model in a poro-elastic medium. Math. Models Methods Appl. Sci. **25**(04), 587–645 (2015). https://doi.org/10.1142/S0218202515500141

Girault, V., Kumar, K., Wheeler, M.F.: Convergence of iterative coupling of geomechanics with flow in a fractured poroelastic medium. Comput. Geosci. **20**(5), 997–1011 (2016). https://doi.org/10.1007/s10596-016-9573-4

Guiducci, C., Collin, F., Radu, J.P., Pellegrino, A., Charlier, R.: Numerical modeling of hydro-mechanical fracture behaviour. NUMOG VIII, pp. 293–299 (2003)

Hanowski, K.K., Sander, O.: Simulation of deformation and flow in fractured. Poroelastic materials. ArXiv e-prints (2016)
Kim, J.M., Parizek, R.R.: Numerical simulation of the noordbergum effect resulting from groundwater pumping in a layered aquifer system. J. Hydrol. **202**(1), 231–243 (1997). https://doi.org/10.1016/S0022-1694(97)00067-X
Kim, J., Tchelepi, H., Juanes, R.: Stability and convergence of sequential methods for coupled flow and geomechanics: drained and undrained splits. Comput. Methods Appl. Mech. Eng. **200**(23), 2094–2116 (2011). https://doi.org/10.1016/j.cma.2011.02.011
Kim, J., Tchelepi, H., Juanes, R.: Stability and convergence of sequential methods for coupled flow and geomechanics: fixed-stress and fixed-strain splits. Comput. Methods Appl. Mech. Eng. **200**(13), 1591–1606 (2011). https://doi.org/10.1016/j.cma.2010.12.022
Martin, V., Jaffr, J., Roberts, J.: Modeling fractures and barriers as interfaces for flow in porous media. SIAM J. Sci. Comput. **26**(5), 1667–1691 (2005). https://doi.org/10.1137/S1064827503429363
Nordgren, R., et al.: Propagation of a vertical hydraulic fracture. Soc. Pet. Eng. J. **12**(04), 306–314 (1972)
Oliver, D.S., Chen, Y.: Recent progress on reservoir history matching: a review. Comput. Geosci. **15**(1), 185–221 (2011)
Ortiz, R,A.E., Renner, J., Jung, R.: Hydromechanical analyses of the hydraulic stimulation of borehole basel 1. Geophy. J. Int. **185**(3), 1266–1287 (2011)
Ortiz, R,A.E., Jung, R., Renner, J.: Two-dimensional numerical investigations on the termination of bilinear flow in fractures. Solid Earth **4**(2), 331–345 (2013)
Peirce, A.P., Siebrits, E.: A dual mesh multigrid preconditioner for the efficient solution of hydraulically driven fracture problems. Int. J. Numer. Methods Eng. **63**(13), 1797–1823 (2005). https://doi.org/10.1002/nme.1330
Perkins, T., Kern, L., et al.: Widths of hydraulic fractures. J. Pet. Technol. **13**(09), 937–949 (1961)
Renner, J., Steeb, H.: Modeling of Fluid Transport in Geothermal Research, pp. 1443–1500. Springer, Berlin (2015)
Renshaw, C.E.: On the relationship between mechanical and hydraulic apertures in rough-walled fractures. J. Geophys. Res. Solid Earth **100**(B12), 24629–24636 (1995)
Rodrigues, J.: The Noordbergum effect and characterization of aquitards at the Rio Maior mining project. Ground Water **21**, 200–207 (1983)
Sandve, T., Berre, I., Nordbotten, J.: An efficient multi-point flux approximation method for discrete fracturematrix simulations. J. Comput. Phys. **231**(9), 3784–3800 (2012). https://doi.org/10.1016/j.jcp.2012.01.023
Segura, J.M., Carol, I.: On zero-thickness interface elements for diffusion problems. Int. J. Numer. Anal. Methods Geomech. **28**(9), 947–962 (2004). https://doi.org/10.1002/nag.358
Segura, J.M., Carol, I.: Coupled HM analysis using zero-thickness interface elements with double nodes. Part II: verification and application. Int. J. Numer. Anal. Methods Geomech. **32**(18), 2103–2123 (2008)
Segura, J.M., Carol, I.: Coupled HM analysis using zero-thickness interface elements with double nodes. Part I: theoretical model. Int. J. Numer. Anal. Methods Geomech. **32**(18), 2083–2101 (2008)
Settgast, R.R., Fu, P., Walsh, S.D., White, J.A., Annavarapu, C., Ryerson, F.J.: A fully coupled method for massively parallel simulation of hydraulically driven fractures in 3-dimensions. Int. J. Numer. Anal. Methods Geomech. **41**(5), 627–653 (2017)
Shen, B., Stephansson, O., Rinne, M.: Hydro-Mechanical Coupling, pp. 77–82. Springer, Dordrecht (2014)
Sneddon, I.N., Elliot, H.A.: The opening of a Griffith crack under internal pressure. Q. Appl. Math. **4**(3), 262–267 (1946)
Taleghani, A.D.: Analysis of Hydraulic Fracture Propagation in Fractured Reservoirs: An Improved Model for the Interaction Between Induced and Natural Fractures. The University of Texas at Austin, Austin (2009)
Tunc, X., Faille, I., Gallouët, T., Cacas, M.C., Havé, P.: A model for conductive faults with non-matching grids. Comput. Geosci. **16**(2), 277–296 (2012). https://doi.org/10.1007/s10596-011-9267-x
Vinci, C.: Hydro-mechanical coupling in fractured rocks: modeling and numerical simulations. Ph.D. thesis. Ruhr-University Bochum (2014)
Vinci, C., Renner, J., Steeb, H.: A hybrid-dimensional approach for an efficient numerical modeling of the hydro-mechanics of fractures. Water Resour. Res. **50**(2), 1616–1635 (2014)
Vinci, C., Steeb, H., Renner, J.: The imprint of hydro-mechanics of fractures in periodic pumping tests. Geophys. J. Int. **202**(3), 1613–1626 (2015)
Wang, H.F.: Theory of Linear Poroelasticity. Princeton University Press, Princeton (2000)

Woodbury, A., Zhang, K.: Lanczos method for the solution of groundwater flow in discretely fractured porous media. Adv. Water Resour. **24**(6), 621–630 (2001). https://doi.org/10.1016/S0309-1708(00)00047-6
Yew, C.H., Weng, X.: Mechanics of Hydraulic Fracturing. Gulf Professional Publishing, Houston (2014)
Zheltov, A.K.: Formation of vertical fractures by means of highly viscous liquid. In: 4th World Petroleum Congress. World Petroleum Congress (1955)

Publisher's Note Springer Nature remains neutral with regard to jurisdictional claims in published maps and institutional affiliations.

Affiliations

Patrick Schmidt[1] · Holger Steeb[1]

Holger Steeb
holger.steeb@mechbau.uni-stuttgart.de

1 Institute of Applied Mechanics, University of Stuttgart, Pfaffenwaldring 7, 70569 Stuttgart, Germany

GEM - International Journal on Geomathematics
https://doi.org/10.1007/s13137-019-0126-6

ORIGINAL PAPER

Comparative verification of discrete and smeared numerical approaches for the simulation of hydraulic fracturing

Keita Yoshioka[1] · Francesco Parisio[1] · Dmitri Naumov[1] · Renchao Lu[1,2] · Olaf Kolditz[1,2] · Thomas Nagel[1,3]

Received: 30 March 2018 / Accepted: 21 January 2019

Abstract
The numerical treatment of propagating fractures as embedded discontinuities is a challenging task for which an analyst has to select a suitable numerical method from a range of options. Since their inception in the mid-80s, smeared approaches for fracture simulation such as non-local damage, gradient damage or more lately phase-field modelling have steadily gained popularity. One of the appeals of a smeared implicit fracture representation, the ability to handle complex topologies with unknown crack paths in relatively coarse meshes as well as multiple-crack interaction and multiphysics, is a fundamental requirement for the numerical simulation of hydraulic fracturing in complex situations which is technically more difficult to achieve with many other methods. However, in hydraulic fracturing simulations, not only the prediction of the fracture path but also the computation of fracture width and propagation pressure (frac pressure) is crucial for reliable and meaningful applications of the simulation tool; how to determine some of these quantities in smeared representations is not immediately obvious. In this study, two of the most popular smeared approaches of recent, namely non-local damage and phase-field models, and an approach in which the solution space is locally enriched to capture a strong discontinuity combined with a cohesive-zone model are verified against fundamental hydraulic fracture propagation problems in the toughness-dominated regime. The individual theoretical foundations of each approach are discussed and differences in the treatment of physical and numerical properties of the methods when applied to the same physical problems are highlighted through examples.

Keywords Phase field method · Non-local damage · Cohesive zone models · Brittle fracture · Hydraulic fracturing · OpenGeoSys · GeomInt · GEMex

✉ Thomas Nagel
thomas.nagel@ifgt.tu-freiberg.de

Extended author information available on the last page of the article

Mathematics Subject Classification 74: Mechanics of deformable solids · 35: Partial differential equations · 65: Numerical analysis

1 Introduction

Simulations of hydraulically driven fracture propagation is still a challenging task in computational mechanics that involves moving boundaries, fluid–structure interaction, strongly discontinuous fields and can in certain cases also entail complex rheological behaviour of the fluids and solid constituents. Which specific approach is most suited to represent the combined problem of pressurization and propagation is still under debate. This contribution aims specifically at comparing two distinct families of numerical approaches: continua with embedded discontinuities (cohesive-zone models) and smeared approaches (phase-field and non-local damage).

The scientific interest in hydraulic fracturing has increased rapidly in the recent past driven by a diverse range of technological applications including productivity enhancements in unconventional fossil fuel reservoirs (Economides and Nolte 2000), the stimulation or enhancement of geothermal reservoirs (Legarth et al. 2005), the assessment of barrier integrity in subsurface energy or waste storage applications (Minkley et al. 2016; Johnson et al. 2004), the safe operation of mines and boreholes (Jiang et al. 2016; Morita et al. 1990), but also the stability of fluid-filled biomaterials undergoing rapid loading (Böger et al. 2017). During the injection of fluid into a fracture, the coupling between hydraulic and mechanical processes spans a wide range of regimes where one or several of the following processes interact: toughness-controlled fracture propagation, dissipative processes in the solid material itself, viscous dissipation within the injected fluid, leak-off into the surrounding porous medium, fluid lag and suction at the crack tip upon fracture propagation, etc. Which of these processes is relevant or even dominant in a particular application depends on material properties, boundary conditions, and the process itself (Detournay 2016). For a low viscosity of the injected fluid and for a very low permeable rock matrix with high toughness, the hypothesis of a toughness-dominated propagation regime can be applied. In this work, we have considered water at room temperature as the injected fluid in a ultra low-permeable and high-toughness metamorphic rock (gneiss) so that the hypothesis of a toughness-dominated regime applies.

Numerical methods are a great asset in unravelling the complex physical mechanisms and in designing controllable stimulation protocols. As previously mentioned, the numerical treatment of crack propagation in general, and with fluid–structure interaction in particular, presents the analyst with a host of numerical challenges that are primarily associated with the creation of new boundary surfaces. The location and time of their creation as well as their topology are generally not known a priori and hence an outcome of the solution process itself. These boundaries can be conceptualised in a number of ways: they can be explicitly modelled by, e.g., remeshing (Bouchard et al. 2000; Meyer et al. 2004; Branco et al. 2015), which presents the additional problem of having to assign boundary conditions to the new surfaces including potential contact formulations and leaves the question of how to model processes that might occur within the newly created fracture space, such as fluid flow. A second approach makes use of the fact that fracture surfaces coincide in the reference configuration and can hence be conceptualized as strong discontinuities in the displacement field. Standard finite element methods have C^0-continuity requirements on primary variables. As such, the ansatz and test function spaces need to be enriched by functions which can capture the displacement jump across the crack tip. Additional enrichments can make use of asymptotic solutions known from fracture mechanical theories to improve numerical performance and decrease mesh sensitivity. An example is crack-tip enrichment by the Westergaard displacement fields (Budyn et al. 2004). These ideas have led to the development of eXtended and Generalized Finite Element Methods (XFEM, GFEM) (Moës et al. 1999; Belytschko et al. 2001, 2009; Gasser and Holzapfel 2005; Fries and Belytschko 2006; Khoei et al. 2012). In their general formulation, the implementational overhead is significant compared to a standard FEM in particular when propagation problems with arbitrary crack topologies are considered in three dimensions where merging and branching are allowed (Belytschko et al. 2001, 2003; Duarte et al. 2007). An extension to multi-physical problems is of course possible (Khoei 2014; Meschke and Leonhart 2015), although it adds additional complexities as the nature of the discontinuity in the various fields relevant for heat transport, multi-phase flow etc. is not always straight-forwardly understood (Zhang et al. 2013; Gordeliy and Peirce 2013; Chessa and Belytschko 2003). In a hydro-mechanical setting, Watanabe et al. (2012) developed a scheme where the local enrichment is introduced by lower-dimensional interface elements (LIE) located at the element boundaries of the full-dimensional domain mesh. Aside from this topological limitation, most advantages of an XFEM scheme are retained while drastically reducing the implementational effort, e.g. by avoiding the need for element splitting and integration routines.

To ease the modelling of the failure process itself and supported by observations in many materials that local failure happens somewhat gradually and covers a fracture process zone, cohesive zone models based on so-called traction-separation laws (TSLs) have been developed (Needleman 1990a, b). They have been successfully applied in the fracture of concrete, composite materials and many others (Nguyen et al. 2001; Elices et al. 2002).

To alleviate some of the difficulties related to these models such as spurious stress transfers, mesh-orientation dependencies, crack merging and branching, discontinuities in multi-physical fields and to be able to utilize standard C^0-continuous FEM software, smeared approaches to fracture have been introduced in which the fracture is not conceptualized as a sharp interface but as a smooth transition zone spanning a characteristic length scale. This transition zone of decreased to null stress transfer can be seen as a purely mathematical concept to regularise a discontinuity but can also be considered to have physical significance in the context of distributed micro-damage coalescing to a macroscopic fracture as well as of fracture process or dissipation zones (Brace et al. 1966; Bažant 1991; Diederichs 2003; Hoek and Martin 2014). Among these methods, gradient-damage (Peerlings et al. 1996; Pham et al. 2011; Alessi et al. 2015; Nedjar 2016) and phase-field methods (Bourdin et al. 2000; Miehe et al. 2010; Kuhn and Müller 2010; Borden et al. 2012; Wheeler et al. 2014; Ambati et al. 2015) are among the most prominent. They are sometimes referred to as weakly non-local, where strongly non-local models (Bazant and Jirasek 2002; He et al. 2015; Silani et al. 2016; Vtorushin 2016) are a third category of smeared models in which the material response at a given point is influenced by long-range interactions with neighbouring points falling into a given radius. These approaches offer the additional benefit of linking the classically separate theories strength of materials and fracture mechanics (Klinsmann et al. 2015; Tanné et al. 2018) and can be interpreted based on common physical origins such that they can be transferred into each other (Kuhl et al. 2000; de Borst and Verhoosel 2016a).

Aside from their numerical advantages and disadvantages, motivating the development of hybrid schemes (Giovanardi et al. 2017), it is imperative that all approaches be able to quantify certain key system properties and their evolution to be able to capture the physics of hydraulic fracturing appropriately. Such quantities include the stress fields and intensities, energy release rates, crack advancement in both direction and increment, etc. Of particular relevance in hydraulic fracturing is also the computation of the fracture opening as it determines the hydraulic properties of the fault and the resulting energy dissipation in the fluid and due to fluid–solid interaction. Discrete approaches are ideally suited for quantifying fracture opening and for including fracture-specific constitutive laws for flow and deformation, but present difficulties when dealing with complex topologies including the propagation/merging/branching of cracks in three-dimensional settings (Roth et al. 2016; Fries et al. 2014; Meschke and Leonhart 2015). In smeared approaches, while adept at handling such complex topologies, it seems less straight-forward to quantify fracture volume, aperture, permeability, and even crack length (Klinsmann et al. 2015). These questions need to be addressed in order to allow the analyst to freely select the most suitable method for a given problem.

In this paper we compare three distinct numerical methods, one discrete and two smeared ones, in their ability to capture hydraulically induced fracture propagation in a toughness-dominated regime and to quantify key variables such as fracture opening displacement, fracture volume, fracture length, and critical propagation pressures. The three methods are: (1) the lower-dimensional interface elements (LIE) with local enrichment developed by Watanabe et al. (2012) and extended here by a cohesive-zone approach based on a bi-linear TSL; (2) a linear-elastic phase-field model for brittle fracture; (3) a non-local elasto-plastic damage model for brittle and ductile failure (Parisio et al. 2018a). All models have been implemented into the open-source finite element framework OpenGeoSys (Kolditz et al. 2012) which can be found at https://github.com/ufz/ogs.

The article is structured as follows: in Sect. 2, the basic mechanical setting is described to set the scene, followed by the introduction of the three methods along with relevant literature in Sects. 3, 4 and 5. In Sect. 6, the parameterization of the different methods based on the same set of experimental data is discussed along with the quantification of key physical properties in the smeared approaches. Fracture volume and fluid mass balance computations are described in Sect. 7 along with the general numerical solution algorithms. Due to the focus on the above mentioned quantities, the simplifying assumption of a toughness-controlled regime with negligible fluid dissipation could be made. As such, fluid pressure is taken as a constant along the fault. Based on this assumption, the numerical solutions are compared with each other and verified against analytical solutions in Sect. 8. Final conclusions are drawn in Sect. 9.

2 General remarks on the mechanical setting

Consider a biphasic setting where a porous solid is fully saturated by a pore fluid and occupies the domain Ω. Embedded in Ω is a domain of co-dimension 1 representing the fracture. The displacement field of the porous solid denoted by $\mathbf{u} \in H^1(\Omega \backslash \Gamma_c; \mathbb{R}^n)$ has a jump across this embedded discontinuity surface Γ_c. Within the matrix, the material's deformation can be expressed by the linearized small strain tensor $\boldsymbol{\epsilon} = \text{sym grad}\,\mathbf{u}$, provided the displacement gradients remain sufficiently small. Following Biot's theory (Biot 1941), the total Cauchy stress tensor $\boldsymbol{\sigma}$ can be decomposed[1] into the effective stress tensor $\boldsymbol{\sigma}'$ and a contribution due to the pore-pressure p weighted by Biot's coefficient α

$$\boldsymbol{\sigma} = \boldsymbol{\sigma}' - \alpha p \mathbf{I}. \tag{1}$$

The effective stress follows from the material's constitutive relation $\boldsymbol{\sigma}' = \mathbf{C}(\boldsymbol{\epsilon})$. In the intact part of the medium, quasi-static equilibrium demands

$$\text{div}\,\boldsymbol{\sigma} + \varrho \mathbf{b} = \mathbf{0} \quad \text{in } \Omega \backslash \Gamma, \tag{2}$$

where $\mathbf{b}$ denotes an applied specific body force and ϱ is the mass density of the porous medium. On the Dirichlet part $\partial_D \Omega$ of the domain boundary $\partial \Omega$, fixed displacement boundary conditions are considered,

$$\mathbf{u} = \bar{\mathbf{u}} \quad \text{on } \partial_D \Omega, \tag{3}$$

while a traction $\bar{\mathbf{t}}$ is applied of the remaining part $\partial_N \Omega := \partial \Omega \backslash \partial_D \Omega$.

$$\boldsymbol{\sigma} \mathbf{n} = \bar{\mathbf{t}} \quad \text{on } \partial_N \Omega, \tag{4}$$

In addition to the equilibrium conditions within the uncracked domain, the stress field across the discontinuity Γ_c has to satisfy mechanical equilibrium as well. Let Γ^+ and Γ^- represent the opposing faces of cracks which coincide in the reference configuration and are therein designated as Γ_c; $\mathbf{n}_\Gamma^\pm$ denote the outer unit normal to Ω on $\Gamma^\pm$ respectively. On $\Gamma^\pm$, the tractions caused by the fluid pressure in the crack p_f are $-p_f \mathbf{n}_\Gamma^\pm$, so that admissible stress fields satisfy the conditions

$$\boldsymbol{\sigma} \mathbf{n}_\Gamma^\pm = -p_f \mathbf{n}_\Gamma^\pm \quad \text{on } \Gamma^\pm. \tag{5}$$

The displacement discontinuity will subsequently be treated by three different approaches: a strong discontinuity approach based on a cohesive zone model using lower-dimensional (i.e. co-dimension 1) interface elements with local enrichment of the displacement field (Sect. 3); and two approaches in which the discontinuity is smeared over a zone characterized by a length-scale parameter: phase-field models of brittle fracture (Sect. 4) and non-local elasto-plastic damage models (Sect. 5). The basic conceptual ideas of the three models are compared in Fig. 1.

[1] Using the sign convention of solid mechanics.

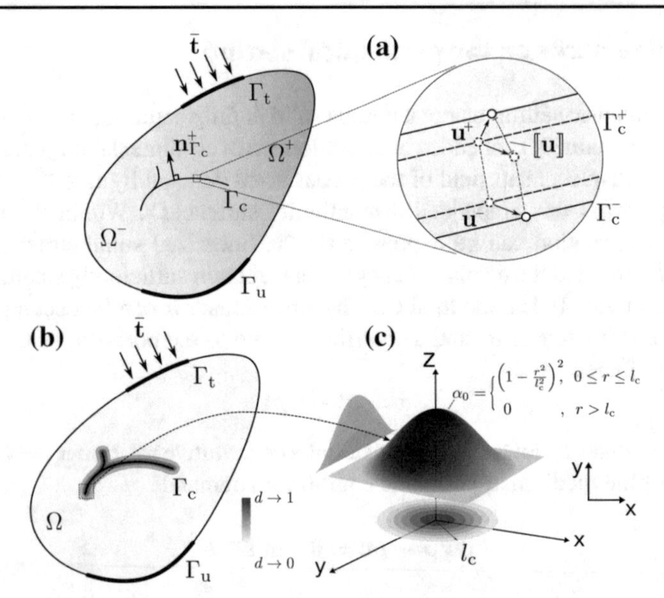

Fig. 1 Conceptual illustration of **a** cohesive zone model using lower-dimensional interface elements with local enrichment to represent a strong displacement discontinuity; **b** phase-field models of brittle fracture in which a crack surface density per unit volume is introduced for regularisation; and **c** non-local elasto-plastic damage models, in which a kernel function with a specified support region is used to characterize a fracture process zone. All models are implemented in OpenGeoSys

The fluid pressure in both matrix and fracture usually follows from the solution of a mass balance equation. In the sequel, the matrix will be considered impermeable and a mass balance only has to be solved for the crack domain. This is done differently for each of the three methods and will be explained in Sect. 7.

3 Lower-dimensional cohesive interface elements with local enrichment

The following implementation is based on Watanabe et al. (2012) and represents a special case of extended finite element methods (Moës et al. 1999; Belytschko et al. 2009, 2001) in that the enrichment is limited to element boundaries. The implementation is here extended to cohesive zone traction separation laws, cf. (Needleman 1990a, b; Nguyen et al. 2001; Elices et al. 2002; Gasser and Holzapfel 2005; Meschke et al. 2007).

3.1 Weak form

The weak form of the static equilibrium equation (2) can be written as the principle of virtual work

$$\int_{\Omega\setminus\Gamma_c} \delta\boldsymbol{\epsilon} : \boldsymbol{\sigma}\,\mathrm{d}\Omega - \int_{\Omega\setminus\Gamma_c} \delta\mathbf{u}\cdot\varrho\mathbf{b}\,\mathrm{d}\Omega - \int_{\partial_N\Omega} \delta\mathbf{u}\cdot\bar{\mathbf{t}}\,\mathrm{d}\Gamma - \int_{\Gamma_c} \underbrace{\left(\delta\mathbf{u}^+ - \delta\mathbf{u}^-\right)}_{\delta[\![\mathbf{u}]\!]}\cdot\mathbf{t}_c\,\mathrm{d}\Gamma = 0, \tag{6}$$

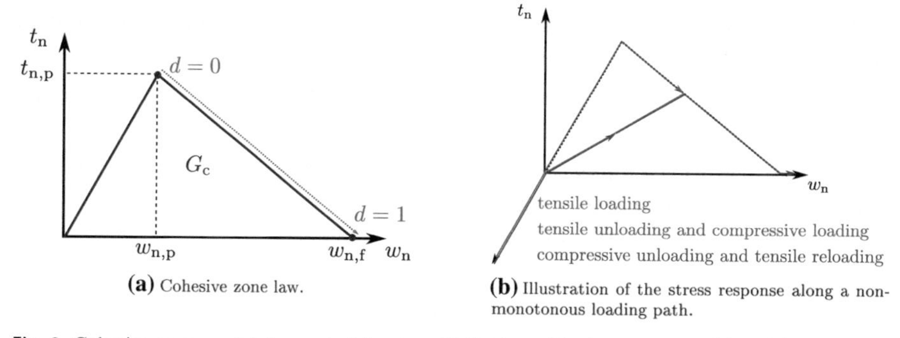

(a) Cohesive zone law.

(b) Illustration of the stress response along a non-monotonous loading path.

Fig. 2 Cohesive zone model for mode I fracture. Definition of basic parameters (**a**) and illustration of cleavage unloading, undamaged compressive loading and damaged reloading

where $\delta\mathbf{u}$ is the first variation of the displacement field (virtual displacements), $\mathbf{u}^+$ and $\mathbf{u}^-$ denote displacements of the opposing fracture phases such that $[\![\mathbf{u}]\!] = \mathbf{u}^+ - \mathbf{u}^-$ is the displacement jump across the interface, $\bar{\mathbf{t}}$ and $\mathbf{t}_c$ are the boundary ($\partial_N \Omega$) and the interface (Γ_c) traction, respectively. While the boundary traction $\bar{\mathbf{t}}$ is given by the applied external forces along the contour, the interface traction $\mathbf{t}_c$ follows from a constitutive response of the interface according to the the relative displacement between the opposing surfaces. Note, that the above definition implies stress continuity between the matrix compartments across the interface, viz. Eq. (5), $\mathbf{t}_c = \boldsymbol{\sigma}(\mathbf{x}_\Gamma)\mathbf{n}_\Gamma^+ = -\boldsymbol{\sigma}(\mathbf{x}_\Gamma)\mathbf{n}_\Gamma^-$.

3.2 Constitutive model

While the matrix material behaves according to linear elasticity and is assumed to be impermeable in the current study, the crack will be fluid saturated and follows an effective stress-type formulation:

$$\mathbf{t}_c = \mathbf{t}_c' - p_f \mathbf{n}_\Gamma, \tag{7}$$

where the fluid pressure in the fault only acts on the normal traction across the fault, not its shear components. For the effective stress, a cohesive-zone law has been implemented based on the following assumptions:

- Damage is driven by fracture opening $[\![\mathbf{u}]\!] \cdot \mathbf{n}_\Gamma$ only, i.e. only mode I fracture propagation is considered.
- The traction-separation law is bilinear (Fig. 2a).
- Cleavage unloading (in contrast to ductile unloading) is considered in view of modelling brittle fracture (Fig. 2b).

The cohesive zone model is characterized by its peak tensile normal traction or normal tensile strength $t_{n,p}$, its initial normal and shear stiffness K_n, K_s, and its fracture toughness/critical energy release rate G_c.

The undamaged elastic fracture constitutive law

$$\mathbf{t}_c' = \mathbf{K}[\![\mathbf{u}]\!] = [K_n(\mathbf{n}_\Gamma \otimes \mathbf{n}_\Gamma) + K_s(\mathbf{I} - \mathbf{n}_\Gamma \otimes \mathbf{n}_\Gamma)]\,[\![\mathbf{u}]\!], \tag{8}$$

remains valid in compression ($w_n = [\![\mathbf{u}]\!] \cdot \mathbf{n}_\Gamma < 0$). During compressive loading, a penalty formulation

$$K_n^{\text{pen}} = K_n \left[1 + \ln^2 \left(\frac{b}{b_0} \right) \right], \tag{9}$$

based on current (b) and initial (b_0) aperture is invoked to prevent fracture face interpenetration at high compressive loads.

In tension, the model is modified to account for damage

$$\mathbf{t} = \mathbf{K}^{\text{d}} [\![\mathbf{u}]\!]. \tag{10}$$

During monotonously increasing loading, damage evolves linearly with normal fault opening between the limiting values

$$d = \begin{cases} 0 & w_n = w_{n,p} \\ 1 & w_n = w_{n,f}, \end{cases} \tag{11}$$

where $w_{n,p} = \frac{t_{n,p}}{K_n}$ and $w_{n,f} = 2\frac{G_c}{t_{n,p}}$.

Damage is implemented as a non-decreasing function according to:

$$d^{t+\Delta t} = \min \left[1, \max \left(\frac{\langle w - w_{n,p} \rangle}{w_{n,f} - w_{n,p}}, d^t \right) \right], \tag{12}$$

where the Macauley brackets have been used.

The new tensile normal stiffness is found via

$$K_n^{\text{d}} = \frac{t_n}{w_n} = \frac{(1-d) t_{n,p}}{\max_{0 \le \tau \le t} w_n(\tau)} = \frac{(1-d) K_n w_{n,p}}{d(w_{n,f} - w_{n,p}) + w_{n,p}}. \tag{13}$$

Accordingly, the entire stiffness tensor is degraded:

$$\mathbf{K}^{\text{d}} = \frac{(1-d) w_{n,p}}{d(w_{n,f} - w_{n,p}) + w_{n,p}} \mathbf{K} = g(d) \mathbf{K}. \tag{14}$$

In case of $d^{t+\Delta t} > d^t$ within the employed incremental-iterative solution scheme, the algorithmic tangent $\mathbf{D}$ is extended by a second term:

$$\mathbf{D} = \mathbf{K}^{\text{d}} + \mathbf{K} [\![\mathbf{u}]\!] \otimes \frac{\partial g(d)}{\partial d} \frac{\partial d}{\partial [\![\mathbf{u}]\!]}, \tag{15}$$

$$\text{with} \quad \frac{\partial d}{\partial w_n} = \frac{1}{w_{n,f} - w_{n,p}} \quad \text{and} \quad \frac{\partial g(d)}{\partial d} = -\frac{w_{n,p} w_{n,f}}{[d(w_{n,f} - w_{n,p}) + w_{n,p}]^2}. \tag{16}$$

Since the matrix was considered impermeable, a fluid pressure was only required in the cracked domain. For that purpose, Eq. (7) was modified to read:

$$\mathbf{t}_{\mathrm{c}} = \mathbf{t}'_{\mathrm{c}} - d p_{\mathrm{f}} \mathbf{n}_{\Gamma} \tag{17}$$

The determination of p_{f} is described in Sect. 7.

3.3 Numerical implementation

The implementation within an extrinsically enriched finite element scheme follows Watanabe et al. (2012). In addition to the usual nodal displacement degrees of freedom for continuous settings, nodal degrees of freedom equivalent to the displacement jump are introduced as additional unknowns. Hence, in contrast to the other methods used in this paper, the displacement jump is explicitly given as a primary solution of the enriched finite element scheme such that the displacement solution displays a strong discontinuity. Non-linearities are linearized using an incremental iterative Newton–Raphson method resulting in a linear system

$$\begin{bmatrix} \mathbf{K}_{\mathbf{uu}} & \mathbf{K}_{\mathbf{ua}} \\ \mathbf{K}_{\mathbf{au}} & \mathbf{K}_{\mathbf{aa}} \end{bmatrix} \begin{Bmatrix} \Delta\mathbf{u} \\ \Delta\mathbf{a} \end{Bmatrix} = \begin{Bmatrix} \mathbf{f}_{\mathbf{u}}^{\mathrm{ext}} \\ \mathbf{f}_{\mathbf{a}}^{\mathrm{ext}} \end{Bmatrix} - \begin{Bmatrix} \mathbf{f}_{\mathbf{u}}^{\mathrm{int}} \\ \mathbf{f}_{\mathbf{a}}^{\mathrm{int}} \end{Bmatrix},$$

which is solved until convergence is achieved. Here, $\Delta\mathbf{u}$ and $\Delta\mathbf{a}$ denote the increments of the regular nodal displacement and the additional degrees of freedom related to the displacement jump across the interface, $\mathbf{K}_{\bullet\bullet}$ are the sub-matrices of the stiffness matrix of the problem and the right-hand-side consists of the out-of-balance (residual) force vectors.

4 Phase-field model of brittle fracture

Phase-field modelling has emerged as one of the most appealing numerical techniques for simulation of fracture in the last two decades. A widely applied approach of phase-field models of fracture was first introduced by Bourdin et al. (2000) as a numerical implementation method to the variational approach to fracture proposed by Francfort and Marigo (1998) using the Γ-convergence properties. Since this inception, the approach has been adapted in many studies ranging from ductile fracture (Ambati et al. 2015; Alessi et al. 2015; Miehe et al. 2016) to hydraulic fracturing (Bourdin et al. 2012; Wheeler et al. 2014; Miehe et al. 2015; Wilson and Landis 2016; Heider and Markert 2016; Chukwudozie 2016; Yoshioka and Bourdin 2016; Santillán et al. 2017) to name a few.

4.1 Regularisation of the total energy functional

Francfort and Marigo (1998) define a total energy functional as the sum of the elastic strain energy, the potential of the external forces and the surface energy as:

$$E(\mathbf{u}, \Gamma_c) := \int_{\Omega\setminus\Gamma_c} \psi(\mathbf{u})\,\mathrm{d}\Omega - \int_{\partial_N\Omega} \bar{\mathbf{t}}\cdot\mathbf{u}\,\mathrm{d}\Gamma - \int_{\Omega\setminus\Gamma_c} \varrho\mathbf{b}\cdot\mathbf{u}\,\mathrm{d}\Omega + G_c\mathcal{H}^{N-1}(\Gamma_c), \tag{18}$$

where $\mathcal{H}$ is the Hausdorff measure. For the extension to hydraulic fracturing, the work done by the fluid pressure on crack faces $\int_{\Gamma_c} p[\![\mathbf{u}]\!]\cdot\mathbf{n}_\Gamma\,\mathrm{d}\Gamma$ is added to the total energy which can now be written as

$$\begin{aligned} E(\mathbf{u}, \Gamma_c) := &\int_{\Omega\setminus\Gamma_c} \psi(\mathbf{u})\,\mathrm{d}\Omega - \int_{\partial_N\Omega} \bar{\mathbf{t}}\cdot\mathbf{u}\,\mathrm{d}\Gamma - \int_{\Omega\setminus\Gamma_c} \varrho\mathbf{b}\cdot\mathbf{u}\,\mathrm{d}\Omega + G_c\mathcal{H}^{N-1}(\Gamma_c) \\ &+ \int_{\Gamma_c} p[\![\mathbf{u}]\!]\cdot\mathbf{n}_\Gamma\,\mathrm{d}\Gamma. \end{aligned} \tag{19}$$

Bourdin et al. (2000) introduced a phase-field variable to regularise the functional adapting the Γ-convergence result obtained in Ambrosio and Tortorelli (1990) and the total energy functional takes the following form (Bourdin et al. 2014; Marigo et al. 2016):

$$\begin{aligned} E(\mathbf{u}, d) := &\int_{\Omega} \psi(\mathbf{u}, d)\,\mathrm{d}\Omega - \int_{\partial_N\Omega} \bar{\mathbf{t}}\cdot\mathbf{u}\,\mathrm{d}\Gamma - \int_{\Omega} \varrho\mathbf{b}\cdot\mathbf{u}\,\mathrm{d}\Omega \\ &+ \frac{G_c}{4c_w}\int_{\Omega}\left(\frac{w(d)}{\ell} + \ell\,|\nabla d|^2\right)\mathrm{d}\Omega + \int_{\Gamma_c} p[\![\mathbf{u}]\!]\cdot\mathbf{n}_\Gamma\,\mathrm{d}\Gamma, \end{aligned} \tag{20}$$

where d is the damage variable that equals 0 when undamaged and 1 for a fully damaged state. $\psi(\mathbf{u}, d)$ is the degraded strain energy density, and w is a dissipated energy function such that $\psi(\mathbf{u}, 0) = \psi(\mathbf{u})$ and $w(0) = 0$ and $\psi(\mathbf{u}, 1) = 0$ and $w(1) = 1$ are ensured. The most popular choice for the degradation function is a quadratic form, $\psi(\mathbf{u}, d) = (1-d)^2\psi(\mathbf{u})$, which is used in this study, while other higher-order polynomial functions were also considered in literature (Karma et al. 2001; Borden et al. 2012). G_c is the fracture surface energy, c_w is a normalization parameter defined as $c_w := \int_0^1 \sqrt{w(s)}\mathrm{d}s$, and ℓ is a regularisation length. Various possible dissipative energy functions w are discussed in Marigo et al. (2016). The most widely used function is a quadratic form (Kuhn and Müller 2010; Miehe et al. 2010; Klinsmann et al. 2015) for which Eq. (20) is given by

$$\begin{aligned} E(\mathbf{u}, d) := &\int_{\Omega} (1-d)^2\psi(\mathbf{u}),\mathrm{d}\Omega - \int_{\partial_N\Omega} \bar{\mathbf{t}}\cdot\mathbf{u}\,\mathrm{d}\Gamma - \int_{\Omega} \varrho\mathbf{b}\cdot\mathbf{u}\,\mathrm{d}\Omega \\ &+ \frac{G_c}{2}\int_{\Omega}\left(\frac{d^2}{\ell} + \ell\,|\nabla d|^2\right)\mathrm{d}\Omega + \int_{\Gamma_c} p[\![\mathbf{u}]\!]\cdot\mathbf{n}_\Gamma\,\mathrm{d}\Gamma. \end{aligned} \tag{21}$$

This regularised Eq. (21) is called *AT2* model in Bourdin et al. (2014) and Tanné et al. (2018) after Ambrosio and Tortorelli (1990, 1992), and this terminology is adopted in the present study. Another choice for $w(d)$ is a linear form called *AT1* introduced in Bourdin et al. (2014) following Braides (1998, Sect. 3.2) for which Eq. (20) is consequently given as

$$E(\mathbf{u}, d) := \int_{\Omega} (1-d)^2 \psi(\mathbf{u}) \mathrm{d}\Omega - \int_{\partial_{\mathrm{N}}\Omega} \bar{\mathbf{t}} \cdot \mathbf{u} \, \mathrm{d}\Gamma - \int_{\Omega} \varrho \mathbf{b} \cdot \mathbf{u} \, \mathrm{d}\Omega \\ + \frac{3G_{\mathrm{c}}}{8} \int_{\Omega} \left(\frac{d}{\ell} + \ell \, |\nabla d|^2 \right) \mathrm{d}\Omega + \int_{\Gamma_{\mathrm{c}}} p[\![\mathbf{u}]\!] \cdot \mathbf{n}_{\Gamma} \, \mathrm{d}\Gamma. \tag{22}$$

The work done by the pressure in the crack is still explicitly dependent on the displacement jump and remains to be cast into a continuous setting. In Bourdin et al. (2012), an approximation $\int_{\Omega} p\mathbf{u} \cdot \nabla d \, \mathrm{d}\Omega$ has been proposed to take into account the pressure work, which Γ-converges to $\int_{\Gamma_{\mathrm{c}}} p[\![\mathbf{u}]\!] \cdot \mathbf{n}_{\Gamma} \, \mathrm{d}\Gamma$. Using this approximation, Eq. (20) can be written in its now fully regularised form

$$E(\mathbf{u}, d, p) = \int_{\Omega} (1-d)^2 \psi(\mathbf{u}) \mathrm{d}\Omega - \int_{\partial_{\mathrm{N}}\Omega} \bar{\mathbf{t}} \cdot \mathbf{u} \, \mathbf{ds} - \int_{\Omega} \varrho \mathbf{b} \cdot \mathbf{u} \, \mathrm{d}\Omega \\ + \frac{G_{\mathrm{c}}}{4c_w} \int_{\Omega} \left(\frac{w(d)}{\ell} + \ell \, |\nabla d|^2 \right) \mathrm{d}\Omega + \int_{\Omega} p\mathbf{u} \cdot \nabla d \, \mathrm{d}\Omega. \tag{23}$$

While it was proposed to split the strain energy term, $\int_{\Omega} (1-d)^2 \psi(\mathbf{u}) \mathrm{d}\Omega$, to take into account the possibility of crack closure into a degraded part and an undegradable (usually compressive) part (Amor et al. 2009; Freddi and Royer-Carfagni 2010; Miehe et al. 2010) as $\int_{\Omega} (1-d)^2 \psi(\mathbf{u}) \mathrm{d}\Omega = \int_{\Omega} [(1-d)^2 \psi^+(\mathbf{u}) + \psi^-(\mathbf{u})] \, \mathrm{d}\Omega$, the choice of the split remains an open question (Li et al. 2016). In the examples shown in the subsequent sections, the symmetric model is used to be consistent with the closed-form solutions.

4.2 Numerical implementation

The solution of Eq. (23) follows the alternate minimisation scheme introduced in Bourdin et al. (2000) with respect to $\mathbf{u}$ and d given a time-evolving volume constraint $V_{\mathrm{inj}} = V_{\mathrm{crack}} (= \int_{\Omega} \mathbf{u} \cdot \nabla d \, \mathrm{d}\Omega)$.[2] The functional is cast into a dimensionless format before application of the alternating minimization scheme (see "Appendix A") to avoid ill-posedness that may be caused by input parameter units Chukwudozie (2016). Thus, the minimisation problem can be stated as

$$(\tilde{\mathbf{u}}, d, \tilde{p})^* = \arg\min_{\begin{cases} \tilde{\mathbf{u}} \in H^1 \\ d \in H^1, d^t \subset d^{t+\Delta t} \\ \tilde{V} = \int_{\Omega} \tilde{\mathbf{u}} \cdot \nabla d \, \mathrm{d}\Omega \end{cases}} \tilde{E}(\tilde{\mathbf{u}}, d, \tilde{p}) \,, \tag{24}$$

where $\tilde{V}_{\mathrm{crack}} = u_0 x_0^{n-1} V$ and irreversibility of d is enforced as in Bourdin et al. (2000). The first variation of the energy functional with respect to $\mathbf{u}$ is

[2] This constraint can also be added to the functional and its Lagrange multiplier turns out to be exactly the pressure (Chukwudozie 2016).

$$\delta \tilde{E}(\mathbf{u}, d, p; \delta\mathbf{u}) = \frac{1}{2}\int_{\Omega} (1-d)^2 \tilde{\mathbf{C}}(\boldsymbol{\epsilon}(\tilde{\mathbf{u}})) : \boldsymbol{\epsilon}(\delta\tilde{\mathbf{u}})\,\mathrm{d}\Omega - \\ - \int_{\partial_{\mathrm{N}}\Omega} \tilde{\mathbf{t}} \cdot \delta\tilde{\mathbf{u}}\,\mathrm{d}\Gamma - \int_{\Omega} \tilde{\varrho\mathbf{b}} \cdot \delta\tilde{\mathbf{u}}\,\mathrm{d}\Omega + \int_{\Omega} \tilde{p}\delta\tilde{\mathbf{u}} \cdot \nabla d\,\mathrm{d}\Omega, \tag{25}$$

where $\tilde{\mathbf{C}}$ is the constitutive relation that satisfies $\psi = \tilde{\mathbf{C}}(\boldsymbol{\epsilon}) : \boldsymbol{\epsilon}/2$. In the present study, isotropic linear elasticity was used: $\psi = \boldsymbol{\epsilon} : \boldsymbol{C} : \boldsymbol{\epsilon}/2$. The first variation of the energy functional with respect to d for the *AT1* model is given as

$$\delta \tilde{E}(\mathbf{u}, d, p; \delta d) = -\int_{\Omega} d\delta d\tilde{\mathbf{C}}(\boldsymbol{\epsilon}(\tilde{\mathbf{u}})) : \boldsymbol{\epsilon}(\tilde{\mathbf{u}})\mathrm{d}\Omega \\ + \frac{3G_{\mathrm{c}}}{8}\int_{\Omega}\left(\frac{\delta d}{\ell} + 2\ell\nabla d \cdot \nabla\delta d\right)\mathrm{d}\Omega + \int_{\Omega} \tilde{p}\tilde{\mathbf{u}} \cdot \nabla\delta d\,\mathrm{d}\Omega, \tag{26}$$

and for the *AT2* model as

$$\delta \tilde{E}(\mathbf{u}, d, p; \delta d) = -\int_{\Omega} d\delta d\tilde{\mathbf{C}}(\boldsymbol{\epsilon}(\tilde{\mathbf{u}})) : \boldsymbol{\epsilon}(\tilde{\mathbf{u}})\mathrm{d}\Omega \\ + G_{\mathrm{c}}\int_{\Omega}\left(\frac{d}{\ell}\delta d + \ell\nabla d \cdot \nabla\delta d\right)\mathrm{d}\Omega + \int_{\Omega} \tilde{p}\tilde{\mathbf{u}} \cdot \nabla\delta d\,\mathrm{d}\Omega. \tag{27}$$

Note that d represents the state of the material constrained in $[1, 0]$ and that this constraint is automatically fulfilled with *AT2* in pure mechanical settings whereas only *AT1* requires the solution of a variational inequality. As can be seen in the above extension to hydraulic fracturing, however, both *AT1* and *AT2* models now require enforcement of the variational inequality ($0 \leq d \leq 1$) due to the presence of the pressure-work term in the energy functional. In this study, a variational inequality non-linear solver of the PETSc library (Balay et al. 1997, 2017a, b) has been applied to satisfy the constraint on d.

5 Non-local integral formulation of elasto-plasticity with damage

5.1 Constitutive model

The theory of continuum damage mechanics has been consistently and successfully applied to the simulation of failure and fracture of brittle and ductile solids in the spirit of the local approach to fracture (Lemaitre et al. 2009; Murakami 2012). Because of the softening response, continuum damage models suffer from spurious mesh-dependency: the rate equation of equilibrium loses ellipticity, the acoustic tensor becomes singular and the problem ill-posed, bifurcated solutions of the static equilibrium might appear and the global dissipated energy decreases upon mesh-element size decrement. In the extreme case, snap-backs are possible and a cusp-catastrophe surface appears in the three-dimensional stress versus strain versus element-size space. Restoring objectivity in the solution with respect to mesh size has often been achieved by introducing an internal length in what is usually referred to as an extended or non-local continuum. In this contribution, we employ an integral-type non-local plastic-damage model, which can successfully prevent ill-posedness of the rate problem for softening damage models (Bazant and Jirasek 2002; Duddu and Waisman 2013; Nguyen et al. 2015; Desmorat et al. 2015; Parisio et al. 2018a). The constitutive equation of the plastic-damage model relates Biot's effective stress to the elastic strain tensor as

$$\boldsymbol{\sigma}' = \tilde{\boldsymbol{C}} : \boldsymbol{\epsilon}_{\text{el}}, \tag{28}$$

where $\tilde{\boldsymbol{C}} = (1-d)\,\boldsymbol{C}$ is the damaged elastic tensor,[3], $0 \leq d \leq 1$ is the damage parameter, $\boldsymbol{\epsilon}_{\text{el}}$ is the elastic strain tensor defined as the difference between total and plastic strain

$$\boldsymbol{\epsilon}_{\text{el}} = \boldsymbol{\epsilon} - \boldsymbol{\epsilon}_{\text{pl}}. \tag{29}$$

According to Eq. (28), the concept of damage effective stress can be defined as the stress that acts on the undamaged part of the material as

$$\tilde{\boldsymbol{\sigma}} = \frac{\boldsymbol{\sigma}'}{1-d} = \boldsymbol{C} : \boldsymbol{\epsilon}_{\text{el}}. \tag{30}$$

The loading surface for this class of plastic-damage model can be chosen as unique for the two dissipative mechanisms. In this case, we will employ a J_2-type failure surface of the Drucker–Prager family formulated in the damage effective stress space $\tilde{\boldsymbol{\sigma}}$:

$$F = \sqrt{J_2} - \beta I_1 + k = 0, \tag{31}$$

with β and k material parameters and the invariants of the stress tensor defined as

$$I_1 = \text{tr}\left(\tilde{\boldsymbol{\sigma}}\right), \; J_2 = \left(\tilde{\mathbf{s}} : \tilde{\mathbf{s}}\right)/2, \tag{32}$$

where $\tilde{\mathbf{s}} = \tilde{\boldsymbol{\sigma}} - \text{tr}\left(\tilde{\boldsymbol{\sigma}}\right)/3\,\mathbf{I}$ is the deviatoric effective stress tensor. The Karush–Kuhn–Tucker loading-unloading conditions write

$$F\left(\tilde{\boldsymbol{\sigma}}\right) \leq 0 \quad \lambda \geq 0 \quad \lambda F\left(\tilde{\boldsymbol{\sigma}}\right) = 0, \tag{33}$$

with λ the plastic multiplier that defines the plastic strain rate as

$$\dot{\boldsymbol{\epsilon}}_{\text{pl}} = \lambda \frac{\partial G}{\partial \tilde{\boldsymbol{\sigma}}}, \tag{34}$$

In the current study, associated plasticity was modelled, i.e. the plastic potential function G was identical to the yield function $G = F$.

[3] Observe the linear degradation of stiffness in comparison to the quadratic formulation used in the phase-field formulation. Consequences of this choice have been discussed in de Borst and Verhoosel (2016b) where gradient-damage and phase-field models were compared. For links on non-local integral formulations and gradient-damage models, the reader is referred to Kuhl et al. (2000) and references therein.

A non-local damage-driving variable $\bar{k}_d$ is obtained by integral averaging of its local counterpart k_d

$$\bar{k}_d(\boldsymbol{x}) = \frac{1}{\int_V \alpha_0(\|\boldsymbol{x} - \boldsymbol{\psi}\|)\,\mathrm{d}\boldsymbol{\psi}} \int_V \alpha_0(\|\boldsymbol{x} - \boldsymbol{\xi}\|)\,k_d(\boldsymbol{\xi})\,\mathrm{d}\boldsymbol{\xi}, \tag{35}$$

where $r = \|\boldsymbol{x} - \boldsymbol{\xi}\|$ is the distance between two points in the continuum and α_0 is the weight function (compare Fig. 1)

$$\alpha_0 = \begin{cases} \left(1 - \dfrac{r^2}{l_c^2}\right) & \text{if } 0 \le r \le l_c \\ 0 & \text{if } l_c \le r \end{cases}. \tag{36}$$

In Eq. (36), l_c is an internal length of the continuum which defines the radius of interaction of the different points in the material. For the theoretical justification and the relation of l_c with internal micro-structural characteristics, the interested reader is addressed to consult the literature (Bazant and Jirasek 2002). The local damage-driving variable k_d of Eq. (35) is in turn a function of the rate of effective plastic strain and is defined in rate form as

$$\dot{k}_d = \dot{\epsilon}_{\mathrm{pl}}^{\mathrm{eff}} \tag{37}$$

with

$$\epsilon_{\mathrm{pl}}^{\mathrm{eff}}(t) = \int_0^t \sqrt{\frac{2}{3}\dot{\boldsymbol{\epsilon}}_{\mathrm{pl}} : \dot{\boldsymbol{\epsilon}}_{\mathrm{pl}}}\,\mathrm{d}\tau. \tag{38}$$

Finally, damage d is an exponential function of the non-local driving variable as $\bar{k}_d$

$$d = \omega\left(\bar{k}_d\right) = 1 - \exp\left(-\frac{\bar{k}_d}{\alpha_d}\right), \tag{39}$$

where α_d is a material parameter controlling the rate of damage growth, and therefore material brittleness (Parisio and Laloui 2017).

As the matrix was considered impermeable, a fluid pressure was only required in the damaged domain. For that purpose, Eq. (1) was modified to read

$$\boldsymbol{\sigma} = \boldsymbol{\sigma}' - dp\mathbf{I}. \tag{40}$$

The determination of p is described in Sect. 7.

5.2 Numerical implementation

The solution of the differential-algebraic system of equations describing the plastic-damage model is performed at each integration point within a fully implicit scheme for non-linear materials (Parisio et al. 2015; Nagel et al. 2016, 2017; Parisio et al.

2018b). Because of the adopted coupling between damage and plasticity, damage can be explicitly computed and updated after the return-mapping algorithm has solved the plastic sub-problem in damage effective-stress space. The Newton–Raphson method for the plastic problem minimizes the residual $\boldsymbol{R}(z)$, which is a function of the vector of state variables z.

The sought state variable vector contains the solution of the plastic problem at time step $t + \Delta t$ in terms of damage effective stress $\tilde{\boldsymbol{\sigma}}^{t+\Delta t}$, plastic strain $\boldsymbol{\epsilon}_{\text{pl}}^{t+\Delta t}$ and plastic multiplier $\lambda^{t+\Delta t}$. The residual to be minimized associated with the plastic state variables contains the equations of stress, plastic strain rate and the yield surface as an algebraic constraint and is expressed as

$$\boldsymbol{R}(z) = \begin{cases} \tilde{\boldsymbol{\sigma}}^{t+\Delta t} - \mathbf{C}\left(\boldsymbol{\epsilon}^{t+\Delta t} - \boldsymbol{\epsilon}_{\text{pl}}^{t+\Delta t}\right) \\ \boldsymbol{\epsilon}_{\text{pl}}^{t+\Delta t} - \boldsymbol{\epsilon}_{\text{pl}}^{t} - \Delta t \lambda^{t+\Delta t} \dfrac{\partial G}{\partial \tilde{\boldsymbol{\sigma}}^{t+\Delta t}} \\ F\left(\tilde{\boldsymbol{\sigma}}^{t+\Delta t}\right) \end{cases}. \tag{41}$$

Linearization of the residual with respect to the state variables yields the following iterative Newton–Raphson scheme

$$z_{n+1}^{t+\Delta t} = z_n^{t+\Delta t} - \left(\boldsymbol{J}_n^{t+\Delta t}\right)^{-1} \boldsymbol{R}_n^{t+\Delta t}, \tag{42}$$

with the Jacobian

$$\boldsymbol{J}_n^{t+\Delta t} = \left.\frac{\partial \boldsymbol{R}^{t+\Delta t}}{\partial z^{t+\Delta t}}\right|_n, \tag{43}$$

which can be computed analytically (as in the present case) or using a numerical perturbation technique. The process is iterated over n until $\|\boldsymbol{R}\| < \theta_{\text{tol}}$. The elasto-plastic tangent matrix $\mathbf{C}_{\text{pl}}$ is extracted from the Jacobian by solving the following system after local convergence

$$\frac{\mathrm{d}z}{\mathrm{d}\boldsymbol{\epsilon}} = -\boldsymbol{J}^{-1} \frac{\partial \boldsymbol{R}}{\partial \boldsymbol{\epsilon}}, \tag{44}$$

where the first entries of z contain $\tilde{\boldsymbol{\sigma}}$ (Nagel et al. 2017). The non-local damage variable $\bar{k}_d$ can be computed explicitly after convergence of the plastic algorithm, and is approximated in the finite element scheme as

$$\bar{k}_{d,i} = \frac{\sum_{j=1}^{n_p} w_j \alpha_0 \left(\left\|\boldsymbol{x}_i - \boldsymbol{x}_j\right\|\right) k_d\left(\boldsymbol{x}_j\right) \det J\left(\boldsymbol{x}_j\right)}{\sum_{k=1}^{n_p} w_k \alpha_0 \left(\|\boldsymbol{x}_i - \boldsymbol{x}_k\|\right) \det J\left(\boldsymbol{x}_k\right)}, \tag{45}$$

where n_p is the number of integration points, $\det J(\boldsymbol{x}_k)$ is the determinant of the Jacobian of the isoparametric element coordinate transformation. The integration scheme is run on the integration points that fall into the non-local interaction radius l_c. From $\bar{k}_{d,i}$, damage at integration points is computed from Eq. (39).

6 Material properties

Before comparing simulation results based on the three model formulations, inter-model consistency in terms of parametrization has to be ensured in face of the different mathematical and physical concepts underlying the models. The calibration procedures for the three models are therefore briefly illustrated in the following sections, relations to spatial discretization discussed, and then applied to represent an anisotropic pink Gneiss from the Erzgebirge mountain range in Saxony, Germany (personal communication from Thomas Frühwirt and Heinz Konietzky; cf. Acknowledgements).

6.1 Rock properties

The properties of the Gneiss are taken only according to one configuration of the foliation planes, so that the material is finally considered isotropic. The density is $\varrho = 2.68\,\mathrm{g\,cm^{-3}}$, Young's modulus $E = 80\,\mathrm{GPa}$, Poisson's ratio $\nu = 0.15$, uniaxial compressive strength $\sigma_\mathrm{c} = 120{-}160\,\mathrm{MPa}$, uniaxial tensile strength $\sigma_\mathrm{t} = 6{-}17\,\mathrm{MPa}$ and fracture energy $G_\mathrm{c} = 20\,\mathrm{Pa\,m}$. Each model is calibrated based on E, ν, and G_c; additional properties to complete the model description are quantified from, e.g., the tensile σ_t and compressive strengths σ_c, as discussed below.

6.2 Cohesive zone traction-separation law

The tensile strength $t_{\mathrm{n,p}}$ and critical energy release rate G_c required to parameterize the failure behaviour of the cohesive zone model, cf. Eq. (11), can be taken directly from the above set of experimentally determined material parameters. The determination of the initial normal stiffness K_n in Eq. (11) may proceed from the intrinsic mechanical properties of the matrix material as well as the geometry of a potential pre-existing fault, see also (Oliver 2000; Oliver et al. 2002). Therefore, the setting of the normal stiffness can be estimated as follows

$$K_\mathrm{n} = \frac{E}{b_0} \geq \frac{t_{\mathrm{n,p}}^2}{2G_\mathrm{c}}. \tag{46}$$

In general and if no pre-existing fault is modelled, however, K_n can be set independently and controls the brittleness of failure and the magnitude of elastic opening before failure. The lower limit corresponds to the extreme case in which sudden brittle breakage occurs once the normal tensile strength is mobilised. For numerical reasons, K_n was herein set to $10^{14}\,\mathrm{MPa\,m^{-1}}$, which is several orders of magnitude higher than the lower limit $K_{\mathrm{n,min}} \approx 10^7\,\mathrm{MPa\,m^{-1}}$ in case of $\sigma_\mathrm{t} = 17\,\mathrm{MPa}$.

The choice of material parameters introduces a cohesive zone process length l_cz characterizing the spread of the failure zone. This length has the form (Hillerborg et al. 1976; Rice 1979)

$$l_\mathrm{cz} = E\frac{G_\mathrm{c}}{t_{\mathrm{n,p}}^2}. \tag{47}$$

In order to resolve the failure process numerically, sufficient discretization is required (Davila et al. 2001; Moës and Belytschko 2002). Turon et al. (2007) suggested the total number of interface elements across the cohesive zone length might be as many as 3.

6.3 Phase-field: effective fracture energy and effective crack length

The phase-field approach in principle only requires E, ν, and G_c in the material description, whereas ℓ was originally introduced as a numerical regularisation parameter in view of Γ-convergence. However, recent studies of Marigo et al. (2016) and Zhang et al. (2017) suggest that ℓ be treated as an internal length, i.e. a material property. Given the uniaxial tensile strength, σ_t, the internal length can be obtained as follows (Tanné et al. 2018)

$$\sigma_t = \begin{cases} \sqrt{\dfrac{3G_c E'}{8\ell}} & \text{for } AT1 \\ \dfrac{3}{16}\sqrt{\dfrac{3G_c E'}{\ell}} & \text{for } AT2 \end{cases} . \tag{48}$$

Using the material properties from Sect. 6.1, the internal length for *AT1* and *AT2* can be computed as 0.017 m and 0.0048 m respectively, which is comparable to the internal length sizes used in this study ranging from 0.00375 to 0.03 m. As investigated by Tanné et al. (2018), as long as the internal length is insignificant with respect to the defect size, which in our case is 0.2 m, the choice of ℓ will have minimum impact on fracture propagation in toughness-dominated settings; similar conclusions were drawn in Klinsmann et al. (2015) and Zhang et al. (2017).

In the phase-field models, the crack "surface" is approximated by

$$S(x, d, \nabla d) := \frac{1}{4c_w} \int_\Omega \left(\frac{w(d)}{\ell} + \ell \, |\nabla d|^2 \right) \mathrm{d}\Omega. \tag{49}$$

The optimal profile of the damage field, d can be obtained by taking $\delta S/\delta d = 0$. Thus,

$$\frac{w'(d)}{\ell} - 2\ell \nabla^2 d = 0. \tag{50}$$

For *AT1*, $w(d) = d$. Therefore, the optimal profile is given by

$$d(x) = \begin{cases} \left(1 - \dfrac{|x - x_0|}{2\ell}\right)^2, & \text{for } |x - x_0| \leq 2\ell \\ 0, & \text{otherwise} \end{cases} \tag{51}$$

and for *AT2*, $w(d) = d^2$, it is

$$d(x) = \exp\left(-\frac{|x - x_0|}{\ell}\right). \tag{52}$$

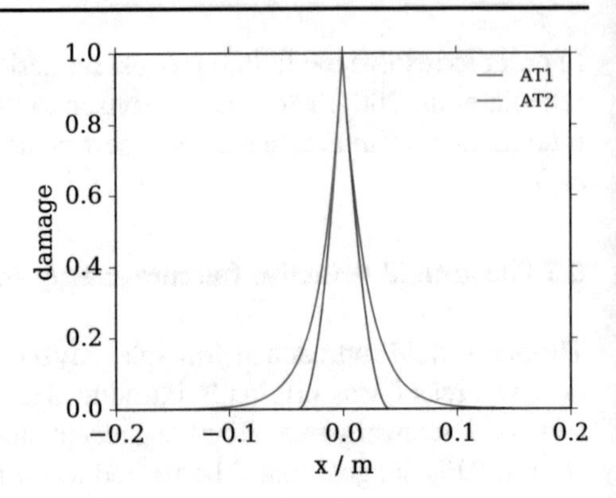

Fig. 3 Optimal damage profiles for *AT1* and *AT2*

where x_0 is the crack location. The optimal profiles of the damage field for *AT1* and *AT2* are plotted in Fig. 3.

Provided that Eq. (49) multiplied by G_c represents the surface energy term, the regularised surface energy terms in a discretised domain can be approximated using the optimal profile [Eqs. (51) or (52)]. We consider a line crack with a length of $2a$ centered at the origin ($-a \leq x \leq a$) which lies in an finite domain. Then, the damage profile for *AT1* is given by

$$d_h(\xi) = \begin{cases} 1, & \text{for } |x| \leq a,\ |y| \leq \frac{h}{2} \\ \left(1 - \frac{\xi}{2\ell}\right)^2, & \text{for } \xi \leq 2\ell \\ 0, & \text{otherwise} \end{cases} \tag{53}$$

where h is the discretisation size and ξ is the distance from the crack. The surface energy term can then be approximated from an areal integral

$$\begin{aligned} &\int_{-\infty}^{+\infty} \int_{-\infty}^{+\infty} \frac{3G_c}{8} \left(\frac{d_h}{\ell} + \ell\, |\nabla d_h|^2 \right) \mathrm{d}x\, \mathrm{d}y \\ &= G_c \left(\frac{3h}{8\ell} + 1 \right) a \left[1 + \frac{\pi \ell}{4a(3h/8\ell + 1)} \right] = G_c^{\text{eff}} a^{\text{eff}}. \end{aligned} \tag{54}$$

The first term on the RHS, $G_c\,(3h/8\ell + 1)$, is known as the effective fracture toughness, G_c^{eff} (Bourdin et al. 2008; Chukwudozie 2016). The second term originates from the damage profile near the crack tip and will be called the effective crack length, a^{eff}, in this study, which can also be estimated using the level-set at certain damage value. Note, that the effective crack length approaches the true crack length a as the crack length (defect size) increases with respect to the internal length, ℓ, unlike the effective fracture toughness which only depends on the discretization with respect to ℓ. Similarly for *AT2*, the discretised damage profile is given by

$$d_h(\xi) = \begin{cases} 1, & \text{for } |x| \leq a_0,\ |y| \leq \frac{h}{2} \\ \exp\left(-\frac{\xi}{\ell}\right), & \text{otherwise} \end{cases} \tag{55}$$

Then the effective fracture toughness and crack length can be obtained from

$$\int_{-\infty}^{+\infty}\int_{-\infty}^{+\infty}\frac{G_\mathrm{c}}{2}\left(\frac{d_h^2}{\ell}+\ell\,|\nabla d_h|^2\right)\mathrm{d}x\,\mathrm{d}y = G_\mathrm{c}\left(\frac{h}{2\ell}+1\right)a\left[1+\frac{\pi\ell}{4a(h/2\ell+1)}\right]$$
$$= G_\mathrm{c}^\mathrm{eff}a^\mathrm{eff}. \tag{56}$$

As can be seen in the above equations, the effective fracture surface energy is augmented in the phase-field model to maintain physically adequate dissipation. Therefore, the equivalent $G_\mathrm{c}^\mathrm{eff}$ needs to be provided in the analysis in order to obtain correct results. For example, for $G_\mathrm{c} = 20$ Pa m, $h = 0.01$ m, and $\ell = 0.03$ m, the values $G_\mathrm{c}^\mathrm{eff} = 17.78$ Pa m and $G_\mathrm{c}^\mathrm{eff} = 17.14$ Pa m have to be provided for *AT1* and *AT2*, respectively, to remain physically equivalent. When comparing crack propagation, the respective formulations for a^eff should be employed to ensure equivalence.

6.4 Non-local plastic-damage: volumetric and surface energy density

For the plastic-damage model, the fracture energy in tension (mode I) G_c can be computed from the fracture energy density g_f as

$$G_\mathrm{c} = g_\mathrm{f} w_\mathrm{t}, \tag{57}$$

where w_t is the fracture width, which in this case corresponds to the width of the damaged zone. The fracture energy density in tension can be computed from the total dissipated energy in a uniaxial tensile test performed to failure

$$g_\mathrm{f} = \int_0^\infty \sigma(\epsilon)\,\mathrm{d}\epsilon = A_1 + A_2, \tag{58}$$

which can be split into two contributions: a first contribution of elastic energy at peak tensile stress (tensile strength; somewhat analogous to Fig. 2a)

$$A_1 = \frac{\sigma_\mathrm{t}^2}{2E}, \tag{59}$$

and a second contribution due to inelastic post-peak dissipation

$$A_2 = \int_{\sigma_\mathrm{t}/E}^{\infty}\sigma(\epsilon)\,\mathrm{d}\epsilon = \int_{\sigma_\mathrm{t}/E}^{\infty}\sigma_\mathrm{t}(1-d)\,\mathrm{d}\epsilon$$
$$= \int_{\sigma_\mathrm{t}/E}^{\infty}\sigma_\mathrm{t}\exp\left[-(\epsilon-\sigma_\mathrm{t}/E)/\alpha_\mathrm{d}\right]\mathrm{d}\epsilon = \alpha_\mathrm{d}\sigma_\mathrm{t}. \tag{60}$$

The fracture energy G_c does not explicitly enter the set of parameters of the non-local plastic-damage. Due to the fact that the non-local operator averages an a-priori unknown field (be that damage, strain energy or, as in this case, k_d), the extent of the damaged zone (i.e., the fracture width w_t) is not clearly defined and will depend upon the problem solved. Furthermore, G_c depends on structural size relative to the internal length, which makes the correlation between global fracture energy in mode I and the internal length l_c a non-trivial task. Often, such correlation is written as a linear dependency (Nguyen and Houlsby 2007; Nguyen et al. 2015)

$$G_c = g_f w_t = g_f \beta_k l_c, \tag{61}$$

where β_k is an empirical coefficient which can take values from 1 to 3 depending on the analysed problem. In the current case, best fit to the fracture energy of $G_c = 20\,\text{Pa m}$ is obtained with a value of $\beta_k = 2.4$. The calibration can be carried out by fixing either σ_t or α_d and the remaining free parameter is obtained by solving Eq. (61) for a fixed G_c. In the current case we have made the assumption that compressive strength is $\sigma_c = 10\sigma_t$, which falls into the range of the experimentally given parameters (Sect. 6.1), so that Drucker–Prager's parameters can be computed as

$$\begin{aligned} \beta &= \frac{1}{\sqrt{3}}\left(\frac{\sigma_c - \sigma_t}{\sigma_t + \sigma_c}\right) \\ k &= \frac{2}{\sqrt{3}}\left(\frac{\sigma_t \sigma_c}{\sigma_t + \sigma_c}\right) \end{aligned}. \tag{62}$$

Given $E = 80\,\text{GPa}$, $\nu = 0.15$ $\sigma_t = 10\,\text{MPa}$ and $l_c = 0.005$, the calibration yields $\alpha_d = 10^{-4}$, $\beta = 0.47$ and $k = 10.50\,\text{MPa}$.

7 Quantification of fracture volume and treatment of the fluid mass balance

The hydraulic fracturing process can either be driven by applying the fluid pressure in the crack directly which can result in unstable crack propagation, or prescribing a constant volume flux. In the latter case, the fluid pressure represents an additional unknown. In order to solve for p or p_f in the system the updated cracked volume V_{crack}, the mass balance $V_{\text{inj}} = V_{\text{crack}}$ is enforced as a constraint where V_{inj} is the injected fluid volume.

While the crack volume is explicitly given in the cohesive zone approach by the displacement jump (cf. Algorithm 1), it is provided by $V_{\text{crack}} = \int_\Omega \mathbf{u} \cdot \nabla d \, \mathrm{d}\Omega$ in the phase-field model from Γ-convergence considerations. This integral can readily be evaluated by standard finite element routines in the phase-field implementation. In the non-local model, d is not a nodal degree of freedom but a state variable given at integration points. Hence, the gradient of damage is not readily available in a finite element setting and would require additional extra- and interpolation procedures which may induce numerical inaccuracies (Kuhn et al. 2015). Hence, the divergence theorem is applied with the additional assumption of undamaged domain boundaries[4] to yield $V_{\text{crack}} = \int_\Omega d\,(\operatorname{div} \mathbf{u})\, \mathrm{d}\Omega$, which is the formulation used subsequently.

[4] This assumption is chosen here for simplicity as it is valid in all subsequent test cases. In the more general case, boundary integrals need to be evaluated.

The schemes to update the pressure during the global Newton iterations differ slightly in each implementation. Since both LIE and the non-local model solve only for $\mathbf{u}$ as a primary variable, its update of p is simply done proportional to the volume defect as in Algorithm 1 (LIE) and Algorithm 2 (non-local damage).

Algorithm 1 Incorporation of the volume constraint in the cohesive-zone model.

1: **while** $V < V_{\text{final}}$ **do**
2: update the injected volume, $V_{\text{inj}}(t_n + \Delta t)$
3: **while** $\|p - p_n\|/p \geq \theta_p$ and "no global equilibrium" **do**
4: calculate crack volume, $V_{\text{crack}} = \int_{\Gamma_c} [\![\mathbf{u}]\!] \cdot \mathbf{n} \, d\Gamma$
5: update pressure $p = (V_{\text{inj}}/V_{\text{crack}})p_n$
6: update total stress, $\mathbf{t}_c = \mathbf{t}'_c - dp\mathbf{n}_\Gamma$
7: solve Eq. (3.3) for $\mathbf{u}$ and $\mathbf{a}$

In LIE (Algorithm 1), the primary variables $\mathbf{u}$ and $\mathbf{a}$ are computed with absolute convergence tolerances of 10^{-8} m and 10^{-4} m for the associated residuals $\|\mathbf{R_u}\|$ and $\|\mathbf{R_a}\|$, respectively, the fluid pressure p being with a relative tolerance of $\theta_p = 10^{-4}$.

Algorithm 2 Incorporation of the volume constraint in the non-local plastic-damage model.

1: **while** $V < V_{\text{final}}$ **do**
2: update the injected volume, $V_{\text{inj}}(t_n + \Delta t)$
3: **while** $\|p_{\text{old}} - p\| / p < \theta_p$ and "no global equilibrium" **do**
4: calculate crack volume, $V_{\text{crack}} = \int_\Omega d\,(\operatorname{div} \mathbf{u})\, d\Omega$
5: update pressure, $\Delta p = p - p_{\text{n}} = \left(V_{\text{crack}} - V_{\text{inj}}\right) K_w$
6: update total stress, $\boldsymbol{\sigma} = \boldsymbol{\sigma}' - dp\boldsymbol{I}$
7: solve for $\mathbf{u}$

In the non-local plastic-damage model (Algorithm 2), the primary variable $\mathbf{u}$ is with an absolute convergence tolerance of $\|\Delta\mathbf{u}\| < 10^{-6}$ m, the fluid pressure's relative tolerance is $\theta_p = 10^{-4}$ and the local stress return algorithm is computed with a relative tolerance of 10^{-10} for the norm of the residuals $\|\mathbf{R}\|$.

Algorithm 3 Incorporation of the volume constraint in the phase-field model.

1: **while** $V < V_{\text{final}}$ **do**
2: update the injected volume, $V_{\text{inj}}(t_n + \Delta t)$
3: **while** $\|d - d_n\| < 10^{-4}$ **do**
4: set $p = 1.0$
5: solve for $\mathbf{u}_1$
6: calculate crack volume, $V_1 = \int_\Omega \tilde{\mathbf{u}}_1 \cdot \nabla d \, d\Omega$
7: calculate pressure, $p = V_{\text{inj}}/V_1$
8: update displacement, $\mathbf{u} = p\mathbf{u}_1$
9: solve for d

The phase-field implementation solves for both $\mathbf{u}$ and d at the nodes of the finite element discretization. Its update of p is performed at the intermittent stage and additionally exploits the linearity of the deformation (Algorithm 3). Let $\mathbf{u}_1$ denote the displacement field under an applied unit pressure $p_1 = 1$: then, due to linearity, the fracture volume is given as $V_{\text{crack}} = p \int_\Omega \tilde{\mathbf{u}}_1 \cdot \nabla d \, \mathrm{d}\Omega$ which can be inverted to obtain p as in Algorithm 3.

8 Numerical examples of fluid injection into a single fault

In the following, two examples will be considered. The first example addresses a static crack under sub-critical pressurization. The pressure is applied directly such that no fluid mass balance needs to be solved. The example tests the ability of all models to reproduce the pressure-dependent fracture opening profile compared to an analytical solution obtained by elasticity theory.

In the second example, water at room temperature is injected into the crack until it propagates to a certain length. The injection rate of $1 \times 10^{-6}\,\mathrm{m}^3\,\mathrm{s}^{-1}$ is chosen so that the corresponding dimensionless viscosity falls in the toughness dominated regime (Garagash 2006; Detournay 2016) where the pressure diffusions are negligible. The injection volume is prescribed as a function of time, requiring the solution of the mass balance. This comparative study thus entails the fracture volume computation, the capturing of the onset of propagation as well as the energy dissipation during propagation.

8.1 Internally pressurised static crack

A widely used classical validation example of an internally pressurised line crack in a two-dimensional plain-strain infinite domain (Ji et al. 2009; Bourdin et al. 2012; Wheeler et al. 2014; Santillán et al. 2017) is considered as a first verification example. The evaluation of the static fracture analysis focuses on the fracture width computation by smeared and discrete approaches. A single crack of length a_0 along the x axis is pressurised by the internal pressure of p in a homogeneous isotropic linear elastic material with the Young's modulus E and Poisson's ratio ν (Fig. 4). Setting $E' = E/(1 - \nu^2)$, the crack opening displacement for $-a_0 \leq x \leq a_0$ (see Sneddon and Lowengrub (1969, Sect. 2.4), for instance) is given by

$$u_y^+(x,0) = \frac{2pa_0}{E'} \left(1 - \frac{x^2}{a_0^2}\right)^{1/2} = -u_y^-(x,0). \tag{63}$$

The representation of the initial crack varies between the different approaches.

In the LIE approach, the crack is discretized by one-dimensional quadratic finite elements having a uniform length of 0.01 m. Note, that the entire crack path is predetermined by defining the interface along which the crack will propagate. An initial crack is set by assigning an initial condition to the internal damage variable, $d = 1.0$, at the integration points of the elements in the cracked domain. Irreversibility is accounted for via Eq. (12).

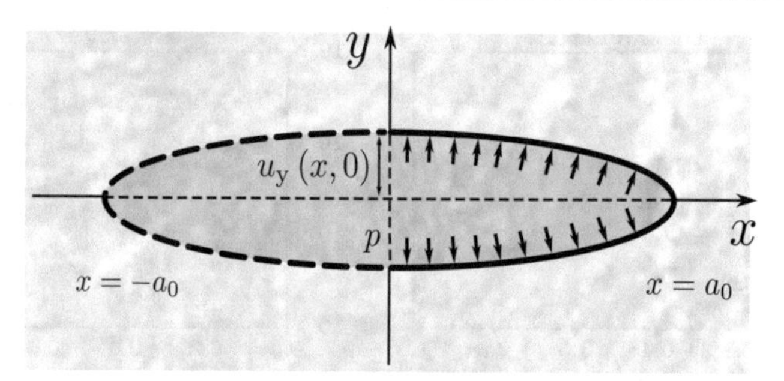

Fig. 4 Internally pressurised crack in 2D inifinite domain

In the phase-field approach, the initial crack is represented by specifying $d = 1.0$ at the nodes of the cracked domain and the effective crack length discussed in the previous section is used to evaluate the computational results.

Concerning the damage model, the crack is represented by a damaged area in which $d = 1.0$ is set at the integration points. Because no regularisation is needed for this case (no damage propagation) and to avoid damage diffusion by integral averaging, we have set $l_c = 0$ (equivalent to a local model).

For $a_0 = 0.1$ m and $p = 10$ MPa, computational results from all three methods are compared against the closed form solution in Fig. 5a and their relative errors are shown in Fig. 5b. The errors from the LIE and the phase-field approaches become worse towards the crack tip while those of the non-local plastic-damage model are almost steady along the crack. The LIE and the phase-field may benefit from higher mesh resolution around the tip though, all three models are in good agreement with the closed form solution. Given the static damage profile or the explicit crack, this is essentially an elastic deformation problem and the computational cost is trivial in all three methods. A quadrilateral structured mesh with 0.01 m resolution is used for all the computations. The crack opening profile is explicitly obtained from the displacement jump **a** in the LIE approach whereas it is a post-processed quantity in the smeared approaches. It is obtained from a line integral of $\mathbf{u} \cdot \nabla d$ in the phase-field modelling and as the differential nodal displacement of the damaged element for the damage model. Although their matches degrade slightly near the tip, all three methods are capable of computing crack opening displacement quite accurately. Damage performs the worst, as with a single damaged element row, strong gradients are present and not captured as well as in the phase-field case. The LIE approach benefits from the direct incorporation of the strong discontinuity across the crack faces into the formulation, and lies in between the two smeared approaches in the present case.

8.2 Internally pressurised propagating crack

The second verification example addresses the propagation of a fracture by hydraulic forces (Dean and Schmidt 2009; Bourdin et al. 2012; Gupta and Duarte 2014; Lee et al. 2016). Since the work done by the internal pressure is $W = pu_y^+(0,0)a_0\pi$, the strain energy ($= -\frac{1}{2}W$) is

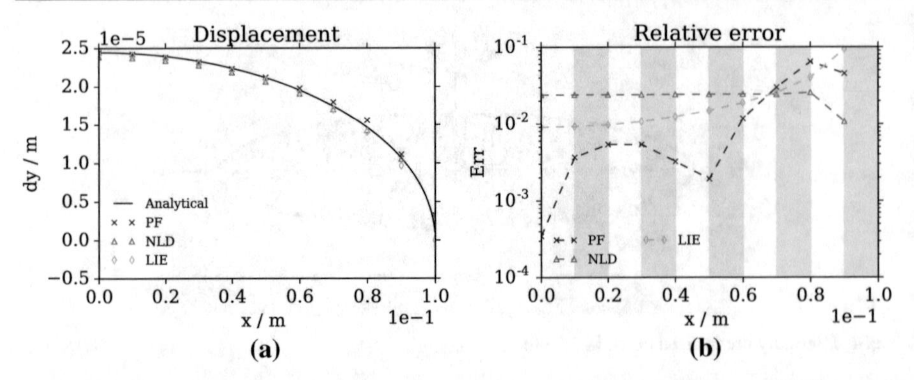

Fig. 5 Comparison of the numerical solution given by the phase-field, non-local damage and LIE methods against the analytical solution (**a**) and relative error of the numerical methods against the analytical solution (**b**). The transparent bands in **b** indicate the size of the finite element discretization for the phase-field and non-local damage model, which has an influence in the computed displacement field

$$E_b = \frac{\pi a_0^2 p^2}{E'}. \tag{64}$$

Given material properties (E' and G_c) and the initial fracture geometry (a_0), the critical propagation pressure p_c can be obtained from Griffith's criterion ($-\partial E_b / \partial (2a_0) = G_c$). Hence, fracture propagation occurs when

$$\frac{\pi a_0 p_c^2}{E'} = G_c. \tag{65}$$

Thus the critical pressure can be obtained as[5]

$$p_c = \left[\frac{E' G_c}{\pi a_0} \right]^{1/2}, \tag{66}$$

Also the critical volume is given by

$$V_c = \left[\frac{4\pi a_0^3 G_c}{E'} \right]^{1/2}. \tag{67}$$

Once the pressure reaches p_c (or, equivalently, the volume V_c), the fracture propagates quasi-statically given a prescribed injection volume. As the propagation pressure decreases with the fracture length following Eq. (66), the fracture length a is determined from the internal fracture pressure p as

$$a = \frac{E' G_c}{\pi p^2}, \tag{68}$$

[5] In terms of the stress intensity factor, $K_{IC} = (E' G_c)^{1/2}$, it is given as $p_c = K_{IC} / \sqrt{\pi a_0}$.

Using the volume, the pressure response can be written as a function of injection volume during fracture propagation as

$$p(V) = \left[\frac{2E'G_c^2}{\pi V}\right]^{1/3}. \tag{69}$$

and the propagating fracture length can also be expressed as a function of injection volume;

$$a(V) = \left[\frac{E'V^2}{4\pi G_c}\right]^{1/3} \tag{70}$$

Comparisons against the closed form solution from all three approaches are shown in Fig. 6. The pressure is slightly overestimated at the peak by the phase-field modelling and converges to the solution as the crack grows (Fig. 6a). Though both fracture toughness and initial crack length are corrected using the approximated effective values [see Eqs. (54) or (56)], the surface energy term may still be slightly overestimated but in an acceptable range ($\sim$2 %). As the crack length increases and becomes dominant over the internal length, ℓ, the computed pressure converges to the theoretical value. The peak pressure match also improves with finer mesh (i.e. smaller internal length at constant h/ℓ) while accurate crack length evolution prediction can be achieved already with a coarser mesh. This relative insensitivity of crack growth to the internal length (mesh size) agrees with the observations made by Klinsmann et al. (2015) and Zhang et al. (2017). Damage initiation is immediate with the *AT2* model while *AT1* exhibits an elastic regime until failure. With the fine mesh used in this example, both models show an almost identical behaviour. In the computation shown in Fig. 6a, an element edge length of 0.00125 m was used for the linear triangular elements.

The non-local damage model was run with a structured mesh made of linear quadrilateral elements of size 0.00125 m. The solution also satisfactorily reproduces the analytical solution as shown in Fig. 6b, with a higher discrepancy observed in the elastic range compared to the other methods. This is related to the fact that strong gradients across a single element (initial condition) are not captured well by the formulation, as was shown in the static crack case. On the other hand, the model performs best in the inelastic dissipation range during crack propagation phase. The initial crack length is slightly over-estimated because of damage diffusion originating from the non-local averaging formulation and during propagation the over-estimated length is around $\sim$5 %, which is still an acceptable value. The discrepancy could also be related to the way in which crack-length is computed, i.e., as an integral of damage along a line passing through the center of the damaged zone and parallel to the crack advancing direction. As such, the distribution of damage at the tip decreases from 1 to 0 over a length of roughly $\sim$8 elements, i.e., 0.01 m (process length). Applying this correction, the predicted crack length results in a better fit with the analytical solution (black diamonds in Fig. 6b). This discrepancy highlights a long debated concept of smeared approaches to simulate crack problems: where is the exact position of the crack boundary? Being smeared, it is implicit that the crack boundary cannot be precisely assessed, as it is instead a "transition" between intact and fully broken material.

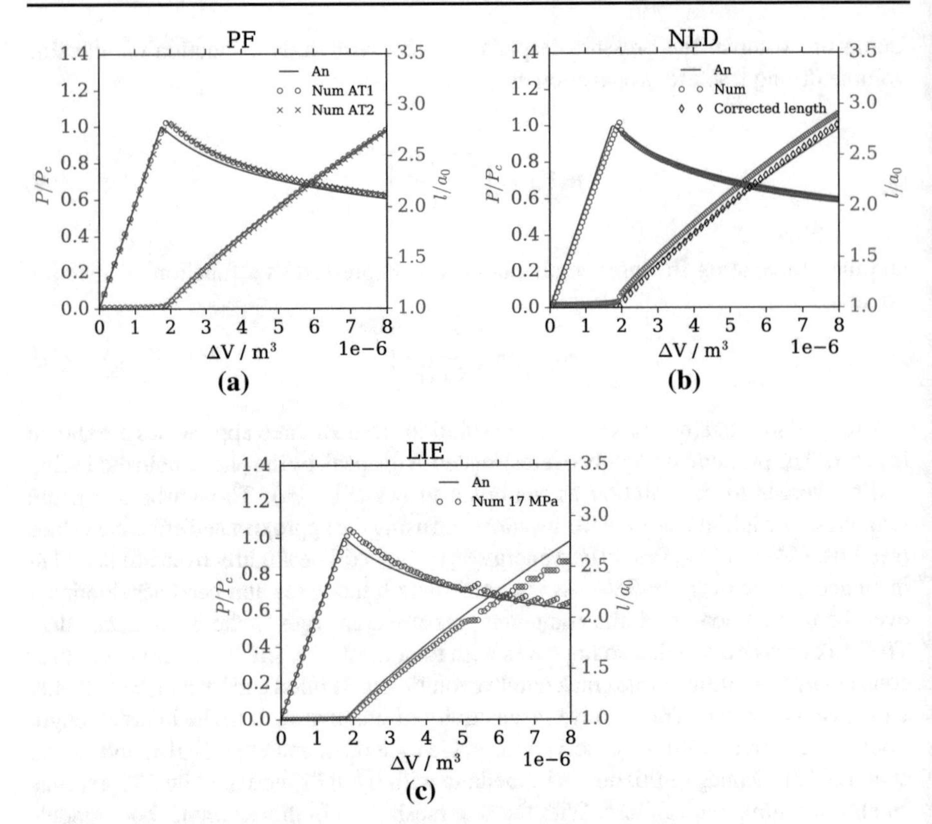

Fig. 6 Analytical and numerical solutions comparison in terms of pressure and crack-length versus injected fluid volume for the phase-field (**a**), the non-local damage (**b**) and the lower-dimensional interface element method (**c**)

In this sense, phase-field methods offer an appealing means to analytically derive an expression for the effective crack length, Eqs. (54) and (56).

Following the guidelines for spatial discretization with respect to the cohesive zone process length ($l_{cz} = 0.00554\,\text{m}$ in case of tensile strength $t_{n,p} = 17\,\text{MPa}$, see Sect. 6.2), a uniform mesh size of 0.00167 m was arranged along the predefined propagation path ahead of the initial crack tip for 0.1 m. After this point, element spacing becomes progressively coarsened with a minimum length of 0.005 m. Once the propagating crack exits the uniformly refined domain, the expected presence of oscillations can be observed in the pressure and crack-length curves (see Fig. 6c). In general, the LIE implementation captures the elastic range very well and shows a slight overestimation of the peak pressure. The solution approaches the analytical curve as the crack propagates.

Figure 7 compares stress profiles at different locations from the three different models. All models reproduce the traction boundary condition on the crack faces, Fig. 7a. Similarly, no appreciable solid stress transfer can be observed for any of the methods across the crack which is a prerequisite for a meaningful stress-field (and displacement) computation. Furthermore, the crack-tip singularity known from linear

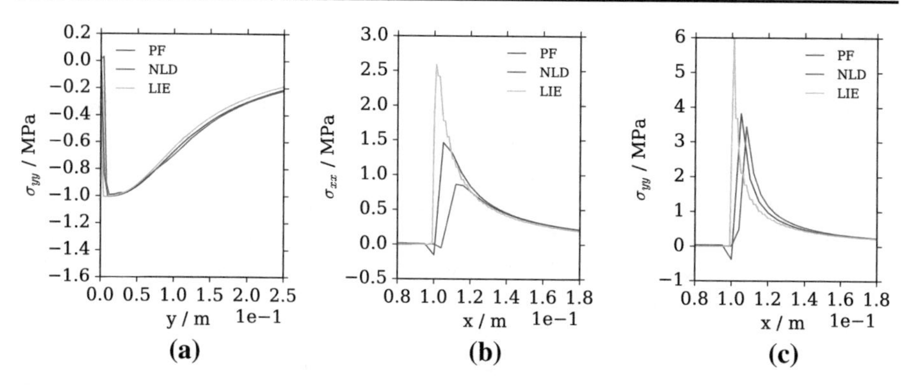

Fig. 7 Stress profiles for the three models: σ_{yy} along a central vertical line (**a**), σ_{xx} along a central horizontal line (**b**), σ_{yy} along a central horizontal line (**c**)

elasticity theory and typically characterised by stress-intensity factors is captured to a varying degree by the different approaches. The high consistency in the stress profile along the central vertical line cross-verifies the successful implementation of the three numerical approaches for the simulation of hydraulic fracturing (see Fig. 7a), while the apparent discrepancies in the stress peak values and peak position (see Fig. 7b, c) show that the LIE approach is most consistent with linear elastic fracture mechanics in terms of stress profiles under the present conditions.

Unlike the other two methods, stresses are not explicitly defined in the phase-field model and the mechanical variables (e.g. displacement and stress) do not have clear physical meanings in the phase transition zone (near the crack) even though its jump set can be quantified from an integral. In the non-local plastic-dmage model, the crack propagates when the available strength is mobilised. Due to the mode I loading, the stress state at the tip is tension dominated such that tensile strength becomes the relevant quantity.

Although this study does not particularly aim to compare the computational efficiencies, a few observations made through this limited set of examples indicate that both the phase-field and non-local damage models are much more computationally efficient than the lower-interface element model and that the mesh sensitivity of the lower-interface element model is greater than those of the phase-field model and the non-local damage. In order for the lower-interface element method to achieve similar accuracy to the other two approaches, it required a finer mesh size and significantly greater number of iterations, both of which contribute to its computational time. Apart from this general remark, however, we will restrict ourselves from further commenting on the computational efficiencies as computation time does not depend only on the method but also on the implementation, and its comparison would require more carefully devised measures, which is beyond the scope of the current study.

9 Conclusion

We have presented three different numerical methodologies to simulate hydraulic fracturing. The cohesive zone model based on locally enriched interface elements treats cracks discretely along element edges. Crack propagation is thus discretization-dependent and, for practical purposes, requires a priori knowledge about the crack propagation path. A mesh-independent generalization of the present approach would lead to well-known extended finite element methods associated with drastically increased implementational effort.

Since cracks are represented implicitly in both phase-field and non-local damage models using a dedicated damage variable, neither special element formulations nor prescribed crack paths are required. In the phase-field model, the damage variable is a solution obtained through successive energy minimisation consistent with linear elastic fracture mechanics whereas it is an internal variable, non-locally averaged over a region of influence, in the non-local damage model.

The cohesive model approximates fracture mechanics using a traction-separation law, which requires two additional parameters (e.g. the stiffness, the cohesive strength, or the critical separation displacement) aside from the fracture toughness. The variational phase-field formulation is originally derived from Griffith's fracture mechanics introducing one additional parameter, the internal length, in its regularised phase-field form. It is shown, however, that the computational result is relatively insensitive to the internal length in a toughness-dominated (or non strength-dominated), i.e. fracture mechanical, regime unless the defect (crack) size is comparable to the internal length. The non-local damage model needs a full definition of the yield surface and flow rule for plasticity and some more additional parameters to describe damage evolution. As such, it can well describe strength evolution at different confinements, dilatancy, rate of softening and mixed-mode fracture propagation (not discussed here). Its extension to ductile fracture may proceed without major model changes, a benefit which has not been exploited in the present study. On the other hand, it is very difficult to relate material parameters to fracture energy, which does not enter the formulation explicitly as strains in the dissipation zone are almost never known a-priori. Though this subject has been addressed in literature before (Jirásek 1998; Nguyen and Houlsby 2007; Nguyen et al. 2015), we feel the question remains open as any calibration will necessarily depend on structural size and remain problem-specific. Capitalizing on the advantage of its versatility, one suggestion is to follow alternative paths in the parameter calibration process (e.g., representing full stress strain response in triaxial strength and fracture mechanics tests, cf. Parisio et al. 2018a). In Sect. 6, we have discussed how each of the models can be set up to simulate a brittle fracture behaviour.

From our comparative analyses, it has been evidenced that all three approaches can successfully represent hydraulic fracture propagation in a toughness-dominated regime. The cohesive zone model has the strong advantage of directly accounting for enhanced displacement field in its formulation, which is not the case for the smeared approaches. In the phase-field model, discrete variables such as crack volume or crack aperture opening can be readily approximated from the gradient of the damage field using the Γ-convergence property. The non-local formulation requires the analyst to make an appropriate choice. Concerning propagation, it is clear that both phase-field and non-local damage are better suited to represent crack propagation when the crack path is not known in advance such as in case of complex stress fields, mixed-mode and rotating crack propagation. In such cases, the LIE implementation would require special treatments such as crack tracking and h-adaptive remeshing algorithms. The implementation effort would be significant especially in three-dimensional settings and if branching or merging crack needs to be considered, criteria for all possible variations would be required explicitly.

Our results have shed light on the advantages and drawbacks of some of the most popular numerical techniques employed to simulate hydraulic-fracture propagation. Furthermore, based on each underlying theoretical model we have shown how to obtain key process quantities as well as how to consistently parametrize each of the methods. While the presented examples are all two dimensional, the implementation of all methods in OpenGeoSys is done for three dimensional problems as well. This is connected to additional coding for the LIE implementation but none at all for the smeared approaches.

The discussions in this paper are limited to hydraulic fracturing scenarios where the assumption of a toughness-dominated regime is valid, i.e. for the cases viscous dissipation and leak-off are irrelevant. Comparing the numerical approaches with respect to their conceptual ease as well as numerical accuracy and efficiency in other settings is the subject of ongoing studies.

Acknowledgements We thank Dr.-Ing. Thomas Frühwirt and Prof. Dr.-Ing. habil. Heinz Konietzky from the Institute of Geotechnics, Chair of Rock Mechanics at the TU Bergakademie Freiberg for providing us with the material properties of the local gneiss. The authors gratefully acknowledge the funding provided by the German Federal Ministry of Education and Research (BMBF) for the GeomInt project, Grant Number 03G0866A, as well as the support of the Project Management Jülich (PtJ). The contribution of F.P. was financed by the GEMex project. The GEMex project is supported by the European Union's Horizon 2020 programme for Research and Innovation under Grant Agreement No 727550.

Appendix

A Rescaling of the phase-field energy functional (non-dimensionalization)

The scaling of Eq. (23) to achieve a non-dimensional format proceeds as follows. Let ψ_0, x_0, u_0, and $\tau_0 = p_0$ be scaling factors, and

$$\psi = \psi_0 \tilde{\psi} \tag{71}$$

$$\mathbf{x} = x_0 \tilde{\mathbf{x}} \tag{72}$$

$$\mathbf{u} = u_0 \tilde{\mathbf{u}} \tag{73}$$

$$p = p_0 \tilde{p} \tag{74}$$

$$\bar{\mathbf{t}} = \tau_0 \tilde{\mathbf{t}} \quad \text{and} \quad \varrho \mathbf{b} = \tau_0 \tilde{\varrho \mathbf{b}} \tag{75}$$

where $\tilde{()}$ is a dimensionless variable. Thus, Eq. (23) becomes

$$E(\mathbf{u}, d, p) = \psi_0 u_0^2 x_0^{n-2} \int_\Omega d^2 \tilde{\psi}^+(\tilde{\mathbf{u}}) + \tilde{\psi}^-(\tilde{\mathbf{u}})\,\mathrm{d}\Omega - \tau_0 x_0^{n-1} u_0 \int_{\partial_\mathrm{N}\Omega} \tilde{\mathbf{t}} \cdot \tilde{\mathbf{u}}\,\mathrm{d}\Gamma - \\ - \tau_0 x_0^{n-1} u_0 \int_\Omega \varrho \tilde{\mathbf{b}} \cdot \tilde{\mathbf{u}}\,\mathrm{d}\Omega + x_0^{n-1} \frac{G_\mathrm{c}}{4c_w} \int_\Omega \left(\frac{w(d)}{\ell} + \ell\,|\nabla d|^2 \right) \mathrm{d}\Omega \\ + p_0 x_0^{n-1} u_0 \int_\Omega \tilde{p} \tilde{\mathbf{u}} \cdot \nabla d\,\mathrm{d}\Omega \tag{76}$$

Dividing both sides by $\psi_0 u_0^2 x_0^{n-2}$,

$$\tilde{E}(\tilde{\mathbf{u}}, d, \tilde{p}) = \int_\Omega d^2 \tilde{\psi}^+(\tilde{\mathbf{u}}) + \tilde{\psi}^-(\tilde{\mathbf{u}})\,\mathrm{d}\Omega - \frac{\tau_0 x_0}{u_0 \psi_0} \int_{\partial_\mathrm{N}\Omega} \tilde{\mathbf{t}} \cdot \tilde{\mathbf{u}}\,\mathrm{d}\Gamma - \\ - \frac{\tau_0 x_0}{u_0 \psi_0} \int_\Omega \varrho \tilde{\mathbf{b}} \cdot \tilde{\mathbf{u}}\,\mathrm{d}\Omega + \frac{G_\mathrm{c} x_0}{4c_w \psi_0 u_0^2} \int_\Omega \left(\frac{w(d)}{\ell} + \ell\,|\nabla d|^2 \right) \mathrm{d}\Omega \\ + \frac{p_0 x_0}{u_0 \psi_0} \int_\Omega \tilde{p} \tilde{\mathbf{u}} \cdot \nabla d\,\mathrm{d}\Omega \tag{77}$$

and setting

$$x_0 = \frac{\psi_0 u_0^2}{G_\mathrm{c}} \tag{78}$$

$$\tau_0 = p_0 = \frac{\psi_0 u_0}{x_0} \tag{79}$$

yields the numerically favourable formulation

$$\tilde{E}(\tilde{\mathbf{u}}, d, \tilde{p}) = \int_\Omega d^2 \tilde{\psi}^+(\tilde{\mathbf{u}}) + \tilde{\psi}^-(\tilde{\mathbf{u}})\,\mathrm{d}\Omega - \int_{\partial_\mathrm{N}\Omega} \tilde{\mathbf{t}} \cdot \tilde{\mathbf{u}}\,\mathrm{d}\Gamma - \\ - \int_\Omega \tilde{\mathbf{t}} \cdot \tilde{\mathbf{u}}\,\mathrm{d}\Omega + \frac{1}{4c_w} \int_\Omega \left(\frac{w(d)}{\ell} + \ell\,|\nabla d|^2 \right) \mathrm{d}\Omega + \int_\Omega \tilde{p} \tilde{\mathbf{u}} \cdot \nabla d\,\mathrm{d}\Omega \tag{80}$$

References

Alessi, R., Marigo, J.J., Vidoli, S.: Gradient damage models coupled with plasticity: variational formulation and main properties. Mech. Mater. **80**(PB), 351–367 (2015). https://doi.org/10.1016/j.mechmat.2013.12.005

Ambati, M., Gerasimov, T., De Lorenzis, L.: Phase-field modeling of ductile fracture. Comput. Mech. **55**(5), 1017–1040 (2015). https://doi.org/10.1007/s00466-015-1151-4. arXiv:1011.1669v3

Ambrosio, L., Tortorelli, V.M.: Approximation of functional depending on jumps by elliptic functional via t-convergence. Commun. Pure Appl. Math. **43**(8), 999–1036 (1990). https://doi.org/10.1002/cpa.3160430805

Ambrosio, L., Tortorelli, V.M.: On the approximation of free discontinuity problems. Boll. Unione Mat. Ital. **7**, 105–123 (1992)

Amor, H., Marigo, J.J., Maurini, C.: Regularized formulation of the variational brittle fracture with unilateral contact: Numerical experiments. J. Mech. Phys. Solids **57**(8), 1209–1229 (2009). https://doi.org/10.1016/j.jmps.2009.04.011

Balay, S., Gropp, W.D., McInnes, L.C., Smith, B.F.: Efficient management of parallelism in object oriented numerical software libraries. In: Arge, E., Bruaset, A.M., Langtangen, H.P. (eds.) Modern Software Tools in Scientific Computing, pp. 163–202. Birkhäuser Press, Basel (1997)

Balay, S., Abhyankar, S., Adams, M.F., Brown, J., Brune, P., Buschelman, K., Dalcin, L., Eijkhout, V., Gropp, W.D., Kaushik, D., Knepley, M.G., May, D.A., McInnes, L.C., Rupp, K., Sanan, P., Smith, B.F., Zampini, S., Zhang, H., Zhang, H.: PETSc users manual. Technical Report ANL-95/11—Revision 3.8, Argonne National Laboratory (2017a). http://www.mcs.anl.gov/petsc

Balay, S., Abhyankar, S., Adams, M.F., Brown, J., Brune, P., Buschelman, K., Dalcin, L., Eijkhout, V., Gropp, W.D., Kaushik, D., Knepley, M.G., May, D.A., McInnes, L.C., Rupp, K., Smith, B.F., Zampini, S., Zhang, H., Zhang, H.: PETSc Web page (2017b). http://www.mcs.anl.gov/petsc, http://www.mcs.anl.gov/petsc

Bazant, P.Z., Jirasek, M.: Nonlocal integral formulations of plasticity and damage: survey of progress. J. Eng. Mech. **128**, 1119–1149 (2002)

Bažant, Z.P.: Why continuum damage is nonlocal: micromechanics arguments. J. Eng. Mech. **117**(5), 1070–1087 (1991)

Belytschko, T., Moës, N., Usui, S., Parimi, C.: Arbitrary discontinuities in finite elements. Int. J. Numer. Methods Eng. **50**(4), 993–1013 (2001)

Belytschko, T., Chen, H., Xu, J., Zi, G.: Dynamic crack propagation based on loss of hyperbolicity and a new discontinuous enrichment. Int. J. Numer. Methods Eng. **58**(12), 1873–1905 (2003)

Belytschko, T., Gracie, R., Ventura, G.: A review of extended/generalized finite element methods for material modeling. Model. Simul. Mater. Sci. Eng. **17**(4), 043001 (2009)

Biot, M.: General theory of three-dimensional consolidation. J. Appl. Phys. **12**(2), 155–164 (1941)

Böger, L., Keip, M.A., Miehe, C.: Minimization and saddle-point principles for the phase-field modeling of fracture in hydrogels. Comput. Mater. Sci. **138**, 474–485 (2017)

Borden, M.J., Verhoosel, C.V., Scott, M.A., Hughes, T.J., Landis, C.M.: A phase-field description of dynamic brittle fracture. Comput. Methods Appl. Mech. Eng. **217–220**, 77–95 (2012). https://doi.org/10.1016/j.cma.2012.01.008

Bouchard, P.O., Bay, F., Chastel, Y., Tovena, I.: Crack propagation modelling using an advanced remeshing technique. Comput. Methods Appl. Mech. Eng. **189**(3), 723–742 (2000)

Bourdin, B., Francfort, G., Marigo, J.J.: Numerical experiments in revisited brittle fracture. J. Mech. Phys. Solids **48**(4), 797–826 (2000). https://doi.org/10.1016/S0022-5096(99)00028-9

Bourdin, B., Francfort, G.A., Marigo, J.J.: The variational approach to fracture. J. Elast. **91**, 5–148 (2008)

Bourdin, B., Chukwudozie, C., Yoshioka, K.: A variational approach to the numerical simulation of hydraulic fracturing. In: Proceedings of the 2012 SPE Annual Technical Conference and Exhibition, vol. SPE 159154 (2012)

Bourdin, B., Marigo, J.J., Maurini, C., Sicsic, P.: Morphogenesis and propagation of complex cracks induced by thermal shocks. Phys. Rev. Lett. **112**, 014301 (2014). https://doi.org/10.1103/PhysRevLett.112.014301

Brace, W., Paulding, B., Scholz, C.: Dilatancy in the fracture of crystalline rocks. J. Geophys. Res. **71**(16), 3939–3953 (1966)

Braides, A.: Approximation of Free-Discontinuity Problems. Springer, Berlin (1998)

Branco, R., Antunes, F., Costa, J.: A review on 3D-FE adaptive remeshing techniques for crack growth modelling. Eng. Fract. Mech. **141**, 170–195 (2015)

Budyn, E., Zi, G., Moës, N., Belytschko, T.: A method for multiple crack growth in brittle materials without remeshing. Int. J. Numer. Methods Eng. **61**(10), 1741–1770 (2004)

Chessa, J., Belytschko, T.: An extended finite element method for two-phase fluids. J. Appl. Mech. **70**(1), 10–17 (2003)

Chukwudozie, C.: Application of the variational fracture model to hydraulic fracturing in poroelastic media. Dissertation, Louisiana State University (2016)

Davila, C., Camanho, P., de Moura, M.: Mixed-mode decohesion elements for analyses of progressive delamination. In: 19th AIAA Applied Aerodynamics Conference, p. 1486 (2001)

de Borst, R., Verhoosel, C.V.: Gradient damage vs phase-field approaches for fracture: similarities and differences. Comput. Methods Appl. Mech. Eng. **312**, 78–94 (2016a) de Borst, R., Verhoosel, C.V.: Gradient damage vs phase-field approaches for fracture: similarities and differences. Comput. Methods Appl. Mech. Eng. **312**, 78–94 (2016b)

Dean, R.H., Schmidt, J.H.: Hydraulic-fracture predictions with a fully coupled geomechanical reservoir simulator. SPEJ (2009). https://doi.org/10.2118/116470-PA

Desmorat, R., Gatuingt, F., Jirásek, M.: Nonlocal models with damage-dependent interactions motivated by internal time. Eng. Fract. Mech. **142**, 255–275 (2015)

Detournay, E.: Mechanics of hydraulic fractures. Annu. Rev. Fluid Mech. **48**, 311–339 (2016)

Diederichs, M.: Manuel rocha medal recipient rock fracture and collapse under low confinement conditions. Rock Mech. Rock Eng. **36**(5), 339–381 (2003)

Duarte, C.A., Reno, L., Simone, A.: A high-order generalized fem for through-the-thickness branched cracks. Int. J. Numer. Methods Eng. **72**(3), 325–351 (2007)

Duddu, R., Waisman, H.: A nonlocal continuum damage mechanics approach to simulation of creep fracture in ice sheets. Comput. Mech. **51**(6), 961–974 (2013)

Economides, M.J., Nolte, E.K.G.: Reservoir Stimulation, vol. 2. Wiley, New York (2000)

Elices, M., Guinea, G., Gomez, J., Planas, J.: The cohesive zone model: advantages, limitations and challenges. Eng. Fract. Mech. **69**(2), 137–163 (2002)

Francfort, G., Marigo, J.J.: Revisiting brittle fracture as an energy minimization problem. J. Mech. Phys. Solids **46**(8), 1319–1342 (1998). https://doi.org/10.1016/S0022-5096(98)00034-9

Freddi, F., Royer-Carfagni, G.: Regularized variational theories of fracture: a unified approach. J. Mech. Phys. Solids (2010). https://doi.org/10.1016/j.jmps.2010.02.010

Fries, T.P., Belytschko, T.: The intrinsic XFEM: a method for arbitrary discontinuities without additional unknowns. Int. J. Numer. Methods Eng. **68**(13), 1358–1385 (2006)

Fries, T.P., Schätzer, M., Weber, N.: XFEM-simulation of hydraulic fracturing in 3D with emphasis on stress intensity factors. In: Oñate, E. Oliver, J., Huerta, A. (eds.) 11th World Congress on Computational Mechanics (WCCM XI), 5th European Conference on Computational Mechanics (ECCM V), 6th European Conference on Computational Fluid Dynamics (ECFD VI) (2014)

Garagash, D.I.: Plane-strain propagation of a fluid-driven fracture during injection and shut-in: asymptotics of large toughness. Eng. Fract. Mech. **73**(4), 456–481 (2006). https://doi.org/10.1016/j.engfracmech.2005.07.012

Gasser, T.C., Holzapfel, G.A.: Modeling 3D crack propagation in unreinforced concrete using PUFEM. Comput. Methods Appl. Mech. Eng. **194**(25–26), 2859–2896 (2005)

Giovanardi, B., Scotti, A., Formaggia, L.: A hybrid XFEM-phase field (xfield) method for crack propagation in brittle elastic materials. Comput. Methods Appl. Mech. Eng. **320**, 396–420 (2017)

Gordeliy, E., Peirce, A.: Coupling schemes for modeling hydraulic fracture propagation using the XFEM. Comput. Methods Appl. Mech. Eng. **253**, 305–322 (2013)

Gupta, P., Duarte, C.A.: Particle shape effect on macro-and micro behaviours of monodisperse ellipsoids. Int. J. Numer. Anal. Methods Geomech. **38**, 1397–1430 (2014). https://doi.org/10.1002/nag.732

He, W., Wu, Y.F., Xu, Y., Fu, T.T.: A thermodynamically consistent nonlocal damage model for concrete materials with unilateral effects. Comput. Methods Appl. Mech. Eng. **297**, 371–391 (2015)

Heider, Y., Markert, B.: Simulation of hydraulic fracture of porous materials using the phase-field modeling approach. Pamm **16**(1), 447–448 (2016). https://doi.org/10.1002/pamm.201610212

Hillerborg, A., Modéer, M., Petersson, P.E.: Analysis of crack formation and crack growth in concrete by means of fracture mechanics and finite elements. Cement Concr. Res. **6**(6), 773–781 (1976)

Hoek, E., Martin, C.: Fracture initiation and propagation in intact rock—a review. J. Rock Mech. Geotech. Eng. **6**(4), 287–300 (2014)

Ji, J., Settari, A., Sullivan, R.: A novel hydraulic fracturing model fully coupled with geomechanics and reservoir simulation. SPE J. (2009). https://doi.org/10.2118/110845-PA

Jiang, L., Sainoki, A., Mitri, H.S., Ma, N., Liu, H., Hao, Z.: Influence of fracture-induced weakening on coal mine gateroad stability. Int. J. Rock Mech. Min. Sci. **88**, 307–317 (2016). https://doi.org/10.1016/j.ijrmms.2016.04.017

Jirásek, M.: Comparison of nonlocal models for damage and fracture. LSC Report 98(02) (1998)

Johnson, L., Marschall, P., Zuidema, P., Gribi, P.: Effects of post-disposal gas generation in a repository for spent fuel, high-level waste and long-lived intermediate level waste sited in opalinus clay. Technical Report, National Cooperative for the Disposal of Radioactive Waste (NAGRA) (2004)

Karma, A., Kessler, D.A., Levine, H.: Phase-field model of mode III dynamic fracture. Phys. Rev. Lett. **87**(4), 3–6 (2001). https://doi.org/10.1103/PhysRevLett.87.045501

Khoei, A.R.: Extended Finite Element Method: Theory and Applications. Wiley, London (2014)

Khoei, A., Moslemi, H., Sharifi, M.: Three-dimensional cohesive fracture modeling of non-planar crack growth using adaptive FE technique. Int. J. Solids Struct. **49**(17), 2334–2348 (2012)

Klinsmann, M., Rosato, D., Kamlah, M., McMeeking, R.M.: An assessment of the phase field formulation for crack growth. Comput. Methods Appl. Mech. Eng. **294**(Supplement C), 313–330 (2015). https://doi.org/10.1016/j.cma.2015.06.009

Kolditz, O., Bauer, S., Bilke, L., Böttcher, N., Delfs, J., Fischer, T., Görke, U., Kalbacher, T., Kosakowski, G., McDermott, C., et al.: OpenGeoSys: an open-source initiative for numerical simulation of thermo-hydro-mechanical/chemical (THM/C) processes in porous media. Environ. Earth Sci. **67**(2), 589–599 (2012)

Kuhl, E., Ramm, E., de Borst, R.: An anisotropic gradient damage model for quasi-brittle materials. Comput. Methods Appl. Mech. Eng. **183**(1), 87–103 (2000)

Kuhn, C., Müller, R.: A continuum phase field model for fracture. Eng. Fract. Mech. **77**(18), 3625–3634 (2010). https://doi.org/10.1016/j.engfracmech.2010.08.009. **(computational Mechanics in Fracture and Damage: A Special Issue in Honor of Prof. Gross)**

Kuhn, C., Lohkamp, R., Schneider, F., Aurich, J.C., Mueller, R.: Finite element computation of discrete configurational forces in crystal plasticity. Int. J. Solids Struct. **56**, 62–77 (2015)

Lee, S., Wheeler, M.F., Wick, T.: Pressure and fluid-driven fracture propagation in porous media using an adaptive finite element phase field model. Comput. Methods Appl. Mech. Eng. **312**, 509–541 (2016). https://doi.org/10.1016/j.cma.2016.02.037

Legarth, B., Huenges, E., Zimmermann, G.: Hydraulic fracturing in a sedimentary geothermal reservoir: results and implications. Int. J. Rock Mech. Min. Sci. **42**(7–8), 1028–1041 (2005)

Lemaitre, J., Chaboche, J.L., Benallal, A., Desmorat, R.: Mécanique des matériaux solides-3eme édition. Dunod (2009)

Li, T., Marigo, J.J., Guilbaud, D., Potapov, S.: Gradient damage modeling of brittle fracture in an explicit dynamic context. Int. J. Numer. Methods Eng. **00**(March), 1–25 (2016). https://doi.org/10.1002/nme

Marigo, J.J., Maurini, C., Pham, K.: An overview of the modelling of fracture by gradient damage models. Meccanica **51**(12), 3107–3128 (2016). https://doi.org/10.1007/s11012-016-0538-4

Meschke, G., Leonhart, D.: A generalized finite element method for hydro-mechanically coupled analysis of hydraulic fracturing problems using space–time variant enrichment functions. Comput. Methods Appl. Mech. Eng. **290**, 438–465 (2015)

Meschke, G., Dumstorff, P., Fleming, W.: Variational extended finite element model for cohesive cracks: influence of integration and interface law. In: IUTAM Symposium on Discretization Methods for Evolving Discontinuities, pp. 283–301. Springer (2007)

Meyer, A., Rabold, F., Scherzer, M.: Efficient finite element simulation of crack propagation. Preprintreihe des Chemnitzer SFB 393 (2004)

Miehe, C., Welschinger, F., Hofacker, M.: Thermodynamically consistent phase-field models of fracture: variational principles and multi-field fe implementations. Int. J. Numer. Methods Eng. **83**(10), 1273–1311 (2010). https://doi.org/10.1002/nme.2861

Miehe, C., Mauthe, S., Teichtmeister, S.: Minimization Principles for the Coupled Problem of Darcy–Biot-Type Fluid Transport in Porous Media Linked to Phase Field Modeling of Fracture, vol. 82. Elsevier, Amsterdam (2015). https://doi.org/10.1016/j.jmps.2015.04.006

Miehe, C., Aldakheel, F., Raina, A.: Phase Field Modeling of Ductile Fracture at Finite Strains: A Variational Gradient-extended Plasticity-damage Theory, vol 84. Elsevier, Amsterdam(2016). https://doi.org/10.1016/j.ijplas.2016.04.011

Minkley, W., Brückner, D., Lüdeling, C.: Tightness of salt rocks and fluid percolation. In: 45. Geomechanik-Kolloqium, Freiberg, Germany (2016)

Moës, N., Belytschko, T.: Extended finite element method for cohesive crack growth. Eng. Fract. Mech. **69**(7), 813–833 (2002)

Moës, N., Dolbow, J., Belytschko, T.: A finite element method for crack growth without remeshing. Int. J. Numer. Methods Eng. **46**(1), 131–150 (1999)

Morita, N., Black ,A.D., Guh, G.F.: Theory of Lost Circulation Pressure. SPE Annual Technical Conference and Exhibition, 23-26 September, New Orleans, Louisiana (1990). https://doi.org/10.2118/20409-MS

Murakami, S.: Continuum Damage Mechanics: A Continuum Mechanics Approach to the Analysis of Damage and Fracture, vol. 185. Springer, Berlin (2012)

Nagel, T., Görke, U.J., Moerman, K.M., Kolditz, O.: On advantages of the kelvin mapping in finite element implementations of deformation processes. Environ. Earth Sci. **75**(11), 1–11 (2016). https://doi.org/10.1007/s12665-016-5429-4

Nagel, T., Minkley, W., Böttcher, N., Naumov, D., Görke, U.J., Kolditz, O.: Implicit numerical integration and consistent linearization of inelastic constitutive models of rock salt. Comput. Struct. **182**, 87–103 (2017)

Nedjar, B.: On a concept of directional damage gradient in transversely isotropic materials. Int. J. Solids Struct. **88**, 56–67 (2016)

Needleman, A.: An analysis of decohesion along an imperfect interface. Int. J. Fract. **42**(1), 21–40 (1990a)

Needleman, A.: An analysis of tensile decohesion along an interface. J. Mech. Phys. Solids **38**(3), 289–324 (1990b)

Nguyen, G.D., Houlsby, G.T.: Non-local damage modelling of concrete: a procedure for the determination of model parameters. Int. J. Numer. Anal. Methods Geomech. **31**(7), 867–891 (2007)

Nguyen, O., Repetto, E., Ortiz, M., Radovitzky, R.: A cohesive model of fatigue crack growth. Int. J. Fract. **110**(4), 351–369 (2001)

Nguyen, G.D., Korsunsky, A.M., Belnoue, J.P.H.: A nonlocal coupled damage-plasticity model for the analysis of ductile failure. Int. J. Plast. **64**, 56–75 (2015)

Oliver, J.: On the discrete constitutive models induced by strong discontinuity kinematics and continuum constitutive equations. Int. J. Solids Struct. **37**(48–50), 7207–7229 (2000)

Oliver, J., Huespe, A.E., Pulido, M., Chaves, E.: From continuum mechanics to fracture mechanics: the strong discontinuity approach. Eng. Fract. Mech. **69**(2), 113–136 (2002)

Parisio, F., Laloui, L.: Plastic-damage modeling of saturated quasi-brittle shales. Int. J. Rock Mech. Min. Sci. **93**, 295–306 (2017)

Parisio, F., Samat, S., Laloui, L.: Constitutive analysis of shale: a coupled damage plasticity approach. Int. J. Solids Struct. **75**, 88–98 (2015)

Parisio, F., Tarokh, A., Makhnenko, R., Naumov, D., Miao, X.Y., Kolditz, O., Nagel, T.: Experimental characterization and numerical modelling of fracture processes in granite. Int. J. Solids Struct. (2018a, in press). https://doi.org/10.1016/j.ijsolstr.2018.12.019

Parisio, F., Vilarrasa, V., Laloui, L.: Hydro-mechanical modeling of tunnel excavation in anisotropic shale with coupled damage-plasticity and micro-dilatant regularization. Rock Mech. Rock Eng. (2018b) https://doi.org/10.1007/s00603-018-1569-z

Peerlings, R.H.J., De Borst, R., Brekelmans, W.A.M., De Vree, J.H.P.: Gradient enhanced damage for quasi-brittle materials. Int. J. Numer. Methods Eng. **39**(19), 3391–3403 (1996)

Pham, K., Amor, H., Marigo, J.J., Maurini, C.: Gradient damage models and their use to approximate brittle fracture. Int. J. Damage Mech. **20**(4), 618–652 (2011). https://doi.org/10.1177/1056789510386852

Rice, J.: The Mechanics of Earthquake Rupture. Division of Engineering, Brown University, Providence (1979)

Roth, S.N., Léger, P., Soulaïmani, A.: Coupled hydro-mechanical cracking of concrete using XFEM in 3D. In: Saouma, V, Bolander, J., Landis, E. (eds.) 9th International Conference on Fracture Mechanics of Concrete and Concrete Structures FraMCoS-9 (2016)

Santillán, D., Juanes, R., Cueto-Felgueroso, L.: Phase field model of fluid-driven fracture in elastic media: immersed-fracture formulation and validation with analytical solutions. J. Geophys. Res. Solid Earth **122**(4), 2565–2589 (2017). https://doi.org/10.1002/2016JB013572

Silani, M., Talebi, H., Hamouda, A.M., Rabczuk, T.: Nonlocal damage modelling in clay/epoxy nanocomposites using a multiscale approach. J. Comput. Sci. **15**, 18–23 (2016)

Sneddon, I., Lowengrub, M.: Crack Problems in the Classical Theory of Elasticity. The SIAM Series in Applied Mathematics. Wiley, London (1969)

Tanné, E., Li, T., Bourdin, B., Marigo, J.J., Maurini, C.: Crack nucleation in variational phase-field models of brittle fracture. J. Mech. Phys. Solids **110**(Supplement C), 80–99 (2018). https://doi.org/10.1016/j.jmps.2017.09.006

Turon, A., Davila, C.G., Camanho, P.P., Costa, J.: An engineering solution for mesh size effects in the simulation of delamination using cohesive zone models. Eng. Fract. Mech. **74**(10), 1665–1682 (2007)

Vtorushin, E.: Application of mixed finite elements to spatially non-local model of inelastic deformations. GEM-Int. J. Geomath. **7**(2), 183–201 (2016)

Watanabe, N., Wang, W., Taron, J., Görke, U., Kolditz, O.: Lower-dimensional interface elements with local enrichment: application to coupled hydro-mechanical problems in discretely fractured porous media. Int. J. Numer. Methods Eng. **90**(8), 1010–1034 (2012). https://doi.org/10.1002/nme.3353/full

Wheeler, M., Wick, T., Wollner, W.: An augmented-lagrangian method for the phase-field approach for pressurized fractures. Comput. Methods Appl. Mech. Eng. **271**(Supplement C), 69–85 (2014). https://doi.org/10.1016/j.cma.2013.12.005

Wilson, Z.A., Landis, C.M.: Phase-field modeling of hydraulic fracture. J. Mech. Phys. Solids **96**, 264–290 (2016). https://doi.org/10.1016/j.jmps.2016.07.019

Yoshioka, K., Bourdin, B.: A variational hydraulic fracturing model coupled to a reservoir simulator. Int. J. Rock Mech. Min. Sci. **88**(Supplement C), 137–150 (2016). https://doi.org/10.1016/j.ijrmms.2016.07.020

Zhang, Z., Guazzato, M., Sornsuwan, T., Scherrer, S.S., Rungsiyakull, C., Li, W., Swain, M.V., Li, Q.: Thermally induced fracture for core-veneered dental ceramic structures. Acta Biomater. **9**(9), 8394–8402 (2013)

Zhang, X., Vignes, C., Sloan, S.W., Sheng, D.: Numerical evaluation of the phase-field model for brittle fracture with emphasis on the length scale. Comput. Mech. **59**(5), 737–752 (2017). https://doi.org/10.1007/s00466-017-1373-8

Publisher's Note Springer Nature remains neutral with regard to jurisdictional claims in published maps and institutional affiliations.

Affiliations

Keita Yoshioka[1] · Francesco Parisio[1] · Dmitri Naumov[1] · Renchao Lu[1,2] · Olaf Kolditz[1,2] · Thomas Nagel[1,3]

Keita Yoshioka
keita.yoshioka@ufz.de

1 Department of Environmental Informatics, Helmholtz Centre for Environmental Research – UFZ, Leipzig, Germany

2 Applied Environmental Systems Analysis, Technische Universität Dresden, Dresden, Germany

3 Chair of Soil Mechanics and Foundation Engineering, Geotechnical Institute, Technische Universität Bergakademie Freiberg, Freiberg, Germany

Printed in the United States
By Bookmasters